DESIGNING EMBEDDED PROCESSORS

Designing Embedded Processors

A Low Power Perspective

Edited by

J.HENKEL
University of Karlsruhe,
Germany

and

S.PARAMESWARAN
University of South Wales,
NSW, Australia

 Springer

A C.I.P. Catalogue record for this book is available from the Library of Congress.

ISBN 978-1-4020-5868-4 (HB)
ISBN 978-1-4020-5869-1 (e-book)

Published by Springer,
P.O. Box 17, 3300 AA Dordrecht, The Netherlands.

www.springer.com

Printed on acid-free paper

Contents

Part IV   Compiler Techniques

13
Compilation Techniques for Power, Energy,   287
    and Thermal Management
Ulrich Kremer

14
Compiler-Directed Dynamic CPU Frequency and Voltage Scaling   305
Chung-Hsing Hsu and Ulrich Kremer

Foreword

Embedded Processors – What is Next?

Jörg Henkel and Sri Parameswaran

These are exciting times for the system level designer/researcher. The world seems to burgeon with embedded systems. Consumers demand superior electronic products more often than ever before. Moore's law continues to be valid 40 years after it was first stated, allowing the adventurous to design with billions of transistors. Demanding managers require products in shorter time than previously. All of this has led to an unending search for superior methods and tools to design embedded systems.

Interminable appetite by consumers for portable embedded systems has continued to churn out chips with ever growing functionality. Soaring non-recurring engineering costs of chips has forced designers towards large scale chips which exhibit computation capability along with communication protocols. These designs are expected to be flexible by being software upgradeable, reduce time to market by being rapidly verifiable, and produced in large volumes to reduce the cost per chip. To truly make such systems ubiquitous, it is necessary to reduce the power consumed by such a system. These often conflicting demands have meant that chips have to exhibit smaller footprint and consume less power. For a long time now, the narrowing feature sizes of chips and continuously reducing supply voltages were sufficient to satisfy the size and power demands. Unfortunately, this trend towards smaller feature sizes and lower supply voltages is slowing due to physical limitations. This has led to looking at system level methods to reduce power in embedded systems.

Unlike circuit level methods to reduce power, system level methods often allow a plethora of techniques to be applied at various levels. Often these techniques are orthogonal to one another, and can be applied simultaneously, assuming that the designer has sufficient time. Some of these techniques are at the architecture level-such as application specific processors, some are run-time techniques-which respond to the workload by switching voltage and frequency, some are at design time-such as compiler techniques which allow lower power consumption of the compiled code.

Time is indeed changing the way we design systems. Reducing design time and the size of a design team are increasingly crucial. Numerous tools and methods are available to educated designer. Many of these are point tools, though several tool vendors work tirelessly towards making these point tools interoperable so that seamless design flows can be created, which are useable by designers, increasing productivity several times. While such design flows from the RTL level down are quite mature, the design tools and flows at the system level are still evolving and will evolve for some time to come. Research in this area is very much alive at this time and will be for the foreseeable future.

This book examines system level design techniques, which allow the automation of system level designs, with a particular emphasis towards low power. We expect researchers, graduate students and system level designers to benefit from this book. The authors of the individual chapters are all well known researchers in their respective fields.

1.   Philosophy of This Book

In order to provide a maximum degree of usability for novices and researchers alike, the book is organized in the following way: each of the individual six sub-topics comprises one section that introduces to the whole area. For example, the section Application Specific Embedded Processors starts with the chapter Designing Application Specific Embedded Processors. That chapter gives an introduction to the field in a textbook-like style along with a mentioning of the specific grand challenges followed by a review of the most relevant related work. Thus, the first chapter of a section introduces a novice to the field. Experts in the respective fields may skip that chapter. The subsequent chapters then pick a research topic and present state-of-the-art research approaches that are representative and most challenging from the current perspective. If too specific, more generally interested readers may skip those chapters. In that sense, the book is useful for both, the expert researcher as well as the novice. Also in that sense, this book provides a new (and hopefully useful) concept.

2.   Contents

Though there are certainly many more aspects in designing embedded processors with respect to low power, we had to restrict the book and eventually identified six main topics (according to the six sections) namely: I. Application

Specific Embedded Processors, II. Embedded Memories, III. Dynamic Voltage and Frequency Scaling, IV. Compiler Techniques, V. Multi-Processors, and VI. Reconfigurable Computing. These topics comprise the most relevant techniques to minimize energy/power consumption from a system-level design point of view.

Starting with Section I, Application Specific Embedded Processors, explores optimization energy/power optimization potentials through architectural customizations ranging from application-specific instruction extension through parameterization etc. After the introductory section by Henkel, Parameswaran entitled Designing Application Specific Embedded Processors, the so-called NISC approach by Gorjiara, Reshadi, Gajski is presented in the chapter Low-Power Design with NISC Technology. It describes a novel architecture for embedded processors that does not use instructions in the traditional way any more. The chapter entitled Synthesis of Instruction Sets for High Performance and Energy Efficient ASIP by Lee, Choi, Dutt focuses on reducing the energy-delay product of instruction-extensible through synthesis. The chapter A Framework for Extensible Processor-based MPSoC Design by Sun, Ravi and Raghunathan with an approach to utilize the energy/power efficiency of extensible processors in the context of a multi processor system. This section concludes with Design and Run-Time Code Compression for Embedded Systems. Parameswaran, Henkel, Janapsatya, Bonny and Ignjatovic show the increase in efficiency for an embedded processor when code compression is used.

Section II, Embedded Memories, addresses energy/power related issues around memories, in specific application memories may account for largest fraction energy/power consumption in an embedded application. It starts with the chapter Power Optimization Strategies Targeting the Memory Subsystem by P. Panda with an overview and review of the challenges in embedded memories. The following chapter Layer Assignment Techniques for Low Energy Multi-Layered Memory Organizations by Brockmeyer, Durinck, Corporaal, Catthoor that exploits layered memory assignments for energy/power efficiency. Finally, the chapter Memory Bank Locality and Its Usage in Reducing Energy Consumption within this section exploits the bank locality property to minimize energy.

Section III is dedicated to Dynamic Voltage and Frequency Scaling one of the most efficient techniques for reducing energy/power consumption as to the relationship given by power, V_{DD} and operating frequency of a switching CMOS transistor. S. Hu et al. give an introduction to the field within the first chapter Power Aware Scheduling followed by Dynamic Voltage and Frequency Scheduling by S. X. Hu et al. presenting new techniques. The chapter Voltage Selection for Time-Constrained Multi-Processor Systems applies the traditional one-processor approach to multi-processors.

The role of the compiler i.e. the design-time phase is content of Section IV Compiler Techniques staring with the introductory chapter Compilation Techniques for Power, Energy, and Thermal Management by U. Kremer. The chapter Compiler-Driven DVFS by Hsu, Kremer presents an approach how DVFS can efficiently be incorporated into the the compiler i.e. how this job that is typically decided upon by the run-time system is part of a design time step (i.e. compilation phase). The contribution Link Idle Period Exploitation for Network Power Management by Li, Chen and Kandemir uses the compiler to optimize power consumption of communication links.

Section V is dedicated to Multi-Processors an efficient means to reduce the energy/power consumption as increasing the operating frequency is not an option for increasing the performance any longer because of the power problem. The first chapter A Power and Energy Perspective on Multi-Processors by G. Martin gives an overview of various architectural alternatives and their characteristics with respect to low power. It is followed by the chapter System-level Design of Network on Chip Architectures by Chatha and Srinivasan focusing on a network-on-chip architecture for multi-processors. Paul and Meyer present a modeling approach in the chapter Power-Performance Modeling and Design for Heterogeneous Multiprocessors. Finally, Section VI deals with reconfigurable computing, a computing paradigm that is very promising in solving many energy/power related issues. The first section by R. Hartenstein gives an overview of reconfigurable computing and its suitability for energy/power efficient systems in the chapter Basic of Reconfigurable Computing. It is followed by the chapter Dynamic Reconfiguration a technique to increase energy/power efficiency. The book concludes with the chapter Applications, Design Tools and Low Power Issues in FPGA Reconfiguration. It focuses on reconfigurable fabrics, namely FPGAs and there usage through design tools when low energy/power is an issue.

We hope that this book helps to increase awareness of the main issues in the design of low power embedded processors.

I

Application Specific Embedded Processors

Chapter 1

Application-Specific Embedded Processors

Jörg Henkel[1], Sri Parameswaran[2], and Newton Cheung[3]

[1] *University of Karlsruhe (TH)*
Germany

[2] *University of New South Wales*
Australia

[3] *VaST System*
Australia

Abstract   Today's silicon technology allows building embedded processors as part of SoCs (systems-on-chip) comprising upto a billion of transistors on a single die. Interestingly, real-world SOCs (in large quantities for mainstream applications) utilizing this potential complexity do hardly exist. Another observation is that the semiconductor industry a couple of years ago experienced an inflection point: the number of ASIC (Application Specific Integrated Circuits) design starts was outpaced by the number of design starts for Application Specific Standard Products (ASSPs). Moreover, we might face a new design productivity gap: the "gap of complexity" (details and references will follow later). Together these observations are reflecting a transition in the way embedded processors are designed. This article reports on and analyzes current and possible future trends from a perspective of embedded system design with an emphasis on design environments for so-called extensible processor platforms. It describes the state-of-the-art in the three main steps when designing an extensible processor, namely *Code Segment Identification, Extensible Instruction Generation, Architectural Customization Selection.*

Keywords: Embedded processors, ESL, design gap

J. Henkel and S. Parameswaran (eds.), Designing Embedded Processors – A Low Power Perspective, 3–23.
© *2007 Springer.*

1.   Introduction and Motivation

Embedded systems are ubiquitous and represent the major source of demand for micro-processors with a 10-digit number of units finding their way into embedded system designs every single year. Examples span from security systems (e.g. video surveillance), control systems (e.g. automotive control), individual health systems (e.g. hearing aids) to main stream consumer products in such areas as personal communication (e.g. cell phones), personal computing (e.g. PDA), entertainment (e.g. MP3 players), video/photo (e.g. digital still/video cameras) and many more.

It can be observed that especially the latter group of main stream consumer products is subject to severe competition between major manufacturers and thus leading to two effects from a consumer's point of view:

- The life cycles of embedded products become increasingly smaller: cell phones represent one of many examples for this trend since they experience the introduction of an average of two new major product lines every year compared to only one years ago.

- The functionality of these products and hence the complexity of the underlying embedded systems is rapidly increasing: staying with cell phones as an example, the observation is that they feature far more functionality beyond their core functionality of providing a wireless voice channel and thus establishing a phone connection. In fact, common functions of cell phones include web browsing capabilities, SMS (Short Message Service), PDA functionalities and even gaming, etc. Almost on a monthly basis there are new features manufacturers of cell phones and service providers announce and eventually integrate into new generations.

Similar scenarios could be discussed in conjunction with other consumer, control, etc. devices as well. The observation would often be an increasing portfolio of functionality combined with decreasing product life cycles. In the following these observations are put in relation with trends in embedded system design.

The consumer's demand for increasing functionality translates directly into increased complexity of embedded systems on a chip.

Indeed, this demand matches well with predictions from the International Technology Roadmap for Semiconductors where within this decade a complexity of 1 billion transistors on a single chip is predicted.

However, already today it can be observed that the maximum possible amount of transistors (silicon-technology-wise) per chip is hardly exploited. This fact is even more obvious when the number of transistors per SOC *without* the embedded memory (i.e. easy to design regular structures) is counted.

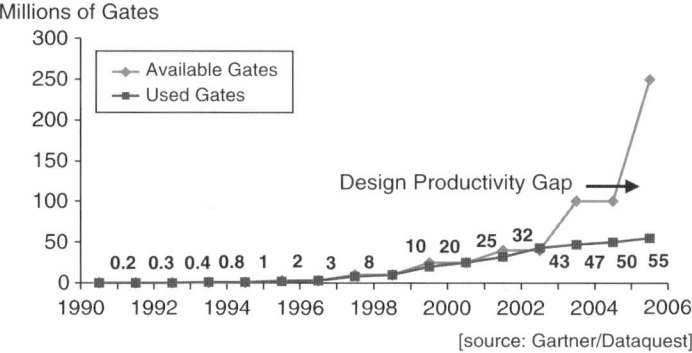

Fig. 1.1   The "Crisis of Complexity" in case no ESL methodologies will be used (G. Smith, Panel at 40ᵗʰ Design Automation Conference, 2003)

In short, real-world SOCs' complexities lag behind the capabilities of current silicon technologies even though there is certainly a demand for higher complexities (i.e. increased functionality) from an application point of view as discussed before.

Figure 1.1 (G. Smith, Panel at 40ᵗʰ Design Automation Conference, 2003) gives a possible answer: shown is the predicted productivity gap. It is measured as the number of available gates per chip for a given silicon technology for SOCs on the one side (red graph) and the number of actually used gates per chip of a given silicon technology on the other side (blue graph). The gap was predicted for the case that there will be *no* ESL (Electronic System Level Design) methodologies deployed for designing future complex SOCs. In other words: the gap might have been avoided if more ESL methodologies would have been deployed in all areas of system level design like specification/modeling, synthesis, simulation/verification, estimation, etc.

Though the need for ESL methodologies has been predicted and solutions have been researched for many years in the according areas, it is just about to start being accepted by system design engineers. An example is the migration to the so-called C-based design methodologies that represent a raised level (up from RTL) of abstraction in designing digital systems. Beyond a raised abstraction level, ESL methodologies also require an extensive IP component reuse through platforms.

Besides the gap of complexity, there is another indicator for the change (or need of change) of designing complex SOCs: the decrease of ASIC design starts. The number of design starts for ASICs were some years ago, outpaced by the number of design starts for ASSPs (Lewis, 2002). Thereby, the definition (as to Gartner) is as follows: an ASIC is an integrated circuit customized for a specific application that comes in form of a gate array or cell-based IC. An

ASSP (Application Specific Standard Product) on the other side includes all other ICs, i.e. ICs that are used in more than one application like DSPs, etc. In 2002 the number of ASIC design starts amounted world-wide to about 3,500–4,000 with a predicted steady decline over the next couple of years whereas ASSP design starts are predicted to stay steady at around 5,000 per year.

A reason for the declining ASIC design start are the extensive NRE costs when migrating to technology nodes far beyond 100 nm.

In order to overcome the problems of demand for increased functionality in embedded systems at constantly shortening product life cycles, more system-level design methodologies combined with the platform-based design paradigm need to be deployed.

1.1   Some Trends in Designing SOCs

Early embedded SOCs were of low complexity as they typically comprised one to two (programmable) processing units (e.g. a micro-controller plus a DSP), some custom hardware, peripherals and memory. However, the number of processing units in today's SOCs is steadily rising. An example is a TCP/IP protocol processing SOC that comprises 10 CPUs on a single chip (NEC, 2003). Each of the CPUs is an extensible processor. The design was preferred over a custom-logic (ASIC) design because it was considered too design-time extensive (and therefore expensive) given the large complexity. On the other side, using general purpose CPUs instead of the actually deployed configurable processors would not have led to the required performance at the same chip complexity (i.e. gate count).

There are other examples for SOCs comprising 10 or more processor including WCDMA chips with close to 15 programmable processing units. Recently (2007) an 80 core chip was announced. So, besides the earlier mentioned increasing functionality of embedded SOCs, there are further trends leading to a rising number of programmable processing units per SOC:

- Complex custom-logic blocks in SOCs are increasingly replaced *in whole* by programmable processing units with high performance. Reasons are the faster SOC development times and thus lower NRE costs at comparable performance values using a new breed of configurable processors (discussed later).

  This general trend has similarities to a trend that ended the hey days of super computing when the world's leading supercomputers were made of custom processors with custom technologies. It turned out that they were too expensive to develop given the relatively short life cycles. Today's supercomputer on the other side consist of off-the-shelf microprocessors.

Similarly, instead of using expensive complex custom logic blocks, off-the-shelf processors with high performance will increasingly be used in future SOCs.

- Certainly, custom-logic blocks will continue to remain on embedded SOCs for various reasons like very high performance, ultra-low power, etc. However, the complexity of these blocks will shrink since *sub-blocks* of these complex custom blocks will most likely be migrated into low complexity processing units where possible as recent designs show. This is accomplished through processors with a very basic instruction set and low architectural complexity of typically only around 10k–20k gates that run a software program to provide the replaced functionality. Hence, these processors further replace parts of complex custom-logic blocks where high performance/low power, etc. is not necessarily required and thus traded in lieu of a higher flexibility (programmable versa hardwired logic), lower NRE and lower per unit costs.

Both trends together increase the number of programmable processing units on SOCs and decrease the allover gate count for custom-logic per SOC and hence contribute partly to the trend of declining ASIC design starts in favor of ASSPs (see above).

The continuation of this trend is what is predicted to evolve into the *"Sea of Processors"* (Rowen, 2003) (*"The SOC Processor is the new Transistor"*, D. Patterson, UC Berkeley). The future of embedded systems is likely to be SOCs with hundreds of heterogeneous processors on it by the end of the decade. Designing these *heterogeneous* multiprocessor SOCs will be a challenge and quite different from designs from the parallel computing domain where processing units tend to be homogeneous.

1.2 Extensible Processor Platforms as a Possible Solution

The solution to overcome the gap of complexity is likely to be the migration of custom hardware into programmable processing units, i.e. the increased usage of off-the-shelf components in form of *extensible processors*.

Since the 1990s ASIPs (Application Specific Instruction-Set Processors) have been researched and developed as a compromise between high performance, high cost and low flexibility hard-wired custom logic on the one end and general-purpose processors (GPP) with a high flexibility (i.e. programmability) and more or less moderate performance and power characteristics on the other end (Figure 1.2). The application-specific instruction sets of ASIPs are domain-specific like instruction sets for multi media, encryption/de-cryption, etc. and thus result in increased performance-per-(chip-)area ratios compared to GPPs. Extensible processor platforms represent the state-of-the-art in ASIP technology. The customization of an extensible processor comprises typically

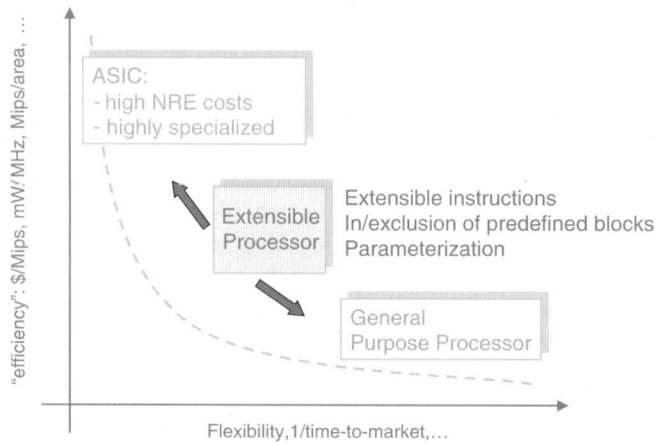

Fig. 1.2   Custom hard-wired logic, extensible processor and general purpose processor

the following three levels *instruction extension, inclusion/exclusion of pre-defined blocks, parameterization*:

- *Instruction Extension*

  The designer has the choice to freely define customized instructions by describing their functionality in a high-level language. This description is used by the platform's design flow as an input and subsequent synthesis steps will generate according custom instructions that are co-existing with the base instruction set of the extensible processor. Practically, restrictions may apply: for example, the complexity of an instruction (in terms of number of cycles for execution) may be limited in order to accommodate the according data path into the pipeline architecture of the base core. Also, the designer of the custom instructions may be responsible for a proper scheduling when instructions require multi-cycling. Other restrictions may constraint the total number of extensible instructions that can be defined and integrated per processor, etc.

- *Inclusion/Exclusion of Predefined Blocks*

  Predefined blocks as part of the extensible processor platform may be chosen to be included or excluded by the designer. Examples are special function registers, built-in self-test, MAC operation blocks, caches etc.

- *Parameterization*

  Finally, the designer may have the choice to parameterize the extensible processor. Examples are: setting the size of instruction/data caches,

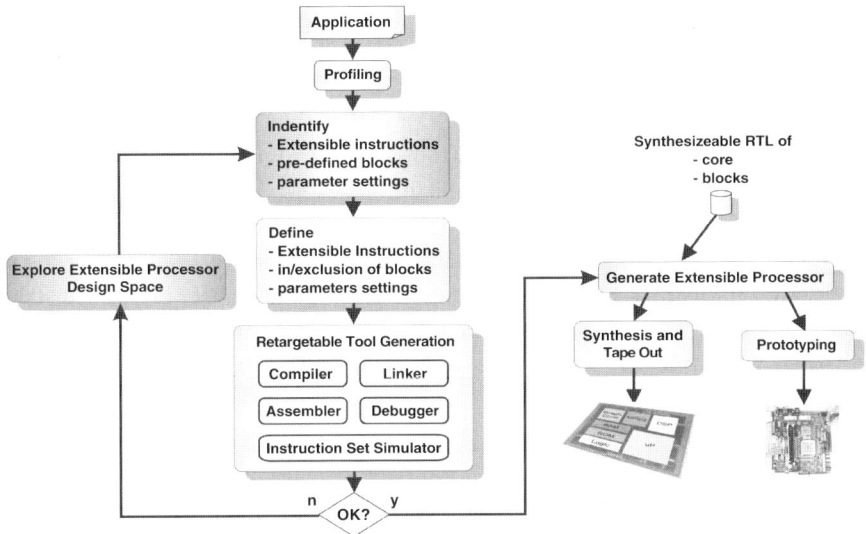

Fig. 1.3   A typical design flow for designing extensible processors

choosing the endianness (little or big endian), choosing the number of registers and many more.

Commercial platforms for extensible processors have been on the market for a while. Major venders of extensible processor platforms include Tensilica, Improv Systems, ARC as well as Coware's LisaTek platform, Target Compiler and others. The latter two pursue a slightly different approach as their focus is directed towards retargetable tool generation (see later) and as they provide means to define/simulate/synthesize a whole new instruction set architecture rather than extending an existing one.

In the following a typical design flow of an extensible processor platform is shortly described[1] with respect to open issues for research.

Figure 1.3 shows the basic steps with the goal to customize the extensible processor to a specific application which might be available in C/C++ description. A profiling by means of, for example, an ISS (Instruction Set Simulator) of the target processor unveils the bottlenecks by cycle-accurate simulation, i.e. it shows which parts of the application represent the most time consuming ones (or, if the constraint is energy consumption, which are the most energy consuming ones). The results of the profiling are the basis for the designer to *identify* (see Figure 1.3) possible instruction extensions, inclusion of pre-defined blocks and parameterization.

---

[1]Please note that is a generic flow only that may or may not cover a specific vendor's platform capabilities. For vendor-specific features please refer to the vendor's web pages.

Finding the right combination of these three major ways of customization is an art and requires an experienced designer. Once, the identification has been completed, the designer starts *defining* (see Figure 1.3). The most important part of this step is the defining of a set of extensible instructions. This is typically done by a custom language that allows to specify an extensible instruction's functionality, scheduling within the pipeline, instruction word format, etc.

After this step is completed, the designer wants to verify that the various customization levels of the extensible processor meet the given constraints (performance, power, etc.) associated with the application. This can be accomplished by retargetable tool generation: simply said, retargetable tool generation is a technique that allows to *retarget* compilation/simulations tools to any architecture. Say, for example, the ISA of an extensible processor is enhanced by a new set of customized instructions. Retargetable techniques then allow to automatically generate a compiler that is aware of the new instruction, i.e. it can generate code and optimize using the recently defined extensible instructions. Accordingly, an ISS is able to cycle-accurate simulate an executable that contains newly defined instructions, etc.

ISS, compiler assembler are then used in the next step to verify in how far the application-imposed constraints can be met using the three discussed customization levels. This is an important step and it can be iterated very fast and thus gives the designer an early feedback through what is called *design space exploration* before synthesis and tape out takes place. In case the designer is satisfied with the simulation results, the specification of instruction definitions and other customization levels are used to *generate* (Figure 1.3) synthesizeable RTL of the extensible processor. This is done by means of a library that contains the synthesizeable RTL of the base processor core. Then, the regular synthesis follow down to the tape out or an evaluation using a rapid prototyping environment follows.

In summary, designing an embedded system using configurable processors as building blocks in a multiprocessor SOC environment represents an efficient way to overcome the complexity gap and will further contribute to the domination of ASSP design starts.

1.3   Open Issues and Key Techniques

The above-mentioned scenario of multiprocessor SOCs replacing traditional domains of custom hardware through extensible processor platforms has still open issues that need more research. Some of them are listed shortly:

- What is the optimum set of extensible instructions for a given application? Currently, the designer of customizable processors does have the means to evaluate instruction extensions on a high level (retargetable

ISSS, etc.) without going through the large turn-around time of implementation first followed by evaluation. However, it is currently entirely up to the designer's experience to identify the kind and granularity of instructions that represent the best return on investment (i.e. chip area) and thus to efficiently explore the extensible processor's design space. Those and similar issues are currently being researched and presented at major design automation conferences like *DAC*, *ICCAD*, *DATE* and others.

- How will all those processors on a SOC communicate with each other? Clearly, traditional on-chip bus systems will not be able to handle the increased traffic when several dozens of heterogeneous processors are integrated on a chip and have to heavily communicate with each other. NOCs not only provide higher bandwidths but they are also scaleable, etc. It is currently an open issue of how an efficient on-chip communication network will look like. Here too, high-level design methodologies are needed to design a custom on-chip network that satisfies the constraints of an embedded application best. The field of architectures and design methodologies for NOCs is currently being researched.

- Retargetable Tool Generation Through several years of research, retargetable tool generation has made significant progresses. However, there are many open issues with respect to extensible processor platforms like, for example, the capability of retargetable compilers to efficiently include extensible (i.e. designer-defined) instructions during the optimization phases etc.

1.4   SOC Design Distinction

Custom-logic hardware blocks in an embedded SOC system are often credited for the SOC's key characteristics that distinguish a certain SOC design of vendor A from a competitive one of vendor B. Those key characteristics include performance, power consumption, etc. With a continued migration of custom logic into programmable processing units, as discussed earlier, design distinction is shifting, too.

Finding the right set of adapted and optimized extensible instructions is an art given the large design space (extensible instruction; inclusion of predefined blocks; parameterization) of extensible processors.

In addition, the processors need to be programmed, hence shifting more and more design expertise into software. An indicator is the increasing size of embedded software: recent designs feature more than a million lines of embedded code tendency rising. Already today, there are many design examples where embedded software takes more resources (engineers) than the hardware design of a SOC.

Conclusion.   The complexity gap will most likely not occur meaning that possible SOC complexities (silicon-technology-wise) of a billion transistors at the end of the decade will be fully exploited by mainstream embedded systems SOCs. Clear indication is an increasing usage of system-level design methodologies and a migration of custom logic into programmable processing units wherever design constraints (performance, power) allow for. A promising approach are extensible processors. In the following, the major challenges when designing an embedded extensible processor are explained.

2.   Challenges in Embedded Extensible Processor Design

Typically, a great amount of expertise is necessay to design extensible processors. The challenging tasks are *code segment identification*, *extensible instruction generation*, *architectural customisation* and *processor evaluation*, which are usually conducted manually. Therefore, recent research has largely focused on these design tasks, and has sought to optimize and automate various steps. This section describes the problems associated with these design tasks and some of the solutions proposed in the literature.

2.1   Code Segment Identification

In order, for example, to increase the performance of an application, code segments can be sped up by generating extensible instructions, including/excluding predefined blocks and setting parameteriations. However, the code segments first need to be identified from the application. While identifying code segments is somewhat supported by profiling tools, it is still a daunting task for large applications, and further complicated when additional constraints (e.g. area and power) must be optimized as well as. Furthermore, the number of code segments for a given application grows large with the program size. It is very common for a function with fewer than one hundred operations to contain several hundred possible code segments.

Recent research can be classified into three categories:

1. Retargetable code generation using matching and covering algorithms

2. Finding patterns in the graph representation (control dataflow graph) of the profiled application

3. High-level extraction from the application

It follows an overview.

Code Generation Using Matching Algorithms.   Code generation using a matching algorithm is a well-known problem, particularly in the fields of technology mapping in logic synthesis (Keutzer, 1987; Cong and Ding, 1992;

Francis et al., 1990) and code generation in compilers (Aho et al., 1989; Corazao et al., 1993; Liem et al., 1994; Shu et al., 1996; Leupers and Marweded, 1998). Matching finds possible instantiations of identical patterns in a structured representation. There are two main approaches to pattern matching: boolean and structural mapping. Boolean matching is often applied to the networks of boolean functions which includes checking the equivalence of functional representations between patterns in the application. This kind of equivalence checking often uses Binary Decision Diagrams (BDDs), which are unsuitable for non-boolean functions. Structural matching focuses on a graph representation, where nodes represent the function. This approach often identifies common patterns with structural rather than functional equivalence. In structural mapping, the type of graph representation can also vary the complexity of the graph. In the early 1990s, most of the matching algorithms revolved around patterns with single acyclic output. However, Arnold (2001) proposed a matching algorithm to identify patterns with multiple outputs which expanded the searching space for possible patterns significantly. Leupers and Marwedel (1995) described an instruction set model to support retargetable compilation and code generation. However, the drawback of these approaches is the lack of application input characteristics (e.g. simulation and profiling information), which may lower the chance to optimize designs for a specific application. In fact, application input provides data-specific information such as estimates for indeterministic loop counts, branch taken or not-taken percentage, the range of input data, etc., which is impossible to ascertain without the use of simulation and input data sets. It is key to select a good coverage of input data sets for an application.

Profiled Graph Representation.   The second category for identifying code segments uses profiling analysis through simulations with a certain set of input data. Using the additional profiling information, a graph representation of the application is created with base instructions as nodes and data dependence between instructions as edges. A code analysis is then performed to identify code suitable segments. Sun et al. (2004) proposed an approach for identifying suitable code segments to implement as extensible instructions in a connected subgraph. First, the application is compiled and profiled using the input data set. Then the program dependence graph is constructed using the profiling instruction and the application with the base instruction as nodes and the dependence between assembly codes as edges. All patterns (e.g. code segments) are identified using a template matching technique. The designer then ranks the patterns from the most frequently executed to the least frequently executed in the application using a priority function. The highly ranked patterns are selected and implemented as extensible instructions. A problem with this approach is that the template patterns need to be pre-defined (which may be best for a particular application) and well constructed in order to maximize the speedup and/or reduce energy consumption.

Clark et al. (2003) proposed a similar pruning technique, called *guide function*, to prune the search design space of code segments in the connected subgraph. Rather than pruning the number of patterns, the authors proposed to prune the searching direction in the graph, thus allowing the possibility that those initial low ranked patterns would amount to a useful pattern in the later stage of the design space. Sun et al., (2003) proposed a scalable approach to extend the existing technique where the matching patterns were not limited to templates. After the patterns are identified, functions can be added or removed from the patterns in order to be well suited to the application. The steps are performed by using a cost function for area and performance tradeoffs.

Atasu et al. (2003) described a binary tree search algorithm that identifies patterns with multiple inputs and outputs in an application dataflow graph, covering an exhaustive design space. This technique achieves maximum speedup and satisfies micro-architectural constraints. It was originally described by Pozzi (2000). Pozzi proposed a generic approach for searching and extracting code segments from an application, where the patterns have multiple inputs and outputs. It is a tree-based searching and extracting algorithm.

Yu and Mitra (2004b) proposed a scalable custom instruction identification method that extracts all possible candidate instructions in a given graph.

High-Level Extraction.   High-level extraction identifies code segments from an application written in high-level language (e.g. C/C++). This approach usually begins with simulation to obtain profiling information. Then, the designer identifies frequently executed sections of the application. Semeria et al. (2002) developed a tool to extract the code segments from C code and generated a functional equivalent of RTL-C and an HDL. Recently, Clarke et al. (2003) introduced an approach for presenting equivalent behavior between C and Verilog HDL which may be used to perform high-level extraction for Yu and Mitra (2004a) described the identification of code segments expoiting the characteristics of the embedded systems application by relaxing constraints.

2.2   Extensible Instruction Generation

Instruction generation involves designing extensible instructions to replace, for example, computationally intensive code segments by specifying new hardware resources and the operations they perform. A typical goal of generating extensible instructions is to maximize performance while satisfying design constraints such as area, power and energy. As mentioned previously, extensible instructions are designed in the execution stage of the processor. If the addition of extensible instructions causes the violation of the base processors clock period, the designer is required to

1. Reduce the amount of computation performed in the instruction

2. Split it into multiple instructions

3. Multi-cycle the execution of the instruction

4. Reduce the clock period of the base processor

Although the instruction generation step is somewhat simplified by specifying extensible instructions at a high level of abstraction, it is still a complex task. Recent research has focused mainly on specific instruction generation on top of a base instruction set which is usually referred to as *instruction generation* or *instruction set extension*. Early research in instruction generation focused on whole custom instruction sets to satisfy design constraints (Holmer, 1994; Huang, and Despain, 1995; Choi et al., 1999a) In 1994, Holmer described a methodology to find and construct the best instruction set on a predefined architecture for an application domain (Holmer, 1994). His method found code segments that are executed in three or four cycles. Huang and Despain (1995) presented an instruction set synthesis for an application in a parameterized, pipelined micro-architecture. This system was one of the first hardware/software systems to be designed for an application with a customized instruction set. The generated instructions are single-cycle instructions. Several years later, Choi et al. (1999a) proposed an approach to generate multi-cycle complex instructions as well as single-cycle instructions for DSP applications. The authors combined regularly executed single-cycle instructions into multi-cycle complex instructions. Recent research has revolved around instruction set extensions (Clark et al., 2003; Biswas et al., 2004; Brisk et al., 2004; Cong et al., 2004) and extensible instruction generation (Lee et al., 2002; Atasu et al., 2003; Goodwin and Petkov, 2003; Sun et al., 2003, 2004).

For instruction set extension, Cong et al., (2004) proposed a performance-driven approach to generate instructions that can maximize application performance. In addition, they allow operations duplication while searching for patterns in the matching phase. The duplication is performed on operations with multiple inputs that are on the critical path of the frequently executed code segments. When the operations are duplicated, parallelism of the code segments may be increased, thus increasing the performance of the application and enhancing design space exploration. The work was evaluated on the NIOS platform, provided by Altera Inc. which is a VHDL reconfigurable embedded processor. Brisk et al. (2004) described an instruction synthesis using resource sharing to minimize the area efficiently. Their approach groups a set of extensible instructions into a co-processor in which common hardware blocks are minimized in the synthesis. The area savings are up to 40% when compared to the original extensible instructions. Biswas et al. (2004) introduced an instruction set extension including local memory elements access. This approach used a hardware unit to enable direct memory access at the execution stage of the processor. In order to enable the local memory access, memory accesses need to be carefully scheduled, or multiple read/write ports are needed. In addition, accessing memory elements in the execution stage potentially increases pipeline hazards, thus increasing the complexity of code optimization. This approach increases the performance of the application at a certain hardware overhead. Clark et al. (2003) described a compiler approach to generate instructions in VLIW architecture without constraining their size or shape. These approaches have largely revolved around maximizing speedup of the application while minimizing area of the processor. None of these approaches focus on the energy consumption of the application.

Extensible instruction generation focuses on generating instructions that satisfy the latency of the base processor while maximizing performance and other constraints (Lee et al., 2002; Atasu et al., 2003; Goodwin and Petkov, 2003; Sun et al., 2003; Sun et al., 2004). (Lee et al., 2002) proposed an instruction encoding scheme for generating complex instructions. The instruction encoding scheme enables tradeoffs between the size of opcode and operands in the instructions to enhance performance and reduce power dissipation. In addition, it contains a flexible approach for creating complex instructions, the combination of basic instructions that regularly appeared in the application, exploring greater design space and achieving improved performance. (Atasu et al., 2003) described a generic method to generate extensible instructions by grouping frequently executed code segments using a tree-based searching and matching approach. The method enables the generation of extensible instructions with multiple inputs and outputs. (Sun et al., 2004) described a methodology to generate custom instructions from operation pattern matching in a template pattern library. The generated instructions increase the application performance by up to 2–5 with a minimal increase in area. (Sun et al., 2003) they described a scalable instruction synthesis that could be adopted by adding and removing operations from custom instructions, further ensuring that the given latency constraint is satisfied. This approach also optimized the area overhead of the instruction while maximizing the performance of the instructions. (Goodwin and Petkov, 2003) described an automatic system to generate extensible instructions using three operation techniques. These operation techniques are (i) Very Long Instruction Word (VLIW) operations – grouping multiple instructions into a single instruction in parallel; (ii) vector operations – parallelizing data and increasing the instruction width; and (iii) fused operations - combining sequential instructions into a single instruction. This system achieved significant speedup for the application while exploring millions of instruction combinations in several minutes. Tensilica Inc. later implemented this system as the Xpress system (Ten, 2007). (Sun et al., 2005) also recently presented a heterogeneous multiprocessor instruction set synthesis using extensible processors to speed up the application. Although these approaches have shown energy reduction, this is achieved by combining computationally intensive code segments into extensible instructions, which reduces execution time significantly (with an incremental increase in power dissipation).

2.3   Architectural Customization Selection

Architectural customization selection involves selecting extensible instructions, pre-defined blocks and parameter settings in the extensible processor to maximize application performance while satisfying design constraints. This process is often referred to as design space exploration. This selection problem can be simplified and formulated as a well-known Knapsack problem, with single or multiple constraints. The single constraint Knapsack problem is defined where an item i has a value v_i and weights w_i. The goal is to find a subset of the n items such that the total value is maximized and the weight constraint is satisfied. In our case, the item is architectural customization, AC, such as extensible instructions, pre-defined blocks and parameter settings. Each customization has a speedup factor, compared with the software code segment that it replaced, and a single design constraint such as area and/or power. In the single design constraint case, the simplified form of the problem is not strongly NP-hard and effective approximation algorithms have been proposed for obtaining a near-optimal solution. Exact and heuristic algorithms are also developed for the single constraint knapsack problem (summarized by Martello and Toth, 1990). A comprehensive review of the single knapsack problem and its associated exact and heuristic algorithms is given there. For a review of the multiple-constraints Knapsack problem, refer to Chu and Beasley (1998).

This section will discuss the literature related to design space exploration and architectural customization selection in extensible processors with single or multiple architectural customizations under single or multiple constraints. Research in extensible processor platform exploration has largely revolved around single architectural customization (either predefined blocks or extensible instructions or parameterizations) under a single constraint. A number of researchers have described methods to include/exclude pre-defined blocks to customize very long instruction word (VLIW) and explicitly parallel instruction computing (EPIC) processors (Aditya et al., 1999; Alippi et al., 1999; Kathail et al., 2002). Choi et al. (1999b) presented a method to select intellectual properties to increase the performance of an application. Gupta et al. (2002) described an automatic method to select among processor options under an area constraint. For extending instructions, Lee et al. (2003) proposed an instruction set synthesis for reducing the energy delay product of application-specific processors through optimal instruction encoding. Various methods to generate extensible instructions automatically from basic, frequently occurring, operation patterns have also been devised (Choi et al., 1999a; Brisk et al., 2002; Atasu et al., 2003; Sun et al., 2003). Parameterizations in the extensible processor platform leads to the setting of the register file size, instruction and data cache sizes, and memory configuration. Jain et al. (2001) proposed a method to evaluate the register file size. Methods to optimize the memory size of the embedded software in a processor have also been proposed (Catthoor et al., 1998; Wolf and Kandemir, 2003). Finally, Abraham and Rau (2000) described a scheme to select instruction and data cache size. They and Lee et al. (2002) presented examples of design exploration with multiple architectural customizations under a single constraint. The PICO system was proposed to explore the design space of the non-programmable hardware accelerators (NPAs) and parameters in memory and cache for VLIW processors. It produces a set of sub-optimal solutions using a divide-and-conquer approach and defers the final constraint tradeoffs to the designer. This is the cornerstone of the system provided by IPflex, Inc. (Lee et al., 2002) proposed a heuristic design space exploration for encoded instructions and parameter settings with the tradeoffs between area overhead and performance.

There is very little work on extensible processor platform exploration under multiple constraints when pre-defined blocks and extensible instructions are involved. Often, research in extensible processor platform exploration only focuses on the area constraint while the energy constraint is neglected (Lee et al., 2002; Sun et al., 2004). This naive assumption is due to the fact that improving performance often also reduces energy consumption of the program running on the custom processor. The selection problem under multiple constraints has been around for more than a decade in other research areas and is often formalized as a multi-dimensional Knapsack problem.

Instruction Set Simulation simulates the performance of the application in the newly configured extensible processor (consisting of the base instruction set, extensible instructions and instructions associated with pre-defined blocks). In order to further reduce time-to-market pressure, research into abstract high-level estimations for extensible processors has been carried out. Gupta et al. (2000) proposed a processor evaluation methodology to quickly estimate the performance improvement when architectural modifications are made. (Jain et al. 2003, 2005) proposed methodologies to evaluate the register file size, register window and cache configuration in an extensible processor design. By selecting an optimum register file size, Bhatt et al. (2002) also proposed a methodology to evaluate the number of register windows needed in processor synthesis. Fei et al. (2004) described a hybrid methodology for estimating energy consumption of extensible processors. This chapter has described a wide range of design approaches for embedded systems and various architectures of application-specific instruction-set processors. It introduced reasons for using extensible processor platforms, and showed that the extensible processor platform is the state-of-the-art design approach for todays embedded systems. This chapter introduced design problems related to extensible processor platform, such as code segment identification, instruction generation, architectural customization selection and processor evaluation and estimation, and described the state-of-the-art work to resolve these issues. In the next chapter, we present our proposed design methodologies to further address these design problems and show how our methodologies advance the existing work.

2.4 Summary

This chapter summarized challenges and state-of-the-art in designing embedded processors. The following four chapters within this section are as follows. The so-called NISC approach by Gorjiara, Reshadi, Gajski is presented in the chapter Low-Power Design with NISC Technology. It describes a novel architecture for embedded processors that does not use instructions in the traditional way any more. The chapter entitled "Synthesis of Instruction Sets for high Performance and Energy efficient ASIP" by Lee, Choi, Dutt focuses on reducing the energy-delay product of instruction-extensible through synthesis. The chapter "A Framework for Extensible Processor-Based MPSoC Design" by Sun, Ravi and Raghunathan with an approach to utilize the energy/power efficiency of extensible processors in the context of a multi processor system. This section concludes with "Design and Run-time Code Compression for Embedded Systems". Parameswaran, Henkel, Janapsatya, Bonny and Ignjatovic show the increase in efficiency for an embedded processor when code compression is used.

References

Abraham, S. G. and Rau, B. R. (2000) Efficient Design Space Exploration in PICO. In *International Conference on Compilers, Architectures, and Synthesis for Embedded Systems*, pp. 71–79.

Aditya, S., Rau, B. R., and Kathail, V. (1999) Automatic Architectural Synthesis of VLIW and EPIC Processors. In *International Symposium on System Synthesis*, pp. 107–113.

Aho, A., Ganapathi, M., and Tjiang, S. (1989) Code Generation using Tree Matching and Dynamic Programming. *ACM Transactions on Programming Languages and Systems*, 11(4):491–561.

Alippi, C., Fornaciari, W., Pozzi, L., and Sami, M. (1999) A DAG-based Design Approach for Reconfigurable VLIW Processors. In *Design, Automation and Test in Europe Conference and Exhibition*, pp. 778–780.

Arnold, M. (2001) *Instruction Set Extension for Embedded Processors*. Ph.D. Thesis, Delft University of Technology.

Atasu, K., Pozzi, L., and Ienne, P. (2003) Automatic Application-Specific Instruction-Set Extensions Under Microarchitectural Constraints. In *ACM/IEEE Design Automation Conference*, pp. 256–261.

Benini, L. and De Micheli, G. (2002) Networks On Chip: A New Paradigm for Systems On Chip Design. In *IEEE/ACM Proc. of Design Automation and Test in Europe Conference (DATE)*, pp. 418–419.

Bhatt, V., Balakrishnan, M., and Kumar, A. (2002) Exploring the Number of Register Windows in ASIP Synthesis. In *International Conference on VLSI Design*, pp. 233–238.

Biswas, P., Choudhary, V., Atasu, K., Pozzi, L., Ienne, P., and Dutt, N. (2004) Introduction of Local Memory Elements in Instruction Set Extensions. In *ACM/IEEE Design Automation Conference*, pp. 729–734.

Brisk, P., Japlan, A., and Sarrafzadeh, M. (2004) Area-Efficient Instruction Set Synthesis for Reconfigurable System-on-Chip Designs. In *ACM/IEEE Design Automation Conference*, pp. 395–400.

Brisk, P., Kaplan, A., Kastner, R., and Sarrafzadeh, M. (2002) Instruction Generation and Regularity Extraction for Reconfigurable Processors. In *International Conference on Compilers, Architecture, and Synthesis for Embedded Systems*, pp. 262–269.

Cataldo, A. (2003) Nec makes series push for asic-lite. In *EE-Times*.

Catthoor, F., DeGreef, E., and Suytack, S. (1998) *Custom Memory Management Methodology: Exploration of Memory Organisation for Embedded Multimedia System Design*. Kluwer Academic Publishers, Norwell, USA.

Choi, H., Kim, J., Yoon, C., Park, I., Hwang, S., and Kyung, C. (1999a) Synthesis of Application Specific Instructions for Embedded DSP Software. *IEEE Transactions on Computers*, 48(6):603–614.

Choi, H., Yi, J., Lee, J., Park, I., and Kyung, C. (1999b) Exploiting Intellectual Properties in ASIP Designs for Embedded DSP Software. In *ACM/IEEE Design Automation Conference*, pp. 939–944.

Chu, P. C. and Beasley, J. E. (1998) A Genetic Algorithm for the Multidimensional Knapsack Problem. *Journal of Heuristics*, 4(1):63–86.

Clark, N., Zhong, H., and Mahlke, S. (2003a) Processor Acceleration through Automated Instruction Set Customization. In *IEEE/ACM International Symposium on Microarchitecture*, pp. 129–140.

Clarke, E., Kroening, D., and Yorav, K. (2003b) Behavioral Consistency of C and Verilog Programs Using Bounded Model Checking. In *ACM/IEEE Design Automation Conference*, pp. 368–371.

Cong, J. and Ding, Y. (1992) An Optimal Technology Mapping Algorithm for Delay Optimization in Lookup-table Based FPGA Designs. In *IEEE International Conference on Computer Aided Design*, pp. 48–53.

Cong, J., Fan, Y., Han, G., and Zhang, Z. (2004) Application-Specific Instruction Generation for Configurable Processor Architectures. In *International Symposium on Field Programmable Gate Array*, pp. 183–189.

Corazao, M., Khalaf, M., Guerra, L., Potkonjak, M., and Rabaey, J. (1993) Instruction Set Mapping for Performance Optimization. In *IEEE International Conference on Computer Aided Design*, pp. 518–521.

Fei, Y., Ravi, S., Raghunathan, A., and Jha, N. (2004) A Hybrid Energy Estimation Technique for Extensible Processors. *IEEE Transactions on Computer-Aided Design of Integrated Circuits and Systems*, 23(5):652–664.

Francis, R. J., Rose, J., and Chung, K. (1990) Chortle: a Technology Mapping Program for Lookup Table-Based Field Programmable Gate Arrays. In *ACM/IEEE Design Automation Conference*, pp. 613–619.

Goodwin, D. and Petkov, D. (2003) Automatic Generation of Application Specific Processors. In *International Conference on Compilers Architectrue and Synthesis for Embedded Systems*, pp. 137–147.

Gupta, T., Ko, R., and Barua, R. (2002) Compiler-Directed Customization of ASIP Cores. In *International Symposium on Hardware/Software Co-Design*, pp. 97–102.

Gupta, T., Sharma, P., Balakrishnan, M., and Malik, S. (2000) Processor Evaluation in an Embedded Systems Design Environment. In *International Conference on VLSI Design*, pp. 98–103.

Holmer, B. (1994) A Tool for Processor Instruction Set Design. In *European Conference on Design Automation*, pp. 150–155.

Holt, D. (2003) Standard cell is broken. In *EE-Times*.

Huang, I. and Despain, A. (1995) Synthesis of Application Specific Instruction Sets. *IEEE Transactions on Computer-Aided Design of Integrated Circuits and Systems*, 14(6):663–675.

Jain, M. K., Balakrishnan, M., and Kumar, A. (2003) Exploring Storage Organization in ASIP Synthesis. In *Euromicro-Symposium on Digital System Design*, pp. 120–127.

Jain, M. K., Balakrishnan, M., and Kumar, A. (2005) Integrated On-Chip Storage Evaluation in ASIP Synthesis. In *International Conference on VLSI Design*, pp. 274–279.

Jain, M. K., Wehmeyer, L., Steinke, S., Marwedel, P., and Balakrishnan, M. (2001) Evaluating Register File Size in ASIP Design. In *International Symposium on Hardware/Software Codesign*, pp. 109–114.

Kathail, V., Aditya, S., Schreiber, R., Rau, B., Cronquist, D., and Sivaraman, M. (2002) PICO: Automatically Designing Custom Computers. *Computer*, 35(9):39–47.

Keutzer, K. (1987) DAGON: Technology Binding and Local Optimization by DAG Matching. In *ACM/IEEE Design Automation Conference*, pp. 617–623.

Lee, J., Choi, K., and Dutt, N. (2002) Efficient Instruction Encoding for Automatic Instruction Set Design of Configurable ASIPs. In *International Conference on Computer Aid Design*, pp. 649–654.

Lewis, B. and Hsu, S. (2002) Asic/soc: Rebuilding after the storm. In *Gartner*.

Lee, J., Choi, K., and Dutt, N. (2003) Energy-Efficient Instruction Set Synthesis for Application-Specific Processors. In *International Symposium on Low Power Electronics and Design*, pp. 330–333.

Leupers, R. and Marwedel, P. (1995) Instruction-Set Modelling for ASIP Code Generation. In *IEEE International Conference on VLSI Design*, pp. 77–80

Leupers, R. and Marwedel, P. (1998) Retargetable Code Generation Based on Structural Processor Descriptions. *Design Automation for Embedded Systems*, 3(1):1–36.

Liem, C., May, T., and Paulin, P. (1994) Instruction-Set Matching and Selection for DSP and ASIP Code Generation. In *European Conference on Design Automation*, pp. 31–37.

Martello, S. and Toth, P. (1990) *Knapsack Problems: Algorithms and Computer Implementations*. John Wiley & Sons, Inc., New York, NY, USA.

NEC (2003) Nec boosts tcp/ip protocol processing with 10 cpu cores on one chip. In *NEC Corporation, Press Release*.

Paulin, P. (2003) Extending SOC Life beyond the 90 nm Wall. In *Talk at DAC Panel, 40th Design Automation Conference (DAC), Anaheim*.

Pozzi, L. (2000) *Methodologies for the Design of Application-Specific Reconfigurable VLIW Processors*. Ph.D. Thesis, Politecnico di Milano.

Rowen, C. (2003) presentation at NEC technology forum.

Semeria, L., Seawright, A., Mehra, R., Ng, D., Ekanayake, A., and Pangrle, B. (2002) RTL C-Based Methodology for Designing and Verifying a Multi-Threaded Processor. In *ACM/IEEE Design Automation Conference*, pp. 123–128.

Shu, J., Wilson, T.C., and Banerji, D.K. (1996) Instruction-Set Matching and GA-Based Selection for Embedded-Processor Code Generation. In *International Conference on VLSI Design*, pp. 73–76.

Smith, G. (2003) DAC panel presentation, 40th Design Automation Conference (DAC), Anaheim.

Sun, F., Raghunathan, A., Ravi, S., and Jha, N. (2003) A Scalable Application Specific Processor Synthesis Methodology. In *International Conference on Computer Aid Design*, pp. 283–290.

Sun, F., Ravi, S., Raghunathan, A., and Jha, N. (2004) Custom-Instruction Synthesis for Extensible-Processor Platforms. *IEEE Transactions on Computer-Aided Design of Integrated Circuits and Systems*, 23(2):216–228.

Sun, F., Ravi, S., Raghunathan, A., and Jha, N. (2005) Synthesis of Application-Specific Heterogeneous Multiprocessor Architectures Using

Extensible Processors. In *International Conference on VLSI Design*, pp. 551–556.

Wolf, W. and Kandemir, M. (2003) Memory System Optimization of Embedded Software. *IEEE*, 91(1):165–182.

Xtensa Processor. Tensilica, Inc. (http://www.tensilica.com) 2007.

Yu, P. and Mitra, T. (2004a) Characterizing Embedded Applications for Instruction-Set Extensible Processors. In *ACM/IEEE Design Automation Conference*, pp. 723–728.

Yu, P. and Mitra, T. (2004b) Scalable Custom Instructions Identification for Instruction-Set Extensible Processors. In *International Conference on Compilers, Architectures, and Synthesis for Embedded Systems*, pp. 69–78.

Chapter 2

Low-Power Design with NISC Technology

Bita Gorjiara, Mehrdad Reshadi, and Daniel Gajski
Center for Embedded Computing System
University of California
Irvine, CA

Abstract   The power consumption of an embedded application can be reduced by moving as much computation as possible from runtime to compile time, and by customizing the microprocessor architecture to minimize number of cycles. This chapter introduces a new generation of processors, called No-Instruction-Set-Computer (NISC), that gives the full control of the datapath to the compiler, in order to simplify the controller hardware, and enable fast architecture customizations.

Keywords: application-specific design; processor; pipelined datapath, low-power; energy; architecture customization; CISC; RISC; NISC; VLIW; ASIP.

1.   Introduction

The power consumption of an embedded application can be reduced by moving as much computation as possible from runtime to compile time, and by customizing the microprocessor architecture to minimize number of cycles. Over the past years, the trend of processor design has been to give compiler more control over the processor. This is more evident in transition from CISC (Complex Instruction Set Computer) to RISC (Reduced Instruction Set Computer) and from Superscalar to VLIW (Very Long Instruction Word) processors. While in CISC, instructions perform complex functionalities, in RISC, the compiler constructs these functionalities from a series of simple instructions. Similarly, while in superscalar, instruction scheduling is done in hardware at runtime, in VLIW, the compiler performs the instruction scheduling statically at compile time.

J. Henkel and S. Parameswaran (eds.), Designing Embedded Processors – A Low Power Perspective,
25–50.
© 2007 *Springer.*

Increasing the role of compiler and its control over the processor has several benefits:

1. The compiler can look at the entire program and hence has a much larger observation window than what can be achieved in hardware. Therefore, much better analysis can be done in compiler than hardware.

2. More complex algorithms (such as instruction scheduling, register renaming) can be implemented in compiler than in hardware. This is because first, the compiler is not limited by the die size and other chip resources; and second, compiler's execution time does not impact the application execution time. In other words, compiler runs at design time, while hardware algorithms run during application execution.

3. The more functionality we move from hardware to compiler, the simpler the hardware becomes, and the less the runtime overhead is. This has a direct effect on power consumption of the circuit.

In general, performance of applications can be improved by exploiting their inherent horizontal and vertical parallelism. Horizontal parallelism occurs when multiple independent operations can be executed simultaneously. Vertical parallelism occurs when different stages of a sequence of operations can be overlapped. In processors, horizontal parallelism is utilized by having multiple functional units that run in parallel, and vertical parallelism is utilized through pipelining. Currently, in VLIW processors, the compiler controls the schedule of parallel independent operations (*horizontal control*). However, in all processors, the compiler has no *vertical control* (i.e. control over the flow of instructions in the pipeline). Therefore, the vertical parallelism of the program may not be efficiently utilized. In Application-Specific Instruction-set Processors (ASIPs), structure of pipeline can be customized for a given application through custom instructions. However, these approaches also have complex design flows and impose limitations on the extent of customizations (see Section 3.1).

On the other hand, in all processors, no matter how many times an instruction is executed, it always goes through an instruction decoder. The instruction decoder consumes power and complicates the controller as well.

The No-Instruction-Set-Computer (NISC) addresses the above issues by moving as much functionality as possible from hardware to compiler. In NISC, the compiler determines both the schedule of parallel independent operations (horizontal control), and the logical flow of sequential operations in the pipeline (vertical control). The compiler generates the control words (CWs) that must be applied to the datapath components at runtime in every cycle. In other words, in NISC, all of the major tasks of a typical processor controller (i.e. instruction decoding, dependency analysis, and instruction scheduling) are done by the compiler statically.

NISC technology can be used for low-power application-specific processor design, because: (a) the compiler-oriented control of the datapath inherently minimizes the need for runtime hardware-based control, and therefore, reduces the overall power consumption of the design; (b) NISC technology allows datapath customizations to reduce total number of cycles and therefore total energy consumption. The extra slack time can also be used for voltage and frequency scaling, which result in more savings.

In the rest of this chapter, an overview of NISC technology is presented in Section 2. NISC is compared with other approaches such as ASIP, VLIW, and High-Level Synthesis (HLS) in Section 3. In Section 4, NISC compilation algorithm is discussed. In Section 6, an overview of power saving techniques in NISC is presented, followed by experimental results in Section 7.

2.   Overview of NISC Technology

A NISC is composed of a pipelined datapath and a pipelined controller that drives the control signals of the datapath components in each clock cycle. The controller has a fixed template and is usually composed of a Program Counter (PC) register, an Address Generator (AG), and a Control Memory (CMem). The control values are stored in a control memory. For small size programs, the control values are generated via logic in the controller. The datapath of NISC can be simple or as complex as datapath of a processor. The controller drives the control signals of the datapath components in each clock cycle. The NISC compiler generates the control values for each clock cycle.

Figure 2.1 shows a NISC architecture with a memory-based controller and a pipelined datapath that has partial data forwarding, multi-cycle and pipelined units, as well as data memory and register file. In presence of controller

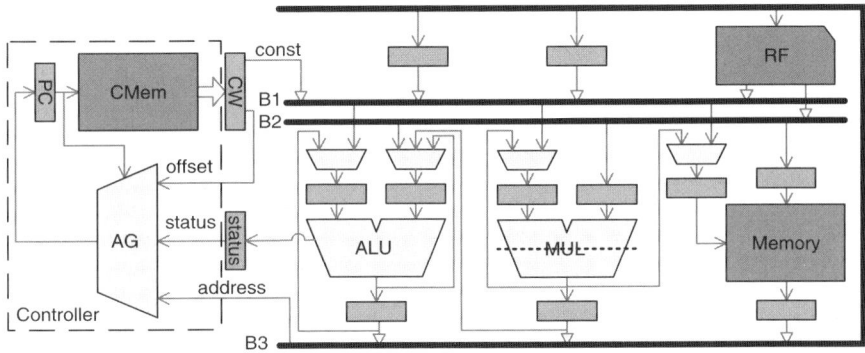

Fig. 2.1   NISC architecture example

pipelining (i.e. CW and Status registers in Figure 2.1), the compiler should also make sure that the branch delay is considered correctly and is filled with other independent operations.

The compiler translates each basic block to a sequence of CWs that run sequentially without interruption. In other words, any pipeline stall or context switch (e.g. interrupt routine call) happens only between basic blocks. This is analogous to traditional processors in which pipeline stalls or context switches happen between instructions. The NISC compiler is called *cycle-accurate compiler* because it decides what the datapath should do in every clock cycle. Compilation algorithm detail is presented in [1,2].

Figure 2.2 shows the flow of designing a custom NISC for a given application. The datapath can be generated (allocated) using different techniques. For example, it can be an IP, reused from other designs, generated by HLS, or specified by the designer. The datapath description is captured in a Generic Netlist Representation (GNR). A component in datapath can be a register, register file, bus, multiplexer, functional unit, memory, etc. The program, written in a high-level language such as C, is first compiled and optimized by a front-end and then mapped (scheduled and bound) on the given datapath. The compiler generates the stream of control values as well as the contents of data memory. The generated results and datapath information are translated to a synthesizable RTL design that is used for simulation and synthesis. After synthesis and Placement-and-Routing (PAR), the accurate timing, power, and area information can be extracted and used for further datapath refinement. For example, the user may add functional units and pipeline registers, or change the bitwidth of the components and observe the effect of modifications on precision

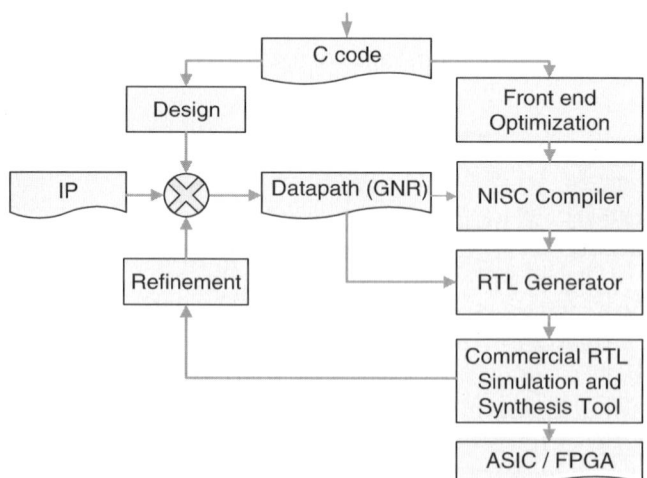

Fig. 2.2   Generating a NISC architecture for an application

of the computation, number of cycles, clock period, power, and area. In NISC, there is no need to design the instruction-set because the compiler automatically analyzes the datapath and extracts possible operations and branch delay. Therefore, the designer can refine the design very fast.

3. NISC Compared to Other Approaches

NISC combines techniques from High-Level Synthesis (HLS), Application-Specific Instruction-set Processor (ASIP), and retargetable compilers. Its execution style is also similar to VLIW and statically scheduled horizontally microcoded machines. This section summarizes the differences between each of these techniques and NISC.

3.1 NISC vs. ASIP

Compiling an application to a customized datapath has also been the goal of Application-Specific Instruction-set Processors (ASIPs) and retargetable compilers.

In ASIPs, functionality and structure of datapath can be customized for an application through custom instructions. At runtime, each custom instruction is decoded and executed by the corresponding custom hardware. Due to practical constrains on size and complexity of instruction decoder and custom hardware, only few custom instructions can be actually implemented in ASIPs. Therefore, only the most frequent or beneficial custom instructions are selected and implemented. Implementing these custom instructions requires: (a) designing custom hardware for each instruction, (b) implementing an efficient instruction decoder, and (c) incorporating the new instructions in the compiler. These steps are complex and usually time consuming tasks that require special expertise.

To automatically incorporate the custom instructions into the compiler, the retargetable compilers use a processor description captured in an Architecture Description Language (ADL). All retargetable compilers rely on high-level instruction abstractions to *indirectly* control the datapath of the processor. They always assume that the processor already has a controller that translates the instructions into proper control signals for the datapath components. In behavioral ADLs, the processor is described in terms of the behavior of its instructions. These ADLs are usually very lengthy because they have to capture all possible configurations of instructions. Furthermore, since no structural information is available in the ADL, the quality of automatically generated RTL for the processor is very low. Structural ADLs try to improve the quality of generated RTL by capturing the controller, instruction decoder, and datapath of the processor. Capturing the instruction decoder significantly complicates these ADLs. Additionally, extracting the high-level instruction behaviors from these ADLs for the compiler is very complex and can only be done for limited architectural features.

The NISC cycle-accurate compiler generates code *as if* each basic block of program is executed with one custom instruction. A basic block is a sequence of operations in a program that are always executed together. Ideally, each basic block should executed by a single instruction that reads the inputs of basic block from a storage unit (e.g. register file) and computes the outputs of basic block and stores them back. The large number of basic blocks in a typical program prevents us from using an ASIP approach to achieve the above goal. To solve this problem, in NISC instruction decoding is moved from hardware to the compiler. In ASIP, after reading the binary of a custom instruction from memory, it is decoded into a set of control words (CWs) that control the corresponding custom datapath and executes the custom functionality. Instead of having too many custom instructions and then relying on a large instruction decoder to generate CWs in hardware, in NISC the CWs are generated in compiler by directly mapping each basic block onto the custom datapath. Therefore, the compiler can construct unlimited number of custom functionalities utilizing both horizontal and vertical parallelism of the input program. If the datapath is designed to improve the execution of certain portions of program, the NISC compiler will automatically utilize it. Since the compiler is no longer limited by the fixed semantics of instructions, it can fully exploit datapath capabilities and achieve better parallelism and resource utilization.

3.2   NISC vs. VLIW

In Very-Long-Instruction-Word (VLIW) machines (also known as Explicitly-Parallel-Instruction-Computing), parallel operations are statically scheduled by the compiler. Removing the instruction scheduling from the controller of the processor makes the hardware simpler and more efficient. As a result, many of the contemporary ASIP and DSP processors are VLIW based. Although compiler's control over the architecture is increased in VLIW, the compiler still has no control over how the instructions go through the pipeline. In other words, compiler can only control the horizontal parallelism and not the vertical parallelism.

In NISC, in addition to utilizing the horizontal parallelism, the compiler can also determine how the operations and their results flow though pipeline. While in VLIW, the instruction word is partitioned to several same-size operation slots, the control word in NISC is not partitioned to predefined sections. Therefore, NISC can achieve higher parallelism using fewer bits.

3.3   NISC vs. Microcoded Architectures

In horizontally microcoded architectures, in each cycle, a set of very simple operations (called microcode) determines the behavior of datapath in that cycle. The assumption in these types of architectures is that a microcode executes in a single cycle. This is a rather limiting assumption for modern architectures. For example consider the execution of a *load* (Memory Read) operation that needs a precisely timed sequence of events. During *load*, the *ChipSelect* control signal of memory becomes 1 at some time (cycle) and then becomes 0 at some other time (cycle). Note that, these events cannot be associated with different microcodes since such microcodes would not have meaningful execution semantics for compiler. Therefore, the microcode operations may require a simple decoder in the architecture that translates them to actual control values. In fact, in most microcoded machines, the compiler only deals with microcodes and the structural details of datapath are often hidden from compiler.

In contrast, in NISC, each low-level action (such as accessing storages, transferring data through busses/multiplexers, and executing operations) is associated with a simple timing diagram that determines the values of corresponding control signals at different times. The NISC compiler eventually schedules these control values based on their timings and the given clock period of the system. Therefore, although a NISC may look similar to a horizontally microcoded architecture, the NISC compiler has much more low-level control over the datapath and hence is closer to a synthesis tool in terms of capability and complexity.

3.4 NISC vs. HLS

Traditional High-Level Synthesis (HLS) techniques [3–6] take an abstract behavioral description of a digital system and generate a register-transfer-level (RTL) datapath and controller. In these approaches, after scheduling, the components are connected (during interconnect synthesis) to generate the final datapath. The generated datapath is in form of a netlist and must be converted to layout for the final physical implementation. Lack of access to layout information limits the accuracy and efficacy of design decisions especially during scheduling. For example, applying interconnect pipelining technique is not easy during scheduling, because wire information is not available yet. It is not also possible to efficiently apply it after generating the datapath because it invalidates the schedule. There have been many attempts [7–11] in the past to predict or estimate the physical attributes of the final datapath layout. Not only these attempts lack the necessary accuracy, they also lead to more complex allocation and scheduling algorithms.

The growing complexity of new manufacturing technologies demands synthesis techniques that support Design-For-Manufacturability (DFM). However, the interdependent scheduling, allocation and binding tasks in HLS are too complex by themselves and adding DFM will add another degree of complexity to the design process. This increasing complexity requires a design flow that provides a practical separation of concerns and supports more aggressive optimizations based on accurate information.

These goals are achieved in NISC by separating the generation of datapath and controller. First the datapath is designed and remains fixed during compilation, then the control word for each clock cycle is generated by mapping (scheduling and binding) the application on the given datapath, and finally the control words are stored in a memory-based controller. In this way, DFM and other layout optimizations are handled independently from compilation/synthesis. Furthermore, accurate layout information can be used by scheduler. The datapath can be generated in several ways. It can be selected from available IPs or reused from previous designs. It can also be designed manually or automatically using techniques from HLS or ASIP design that analyze the application behavior and suggest a custom datapath. Such datapath can be iteratively refined and optimized. Traditional HLS techniques usually support a subset of a high-level language such as C. By adding proper components and structures to the datapath of NISC, it can cover all language features and target a broader application domain.

NISC and HLS can be compared from another point of view too. Resource-constrained HLS algorithms have been developed in order to improve the quality of generated results. However, the focus of these algorithms is only functional unit and storage element resources. In current implementation technologies, the wires have considerably more delay than before, and are major contributors to the design complexity. The next natural step in progress of resource-constrained algorithms is to consider wire constraints as well. In other words, in addition to the number and types of functional units and storage elements, the input to such algorithms must also include connectivity of components. Hence the datapath is fully allocated and the controller is generated by performing scheduling and binding on the operations of the application. This approach eventually leads to NISC.

In summary, in contrast to HLS, NISC design approach separates the design concerns and allows iterative refinement and optimizations, enables use of accurate layout information during design, and enables design reuse.

4.   Overview of the Compilation Algorithm

In this section, the basis of NISC scheduling and binding algorithm is illustrated using an example. The input of algorithm is the CDFG of an application, netlist of datapath components, and the clock period of the system. The output is an FSM in which each state represents a set of Register Transfer Actions (RTAs) that execute in one clock cycle. An RTA can be either a data transfer through buses/multiplexers/registers, or an operation executed on a functional unit. The set of RTAs are later used to generate the control bits of components.

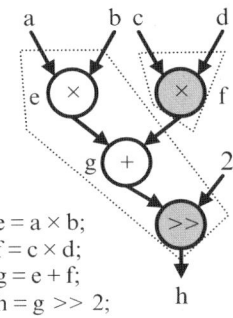

e = a × b;
f = c × d;
g = e + f;
h = g >> 2;

Fig. 2.3 A sample DFG

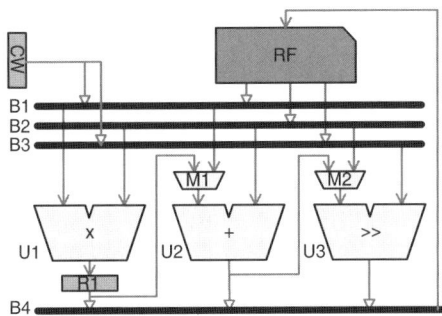

Fig. 2.4 A sample datapath

As opposed to traditional HLS, in NISC, operations cannot be scheduled merely based on the delay of the functional units. The number of control steps between the schedule of an operation and its successor depends on both the binding of operations to functional units (FU) and the delay of the path between corresponding FUs. For example, suppose we want to map DFG of Figure 2.3 on datapath of Figure 2.4. Operation shift-left ($\gg$) can read the result of operation $+$ in two ways. If we schedule operation $+$ on $U2$ and store the result in register file RF, then operation $\gg$ must be scheduled on $U3$ in next cycle to read the result from RF through bus $B2$ and multiplexer $M2$. Operation $\gg$ can also be scheduled in the same cycle with operation $+$ and read the result directly from $U2$ through multiplexer $M2$. Therefore, selection of the path between $U2$ and $U3$ can directly affect the schedule. Since knowing the path delay between operations requires knowing the operation binding, the scheduling and binding must be performed simultaneously.

The basic idea in the algorithm is to schedule an operation and all of its predecessors together. An *output operation* in the DFG of a basic block is an operation that does not have a successor in that basic block. The algorithm

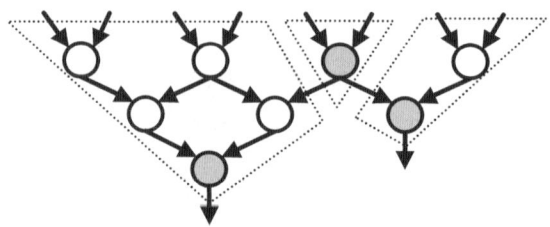

Fig. 2.5   Partitioning a DFG into output sub-trees

starts from output operations and traverse the DFG backward. Each operation is scheduled after all its successors are scheduled. The scheduling and binding of successors of an operation determine when and where the result of that operation is needed. This information can be used for: utilizing available paths between FUs efficiently, avoiding unnecessary register file read/writes, chaining operations, etc.

The DFG of the basic block is partitioned into sub-trees. The root of a subtree is an output operation. The leaves are input variables, constants, or output operations from other basic blocks. If the successors of an operation belong to different sub-trees, then that operation is considered as an *internal output* and will have its own sub-tree. Such nodes are detected during scheduling. Figure 2.5 shows an example DFG that is partitioned into three sub-trees. The roots of the sub-trees (shown with shaded nodes) are the output operations. The algorithm schedules each sub-tree separately. If during scheduling of the operations of a sub-tree, the schedule of an operation fails, then that operation is considered an internal output and becomes the root of a new sub-tree. A sub-tree is available for schedule as soon as all successor of its root (output operation) are scheduled. Available sub-trees are ordered by the mobility of their root. The algorithm starts from output nodes and schedules backward toward their inputs; therefore, more critical outputs tend to be generated towards the end of the basic block.

Consider the example DFG of Figure 2.3 to be mapped on the datapath of Figure 2.4. Assume that the clock period is 20 units and delays of $U1$, $U2$, $U3$, multiplexers, and busses are 17, 7, 5, 1, and 3 units, respectively. The operations of the basic block are scheduled so that all results are available before last cycle, i.e. 0; therefore, the RTAs are scheduled in negative cycle numbers. In each step, we try to schedule the sub-trees that can generate their results before a given cycle *clk*. The *clk* starts from 0 and is decremented in each step until all sub-trees of a basic block are scheduled.

During scheduling, different types of values may be bound to different types of storages (variable binding). For example, global variables may be bound to memory, local variables to stack or register file, and so on. A constant is bound to memory or control word (CW) register, depending on its size. A control word may have limited number of constant fields that are generated in each cycle along with the rest of control bits. These constant fields are loaded into the CW register and then transferred to a proper location in datapath. The NISC compiler determines the values of constant(s) in each cycle. It also schedules proper set of RTAs to transfer the value(s) to where it is needed.

When scheduling an output sub-tree, first step is to know where the output is stored. In our example, assume h is bound to register file RF. We must schedule operation $\gg$ so that its result can be stored in destination RF in cycle -1 and be available for reading in cycle 0. We first select a FU that implements $\gg$ (operation binding). Then, we make sure that a path exists between selected FU and destination RF and all elements of the path are available (not reserved by other operations) in cycle -1 (interconnect binding). In this example, we select $U3$ for $\gg$ and bus $B4$ for transferring the results to RF. Resource reservation will be finalized if the schedule of operands also succeeds. The next step is to schedule proper RTAs in order to transfer the value of g to the left input port of $U3$ and constant 2 to the right input port of $U3$. Figure 2.6 shows the status of schedule after scheduling the $\gg$ operation. The figure shows the set of RTAs that are scheduled in each cycle to read or generated a value. At this point, $B3$ and $M2$ are considered the *destinations* to which values of 2 and g must be transferred in clock cycle -1, respectively.

In order to read constant 2, we need to put the value of CW register on bus $B3$. As for variable g, we schedule the $+$ operation on $U2$ to perform the addition and pass the result to $U3$ though multiplexer $M2$. Note that delay of reading operands of + operation and executing it on $U2$, plus the delay of reading operands of $\gg$ operation and executing it on $U3$ and writing the results to RF is less than one clock cycle. Therefore, all of the corresponding RTAs are scheduled together in clock cycle -1. The algorithm chains the operations in this way, whenever possible. The new status of scheduled RTAs is shown in Figure 2.7. In the next step, we should schedule the $\times$ operations to deliver their results to the input ports of $U2$.

The left operand (e) can be scheduled on $U1$ to deliver its result through register $R1$ in cycle -2 and multiplexer $M1$ in cycle -1. At this point, no other multiplier is left to generate the right operand (f) and directly transfer it to the

clock→ operation↓	-3	-2	-1
g			M2=?;
2			B3=?;
h			B4=U3(M2.B3);RF(h)=B4;

Fig. 2.6 Schedule of RTAs after scheduling $\gg$ operation

clock→ operation↓	-3	-2	-1
e			M1=?;
f			B2=?;
g			M2=U2(M1, B2);
2			B3=CW;
h			B4=U3(M2, B3); RF(h)=B4;

Fig. 2.7   Schedule of RTAs after scheduling h sub-tree

clock→ operation↓	-3	-2	-1
a		B1=RF(a);	
b		B2=RF(b);	
e		R1=U1(B1, B2);	M1=R1;
f			B2=RF(f);
g			M2=U2(M1, B2);
2			B3=CW;
h			B4=U3(M2, B3); RF(h)=B4;

Fig. 2.8   Schedule of RTAs after scheduling h sub-tree

clock→ operation↓	-3	-2	-1
c	B1=RF(c);		
d	B2=RF(d);		
a		B1=RF(a);	
b		B2=RF(b);	
e		R1=U1(B1,B2);	M1=R1;
f	R1=U1(B1,B2);	B4=R1;RF(f)=B4;	B2=RF(f);
g			M2=U2(M1,B2);
2			B3=CW;
h			B4=U3(M2,B3);RF(h)=B4;

Fig. 2.9   Schedule of RTAs after scheduling all sub-tree

right input port of $U2$. Therefore, we assume that f is stored in the register file and try to read it from there. If the read is successful, the corresponding $\times$ operation (f) is considered as an internal output and will be scheduled later. Figure 2.8 shows the status of schedule at this time. The sub-tree of output h is now completely scheduled and the resource reservations can be finalized.

The sub-tree of internal output f must generate its result before cycle -1 where it is read and used by operation $+$. Therefore, the corresponding RTAs must be scheduled in or before clock cycle -2 and write the result in register file RF. The path from $U1$ to RF goes through register $R1$ and hence takes more than one cycle. The second part of the path (after $R1$) is scheduled in cycle -2 and the first part (before $R1$) as well as the execution of operation $\times$ on $U1$ is scheduled in cycle -3. The complete schedule is shown in Figure 2.9.

The above example shows how the DFG is partitioned into sub-trees during scheduling, and how pipelining, operation chaining, and data forwarding are performed during scheduling.

5. Power Optimizations in NISC

This section presents general techniques for reducing total energy consumption of a processor, and discusses how NISC can facilitate such techniques. Dynamic power consumption of CMOS circuits is proportional to $C \times V^2 \times F$, where C is the average switching capacitance, V is the supply voltage, and F is the switching frequency. The energy consumption of the circuit is calculated by

$$E = \frac{P \times N}{F} \tag{2.1}$$

where P is the total power consumption and N is the total number of cycles. In order to reduce total energy consumption of the circuit, the switching capacitance, supply voltage, and/or number of cycles must be reduced. In order to avoid soft errors, supply voltage scaling must be accompanied by frequency reduction as well. Therefore, it adversely affects the performance of the design. In NISC, we leverage the performance gain from architecture customizations to compensate for the performance loss of voltage scaling. In other words, architecture customizations can reduce number of cycles and create some slack time that is used for voltage scaling. In the following, we discuss the techniques for reducing switching capacitance and number of cycles in NISC.

5.1 Reducing the Switching Capacitance

Common techniques for reducing switching capacitance of the processors include *signal gating*, *clock gating*, and memory hierarchy design. Since memory hierarchy design in NISC is the same as other processors, in this section we focus only on signal and clock gating.

Gating the input signals of unused components is called signal gating. The goal of signal gating is to prevent unnecessary switching activities from propagating in the circuit, and it can be implemented using pipeline registers (Figure 2.10b), or AND gates (Figure 2.10c).

In the pipelining approach, the registers are loaded whenever the ALU is used. Otherwise, they maintain their last values. This significantly reduces the dynamic power of the ALU. Pipelining usually reduces the critical path and therefore reduces the cycle time of the circuit as well. Also, pipelining may increase or decrease number of cycles depending on the amount of parallelism in the application code. Pipeline registers also increases the capacity of the clock tree and therefore, increase the clock power.

An alternative approach is to replace pipeline registers with AND gates that are controlled by a *gate signal* called *lock*. In Figure 2.10c, whenever the ALU

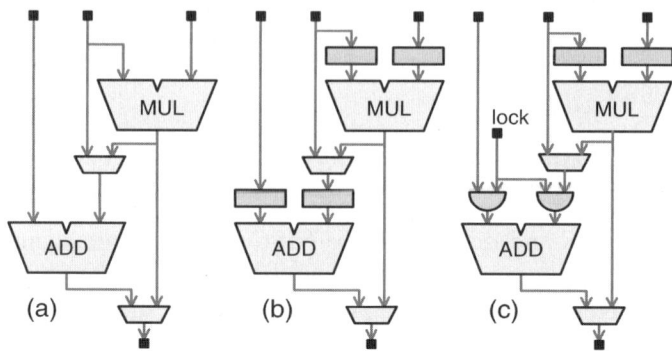

Fig. 2.10 (**a**) No signal gating, (**b**) signal-gating by pipelining, and (**c**) signal gating using AND gates

is not used, the gate signal becomes '0', and as a result, the inputs of the ALU become '0', as well. While this approach does not affect number of cycles, it slightly increases the cycle time because of the delay of the AND gates. Also, unlike the pipelining approach, switching to idle state has some power overhead due to transition of input values to '0'.

In general, depending on application characteristics and data profile, one or both of these techniques may be used. For example, if data profiling reveals that a specific functional unit only operates on small negative numbers, then AND gating is not beneficial due to high power overhead of switching from negative numbers to '0'. For such functional units, pipelining is a better option. NISC technology allows such processor customizations to meet the application requirements.

In pipelined microprocessors, clock power accounts for 20% to 40% of the total power consumption depending on the clock frequency and number of clocked components. Clock gating of unused components is a common technique for reducing clock power, as well as interconnect and logic power. Clock gating of pipeline registers locks the input values of their corresponding functional units and therefore acts as signal gating as well. To identify idle components, in [12], authors suggest an Observability Don't Care (ODC)-based algorithm that conservatively extracts the idle components. In [13], the Deterministic Clock Gating (DCG) technique is proposed that exploits the knowledge of pipeline structure and the behavior of instructions to synthesize decoder logic for determining idle components. These algorithms are implemented in the decoding logic and execute at runtime. In NISC, there is no need for such extra decoding logic because compiler can optimally extract the idle components and generate the proper values of the gate signals for the given architecture and application.

Nevertheless, in NISC, signal gating and clock gating increase the number of control signals in the architecture and hence, they increase the size of the control memory. To balance power and size of control memory vs. that of data-path, the clock gating and signal gating must be applied properly.

5.2 Reducing Number of Cycles

As mentioned earlier, reducing number of cycles can enable voltage scaling and result in significant power and energy savings. NISC reduces number of cycles by exploiting horizontal and vertical parallelism in the given architecture and application. Datapath customizations such as operation chaining and data forwarding can reduce number of cycles by eliminating the need for frequent register-file access. Depending on the application behavior, different operations may be chained to maximize the savings. Although operation chaining and data forwarding can reduce cycle count, they usually increase the number of interconnects and their power consumption. Therefore, they must be tuned to the application needs in order to gain significant energy savings.

6. Experiments

In this section, two sets of experiments are presented. In Section 6.1, a set of general-purpose NISC with different pipeline structure is used for different benchmarks. The effect of the pipeline structure on the performance and power consumption is discussed in this section. In Section 6.2, different datapaths customizations are suggested for improving performance and energy consumption of one of the benchmarks, namely the 2D DCT.

6.1 The Effect of Pipeline Structure on Power and Performance

In general, pipelining improves the clock frequency as well as logic- and interconnect-power consumption. However, it may increase total number of cycles and clock power. On the other hand, data forwarding usually decreases number of cycles, but, it decreases the clock frequency, and increases the interconnect power. The effect of pipelining and data forwarding on total execution time and power consumption depends on the behavior of a given application.

In this section, the following benchmarks are used: *bdist2* function (from MPEG2 encoder), *DCT* 8×8, *FFT*, and *bubble sort* function. The *FFT* and *DCT* benchmarks have data-independent control graphs. The *bdist2* benchmark works on a $16 \times h$ block, where h is 10 in these experiments. The *sort* benchmark, sorts 100 elements stored in an array. Among these benchmarks, *FFT* has the most parallelism and *sort* is a fully sequential code.

Four general-purpose architectures are used in these experiments: The first architecture has only controller pipelining (CP), as shown in Figure 2.11; the second architecture has controller pipelining with signal gating in datapath

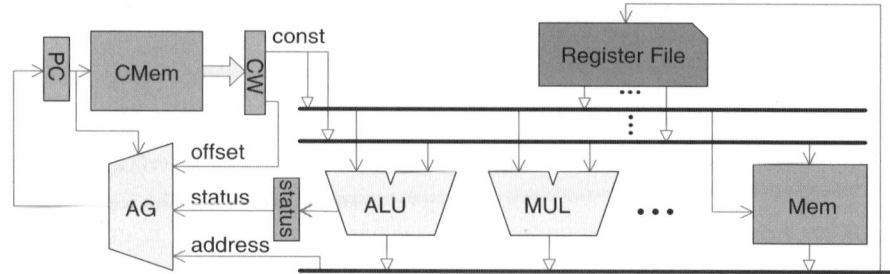

Fig. 2.11   NISC architecture with controller pipelining (CP)

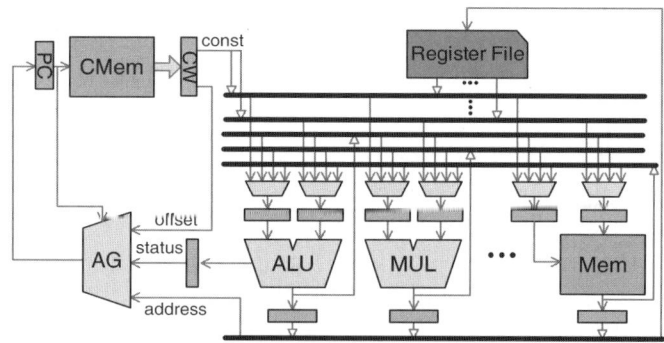

Fig. 2.12   NISC architecture with controller and datapath pipelining and data forwarding
              (CPDPF)

(CPDG). In this architecture the AND gates are added at the input of the
multiplier and ALU; the third architecture has both controller pipelining and
datapath pipelining (CPDP). Pipeline registers are added both at the input
and at the output of the functional units and the memory; the fourth architecture has controller and datapath pipelining with data forwarding (CPDPF),
as shown in Figure 2.12. The bit-width of these datapaths is 32 bits. All benchmarks are compiled on all architectures and are synthesized on Xilinx FPGA
package Virtex2V250-6 using Xilinx ISE 8.3 tool. Two synthesis optimization of retiming, and buffer-to-multiplexer conversion are enabled. The clock
frequency of the architectures after synthesis and placement-and-routing is
reported in Table 2.1. This table shows that pipelining has increased the clock
frequency (from 60 MHz to almost 90 MHz), and data forwarding and signal
gating has decreased the frequency by a few MHz, as expected.

The number of cycles and total execution time of each benchmark on different architectures are shown in Table 2.2. While adding pipelining increases
the clock frequency, it may increase the cycle counts especially if there is

Table 2.1 Clock frequency of architectures after synthesis

Architecture	CP	CPDG	CPDP	CPDPF
Clock frequency (MHz)	60	57.5	89	88

Table 2.2 Execution cycles counts of benchmarks

	Cycle count				Total execution time (μs)			
	CP	CPDG	CPDP	CPDPF	CP	CPDG	CPDP	CPDPF
bdist2	6,326	6,326	7,168	5,226	105.64	110.05	80.35	**59.05**
DCT	11,764	11,764	14,292	13,140	196.46	204.65	160.21	**148.48**
FFT	220	220	218	166	3.67	3.83	2.44	**1.88**
Sort	35,349	35,349	84,161	74,162	**590.33**	614.93	943.44	838.03

Table 2.3 Power breakdown of the NISCs and total energy consumption of DCT

	Clock power (mW)	Logic power (mW)	Interconnect power (mW)	Total power (mW)	Total energy (uJ)	Energy $\times$ Delay (fJs)
CP	35.40	89.25	199.69	324.34	63.72	12.52
CPDG	30.00	63.00	150.00	243.00	49.73	10.18
CPDP	54.65	50.10	61.23	165.98	26.59	4.26
CPDPF	59.81	60.28	92.81	212.90	31.61	4.69

not enough parallelism in the benchmark. Except for the sort algorithm, the minimum execution delay is achieved using CDPF datapath. For sort, a less pipelined architecture (CP) performs the best. The minimum execution times of all the benchmarks are highlighted in Table 2.2.

Table 2.3 columns 2–4 show the power breakdown of the NISC architectures (collected for DCT benchmark), in terms of clock, logic, and interconnect power. Compared to CP, the CPDG, CPDP, and CPDPF architectures save 30%, 44%, and 32% logic power, respectively. They also save 25%, 70%, and 53% interconnect power, respectively. However, in CPDP and CPDPF, the clock power increases by 54% and 68% compared to the clock power of CP. Considering all the three power elements, the total power saving of CPDG, CPDP, and CPDPF compared to the CP architecture is 25%, 49%, and 34% respectively. Columns 5–7 show the total power, energy, and energy-delay product of the DCT benchmark. The minimum power consumption is achieved by CPDP (165.98 mW), which shows a 49% savings compared to CP. According to Table 2.2, DCT achieves the maximum performance on CPDPF, and the minimum power and energy consumption on CPDP. In such cases, depending on the importance of each of these metrics for a specific application, the most suitable architecture is selected.

6.2   Custom Datapath Design for DCT

In this section, different custom pipelined datapaths for DCT algorithm are designed, in order to further improve the performance and power consumption. More details about this example can be found in [16]. The definition of Discrete Cosine Transform (DCT) [15] for a 2-D $N \times N$ matrix of pixels is as follows:

$$F[u,v] = \frac{1}{N^2} \sum_{m=0}^{N-1} \sum_{n=0}^{N-1} f[m,n] \cos \frac{(2m+1)u\pi}{2N} \cos \frac{(2n+1)v\pi}{2N}$$

where u, v are discrete frequency variables ($0 \leq u$, $v \leq N - 1$), $f[i, j]$ gray level of pixel at position (i, j), and $F[u,v]$ coefficients of point (u, v) in spatial frequency. Assuming $N = 8$, matrix C is defined as follows:

$$C[u][n] = \frac{1}{8} \cos \frac{(2n+1)u\pi}{16}$$

Based on matrix C, an integer matrix $C1$ is defined as follows: $C1 = $ round($factor \times C$). The $C1$ matrix is used in calculation of DCT and IDCT: $F = C1 \times f \times C2$, where $C2 = C1^T$. As a result, DCT can be calculated using two consecutive matrix multiplications. Figure 2.13a shows the C code of multiplying two given matrix A and B using three nested loops.

To compare the results of custom architectures with a general-purpose architecture, a NISC-style implementation of MIPS M4K datapath [14] (called NMIPS) is used. The bus-width of the datapath is 16-bit for a 16-bit DCT precision, and the datapath does not have any integer divider or floating point unit. The clock frequency of 78.3 MHz is achieved after synthesis and

```for(inti=0;i<8;i++) for(intj=0;j<8;j++){ sum=0; for(intk=0;k<8;k++) sum=sum+A[i][k]×B[k][j]; C[i][j]=sum; } ```	```ij=0; do{ i8 = ij&0xF8; j = ij&0x7; aL =*(A+(i8|0)); bL =*(B+(0|j)); sum = aL×bL; aL =*(A+(i8|1));bL =*(B+(8|j)); sum+ = aL×bL; aL =*(A+(i8|2));bL =*(B+(16|j)); sum+ = aL×bL; aL =*(A+(i8|3));bL =*(B+(24|j)); sum+ = aL×bL; aL =*(A+(i8|4));bL =*(B+(32|j)); sum+ = aL×bL; aL =*(A+(i8|5));bL =*(B+(40|j)); sum+ = aL×bL; aL =*(A+(i8|6));bL =*(B+(48|j)); sum+ = aL×bL; aL =*(A+(i8|7));bL =*(B+(56|j)); *(C+ij) = sum+(aL×bL); } while(++ij!=64); ```
(a)	(b)

Fig. 2.13   (a) Original and (b) transformed matrix multiplication

Placement-and-Routing (PAR). Two synthesis optimizations of retiming and buffer-to-multiplexer conversions are applied to improve the performance.

In general, customization of a design involves both software and hardware transformations. In Figure 2.13b, the application parallelism is increased by unrolling the inner-most loop of the matrix multiplication, merging the two outer loops, and converting some of the costly operations such as addition and multiplication to OR and AND. In DCT, the operation conversions are possible because of the special values of the constants and variables.

Initial Custom Datapath: CDCT1.   By looking at the body of loop in Figure 2.13b, four steps of computation can be identified: (1) calculation of the memory addresses of the relevant matrix elements; (2) loading the values of those elements from data memory; (3) multiplying the two values; (4) accumulating the multiplication results. Therefore, a custom datapath can be designed so that each of these steps is a pipeline stage. Figure 2.14a shows such datapath (CDCT1). The datapath includes four major pipeline stages that are marked in the figure. This datapath uses operation chaining to reduce RF accesses and decrease register pressure. Chaining the operations improves the energy consumption and performance. The OR and ALU are chained, as well as the Mul and Adder. Note that the chaining of multiply and add forms a MAC unit in the datapath. After compilation, the total number of cycles of the DCT is 3080, and the maximum clock frequency is 85.7 MHz.

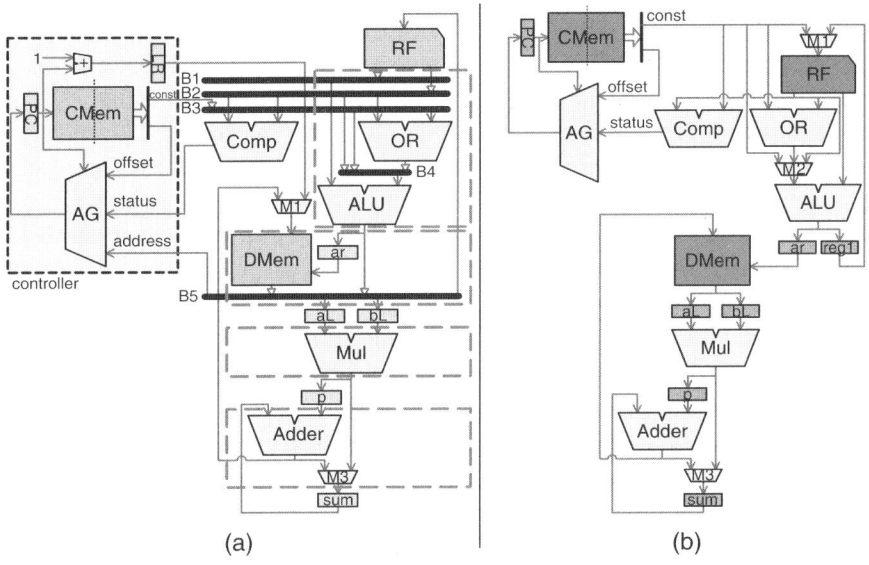

Fig. 2.14   Block diagram of (a) CDCT1 and (b) CDCT2

Table 2.4 Critical-path delay breakdown of CDCT1

Component	CMem + CW	RF + RF_o	ALU + ALU_o	RF setup time
Delay (ns)	3.28	2.39	5.4	0.58

Table 2.5 Critical-path delay breakdown of CDCT2

Component	CMem + CW	RF + RF_o	Comp + comp_o	AG + PC setup
Delay (ns)	2.93	2.45	3.726	2.06

Table 2.4 shows the critical-path breakdown of CDCT1. Each column in the table shows the sum of a component delay and its output-interconnect delay. The critical path goes through CMem, RF, B2, B4, ALU, B5, and back to RF.

CDCT2: Bus Customization and Adding a Pipeline Register to the Datapath. According to Table 2.4, ALU and the wire that connects ALU to RF are in the critical path. To reduce the critical-path delay, an additional pipeline register (called $reg1$) is inserted in the output of the ALU, and also all global buses, including B5, are replaced with point-to-point connections. Only the connections that are used by the DCT application are kept. The new design, CDCT2, is shown in Figure 2.14b. The NISC compiler automatically analyzes the new datapath and regenerates the control words to correctly handle the flow of the data. CDCT2 runs the DCT algorithm in 2952 cycles at the maximum clock frequency of 90 MHz. The reduction in number of cycles is due to additional parallelism created by the separation of interconnects. Table 2.5 shows the breakdown of the critical path of CDCT2. Note that, in CDCT2, the critical path goes through the comparator instead of the ALU.

CDCT3: Eliminating the Unused Parts of ALU, Comparator, and RF. Next, the ALU and comparator are simplified by eliminating the operations not used in the DCT application. In Figure 2.13b, only Add, And, Multiply, and Not-equal (!=) operations are used. The first two operations are executed by ALU, the third by Mul, and the last by Comp. NISC compiler allocates and uses nine registers in RF. Therefore, number of registers in RF is reduced from 32 to 16. The new architecture (CDCT3) runs much faster at the clock frequency of 114.4 MHz. The breakdown of critical-path delay Table 2.6 shows a considerable reduction in the delay of the comparator (i.e. from 3.726 ns to 2.29 ns). Also, the number of fan-outs of RF output wires is reduced, and hence its interconnect delay is reduced (i.e. from 2.45 ns to 1.64 ns). These modifications, also, reduce the area and power consumption significantly.

Table 2.6 Critical-path delay breakdown of CDCT3

Component	CMem + CW	RF + RF_o	Comp + comp_o	AG + PC setup
Delay (ns)	2.76	1.64	2.29	2.06

Table 2.7 Critical-path delay breakdown of CDCT4

Component	CMem + CW	RF + RF_o	Comp + comp_o	AG + PC setup
Delay (ns)	1.39	1.6	1.74	2.06

Table 2.8 Critical-path delay breakdown of CDCT5

Component	bL + bL-o	Mul + Mul-o	P setup time
Delay (ns)	1.29	4.25	0.3

CDCT4 and CDCT5: Controller Pipelining.   Looking at the critical paths of the architectures, it is evident that the controller contributes to a major amount of the delay. The CMem, CW, and Address Generator (AG) delays are part of the critical path of CDCT3. To reduce the effect of the controller delay, one pipeline register (i.e. CW register) is inserted in front of the CMem. The new architecture (CDCT4) can run much faster at the clock frequency of 147 MHz. Table 2.7 shows a reduction in the critical-path delay, where CMem+CW delay is reduced from 2.76 ns to 1.39 ns. Also, due to retiming optimization, the delay of comparator is reduced from 2.29 ns to 1.74 ns. On the downside, however, the number of cycles of DCT increases to 3080 because of an extra branch delay cycle. Note that the NISC compiler automatically analyzes the datapath and notices the extra branch delay. So, the user does not need to change the compiler manually.

To further reduce the effect of controller's delay on the clock cycle, another pipeline register (called status register) is inserted at the output of the Comp. This register eliminates the AG's delay from the critical path. Table 2.8 shows the breakdown of the critical-path delay of the new architecture (CDCT5). In CDCT5, the critical path goes through the multiplier. Note that, CDCT5 has a branch delay of two and runs at the clock frequency of 170 MHz. The total number of cycles of DCT increases to 3208.

CDCT6: Bit-Width Reduction.   To improve the area of the design, the bit-width of some of the components can be reduced without affecting the precision of the calculations. The address values in DCT are in the range of 0 to 255. Therefore, the bit-width of the address-calculation pipeline stage (i.e. RF, OR, ALU, and Comp) can be reduced to 8 bits. In this case, the clock frequency remains fixed at 170 MHz.

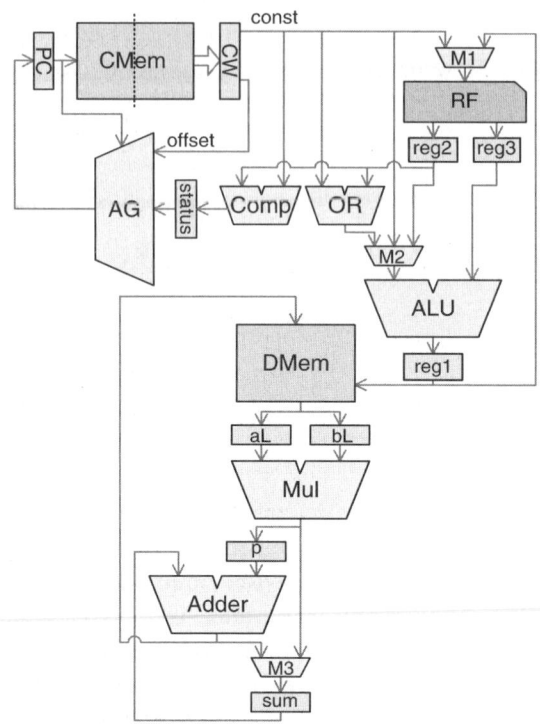

Fig. 2.15   Block diagram of CDCT7

CDCT7: Using Multi-Cycle Paths.   In the final optimization, the clock frequency is further improved by considering Mul as a two cycle unit and adding two pipeline registers at the output of RF. To synthesis the new design correctly, the proper timing constraint for the two cycle path is specified for the synthesis tool. The clock frequency of CDCT7 is 250 MHz and it takes 3460 cycles to run DCT. Figure 2.15 shows final design (CDCT7) after all the transformations.

Comparing Performance, Power, Energy and Area of the NISCs.   Table 2.9 compares the performance, power, energy, and area of all NISC architectures. The third column shows the maximum clock frequency after Placement-and-Routing. The fourth column shows the total execution time of the DCT algorithm calculated based on number of cycles and the clock frequency. Note that although in some cases (such as CDCT4 and CDCT5) the number of cycles increases, the clock frequency improvement compensates for that. As a result, the total execution delay maintains a decreasing trend.

Column 5 shows the average power consumption of the NISC architectures while running the DCT algorithm. All the designs are stimulated with the same

Table 2.9 Performance, power, energy, and area of the DCT implementations

	No. of cycles	Clock freq	DCT exec. time (μs)	Power (mW)	Energy (μJ)	Normalized area
NMIPS	10772	78.3	137.57	177.33	24.40	1.00
CDCT1	3080	85.7	35.94	120.52	4.33	0.81
CDCT2	2952	90.0	32.80	111.27	3.65	0.71
CDCT3	2952	114.4	25.80	82.82	2.14	0.40
CDCT4	3080	147.0	20.95	125.00	2.62	0.46
CDCT5	3208	169.5	18.93	106.00	2.01	0.43
CDCT6	3208	171.5	18.71	104.00	1.95	0.34
CDCT7	3460	250.0	13.84	137.00	1.90	0.35

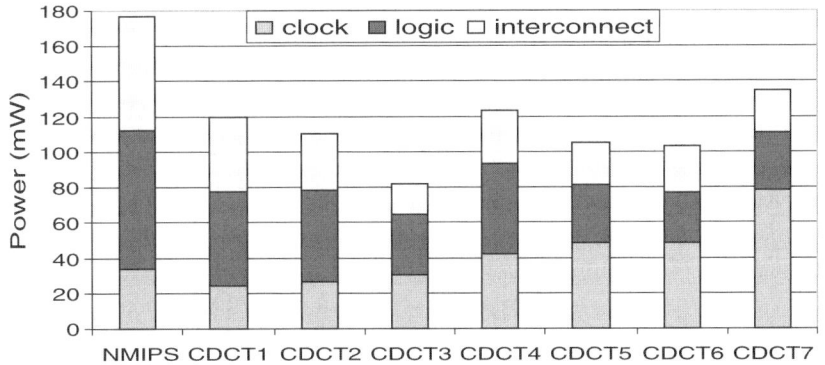

Fig. 2.16 Power breakdown of the DCT implementations

data values, and Post-Placement-and-Routing simulation is used to collect the signal activities. To compute the total power consumption, Xilinx XPower tool is used. Figure 2.16 shows the power breakdown of different designs in terms of the clock, logic, and interconnect power. Column 6 shows the total energy consumption calculated by multiplying power and execution time.

In these experiments, CDCT1 consumes less power than NMIPS due to its different pipeline structure and better signal gating. CDCT2 consumes less power compared to CDCT1 because of replacing shared bus B5 with short point-to-point connections with lower bus capacitance. The diagram of Figure 2.16 shows the reduction in interconnect-power consumption of CDCT2.

Power consumption of CDCT3 is lower than CDCT2 because of the elimination of unused operations in ALU and comparator. Elimination of operations reduces number on fan-outs of the RF output wires. Therefore, reduction in interconnect power, as well as logic power is achieved. The power breakdown of

CDCT3 confirms this fact. Note that as the clock frequency goes up, the clock power gradually increases.

In CDCT4, the power consumption further increases, because of: (1) higher clock power due to higher clock frequency and higher number of pipeline registers; and more importantly, because of (2) the power consumption of logic and interconnects added by retiming algorithm. Since the difference between the delays of the two pipeline stages located before and after CW register is high, the retiming works aggressively to balance the delay. As a result it adds extra logic to the circuit. In CDCT5, the status register is added to the output of Comp to reduce the critical path. In this case, the retiming algorithm works less aggressively because the delays of the pipeline stages are less imbalanced. This reduces logic and interconnect power. In CDCT7, the logic and interconnect power remain the same as CDCT5 but the clock power increases due to higher clock frequency.

The last column of Table 2.9 shows the normalized area of different designs calculated based on the number of FPGA slices that each design (including memories) occupies. The area trend also confirms the increase in area in CDCT4 followed by a decrease in CDCT5, evidently due to retiming.

Figure 2.17 shows the performance, power, energy, and area of the designs normalized against NMIPS. The total execution delay of DCT algorithm has a decreasing trend, while the power consumption decreases up to CDCT3 and then increases. The energy consumption significantly drops at CDCT1, because of the reduction in number of cycles and power consumption. From CDCT1 to CDCT7, the energy decreases gradually in a slow paste.

As shown in Figure 2.17, CDCT7 is the best design in terms of delay and energy consumption, while CDCT3 is the best in terms of power, and CDCT6 is the best in terms of area. As a result, CDCT3, CDCT6, and CDCT7 are

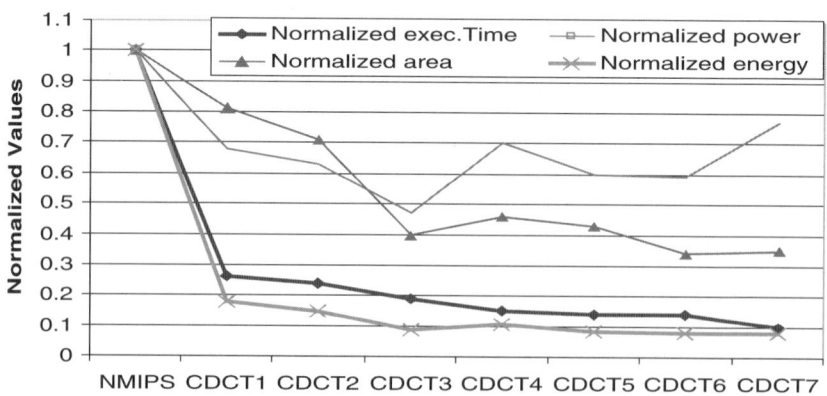

Fig. 2.17   Comparing different DCT implementations

considered the pareto-optimal solutions. Note that minimum energy and minimum power are achieved by two different designs: CDCT7 and CDCT3, respectively. Compared to NMIPS, CDCT7 runs 10 times faster, consumes 1.3 times less power, and 12.8 times less energy. Also, it occupies 2.9 times less area than NMIPS.

In summary, designing a custom datapath for a given application by properly connecting functional units and pipeline registers is the key to reducing number of cycles and energy consumption. Also, eliminating the unused logic and interconnects, adjusting the bus-width of the datapath to the application requirement, signal gating, and clock gating are the key to reducing power consumption. The NISC technology makes these customizations very easy to apply.

7.   Conclusion

This chapter presented No-Instruction-Set-Computer (NISC) technology. In NISC, architecture is both horizontally and vertically controlled by compiler, and therefore, has less hardware and runtime overhead for controlling the pipeline. NISC technology allows fast datapath customizations for a given application. In NISC, properly connecting functional units and pipeline registers is the key to reducing number of cycles and energy consumption of a given application. Also, eliminating the unused logic and interconnects, adjusting the bus-width of the datapath to the application requirement, signal gating, and clock gating are the key to reducing power consumption. Moreover, in NISC, reducing number of cycles by architecture customization creates some slack time that can be used for voltage scaling. Because of these features the NISC technology is very powerful for IP and embedded system design.

References

[1] M. Reshadi, B. Gorjiara, D. Gajski, Utilizing Horizontal and Vertical Parallelism Using a No-Instruction-Set Compiler and Custom Datapaths, in *Proceedings of the International Conference on Computer Design* (ICCD), pp. 69–76, October 2005.

[2] M. Reshadi, D. Gajski, A Cycle-Accurate Compilation Algorithm for Custom Pipelined Datapaths, in *Proceedings of the International Symposium on Hardware/Software Codesign and System Synthesis* (CODES + ISSS), pp. 21–26, September 2005.

[3] P.G. Paulin, J. Knight, Algorithms for High-Level Synthesis, *IEEE Design & Test of Computers*, 1989.

[4] P.G. Paulin, J.P. Knight, Force-Directed Scheduling for the Behavioral Synthesis of ASIC's, *IEEE Transactions on Computer-Aided Design*, 1989.

[5]   R. Camposano, Path-Based Scheduling for Synthesis, *IEEE Transactions on Computer-Aided Design*, 1991.

[6]   D. Gajski, N. Dutt, A. Wu, S. Lin, *High-Level Synthesis Introduction to Chip and System Design*, Kluwer Academic Publishers, The Netherlands, 1994.

[7]   M. Xu, F.J. Kurdahi, Layout-Driven High Level Synthesis for FPGA Based Architectures, *DATE*, 1998.

[8]   S.Y. Ohm, F.J. Kurdahi, N. Dutt, M. Xu, A Comprehensive Estimation Technique for High-Level Synthesis, *International Symposium on Systems Synthesis*, 1995.

[9]   D. Kim, J. Jung, S. Lee, J. Jeon, K. Choi, Behavior-to-Placed RTL Synthesis with Performance-Driven Placement, *International Conference Computer Aided Design*, 2001.

[10] J. Zhu, D. Gajski, "Soft Scheduling in High Level Synthesis, *Design Automation Conference*, 1999.

[11] W.E. Dougherty, D.E. Thomas, Unifying Behavioral Synthesis and Physical Design, *Design Automation Conference*, 2000.

[12] H. Kapadia, L. Benini, G.D. Micheli. Reducing Switching Activity on Datapath Buses with Control-Signal Gating. *IEEE Journal of Solid-State Circuits*, 34(3):405–414, 1999.

[13] H. Li, S. Bhunia, Y. Chen, T.N. Vijaykumar and K. Roy. Deterministic Clock Gating for Microprocessor Power Reduction, in *Proceedings of the International Symposium on High-Performance Computer Architecture (HPCA)*, pp. 113–122, 2003.

[14] MIPS32® M4K™ Core, http://www.mips.com

[15] N. Ahmed, T. Natarajan, K.R. Rao, Discrete Cosine Transform, *IEEE Transactions on Computers*, vol. C-23, 1974.

[16] B. Gorjiara, D. Gajski, Custom Processor Design Using NISC: A Case-Study on DCT Algorithm, in *Proceedings of Workshop on Embedded Systems for Real-time Multimedia* (ESTIMEDIA), 2005.

Chapter 3

Synthesis of Instruction Sets for High-Performance and Energy-Efficient ASIP

Jong-Eun Lee[1], Kiyoung Choi[2], and Nikil D. Dutt[3]

[1]*SoC R&D Center*
System LSI
Samsung Electronics
Korea

[2]*School of Electrical Engineering and Computer Science*
Seoul National University
Seoul, Korea

[3]*Donald Bren School of Information and Computer Sciences*
University of California
Irvine, CA

Abstract   Several techniques have been proposed to reduce the energy consumption of ASIPs (Application-Specific Instruction set Processors). While those techniques can reduce the energy consumption with minimal change in the instruction set (IS), they often fail to exploit the opportunity of designing the entire IS from the energy-efficiency perspective. In this chapter we present an energy-efficient IS synthesis that can comprehensively reduce the energy-delay product (EDP) of ASIPs through optimal instruction encoding, considering both the instruction bitwidth and the dynamic instruction fetch count. Experimental results with a typical embedded RISC processor show that the proposed energy-efficient IS synthesis technique can generate application-specific ISs that are up to 40% more energy-efficient over the native IS for several application benchmarks.

Keywords: instruction set synthesis; CISC (Complex Instruction Set Computer); instruction encoding; low-power ASIP (Application-Specific Instruction set Processor).

J. Henkel and S. Parameswaran (eds.), Designing Embedded Processors – A Low Power Perspective,
51–64.

1.   Introduction

It is well known that CISC (Complex Instruction Set Computer) ISs (Instruction Sets) are more energy-efficient than RISC (Reduced Instruction Set Computer) ISs for the same microarchitecture (Bunda et al., 1995). However, it is not well known how one can utilize this observation to generate more energy-efficient ISs, especially when one is given the freedom to modify an IS on an application or application domain basis. With the recent development in soft IPs (Intellectual Properties) and configurable processors, IS customization has become possible and even necessary to make differentiation in today's competitive markets. Nonetheless, previous work on low-power ASIPs (Application-Specific Instruction Set Processors) has not been so ambitious as to fully exploit the flexibility of ASIPs and redesign the IS from the energy-efficiency perspective; most techniques (Benini et al., 1998; Kim and Kim, 1999; Glokler and Bitterlich, 2001; Inoue et al., 2002) are concerned with optimizing only the bit pattern assignment while some techniques (Dougherty et al., 1998) have considered only to remove less useful instructions from a given set of instructions.

In this chapter, we present an energy-efficient IS synthesis approach for application-specific processors. Unlike the previous low-power opcode encoding techniques (Benini et al., 1998; Kim and Kim, 1999), the energy-efficient IS synthesis proposed in this chapter is not limited to opcode reassignment but provides a comprehensive method to synthesize ISs optimized for given applications. Specifically, we optimize ISs under given microarchitectural constraints, as the design change in the data path may incur significant engineering cost and thus not be desirable. With a fixed microarchitecture, specialization can be made in such areas as instruction encoding, the number of instructions, and the instruction bitwidth, all of which can be considered as instruction encoding in a broad sense. Thus, the objective is to find the best instruction encoding (manifested by RISC vs. CISC) that leads to the maximal energy-efficiency through fewer number of instructions fetched (reducing the instruction memory energy) or fewer number of execution cycles (reducing the processor core energy) or a balance of the two.

One of the critical elements of the proposed energy-efficient IS synthesis is reducing the instruction fetch energy through multiple dimensions of the *code volume* (i.e., the number of instructions fetched multiplied by the instruction bitwidth). While some previous low-power techniques can also have similar effects of reducing the code volume, only one dimension has typically been considered. For example, low-power instruction compression schemes (Benini et al., 1999; Chander et al., 2001) try to reduce the code volume by focusing only on the bitwidth of frequently occurring binary instruction patterns (thus not changing the number of instructions whether static or dynamic). Likewise, code size reduction techniques (Liao et al., 1999) aim to reduce the static code size, thus may not be effective for reducing the dynamic instruction count.

Our IS synthesis technique, on the contrary, addresses the multiple dimensions of the code volume and provides a comprehensive optimization framework for energy-efficient ISs. Also, our scheme generates a single IS as opposed to dual ISs (compressed, uncompressed) as the code compression techniques do; thus, it avoids the problems of dual ISs such as requiring an instruction decompressor (or re-map table) and, for some techniques, having to take care of the changes in the branch target addresses. It should be noted, however, that the energy-efficient IS synthesis assumes minor architectural changes in the data path, allowed by the given architectural constraints, such as inserting additional muxes in front of functional units, as well as changing the instruction decoder logic. Our experimental results show that the proposed energy-efficient IS synthesis can generate application-specific ISs that outperform the native IS of a typical embedded RISC processor, not only in performance but also in energy and the EDP, up to about 40% with each metric.

The rest of the chapter is organized as follows. Section 2 briefly discusses the previous work for energy-efficient ASIPs and Section 16.2 highlights the key elements of the encoding-oriented IS synthesis for ASIP customization. Section 4 derives the contribution of each instruction to the overall energy-efficiency based on an ASIP energy consumption model. Finally, experimental results are presented in Section 5 and the chapter is concluded with Section 6.

2.   Related Work

Previous work on low-power techniques for ASIPs has mostly concentrated on bit pattern assignment of instructions, without changing the number of instructions in the IS. To reduce the switching activity and the dynamic energy consumption in the IF (Instruction Fetch) registers of ASIPs, it was proposed to re-encode the opcode part of instructions so that the most frequent opcode sequences (in a typical application execution) can have the smallest Hamming distances (Benini et al., 1998; Kim and Kim, 1999). In other cases, exploiting the asymmetric energy consumption of some memory devices (with some ROMs, reading the value "1" does not require any switching activity while reading "0" does), even the whole instruction can be re-encoded (frequently used opcodes or constant values are encoded with as many 1's as possible) (Glokler and Bitterlich, 2001; Inoue et al., 2002).

As for a more aggressive approach, removing unused or less useful instructions from the IS has been suggested. By implementing only the subset of instructions decided to be used for the applications at hand (*instruction subsetting*), one may reduce the area, critical path, and power dissipation, while partially retaining its programability (Dougherty et al., 1998).

While these techniques may fit well where only minimal changes are allowed in the architecture, they are too conservative when a more aggressive IS redesign is preferred to seize the opportunity afforded by configurable processors. Contrastingly, the proposed IS synthesis approach considers redesigning the IS from the energy-efficiency perspective to better exploit the flexibility of configurable processors. The proposed technique is based on the encoding-oriented IS synthesis flow described in Lee et al. (2002) but extends it considering various optimization goals and their efficacy.

3.   Synthesizing Instruction Sets

The encoding-oriented IS synthesis flow (Lee et al., 2002) assumes *basic instructions*, which are provided by users once for each processor. The existing IS of a processor, called *native instruction set*, is normally optimized for general applications and includes many complex instructions as well as simple ones. To facilitate automatic synthesis of application-specific instructions, we start from the basic IS. A basic instruction includes only one operation, so that more complex instructions can be easily created by combining multiple basic instructions that appear frequently in the application program. The application-specific instructions created this way are called *C-instructions*, meaning compound-operation instructions. C-instructions can be generated for each application or a set of applications representing an application domain. The C-instructions together with the basic instructions comprise a *synthesized instruction set*. The relationship between the native IS, basic IS, and synthesized IS are illustrated in Figure 3.1.

Figure 3.2 illustrates the process of synthesizing an IS. First the basic IS and the data path resource description (e.g., the number and types of functional units, and their timing in the pipeline during each instruction execution) are given as input. The application is compiled using a retargetable compiler targeted for the basic IS. This preliminary assembly code is used in the rest of the IS synthesis process. The core synthesis process consists of two phases: candidate C-instruction generation and C-instruction selection. In the generation phase, a group of C-instructions are created for every sequence of up to N basic instructions appearing in the assembly code, where N is a design

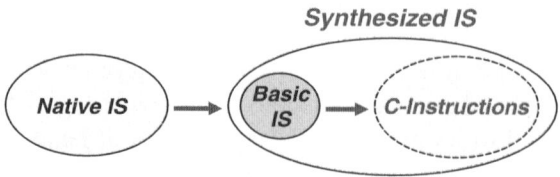

Fig. 3.1   Relationship between the native, basic, and synthesized instruction sets

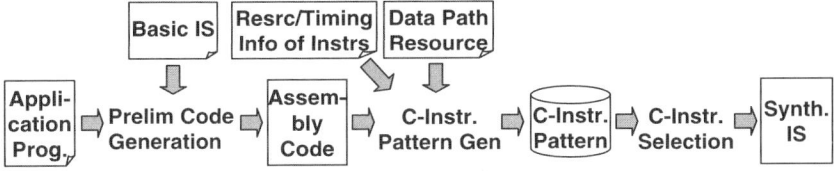

Fig. 3.2   Instruction set synthesis flow

parameter. Since only those appearing in a linear sequence can become candidates for C-instruction creation, this method has an advantage of low computational complexity (linear in the number of instructions in the application assembly code).

Though every C-instruction can contribute to a better code quality such as smaller code size and fewer execution cycles, not all of them can be included in the synthesized IS due to the instruction encoding constraint. There can be at most 2^{IW} (IW is the instruction bitwidth) number of distinct bit patterns that can be assigned to the synthesized instructions (including the basic instructions). Therefore if a C-instruction i uses W_i bits to encode all its operands, 2^{W_i} number of bit patterns (called the *code space* used by i) should be reserved for instruction i. Thus, we need to select the best set of C-instructions that will bring the most benefit (e.g., code size reduction, execution cycle count reduction) while meeting the code space constraint.

However, the formulation is rather complicated because of the nonlinearity of the cost (as well as benefit) function. For instance, the code space (representing the cost) of two C-instructions may not be equal to the sum of the code space of each if one is a superset of the other. As a result, the benefit and the cost of a C-instruction depend on what other C-instructions are already selected. Lee et al. (2002) gives a more detailed discussion on the problem, providing an ILP (Integer Linear Programing) formulation for the C-instruction selection problem, as well as an efficient heuristic algorithm, since solving an ILP problem may take a prohibitive amount of computation resources.

4.   Optimizing for Energy-Efficiency

By redesigning the entire IS for specific applications, the code volume can be reduced in all its dimensions, generating more effective IF reduction than has been possible with previous approaches. On the other hand, simultaneously considering all the factors affecting the code volume requires the non-trivial task of quantifying energy (and EDP) changes during the IS synthesis. In this section, we analyze the energy and EDP changes during the IS synthesis based on an ASIP energy model, and derive a quantitative energy-efficiency criterion which can be used for selecting C-instructions.

4.1   ASIP Energy Model

Figure 3.3 illustrates a simple ASIP chip including a processor core and on-chip memory. From the behavioral point of view, the same ASIP chip (the part that covers the processor core and on-chip memory) can be viewed, at the cycle level, as a pipeline of IF, ID, and EX stage operations. The CMOS *dynamic* energy of the whole ASIP can then be seen as $E_{\text{IF}} + E_{\text{ID}} + E_{\text{EX}}$, where E_{IF}, E_{ID}, and E_{EX} are the energy consumed by the operations in IF, ID, and EX stages, respectively. Also, the energy-delay product (EDP) of the ASIP, for a given clock frequency, can be defined as

$$\text{EDP} = (E_{\text{IF}} + E_{\text{ID}} + E_{\text{EX}}) \cdot N_{cyc} \tag{3.1}$$

where N_{cyc} is the number of execution cycles of an application.

The energy consumed in each stage can be modeled, simplifying interrupts, pipeline flush, etc. as

$$E_{\text{IF}} + E_{\text{ID}} + E_{\text{EX}}$$
$$= [N_{ins} \cdot (e_{IM} + e_{IBus} + e_{\text{IF}})] + [N_{ins} \cdot e_{\text{ID}}] + \left[\sum_{op} N_{op} \cdot e_{op}\right] \tag{3.2}$$

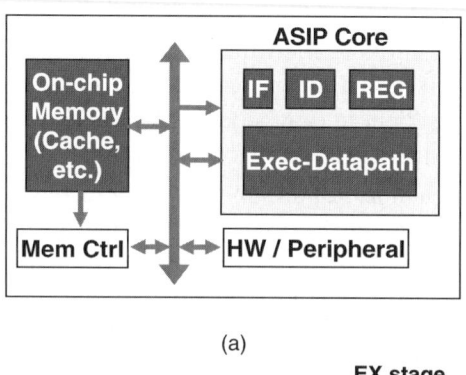

(a)

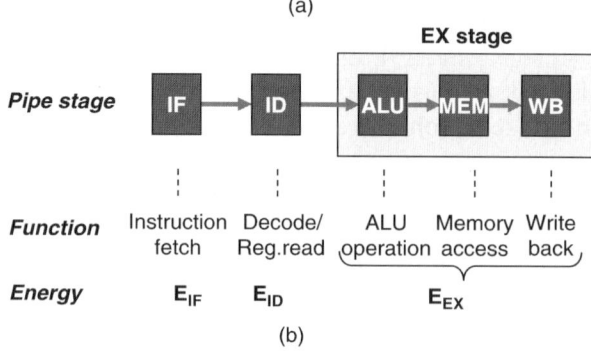

(b)

Fig. 3.3   An example ASIP. (a) The structural view. Major components include ASIP core, on-chip memory, and on-chip bus. (b) The behavioral view. Major functions inlcude IF (instruction fetch), ID (instruction decode and register read), and EX (execution) stages

where N_{ins} is the dynamic instruction count; e_{IM}, e_{IBus}, e_{IF}, and e_{ID} are the per-access energy consumption in the instruction memory, instruction bus, instruction fetch unit, and instruction decoder, respectively; and N_{op} and e_{op} are the number of operations and the per-operation energy consumption in the EX stage for each operation op, respectively.

4.2   EDP Change due to IS Customization

Let us consider the energy-efficiency (defined by EDP) when using the basic IS (denoted by B) vs. a synthesized IS (denoted by C), where C is equal to B plus selected C-instructions. There are a number of different C's and we have to choose the best C through C-instruction selection. Now for two ISs B and C, let N_{cyc}^B and N_{cyc}^C be the numbers of execution cycles, and N_{ins}^B and N_{ins}^C be their dynamic instruction counts, respectively. Note that N_{cyc}^B and N_{ins}^B are fixed for a given application whereas N_{cyc}^C and N_{ins}^C are subject to optimization.

The advantage that a C-instruction has over basic instructions is essentially that it can express a number of basic instructions with a single instruction. This leads first to (static) instruction count reduction (or code size reduction) and also in many cases to execution cycle count reduction (which may be called dynamic instruction count reduction). A typical example is a C-instruction combining an ADD instruction with a LOAD instruction (Lee, 2004). Let R_i^{ins} and R_i^{cyc} be the reduction in the number of instructions and in the number of execution cycles per each use of a C-instruction i instead of the corresponding basic instruction sequence. R_i^{ins} can be easily calculated, i.e., the length of the basic instruction sequence minus one. To calculate R_i^{cyc}, one has to schedule the operations of the corresponding basic instructions considering the data path resource constraint, which is similar to running simple high-level synthesis (Gajski et al., 1992; Lee, 2004).

From the IS synthesis flow of Figure 3.1, the following relationship can be obtained:

$$N_{cyc}^C = N_{cyc}^B - \sum_i R_i^{cyc} \cdot d_i \cdot \chi_i \qquad (3.3)$$

$$N_{ins}^C = N_{ins}^B - \sum_i R_i^{ins} \cdot d_i \cdot \chi_i \qquad (3.4)$$

where d_i is the dynamic matching count, by which times C-instruction i is used in the application, and χ_i is a binary variable with the value 1 indicating that C-instruction i is selected.

Now, with the following observations we can derive the EDP difference (ΔEDP) of using a synthesized IS (C) instead of the basic IS (B). First, the EX stage operations such as ALU operations and memory operations can be considered the same in terms of energy consumption for both B and C, assuming that clock gating is extensively used during synthesis to avoid clocking for unnecessary computation. Therefore E_{EX} is the same for both ISs. Second, e_{IM}, e_{IBus}, and e_{IF} can be considered the same for both B and C. Though the two ISs will have different binary encoding patterns (in terms of assigning a binary code to each instruction) resulting in different bit patterns on the bus and memory and thus slightly different energy consumption, it can be reasonably assumed that the average energy consumption in this part is more or less the same for the two ISs. The instruction bitwidth is a given parameter of IS synthesis and thus constant during IS synthesis. Third, e_{ID}^C (for C) is likely to be larger than e_{ID}^B (for B), due to the increased number of instructions in C. Assuming, however, that IF energy is much larger than ID energy *increase* (i.e., $e_{IM} + e_{IBus} + e_{IF} \gg e_{ID}^C - e_{ID}^B$), which is also very reasonable, the EDP difference can be approximated as

$$\Delta\text{EDP} \approx \sum_i (a \cdot R_i^{cyc} + b \cdot R_i^{ins}) \cdot d_i \cdot \chi_i \qquad (3.5)$$

where constants a and b are defined as $a = E_{IF}^B + E_{ID}^B + E_{EX}$ and $b = E_{IF}^B + E_{ID}^B$.

4.3   Modifying the Selection Algorithm

The EDP difference of (3.5) means that the contribution of a C-instruction i to the reduction of EDP, which defines Ben_i or the benefit of i, can be written (approximately) as

$$Ben_i = a \cdot R_i^{cyc} + b \cdot R_i^{ins} \qquad (3.6)$$

Note that this expression of benefit renders it very easy to extend the heuristic algorithm of (Lee et al. 2002) for the enery-efficiency context. In fact, (3.6) can also be used to represent the performance improvement or the energy reduction simply by changing the parameters: performance improvement if $a = 1$ and $b = 0$ and energy reduction if $a = 0$ and $b = 1$. Thus instead of using R_i^{cyc}, which is the form that is originally used, we can use (3.6) and extend the heuristic algorithm easily. More details of the modified algorithm can be found in (Lee 2004).

5.   Experiments

We now present the experimental results applying the energy-efficient IS synthesis technique to improve the native IS of the MIPS processor, and also show the effects of the instruction bitwidth variation. To drive the IS synthesis, a number of realistic benchmark applications are used covering multimedia (e.g., h.263 decoder, JPEG encoder), control-intensive (e.g., ADPCM coder/decoder), and cryptography (e.g., DES) domains. After describing the experimental setup, we first show the effectiveness of the synthesized IS over the native IS, comparing the results produced by different optimization criteria. Next, the effects of the instruction bitwidth variation are also presented.

5.1   Experimental Setup

For our experiments, we used the MIPS microprocessor architecture (Patterson and Hennessy, 1997), from which a basic IS was defined annotated with resource and timing information. The benchmark applications were pre-processed using the EXPRESS retargetable compiler (Halambi et al., 2001) targeting the basic IS to generate preliminary assembly code, which was used for the rest of the IS synthesis process.

While the MIPS architecture has 32-bit instructions, the native IS of the MIPS uses the code space of only about 2^{30}, meaning that the native IS essentially uses only 30 bits, reserving the rest of the code space for future versions. Since the basic IS for the MIPS was defined with the code space of about 1.42×10^8, we generated C-instructions into the remaining code space of $2^{30} - 1.42 \times 10^8$, with 30-bit bitwidth, so that the synthesized IS should take no more code space than the native IS uses. In the second set of experiments, where the instruction bitwidth was varied, the code space of $2^{29} - 1.42 \times 10^8$ and $2^{28} - 1.42 \times 10^8$ were used for 29-bit and 28-bit IS synthesis, respectively.

After the IS synthesis process, the application was recompiled using the same retargetable compiler targeted for the synthesized IS. The execution cycle counts and the instruction fetch counts were obtained through cycle-accurate simulation (within basic blocks) and profiling (at the global level). It is assumed that the ID energy is the same for all ISs, since the comparison is between the synthesized ISs and the native IS, the latter of which also exhibits many "compound-operation" instructions.

For the energy consumption estimation, the following assumptions are made on the architecture. It is assumed that the instruction memory is large enough to fit each application so that there is no off-chip memory access for instruction fetch. To further simplify, it is assumed that there is no instruction cache; thus, instruction memory energy consumption only depends on the number of instruction fetch. For the energy consumption ratio between IF, ID, and EX stages, it is assumed that the ratio between $E_{IF}^B + E_{ID}^B$ and E_{EX} is 1:1 regardless of the application.[1] This greatly simplifies the EDP comparison between different ISs, i.e.,

$$r_{edp} = [1 - 0.5\,(1 - r_{if})] \cdot r_{cyc} \qquad (3.7)$$

---

[1] Our architecture-level power simulation using a typical processor power simulation framework did not show enough fidelity; the power breakdown was the same for all applications. Thus, we estimated the ratio based on the ARM920T power analysis result in (Segars 2001) taking into account different activation factors of components.

where r_{edp}, r_{if}, and r_{cyc} are the ratios of the EDP values, the IF counts, and the cycle counts, respectively, of two ISs. r_{if} itself represents the energy consumption ratio between two ISs, provided that we use the earlier assumption that the number of EX operations does not change by different ISs. Lastly, r_{cyc} is the performance ratio, assuming the same clock speed between two ISs.[2]

5.2   Improvement through IS Synthesis

Recall that the strategy employed by the IS synthesis framework improves the native IS in two steps: (1) extract the basic IS from the native IS; and (2) build C-instructions on top of the basic IS. Therefore, the synthesized ISs always generated far better results compared to the basic IS in all the experiments performed, and these trivial results are not shown here. More importantly, in most cases the synthesized IS generated better results even compared to the native IS, as shown in Figure 3.4.

Figure 3.4 shows the results generated by the synthesized ISs, in terms of the performance, the number of IF (representing energy), and the EDP value, normalized to those of the native IS. To see the effects of different optimization goals, i.e., performance, energy, and EDP, three ISs were synthesized for each benchmark application; thus, the three bars for each application represents the results of the three different synthesized ISs. The graphs show that the synthesized ISs can generate performance improvements of up to 43% or the IF reduction of up to 44%, though the improvements vary depending on the application as well as the optimization goal used. Also, when translated into EDP, the synthesized ISs can reduce the EDP by up to 42% (compared to the native IS), and 25% on average for all the applications using the EDP optimization. These results clearly show that the proposed technique can generate energy-efficient ISs for many applications in various domains.

From the figure, it is clear that optimizing for one metric does not necessarily lead to optimal results for other metrics as well, which confirms the need to consider the energy-efficiency metric more explicitly. Also, as expected, the best results for a metric were obtained by directly optimizing for the metric in most cases. There were minor exceptions, however, for the ADPCM benchmark the greatest energy reduction (the lowest IF count) was achieved by optimizing for EDP, and for the h.263 benchmark the greatest EDP reduction was achieved by optimizing for performance. This phenomenon is most likely due to

---

[2]We suppose that the performance of the native and the synthesized ISs can be compared directly in terms of the cycle counts, since the native IS also has its own "compound-operation" instructions often with elaborated encoding schemes.

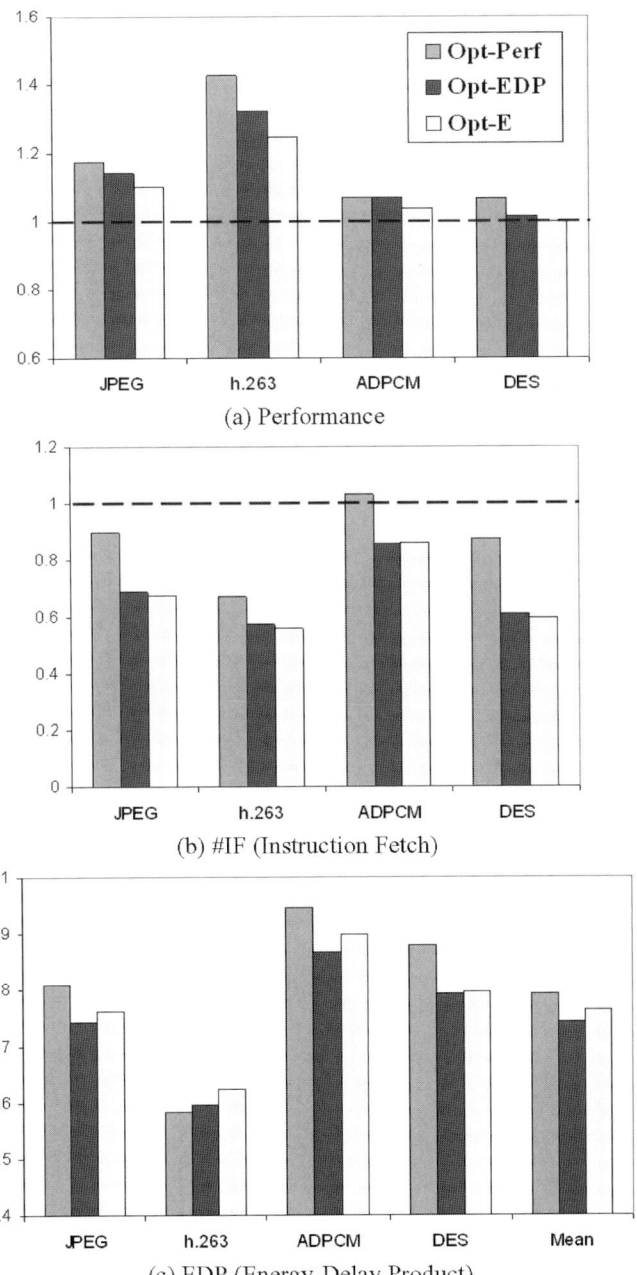

Fig. 3.4   Performance, #IF, and EDP of the synthesized ISs, normalized to that of the native
IS. The three bars per each application correspond to the three ISs synthesized with
different optimization goals: performance, EDP, and energy (in order). The rightmost
column in (c) is the geometric mean of the four benchmark results on EDP

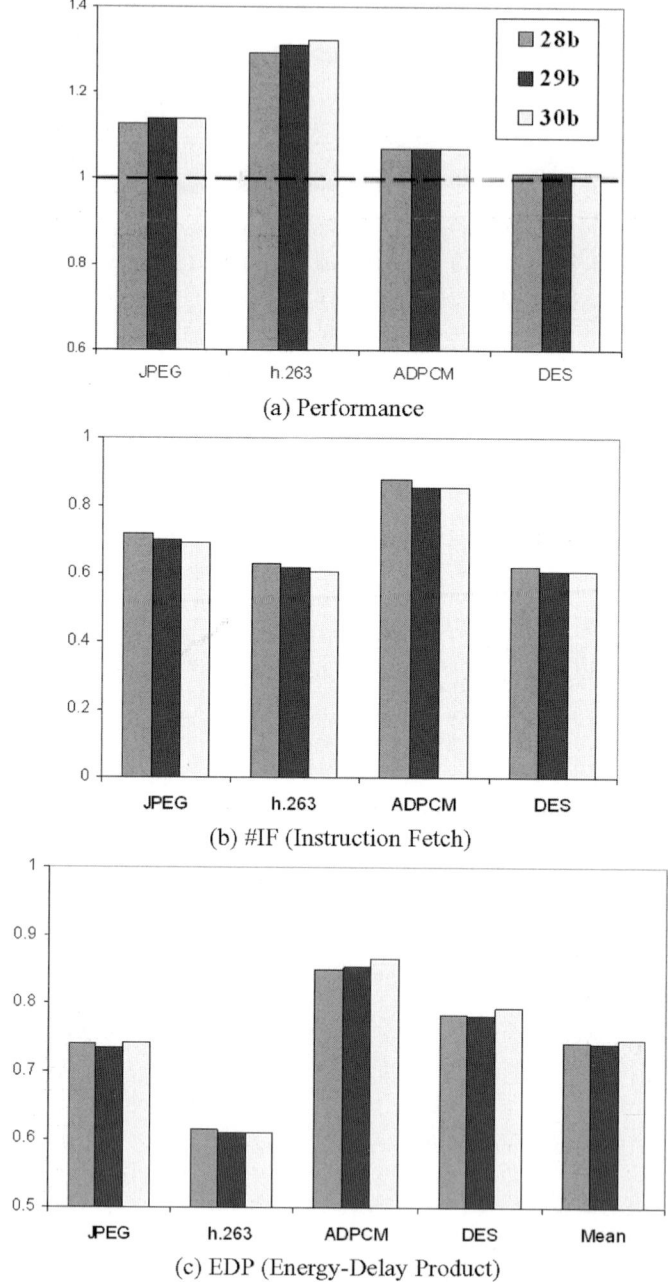

Fig. 3.5 Performance, #IF, and EDP of the synthesized ISs, normalized to that of the native IS. The three bars per each application correspond to the ISs with different instruction bitwidths: 28, 29, and 30 bits. The rightmost column in (c) is the geometric mean of the four benchmark results on EDP

the suboptimality of the instruction selection heuristic algorithm. In the case of the EDP optimization, the approximation such as (3.5) is also responsible.

5.3   Effects of Bitwidth Variation

The proposed technique can effectively reduce the code volume through a careful selection of instruction encoding, considering the dynamic instruction fetch count as well as the code space constraint for the best use of the instruction bitwidth. But another possibility for reducing the code volume is simply reducing the bitwidth. Reducing the instruction bitwidth can affect the IF power directly, resulting in, at the most, a linear decrease of the IF power with respect to the bitwidth reduction. On the other hand, the reduced bitwidth may not permit enough code space for the C-instructions that are essential to boost the performance.

To see this aspect of code volume reduction, another set of IS synthesis experiments is performed varying the bitwidth (and the code space constraint, consequently) while optimizing for the EDP reduction. Figure 3.5 summaries the results, in which it is aggressively assumed that the IF energy consumption is proportional to the instruction bitwidth. Though the difference is small, the lowest EDP values are found at 28-bit (ADPCM), 29-bit (JPEG, DES), and 30-bit (h.263) instructions, which is in accordance with the intuition that smaller applications will have less number of application-specific instructions and hence require smaller instruction bitwidth. Though it is possible and might be advantageous to fine-tune the instruction bitwidth for the application, very small EDP differences with different bitwidths suggest that the energy-efficient IS synthesis technique can make a very good use of the given bitwidth, so that instruction bitwidth fine-tuning does not look very much necessary.

6.   Conclusion

We have presented an energy-efficient IS synthesis technique that is based on an encoding-oriented IS synthesis framework. To comprehensively reduce the code volume, our technique optimizes the instruction encoding, considering both the instruction bitwidth and the dynamic instruction count. To apply the IS synthesis algorithm for energy-efficiency optimization, we formulated the energy and EDP change due to IS customization and derived the contribution of a C-instruction. The experimental results show that the proposed IS synthesis technique can generate application-specific ISs that outperform the native IS of a typical embedded RISC processor, in performance, energy, and EDP, up to about 40% with each metric.

References

Benini, L., De Micheli, G., Macii, A., Macii, E., and Poncino, M. (1998) Reducing power consumption of dedicated processors through instruction set encoding. In *Great Lakes Symposium on VLSI (GLSVLSI)*.

Benini, L., Macii, A., Macii, E., and Poncino, M. (1999) Selective instruction compression for memory energy reduction in embedded systems. In *International Symposium on Low Power Electronics and Design (ISLPED)*.

Bunda, J., Fussell, D., and Athas, W. C. (1995) Energy-efficient instruction set architecture for CMOS microprocessors. In *Twenty-Eighth Hawaii International Conference on System Sciences*.

Chandar, S., Mehendale, M., and Govindarajan, R. (2001) Area and power reduction of embedded dsp systems using instruction compression and re-configurable encoding. In *International Conference on Computer Aided Design (ICCAD)*.

Dougherty, W., Pursley, D., and Thomas, D. (1998) Instruction subsetting: Trading power for programmability. In *Workshop on System Level Design*.

Gajski, D., Dutt, N., Wu, A., and Lin, S. (1992) *High-Level Synthesis: Introduction to Chip and System Design*. Kluwer Academic Publishers.

Glokler, T. and Bitterlich, S. (2001) Power efficient semi-automatic instruction encoding for application specific instruction set processors. In *International Conference on Acoustics, Speech, and Signal Processing (ICASSP)*, pp. 1169–1172.

Halambi, A., Shrivastava, A., Dutt, N., and Nicolau, A. (2001) A customizable compiler framework for embedded systems. In *Workshop on Software and Compilers for Embedded Systems (SCOPES)*.

Inoue, K., Moshnyaga, V., and Murakami, K. (2002) Reducing power consumption of instruction ROMs by exploiting instruction frequency. In *Asia-Pacific Conference on Circuits and Systems*.

Kim, S. and Kim, J. (1999) Opcode encoding for low-power instruction fetch. *IEE Electronics Letters*, Vol. 35, no. 13, pp. 1064–1065.

Lee, J.-E. (2004) *Architecture Customization for Configurable Processors and Reconfigurable ALU Arrays*. PhD thesis, Seoul National University.

Lee, J.-E., Choi, K., and Dutt, N. (2002) Efficient instruction encoding for automatic instruction set design of configurable ASIPs. In *International Conference on Computer Aided Design (ICCAD)*.

Liao, S., Devadas, S., and Keutzer, K. (1999) A text-compression-based method for code size minimization in embedded systems. *ACM Transactions on Design Automation of Electronic Systems*, Vol. 4, no. 1, pp. 12–38.

Patterson, D. and Hennessy, J. (1997) *Computer Organization and Design: The Hardware/Software Interface*, 2nd ed. Morgan Kaufmann Publishers.

Segars, S. (2001) Low power design techniques for microprocessors. In *International Solid State Circuits Conference (ISSCC)*.

Chapter 4

A Framework for Extensible Processor Based MPSoC Design

Fei Sun[1], Srivaths Ravi[2], Anand Raghunathan[3], and Niraj K. Jha[4]

[1] *Tensilica Inc., 3255-6 Scott Blvd.*
Santa Clara
CA 95054

[2] *NEC Laboratories America*
4 Independence Way
Princeton
NJ 08540

[3] *NEC Laboratories America*
4 Independence Way
Princeton
NJ 08540

[4] *Department of Electrical Engineering, Princeton University*
Princeton
NJ 08544

Abstract   Multiprocessor system-on-chip (MPSoC) architectures have emerged as a popular solution to the ever-increasing performance requirements of embedded systems. MPSoC architectures that are customized to a specific application or domain have the potential to achieve very high performance, while also requiring low power consumption. The recent emergence of extensible processors has greatly facilitated the design of efficient yet flexible application-specific processors, making them a promising building block for MPSoC architectures. However, the inter-dependent multiprocessor, co-processor, and custom instruction design problems result in a huge design space. Therefore, efficient tools are needed that assist designers to create high-quality architectures in limited time.

J. Henkel and S. Parameswaran (eds.), Designing Embedded Processors – A Low Power Perspective,
65–95.

In this chapter, we describe a framework that generates extensible processor based MPSoC architectures for a given application, by synergistically exploring custom instruction, co-processor, and multiprocessor optimizations. The framework automatically maps embedded applications to MPSoC architectures, aiming to minimize application execution time and energy consumption, while the overall area for the MPSoC is kept within a given budget.

Keywords: MPSoC; ASIP; custom instruction; extensible processor; co-processor; hardware accelerator; hardware–software co-design.

1.   Introduction

Trends in embedded system applications, features, and complexity have translated into increasingly stringent performance requirements for the underlying architectures, which must be met at low power consumption and low cost. For example, let us consider video playback on handheld devices such as cell phones, media players, etc. Screen resolutions have increased from QCIF to CIF to D1, which translates to $16\times$ more computation and I/O throughput. The implementation of an H.264 video codec requires $5\times$ to $10\times$ more computation than the previous H.263 codec. As a result, the handheld device needs to provide $80\times$ to $160\times$ more real-time computation capability with little or no impact on battery life. At the same time, embedded systems need to be flexible enough so that the design can be re-used between different product variants or versions, and easily modified in response to bugs, market shifts, or user requirements, during the design cycle and even after production.

A good trade-off between efficiency and flexibility in SoC design can be obtained through the use of configurable and extensible processors. The processor's instruction set can be extended, and the underlying micro-architecture configured, for a specific application in order to improve efficiency, while the basic instructions of the processor leave room for software upgrades. Many companies have provided commercial configurable and extensible processors (e.g., Xtensa from Tensilica [1], MeP from Toshiba [2], ARCtangent from ARC [3], MicroBlaze from Xilinx [4], and Nios from Altera [5]).

The complexity of embedded software applications and their performance requirements have reached a point where they can no longer be supported by conventional embedded system architectures. The energy efficiency (MIPS/mW) of monolithic general-purpose processor architectures scales poorly with the addition of advanced micro-architectural features. As a result, traditional uni-processor architectures have given way to multiprocessor system-on-chip (MPSoC) architectures for many embedded applications. Many commercial SoC products have successfully integrated tens to even hundreds of processors on a chip. For example, the Cisco CRS-1 network router uses 192 processors per chip [6]; the cell processor used by Sony's Playstation 3 includes 9 cores [7]; and Intel has recently prototyped an 80-core chip delivering teraflop performance [8].

Configurable and extensible processors can be tailored to the specific task they perform, thus, they could be ideal building blocks for modern MPSoCs. However, realizing this potential requires the development of supporting tools and methodologies so that the design turnaround times are kept short and comparable to software solutions. Current extensible processor design flows automate several critical, but tedious steps, such as automatic generation of register-transfer level (RTL) description of the custom processor, and automatic generation of re-targetable software tool chains, and automatic generation of synthesis, verification, test and physical design scripts. However, several key questions remain to be answered: how to map the applications to a multiprocessor architecture, how to add hardware engines (co-processors) for each processor to speed up critical functions, how to customize each processor to fit the functions it performs using custom instructions, and how to explore the best trade-offs between hardware added, performance, and power.

In this chapter, we answer these questions by presenting a framework for automating these steps in the design of extensible processor based MPSoC architecture.

2. Overview of MPSoC Synthesis

In this section, we provide an overview of the extensible processor based MPSoC synthesis problem. Given an application and an extensible processor platform that can be used as a building block, the objective is to create an architecture that consists of multiple processor instances, where each processor instance is further customized through the addition of co-processors and custom instructions. The objective is to minimize the execution time and/or energy consumption of the application, while the total area of the MPSoC is kept within a fixed budget. This problem contains inter-dependent sub-problems that address architectural optimizations at different levels of granularity – how many processors to use and how to partition the application among them (coarse-grained), which functions to map to co-processors (medium-grained), and how to extend each processor with custom instructions (fine-grained).

Complex applications are frequently represented as a set of communicating tasks, or *task graph* during system-level design. In order to effectively perform optimizations at the three different levels of granularity, we utilize hierarchical task graphs, where each task may itself correspond to a sub-graph at a lower level of the hierarchy. The tasks that correspond to leaf nodes in the task graph are represented using program dependence graphs, which are commonly used in software compilers.

Following the definition by Girkar [9], an HTG is a directed acyclic graph (HV, HE), where each node in HV can be one of the following types: (i) *simple* task node representing a task that has no subtasks, (ii) *compound* task node representing a task that consists of other tasks in an HTG, or (iii) *loop* task node representing a task that is a loop whose iteration body is an HTG. Girkar also provides a method to automatically extract HTGs from ordinary sequential programs [9].

Example 4.1 Figure 4.1 is part of an HTG for the MPEG-2 decoder application taken from the MediaBench benchmark suite [10]. There are only two tasks at the first level. *Frame body decoder* executes after *Frame header decoder*. Note that *Frame body decoder* is a loop node. Its iteration body is another HTG, which has three tasks: *For header*, *Slice header decoder*, and *Slice body decoder*. Task *Slice body decoder*, which decodes all the macro-blocks inside a slice, is in turn a loop node. In its HTG, after *For header* and *Variable length decoder* are executed, task *Motion compensation* can be executed in parallel with tasks *Inverse quantization* and *Inverse DCT*. Finally, *Add block* combines the data generated by the previous tasks and updates the frame buffer. *Inverse quantization*, *Inverse DCT*, and *add block* can be further divided into loops at the block level.

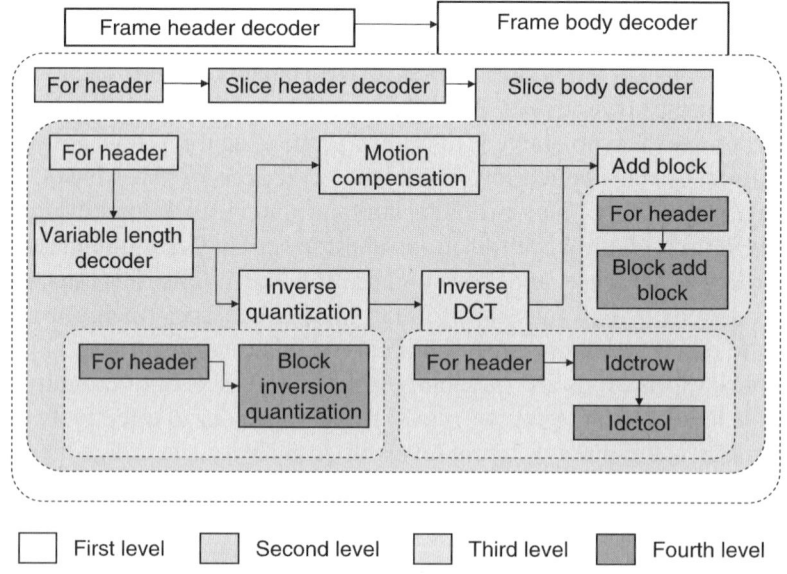

Fig. 4.1   HTG for the MPEG-2 decoder

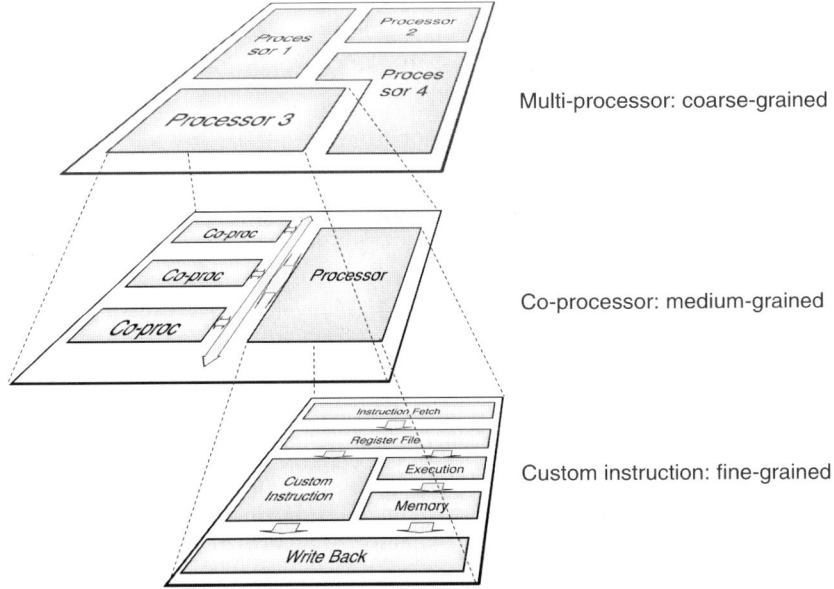

Fig. 4.2 Granularities of architectural optimization in MPSoC design

Figure 4.2 illustrates the different granularities of architectural optimization that are addressed in our framework. A complex SoC may consist of several processors, which exploit coarse-grained parallelism across tasks. Tasks at any level in an HTG may be mapped to a processor to be executed as embedded software, or offloaded to dedicated hardware (co-processors). The leaf tasks in an HTG may still be complex sub-programs, and are represented using program dependence graphs. If they are mapped to a processor, they can be further optimized through custom instructions.

We describe different parts of our framework in the following three sections. First, we describe how to generate custom instructions in order to speed up a program or sub-program that is mapped to an extensible processor. Next, we describe how to simultaneously generate co-processors and custom instructions, while exploring their inter-dependencies. Finally, we describe how a complete application can be mapped to an efficient multiprocessor architecture, while also optimizing the constituent processors by selecting a combination of co-processors and custom instructions.

3.   Custom Processor Synthesis

In this section, we describe how to automatically generate custom instructions to speed up the execution of a program on an extensible processor.

3.1   Motivation

Current extensible processor design flows automate several tedious steps. However, most designers still manually design the custom instructions that are used to augment the base processor platform. This involves profiling the program source code on the base processor, identifying the most performance-critical sections of the program (the "hot spots"), formulating custom instructions by specifying new hardware resources and the operations they perform, and re-writing the source code of the program to directly invoke the custom instructions. While these steps are somewhat simplified by profiling tools, and through the specification of the custom instruction hardware at a high level of abstraction, they are still daunting tasks for large programs, and are further complicated when performance or energy consumption needs to be optimized subject to constraints on hardware overhead.

The number of candidate custom instructions for a program grows exponentially with the program size. It is not uncommon for functions with a few tens of operations to contain several hundred instruction candidates. For example, consider the C code for the BYTESWAP() function shown in Figure 4.3a. We used the tools developed in our work to identify all possible (unique) instruction candidates for this function. The function, although quite small, contains 482 potential custom instructions. In more realistic programs, a *combination of several custom instructions* may be necessary to achieve the desired performance, further increasing the design space.

Example 4.2 In order to illustrate the variation of speedup over the set of all candidate instructions (the "design space"), we generated and evaluated

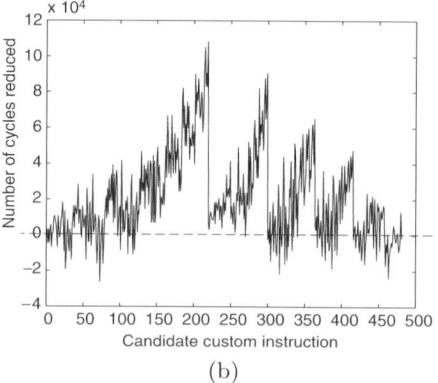

```
/* This function swaps the order of bytes in s if
   argument do_swap is non-zero */
static unsigned
BYTESWAP(unsigned s, unsigned char do_swap) {
    unsigned ss, s1, s2, s3, s4,
             s5, s6, s7, s8, result;
    s1 = s<<24;
    s2 = s<<8;
    s4 = s>>8;
    s6 = s>>24;
    s3 = s2 & 0xff0000;
    s5 = s4 & 0xff00;
    s7 = s1 | s3;
    s8 = s5 | s6;
    ss = s7 | s8;
    /* Global count of #words processed */
    if (do_swap) SWAPPED_COUNT++;
    result = do_swap? ss:s;
    return result;
}
```

(a)   (b)

Fig. 4.3   An example used to demonstrate the size and complexity of the custom instruction design space: (a) example function BYTESWAP() and (b) performance variation across the design space

all possible custom instructions for the `BYTESWAP()` function shown in Figure 4.3a. The function iterates 10,000 times and consumes 130 K cycles on the base processor core. Figure 4.3b plots the execution cycle savings resulting from each of the 482 candidate instructions. Figure 4.3b indicates that there is a large variation in quality across the custom instruction design space. Some custom instructions even increase the execution time of the program. The figure shows that the nature of the design space is quite complex, underlining the need for automatic exploration techniques.

3.2   Extensible Processor Synthesis Methodology

The design flow of automatic custom instruction generation is outlined in Figure 4.4. It takes as input the program source code to be optimized (in C), and outputs the selected custom instructions and a modified C program that calls them.

Given a C program as input, Step **1** generates the *program dependence graphs* [11]. At the same time, the program is simulated and then profiled both at the function level and line level, to determine where the hot spots are (Step **2**). Step **3** ranks the control blocks of the program in descending order

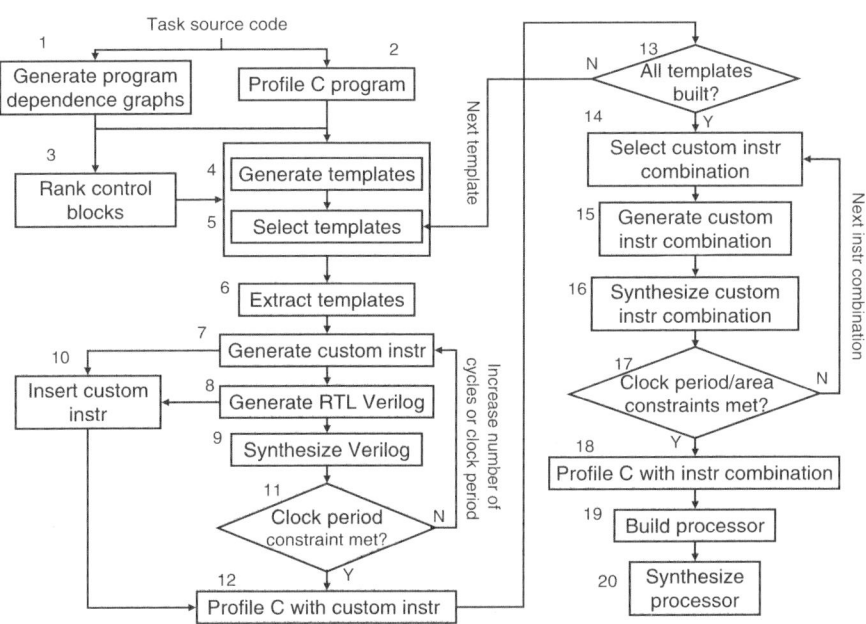

Fig. 4.4   Automatic custom instruction generation flow

of potential for improvement. The ranking criteria may include performance (from a profiler), energy (from an energy estimator) or energy-delay product (from both), depending on the optimization objective.

Steps **4** and **5** generate and select custom instruction templates, respectively. Although these two steps are explained separately for the sake of clarity, their implementation may be combined for the sake of efficiency. A *template* is a set of program statements that is a candidate for implementation as a custom instruction. Since the number of templates grows exponentially with program size, it is necessary to prune the search space.

Each promising candidate selected in Step **5** is extracted from the C program (Step **6**), and transformed to a custom instruction format that describes the opcode, operand, states, user-registers, computations, etc. (Step **7**). It is then compiled to get the Verilog RTL description of the additional hardware that will augment the base processor (Step **8**). The RTL description is synthesized to get the timing and area information (Step **9**). If the new instruction cannot be fit in the base processor core's clock period, either the number of cycles used by the new instruction is increased, or the clock period is increased, and the custom instruction generation phase is repeated (Step **11**). Hence, for a single custom instruction template, there may be several versions with different clock periods and numbers of cycles. At the same time, the original C code is transformed by replacing the appropriate statements with a call to the custom instruction (Step **10**). Then, for each version of the custom instruction, the new C program is compiled and profiled using a cycle-accurate instruction set simulator (ISS) to get the performance improvement (Step **12**). Steps **5** to **12** are iterated for every selected template.

After each individual custom instruction has been verified, a subset (combination) of instructions is chosen to get the maximum performance improvement under the given area constraint, depending on the selection criteria (Step **14**). The hardware corresponding to the selected custom instruction combination is built and synthesized (Steps **15** and **16**). If the timing and/or area constraint is not satisfied, the next best custom instruction combination is selected (Step **17**). Otherwise, the modified C program is compiled and profiled again to get the final performance improvement and/or energy reduction (Step **18**). After having selected the custom instruction combination, the whole processor is built and synthesized (Steps **19** and **20**).

3.3   Template Generation

Although template generation and selection are represented as distinct steps in Figure 4.4, in our implementation, they are interleaved in order to improve efficiency. In the generation process, some templates, which have low potential for performance improvement, are not generated.

We propose to generate templates in three phases. In the first phase, we generate *basic templates*. A basic template consists of a single node in the program dependence graphs that satisfies the given selection (pruning) criteria. In the second phase, we generate *dependent templates*. A dependent template is a fully connected sub-graph of the data dependence graph. Hence, each node of a dependent template is connected to some other node in the template through a variable. Dependent templates are generated by using a basic template as a seed, checking data dependencies of the basic template, and including combinations of data dependence predecessors and successors if they satisfy the selection criteria. In the third phase, we generate *independent templates*. In this step, we use both basic and dependent templates as seeds, and add nodes that are independent of the seed template. The following example illustrates the template generation process.

Example 4.3 Figure 4.5a shows a small fragment of C code corresponding to a single control block, and its data dependence graph. Each node represents a single statement in the C program. Each node also has a weight that represents the fraction of the total program execution time spent in that node. The dotted lines in Figure 4.5a indicate data dependencies with operations that belong to other control blocks. Figure 4.5b shows all possible templates that can be generated from this graph. Nodes 1–4 form basic templates. Templates 5–7 are generated in the dependent template generation phase. The independent template generation phase generates templates 8 to 13.

In Example 4.3, we enumerated all possible templates for the sake of clarity. In practice, the number of candidate templates may potentially be very large,

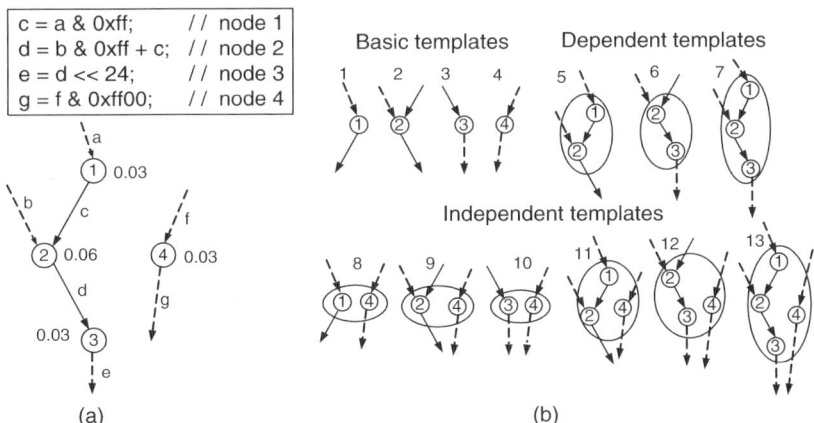

Fig. 4.5 Illustration of the template generation process: (a) code fragment and its data dependence graph and (b) generated templates

even for programs of moderate size. Hence, it is necessary to use pruning criteria to select good templates while discarding less promising ones. Any metric to evaluate the templates should consider the following factors:

- Amdahl's Law [12] suggests that the fraction of the original program's execution time that a template accounts for presents a bound on the performance improvement achievable when it is converted into a custom instruction. Hence, templates that have a larger cumulative weight are more desirable.

- Given two templates that account for the same fraction of total execution time on the original processor, the number of cycles required to execute them when implemented as a custom instruction is an indicator of their optimization potential. This conforms to the well-known principle that operations that require a large execution time on the base processor, but can be very efficiently performed by special hardware units, are good candidates for custom implementation.

- Many extensible processors, including the Xtensa processor, impose a limit on the number of operand fields that can be specified in the instruction format. Also, the general-purpose register file in the processor has a specific number of read and write ports, imposing a limit on the number of general-purpose registers that can be used in a custom instruction. This bottleneck can be overcome by defining custom registers (called *state* or *user-defined registers*) whose use is hardwired into the instruction. However, the use of state registers imposes an additional overhead. When other computations generate (use) data that are used (generated) by the custom instruction, the contents of the state registers need to be written to or read from either memory or the processor's general-purpose registers. The overhead for data transfer is determined by the number of "excess" input and output variables of a given instruction template.

Considering all the above factors, we use the following equation to rank candidate templates:

$$Priority = \frac{OriginalTime}{\max(In - \alpha, 0) + \max(Out - \beta, 0) + \gamma} \tag{4.1}$$

In the above equation, $OriginalTime$ is the fraction of the total execution time of the original program spent in the template, In and Out are the number of inputs and outputs of the template, respectively, α is the number of inputs that can be encoded in the instruction, β is the number of outputs that can be encoded, and γ is the number of cycles required by the template when implemented as a custom instruction. The numerator of Equation (4.1) is the time spent in the original sub-program corresponding to the template, which is automatically computed from the line-by-line profile information. The denominator is an estimate of the number of cycles required by the custom instruction in each invocation. Since one instruction can have at most α inputs and β outputs specified in the instruction (the exact values of α and β are dependent on the processor architecture), if the number of inputs is greater than α, a cycle is needed for each additional input to load it into a user-defined state register, so that the new custom instruction can implicitly use the data in corresponding states to compute the final result. If the number of inputs is less than α, this term is zero. A similar explanation holds for the number of additional outputs.

Since templates having higher values of the $Priority$ metric are likely to get more performance speedup or energy reduction, we first consider those templates as seeds when generating new templates. In order to achieve this, we preserve a ranked index of the templates, and traverse the list in decreasing order of $Priority$ when choosing a seed. Low priority templates may either have a large number of inputs or outputs, or consume little time so they are not worth implementing as custom instructions, and thus may be discarded. Hence, we set a priority threshold to determine the templates that are considered for further analysis. Any template with priority below this threshold is discarded. For the sake of computational efficiency, the threshold mechanism is dynamically enforced during template creation itself (rather than as a post-processing step). While generating templates, we preserve the highest priority ($Priority_{highest}$) template obtained thus far. After we generate a new template, we compute its priority and compute the ratio $\frac{Priority_{current}}{Priority_{highest}}$. If the ratio is below the threshold, we do not add it to the template list. In our experiments, we found that setting the threshold ratio between 0.1 and 0.15 achieved reasonable reductions in the number of templates generated, with negligible or no impact on quality.

3.4 Experimental Results

We have implemented the flow described in Figure 4.4 by integrating several commercial and public-domain tools with our custom tools. Our tool takes a task program (written in C) as input and outputs custom instructions and the modified task in C program. The GNU-based compiler, simulator, and profiler tools provided by Tensilica are used to simulate the program and gather information about execution cycles (Steps **2**, **12**, and **18** in Figure 4.4). We use the Aristotle analysis system [11] to generate the program dependence graphs (Step **1** in Figure 4.4). We use Synopsys Design Compiler [13] to synthesize the RTL circuit and map it to NEC's commercial $0.18\,\mu$m technology library [14] (Steps **9** and **16** in Figure 4.4). The area and clock period information extracted from the synthesized, mapped netlists are used to drive the selection of the final instruction combination that is used to augment the processor.

We evaluated the proposed techniques using six example benchmarks. BYTESWAP is a function to swap the order of bytes in a word. It is mostly used for little-endian to big-endian conversion and vice versa. Add4 adds the value of four bytes in one word and returns the sum. RGBtoCMYK is a color conversion program. Alphablend blends two 24-bit pixels. PopCount implements the population count function, which counts the number of 1's in a word. Rand is a function for ISAAC (indirection, shift, accumulate, add, and count), which is used as a fast cryptographic random number generator. Our experiments are run on a 440 MHz SUN Ultra10 workstation with 1 GB main memory. The area constraint is set to 10% of the original processor's total area in all experiments.

Table 4.1 summarizes the results of our experiments. It compares the execution time, energy, and energy-delay product of the benchmark programs, running on a base processor (without any custom instruction extensions), and on the customized processors generated by our tool. We also report the area overheads incurred due to the addition of extra hardware to the processor. The results in Table 4.1 are based on: (i) execution cycle count reported by the cycle-accurate ISS, (ii) clock period and area information derived from the synthesized, mapped netlists of the complete processor cores, and (iii) power estimates provided by running the commercial tool PowerTheater from Sequence Design Inc. [15]. The results indicate that processors customized using instructions automatically generated by our tool can achieve a performance improvement of up to 5.4× (average of 3.4×) over the base processor cores. Energy consumption is reduced by up to 4.5× (average of 3.2×). The energy-delay product is reduced by up to 24.2× (average of 12.6×), while the average area increase is only 1.8%.

Table 4.1 Area, performance, and energy results for processors generated by the proposed tool

Program		Time (ms)	Energy (μJ)	Energy•Delay (ms•μJ)	Area (grids)	Speedup	Energy•Delay reduction
BYTESWAP	Original	0.958	101.5	97.2	435347		
	New	0.397	42.1	16.7	432496	2.4 ×	5.8 ×
Add4	Original	0.532	63.8	33.9	435347		
	New	0.327	35.0	11.4	445216	1.6 ×	3.0 ×
RGBtoCMYK	Original	2.073	193.0	400.1	435347		
	New	0.387	42.6	16.5	446314	5.4 ×	24.2 ×
Alphablend	Original	2.728	298.4	814.0	435347		
	New	0.531	73.4	39.0	458953	5.1 ×	20.9 ×
PopCountp	Original	0.901	90.1	81.2	435347		
	New	0.217	20.0	4.3	438346	1.2 ×	18.9 ×
Rand	Original	2.063	253.7	523.4	435347		
	New	1.277	159.6	203.8	436995	1.6 ×	2.6 ×

4. Hybrid Custom Instruction and Co-Processor Synthesis

Custom instructions are ideal to explore parallelism inside a small section of code, *e.g.*, a basic block. However, a compound task or loop task in an HTG may sometimes be amenable to hardware implementation in the form of co-processors. In this section, we describe a hybrid custom instruction and co-processor synthesis methodology, and demonstrate that a combination of custom instructions and co-processors could improve performance better than when only one of them is considered.

4.1 Methodology Overview

The hybrid custom instruction and co-processor synthesis problem is formally stated below:

Problem 1. Suppose an HTG with n tasks is given. Any task node t_i has l_i custom instruction versions and m_i co-processor versions. A version requires $Cycle_{ij}$ cycles to execute and consumes $Area_{ij}$ area, where $1 \leq j \leq l_i$ for custom instructions, $1 \leq j \leq m_i$ for co-processors, and $1 \leq i \leq n$. Area for memory may be shared across custom instructions or co-processors. Find all non-dominated custom instruction and co-processor combinations in terms of execution time and area.

We propose a hybrid custom instruction and co-processor synthesis methodology for HTGs (Figure 4.6) to solve Problem 1. It takes an application as input and outputs a number of different custom instruction and co-processor combinations, where each combination has a non-dominated execution time and area.

Given an application program, first the HTG is generated (Step **1**). The method described by Girkar [9] may be used to automatically generate the HTG. The application is then profiled (Step **2**). The HTG is annotated with the profiling statistics (e.g., execution time, number of loop iterations). All the tasks in the HTG are ranked from 1 to N, depending on their hierarchy level (Step **3**). The tasks in the topmost task graph is given rank 1. For all rank 1 compound and loop tasks, all tasks contained in them are given rank 2. This process continues until every task is assigned a level.

Then, the algorithm starts from the highest level, N (Step **4**), and iterates until it reaches the lowest level (Step **5**). At each level, a task graph with all its tasks at that level, G, is selected (Step **6**). In task graph G, one task T is selected (Step **7**).

Then, co-processor implementations of task T are generated (Step **8**). If task T is a compound or loop task, all tasks contained in it are implemented in the co-processor. Different functional unit constraints may be applied to the task, so that the generated co-processors may have different execution time and

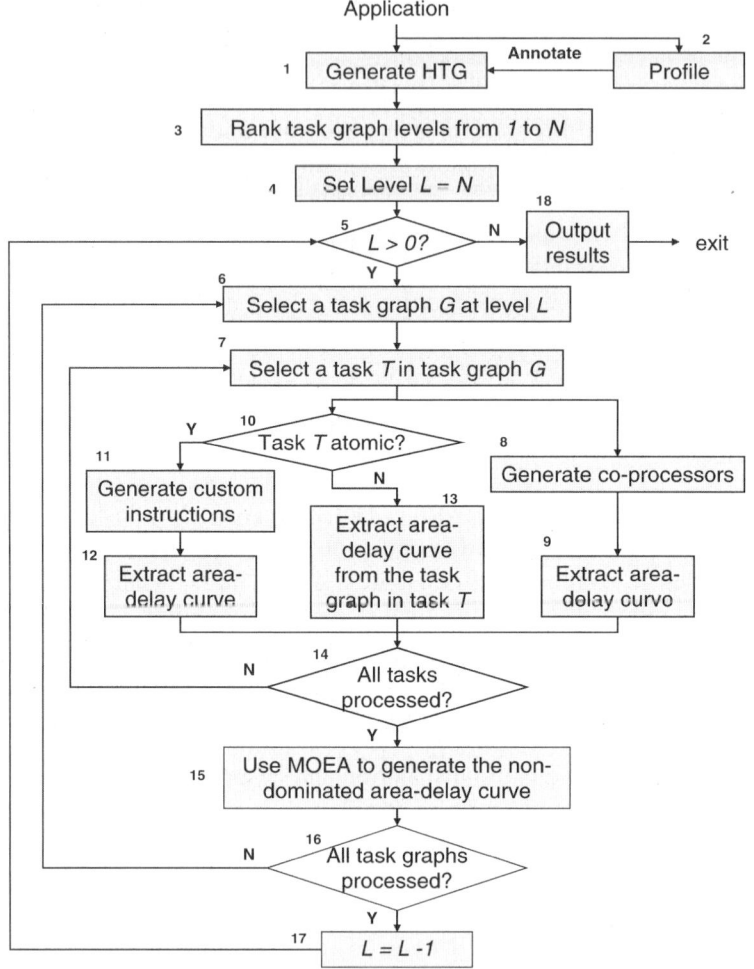

Fig. 4.6   Design flow for hybrid synthesis of custom instructions and co-processors

area characteristics. The execution times and areas are extracted from synthesis and simulation results (Step **9**). The execution time of a co-processor may be obtained by running the cycle-accurate C model for the co-processor, or simulating the generated hardware description language (HDL) code. The area can be obtained by synthesizing the HDL code.

Custom instruction implementations of task T are also generated. If T is a simple node, custom instruction versions with different execution times and areas are generated (Step **11**). The method described in Section 3 may be used to generate custom instructions automatically. Similar to co-processors, execution times and areas are then extracted (Step **12**). If the task is a compound or loop node, the task graph embedded in it should have already been processed (since the embedded tasks have a deeper level number). If the task being currently processed is a compound node, execution times and areas for the embedded task graph can be used directly. If the task being currently processed is a loop node, its execution times are the execution times of the embedded task graph multiplied by the iteration number, which may be obtained from compiler analysis or annotated profile information. Its areas are the same as those for the embedded task graph. Hence, for compound and loop task nodes, execution times and areas can also be extracted (Step **13**).

Steps **7** to **13** are repeated for every task in task graph G (Step **14**). After execution times and areas for both custom instructions and co-processors are obtained for each task in the task graph, the non-dominated curve for total execution times and areas are obtained for the task graph by using a multiobjective evolutionary algorithm (MOEA). (Step **15**).

After all task graphs at the current level are processed (Step **16**), the current level is reduced (Step **17**). The same process (Steps **6** to **17**) is repeated at a lower level until the root task graph is reached. Execution times and areas obtained at the root task graph are the final solutions for the application task. They give the designers the freedom to make execution time and area trade-offs (Step **18**).

4.2   Multiobjective Evolutionary Algorithm

A key step in the design flow is to find non-dominated execution times and areas of a task graph, given execution times and areas of each individual task (Step **15** in Figure 4.6). Since we need to optimize both the execution time and the area of the task graph, it is a multiobjective optimization problem. We have based our MOEA implementation on the NSGA-II framework [16] to find non-dominated solutions. In the following subsections, we present our encoding, fitness function computation, crossover and mutation methods.

Encoding.   A possible solution can be represented in terms of the schedule of the task graph and selection of custom instruction or co-processor versions. It needs to be encoded into a chromosome, which is usually represented as a string. We encode the selected custom instruction or co-processor version as an attribute of the task, and the schedule of a task graph as a topological sort. A sort is topological if for any task, all its predecessors are before it and all its successors are after it. The schedule is implicit in the topological sort and selection of custom instruction or co-processor version. The execution time of the task graph can be obtained by applying the as-soon-as-possible (ASAP) scheduling algorithm to the task graph. The area can be obtained by adding the areas required for each task in the task graph. The following example illustrates the encoding method.

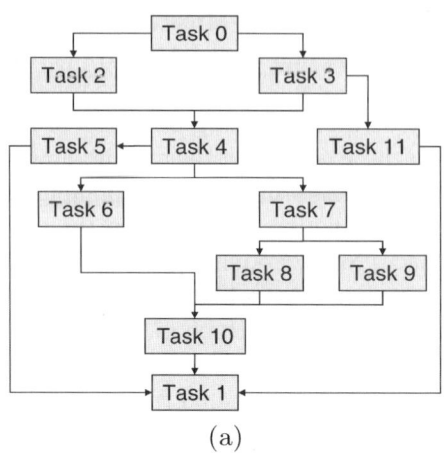

Task 0	N/A
Task 1	N/A
Task 2	I0, I1, C0, C1
Task 3	I0, I1, C0
Task 4	I0, I1, I2
Task 5	I0, C0
Task 6	I0, I1, I2, C0
Task 7	I0, I1, C0, C1
Task 8	I0, I1, C0, C1
Task 9	I0, I1, I2, I3
Task 10	I0, I1, I2
Task 11	I0, I1, C0, C1

(a)   (b)

Fig. 4.7   An example to illustrate the MOEA: (a) task graph and (b) custom instruction and
co-processor versions for each task

T0 → T3 → T2 → T4 → T7 → T5 → T11 → T6 → T8 → T9 → T10 , T1
     C0   I0   I2   C1   C0   C0   I2   C1   I3   I2

Fig. 4.8   Topological encoding of a solution in Figure 4.7

Example 4.4 Figure 4.7a shows an example task graph. Figure 4.7b lists the
custom instruction or co-processor versions for each task. Task 0 (T0)
and Task 1 (T1) are two artificial task nodes. T0 is the predecessor of all
tasks. T1 is the successor of all tasks.

The task graph can be encoded as shown in Figure 4.8. The schedule
can be implicitly obtained by applying the ASAP scheduling algorithm
to the topological sort. Tasks with co-processor versions can start executing immediately after all predecessors have finished execution. Tasks
with custom instruction versions can start executing after all predecessors have finished execution and the processor is available.

Crossover.   The crossover operator operates on two parent chromosomes
and generates two children chromosomes by exchanging parts of the parents'
characteristics. The offspring's characteristics may be better than those of both
parents by inheriting the good characteristics of the parents.

The topological sort maintains the precedence constraints of the task graph.
When the crossover operator is applied to chromosomes, the topology needs to
be maintained in the children chromosomes. We apply the crossover operator
using the following rules: (i) randomly generate a crossover point for the parent
chromosomes to break them into two parts, as shown in Figure 4.9a, (ii) copy
the first part of `Parent1` to `Child1`, (iii) scan `Parent2` from head to tail,

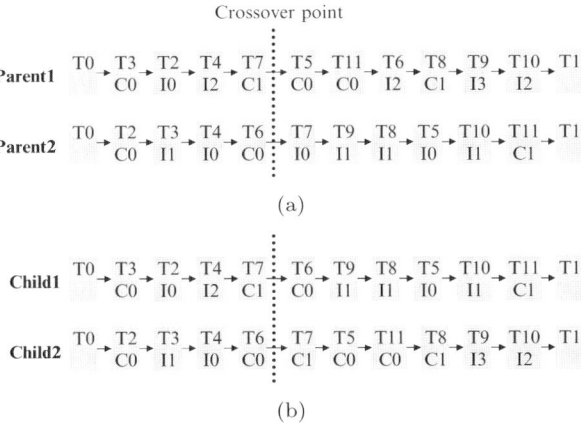

Fig. 4.9 Crossover example: (a) parent chromosomes, (b) offspring generated by applying the crossover operator on the parents.

append the gene to `Child1` if it is not in `Child1` already, and (iv) perform a similar computation to generate `Child2`. The resulting offspring are shown in Figure 4.9b. It is obvious that the offspring maintain the topology of the task graph.

Mutation. Mutation randomly changes parts of the characteristics of a chromosome. We implement two mutation operators and the algorithm randomly chooses between them. The first one randomly selects a task and then alters the custom instruction or co-processor version. The second one randomly selects a task, computes the position range it can move to without violating the topological constraint, and moves the task to a new allowed position. The possible positions of the task can be calculated by computing the position of its latest predecessor and earliest successor. The task can be placed at any position between them.

Initialization, Fitness Function Computation, and Selection. The initial pool of chromosomes is generated by first generating random topological sort strings and then randomly selecting the custom instruction or co-processor versions for each task. The fitness function is based on non-domination rank and crowding distance [16]. We use the roulette wheel selection method to improve convergence speed.

4.3   Experimental Results

We have implemented the hybrid custom instruction and co-processor synthesis methodology and evaluated its efficacy. The custom instructions are based on the Xtensa [1] platform from Tensilica. The co-processors are generated by the Cyber behavioral synthesis tool from NEC [17].

In our MOEA, we fix the mutation rate to 0.1 and the crossover rate to 1, because NSGA-II is an elitist algorithm [16]. The mutation rate and crossover rate are the probability of applying mutation operators and crossover operators on a chromosome. The algorithm exits if it converges, which is defined as no variance in the set of non-dominated solutions for five consecutive generations.

We used MPEG-2 decoder and GSM encoder programs as our bench-marks. The MPEG-2 decoder is a video decompression program. GSM is a lossy speech compression program. They were obtained from the Media-Bench benchmark suite [10]. Figure 4.10 shows the non-dominated solutions for the MPEG-2 decoder example found by our evolutionary algorithm. The execution time is the number of cycles to decode one macroblock. The area is the total system area, which includes the base processor, custom instruc-tions, co-processors, and on-chip memory, if they are present. Three different implementations are shown in Figure 4.10. They are custom instruction only, co-processor only, and custom instruction/co-processor architectures, respec-tively. It is clear from the figure that simultaneous custom instruction and co-processor synthesis gives much better performance/area trade-off than when using only one method. Simultaneous custom instruction and co-processor synthesis results in a 2.28× performance improvement over the pure software implementation.

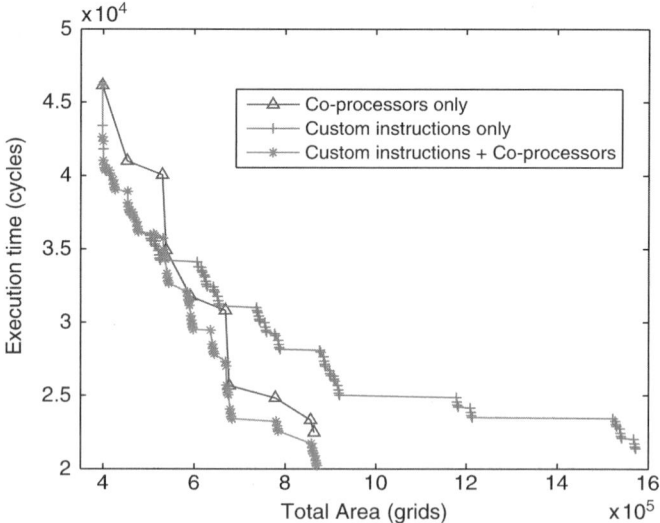

Fig. 4.10   Non-dominated execution time and area trade-offs for the MPEG-2 decoder appli-cation

Table 4.2 Selected completion times and areas for benchmarks and randomly generated task graphs

Name	#Tasks	#Solutions	Solution	Cycles	Area
			SW	20250	869373
MPEG-2	7	74	best-perf	31764	585896
			median-perf	46148	398343
			SW	56918	2739982
GSM	7	50	best-perf	130361	1293786
			median-perf	304771	398343
			SW	85091	442026
TG0	7	7	best-perf	159757	418716
			median-perf	246729	398343
			SW	494446	555633
TG1	17	118	best-perf	619296	468658
			median-perf	1494467	398343
			SW	953057	1003430
TG2	42	879	best-perf	1219268	692397
			median-perf	4183472	398343
			SW	257959	497265
TG3	11	21	best-perf	322205	417735
			median-perf	610489	398343
			SW	736902	584639
TG4	29	111	best-perf	859967	464981
			median-perf	2723501	398343

In Table 4.2, the pure software, best performance, and median performance point in the performance/area trade-off curve are shown for the MPEG-2 decoder. The total number of non-dominated solutions is also shown. The designer can choose the specific implementation based on the desired area or performance requirement.

Similar to MPEG-2, three different implementations are also shown for GSM in Table 4.2. Simultaneous custom instruction and co-processor synthesis results in a $5.35\times$ performance improvement over the pure software implementation. In between, 48 non-dominated solutions are also found. In the experiments, the clock period of the system is fixed at 7 ns. Synopsys Design Compiler [13] is used to synthesize the RTL Verilog code of the processor, custom instructions, and co-processors, and map them to an NEC 0.18 µm technology library [14].

We also used TGFF [18] to generate a number of random task graphs to analyze the scalability of our algorithm. Since each task may have several custom instruction or co-processor versions, each with a different associated execution time and area, the data transfer time, task execution time, and non-shared and shared area were also randomly generated for each version. Table 4.2 shows the task graph completion times and corresponding areas for five randomly generated task graphs: TG0–TG4. It shows that the evolutionary algorithm scales well with regard to the task graph size and yields many non-dominated solutions.

5.   Heterogeneous Multiprocessor Synthesis

In this section, we present our heterogeneous multiprocessor synthesis methodology, which assigns and schedules the application tasks to a number of processors, while customizing each processor with co-processors or custom instructions. It uses the procedures described in the previous two sections for custom instruction and co-processor generation.

5.1   Basic Synthesis Methodology

The heterogeneous multiprocessor synthesis problem is stated below.

Problem 2. Suppose a task graph with n tasks is given where task t_i has m_i custom instruction/co-processor versions. A version requires $Cycle_{ij}$ cycles to execute and consumes $Area_{ij}$ area for hardware accelerators ($1 \leq i \leq n$, $1 \leq j \leq m_i$). Given p initially homogeneous processors in a multiprocessor system, and a total area budget AB for all hardware accelerators, assign and schedule the tasks on these processors and customize each processor with a set of hardware accelerators such that the total execution time of the task graph is minimized while the total area of all the hardware accelerators is within AB.

Note that the above problem specification fixes the number of processors, p, but it can be iteratively solved for different values of p.

We propose a heterogeneous multiprocessor synthesis methodology to solve Problem 2, which is shown in Figure 4.11. It takes an application program as input and outputs the selected hardware accelerators, assignment, and schedule for each task in the program. The boxes with a solid boundary denote steps in the basic methodology, whereas the boxes with a dashed boundary denote the steps added in an enhanced version of the methodology (that uses expected execution times). We focus on the basic methodology first.

At first, the application program is divided into a task graph by functionality (Step **1** in Figure 4.11). Next, custom instructions and co-processors are generated for each task separately (Step **2**). The algorithm described in Section 4 is used to generate a number of non-dominated execution times and areas for each task. Those non-dominated execution times and areas form an area-delay curve in Step **3**. During the subsequent steps of our methodology, several hardware accelerator configurations may be explored for each task. However, in the

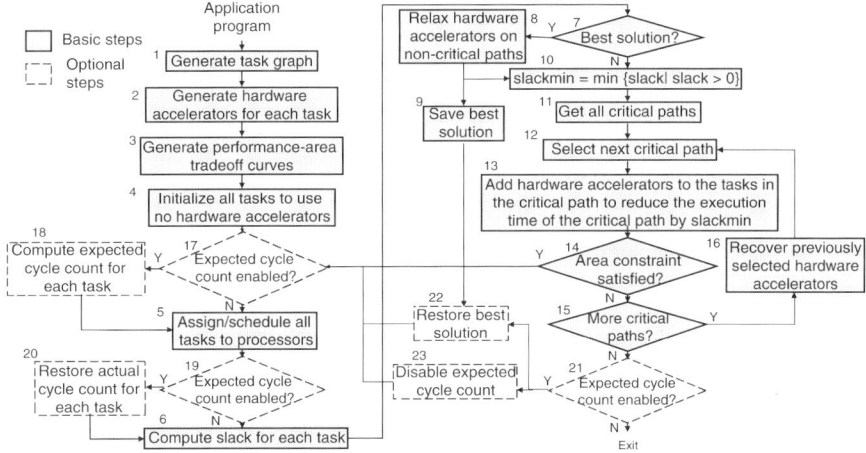

Fig. 4.11   Synthesis methodology for heterogeneous multiprocessors

final custom multiprocessor architecture, only one version is selected for each task.

We use an iterative improvement algorithm to solve Problem 2. Initially, the processors in the multiprocessor architecture are homogeneous. Only later, when each processor gets customized by its own set of hardware accelerators, do the processors become heterogeneous. An iterative improvement algorithm is well suited to this problem. It iterates between two main steps: assigning and scheduling all the tasks on the processors (Step **5**), and selecting hardware accelerators for each task on a given processor to reduce the overall execution time of the task graph (Step **13**). The algorithm iterates until the completion time of the task graph cannot be improved any further.

The basic iterative improvement algorithm starts with a pure software implementation of each task (Step **4**). All the tasks are assigned to and scheduled on the processors (Step **5**). Then the slack for each task is updated (Step **6**). The slack of a task is the time by which the execution time of the task can be increased without increasing the overall task graph execution time on the multiprocessor system. If the current schedule results in the best solution, i.e., one that gives the least completion time (Step **7**), the hardware accelerators of the tasks on the non-critical paths are relaxed (Step **8**) and the solution is saved (Step **9**). Relaxing of a hardware accelerator just implies that for these tasks, a version with less area and more delay may be acceptable. Since the slack of the critical path in the schedule is zero, the minimum slack $slack_{min}$ of all the non-critical paths is extracted (Step **10**). Then, all the critical paths are obtained (Step **11**). From them, one critical path is randomly selected (Step **12**). The hardware accelerators in the tasks along this critical path are changed (based on the many selections available from the area/delay curves of these tasks) so that the execution time along the critical path is reduced by at least $slack_{min}$ (Step **13**). Reducing the execution time of this path further may not be useful, since a different path may become critical. Next, the cumulative area of the hardware accelerators is checked to see whether it exceeds the area budget (Step **14**). If the area constraint is satisfied, the algorithm starts another iteration. If it is not satisfied, but there exists another critical path (Step **15**), the previously selected custom instructions are recovered (Step **16**), a new critical path selected (Step **12**) and the custom instructions recomputed (Step **13**). Otherwise, the algorithm exits.

The completion time of the task graph can only be reduced if the critical path of the task graph (after assignment and scheduling) is reduced. Hence, in Step **13**, hardware accelerators are only added to the tasks in the critical path, in order to reduce its overall execution time. All other tasks have slacks greater than 0, with their minimum slack being $slack_{min}$. When the execution time of the critical path is reduced by at least $slack_{min}$, the critical path of the task graph may change. Hence, the tasks need to be re-assigned and re-scheduled to find the new critical path (Step **5**).

Task Assignment and Scheduling.   Assignment and scheduling of a task graph on the given set of processors is one of the two main steps in the proposed methodology (Step **5**). In our methodology, the number of processors is fixed beforehand. The scheduling algorithms in this category are called BNP (bounded number of processors). Kwok and Ahmad compared the performance of most widely used multiprocessor scheduling algorithms in [19]. Of the six widely used BNP scheduling algorithms Kwok discussed, the modified critical path (MCP) algorithm [20] gives the best performance. Hence, in our iterative improvement algorithm, we use the MCP scheduling algorithm for task assignment/scheduling. Note, however, the selection of the assignment/scheduling algorithm is independent of our design flow. Our methodology can easily be combined with other assignment/scheduling algorithms.

Custom Instruction Selection.   After $slack_{min}$ is computed (Step **10**), and a critical path is selected (Step **12**), Step **13** tries to reduce the execution time of the selected critical path by at least $slack_{min}$ by changing the selection of hardware accelerators for tasks on the critical path. The detailed structure of the hardware accelerators is already known from the pre-processing step (Step **3**). In Step **13**, only the area-delay curve of the various hardware accelerator versions is visible. We use a greedy algorithm and reduce the execution time iteratively. In each iteration, each task on the critical path is examined, and a version that has the highest ratio of number of cycles reduced per unit area increase among all tasks is chosen. Here the selected version may not be the neighboring area-delay entry of the current version, but the version that can give the best overall ratio. This process repeats until $slack_{min}$ cycles are reduced, or no more hardware accelerators can be found to reduce the critical path.

5.2 Enhanced Synthesis Methodology

The above basic steps are enhanced by adding Steps **17 to 23** to the methodology (boxes with a dashed boundary in Figure 4.11). Since only the hardware accelerators for tasks on the critical path are changed, the algorithm delays scheduling of tasks that can be heavily customized so that they are more likely to appear on the critical path, and schedules the tasks that do not have any hardware accelerator as soon as possible to avoid their appearance on the critical path. The enhanced methodology can avoid local minima better than the basic one.

Before the task graph is assigned and scheduled on the processors, the expected cycle count of each task is computed (Step **18**). The expected cycle count of a task is the weighted sum of the number of cycles required for the task under all possible hardware accelerator configurations, considering the possibility of choosing each version. The assignment and scheduling process uses the expected cycle count instead of the actual cycle count as the execution time of the task. After the scheduling step, the actual cycle count is restored (Step **20**). The hardware accelerators are selected based on the actual cycle count of the tasks. If the critical path cannot be improved any further, the best solution found so far is restored (Step **22**). Finally, the expected cycle count computation is disabled (Step **23**), and the entire methodology is executed again using the actual cycle count in the assignment/scheduling step.

Expected Cycle Count Computation. Since the hardware accelerator selection step only selects hardware accelerators on the critical path for a given assignment/scheduling, it is important to schedule the tasks on the critical path whose execution time can be reduced. Otherwise, the methodology exits and the design space cannot be explored any further. Thus, highly customizable tasks should be scheduled on the critical path as much as possible, such that the hardware accelerator selection step has more flexibility in selecting hardware accelerators to reduce the completion time of the task graph. This is achieved by computing the expected cycle count of the tasks (Step **18**) and using the expected cycle count in place of the actual cycle count during scheduling. When scheduling tasks in some iteration of the methodology, the execution time of the task is not chosen to be that of the current custom instruction version, but the weighted sum of the possible execution times of all the hardware accelerator versions, as shown in the following equation:

$$cycle = \frac{1}{\sum_{j} w[j]} \times \sum_{i} w[i] \times v[i] \cdot cycle \tag{4.2}$$

Here $w[i]$ is the weight and $v[i]$ is the version of hardware accelerators.

Several considerations need to be taken into account when constructing the equation to obtain the weights: (i) If no area can be allocated to a specific version, it should be assigned zero weight, (ii) For a given task, a version should be given a higher weight if it has a smaller execution time, (iii) A version should be given a smaller weight if its area is much larger than that of other versions, and (iv) A version should be given a smaller weight if the remaining area in the area budget is only slightly larger than the area the version consumes. In other words, the version should be penalized if the remaining area after choosing the version is so small that there is little opportunity to choose better versions for other tasks.

Based on the above considerations, the following equation is proposed for obtaining the weights:

$$w[i] = \begin{cases} 0 \left(\dfrac{v[i] \cdot area - v[cur] \cdot area}{remArea} \le 0 \ or \ \dfrac{v[i] \cdot area - v[cur] \cdot area}{remArea} \ge 1 \right) \\[2ex] \dfrac{v[cur] \cdot cycle}{v[i] \cdot cycle} \times \left(1 - \dfrac{v[i] \cdot area - v[cur] \cdot area}{remArea} \right) \qquad (otherwise) \end{cases}$$

$$\tag{4.3}$$

Here $w[i]$ is the weight, $v[i]$ is the version under consideration and $v[cur]$ is the version currently selected, and $remArea$ is the remaining area for the custom instructions. Assigning $w[i]$ to zero in Equation (4.3) reflects the first consideration above. In the other formula, the first factor reflects the second consideration above. The third consideration is reflected in the numerator of the second factor. The denominator of the second factor reflects the last consideration.

Note that the expected cycle count is only used in the task assignment/scheduling step (Step **5**). In the hardware accelerator selection step, the actual cycle count of the task is used. This is made possible by restoring the actual cycle count of the task (Step **20**).

5.3   Experimental Results

We have integrated automatic custom instruction generation, co-processor synthesis, and heterogeneous multiprocessor synthesis into a unified framework. We evaluated the performance of our tool by generating heterogeneous multiprocessor architectures for some real-life applications based on the Xtensa [1] platform from Tensilica.

In our experiments, we simulated several Xtensa processor cores together using instruction-set models of each core. The clock periods for all processors are the same. Each processor was extended by adding the custom instructions and co-processors selected using our methodology described in Section 4.4. Our tool takes a global area budget for custom instructions and co-processors. In our experiments, we increased the area budget for each benchmark until further improvements resulted in less than 1% improvement in performance. This was done in order to obtain close to the best speedups without excessive area overheads. We used the shared memory model for inter-processor communication. Semaphores were used to synchronize access to shared memory and enforce critical sections.

We evaluated our methodology using four benchmarks, as shown in Table 4.3. MPEG2DEC refers to an MPEG-2 decoder. The task graph delay is the average number of cycles elapsed between outputting of two successive macroblocks of a total of 192 macroblocks. PGPENC is a modified version of the pretty-good-privacy (PGP) encryption utility. The advanced encryption standard (AES) is used as the block-cipher in this version. The task graph delay is defined as the average number of cycles to process six datasets, with file sizes of 5K, 10K, 15K bytes, and RSA key sizes of 512 bits and 1024 bits, respectively. MPEG2DEC and PGPENC are derived from the MediaBench benchmark suite [10]. PicEn modifies a video frame before it is fed to the encoder. It performs color shifts, byte order switches, and color space conversion. EdgeEn is an example taken from [21]. It enhances the edges of a gray scale picture while preserving low noise. The task graph delays of PicEn and EdgeEn are the average number of cycles elapsed between outputting of two successive common image format images. The hardware accelerator area for the two processors is represented as a tuple. The first and second elements of the tuple represent the hardware accelerator area for the first and second processor,

Table 4.3 Performance comparisons using our tool for Xtensa based custom processors

Programs	Number of processors	Task graph delay		Speedup	Custom instruction area (grids)
		homo	hetero		
MPEG2DEC	1	262K	163K	1.6×	723K
	2	206K	144K	1.4×	13.3K, 705K
PGPENC	1	16.5M	10.5M	1.6×	297K
	2	12.8M	7.7M	1.7×	0, 297K
PicEn	1	37.5M	20.3M	1.8×	9.9K
	2	20.0M	10.9M	1.8×	9.5K, 0
EdgeEn	1	94.1M	32.9M	2.9×	67K
	2	55.3M	18.8M	2.9×	40K, 36K

respectively. Synopsys Design Compiler [13] is used to synthesize the RTL Verilog code of the hardware accelerators and map them to an NEC 0.18 μm technology library [14]. In all benchmarks, the clock period for both base processors and processors augmented with custom instructions is kept at 6.1 ns. The results indicate that the processors in the multiprocessor system can achieve significant performance improvement (average of 2.0×, up to 2.9×) if they are augmented by the custom instructions selected by our methodology. The performance improvements are with respect to homogeneous multiprocessor systems with well-optimized task assignment and scheduling. For the single-processor system, the processors augmented with hardware accelerators can also achieve a significant performance improvement (average of 2.0×, up to 2.9×) over the base processor. The results also indicate that the heterogeneous two-processor system achieves a significant performance improvement (average of 1.6×, up to 1.9×) over the single-processor system augmented with hardware accelerators.

We also used TGFF [18] to generate a number of random task graphs to analyze the scalability of our methodologies and to compare the basic and enhanced methodologies. Since each task may have several versions, each with a different associated execution time and area for different hardware accelerators, the data transfer time, task execution time, and hardware accelerator area were also randomly generated for each version.

In order to test the effectiveness of the proposed methodology, 50 random task graphs with various sizes were generated. Average normalized completion times were compared under different area budgets and number of processors, as illustrated in Table 4.4. All task graph completion times were normalized against the case of one processor with no hardware accelerators. Each entry in Table 4.4 represents the average of the entry for all 50 random task graphs. As seen in Table 4.4, when the number of processors or hardware accelerator area budget is increased, (i) the completion times of task graphs decrease, and (ii) the rate of decrease reduces. This shows that our methodologies scale well with regard to task graph size, the number of processors, and hardware accelerator area budget.

Table 4.4 Average normalized finish times under different area budgets and number of processors

# of processors	Area budget (grids)					
	0	10000	20000	30000	40000	50000
1	1.000	0.838	0.796	0.766	0.748	0.734
2	0.625	0.522	0.492	0.474	0.461	0.455
3	0.548	0.458	0.428	0.413	0.401	0.394
4	0.530	0.440	0.411	0.395	0.387	0.380

Table 4.5 Performance/area comparisons of the basic and enhanced methodologies under different area budgets and number of processors

# of processors		\multicolumn{6}{c}{*Area budget (grids)*}
		0	10,000	20,000	30,000	40,000	50,000
1	Act	0.0	0.0	0.0	0.0	0.0	0.0
	Exp	0.0	0.0	0.0	0.0	0.0	0.0
2	Act	0.0	4.17	5.19	1.17	4.22	7.17
	Exp	0.0	24.11	21.7	27.11	26.9	22.12
3	Act	0.0	7.8	2.10	5.14	4.14	2.15
	Exp	0.0	11.12	17.10	15.9	17.8	22.10
4	Act	0.0	3.8	3.9	2.11	0.11	2.12
	Exp	0.0	4.4	7.4	10.3	12.4	10.3

In Table 4.5, the task graph completion times and the total area required for hardware accelerators, when the basic methodology is used, are compared against the case when the enhanced methodology is used. The same 50 random task graphs are used as before. Rows labeled Act refer to the case when the actual cycle count is used in the assignment/scheduling step. Rows labeled Exp refer to the case when the expected cycle count is used instead. The comparison results are presented as a tuple. The first element of the tuple indicates the number of task graphs that require shorter completion times using one methodology compared to the other. The second element of the tuple indicates the number of task graphs that consume a smaller area for hardware accelerators for one methodology compared to the other. For example, consider the entry for three processors and 50,000 grids area budget. Among the 50 task graphs, the basic methodology (using actual cycle count) performs better than the enhanced methodology (using expected cycle count) for two task graphs, whereas the enhanced methodology performs better for 22 graphs. For the other 26 graphs, the two methodologies have the same performance. The basic methodology results in implementations with a smaller area for hardware accelerators than the enhanced methodology for 15 task graphs, whereas the enhanced methodology results in implementations with a smaller area for 10 task graphs. For the remaining 25 graphs, the two methodologies require the same area. The basic methodology, in general, results in smaller-area implementations since these implementations have longer completion times. However, by subtracting the number of cases when the basic methodology outperforms the enhanced methodology (i.e., 2) from the number of cases when the enhanced methodology yields implementation with smaller area (i.e., 10), the enhanced methodology yields smaller areas while the corresponding completion times are smaller than or equal to those resulting from the basic methodology in the resulting eight cases $(10-2)$. Similarly, by subtracting 15 from 22, in the resulting seven cases, the enhanced methodology gives better performance while the corresponding area is less than or equal to that resulting from the basic methodology.

From the results in Table 4.5, it is clear that the enhanced methodology yields implementations with much better performance, and sometimes better area too. This confirms the advantage of using expected cycle counts in the assignment/scheduling step.

6.   Related Work

Synthesizing an MPSoC, considering all combinations of custom instructions, co-processors, and multi processors, is a very complex problem. In the past, various research efforts have been focused on different aspects of this problem, a comprehensive review of which is beyond the scope of this chapter. We restrict our attention to a few closely related research efforts.

Synthesis of multiprocessors and co-processors has been extensively studied in the hardware–software co-design area. If the target architecture has only one processor and one co-processor, it is called hardware–software partitioning. If the target architecture may have multiple processors and co-processors, it is called hardware-software co-synthesis. Hardware–software co-design has been studied for a long time. Wolf surveyed the development of hardware-software co-design in the past decade and concluded that co-design is becoming a mainstream technology [22]. In [23], De Micheli et al. also surveyed the research developments in hardware-software co-design since its emergence in the early 1990s.

Automatic custom instruction generation for extensible processors has been studied in the ASIP design area. Yu and Mitra studied the impact of different architectural constraints on custom instructions [24]. Automatic custom instruction generation methodologies were also studied in various other research groups [25; 26; 27; 28; 29].

As far as we know, our work on heterogeneous MPSoC synthesis is the first to address the problem of simultaneously synthesizing processor, co-processor, and custom instructions at different granularities. The closest work may be the one presented in [30], where partitioning granularities may range from basic blocks to functions, and the hardware accelerators are implemented as co-processors. In our work, custom instructions may be used to speed up parts of a basic block, and the offloaded tasks may be mapped to either processors or co-processors.

7. Conclusions

MPSoC architectures offer great potential to meet the performance demands of embedded applications. In this chapter, we presented a unified methodology to automatically generate MPSoC architectures using extensible processors. The methodology maps and schedules the applications onto multiple processors, while also generating custom instructions and co-processors that can best improve the performance of the application under a limited area budget. We implemented the methodology and evaluated it using a commercial extensible processor and several embedded benchmarks, demonstrating its capability to automate the design of highly optimized MPSoC architectures.

References

[1] *Xtensa™ microprocessor.* Tensilica Inc. (http://www.tensilica.com).

[2] *MeP™.* Toshiba Semiconductor Company (http://www.semicon.toshiba.co.jp/eng/).

[3] *ARCtangent™ processor.* Arc International (http://www.arc.com).

[4] *MicroBlaze™.* Xilinx Inc. (http://www.xilinx.com).

[5] *Nios™ II.* Altera Corp. (http://www.altera.com).

[6] W. Eatherton, "The push of network processing to the top of the pyramid," in *Symp. Architectures for Networking and Communications Systems*, Oct. 2005.

[7] J. A. Kahle, M. N. Day, H. P. Hofstee, C. R. Johns, T. R. Maeurer, and D. Shippy, "Introduction to the cell multiprocessor," *IBM J. Research and Development*, vol. 49, no. 4/5, pp. 589–604, July 2005.

[8] S. Vangal, J. Howard, G.Ruhl, S. Dighe, H. Wilson, J. Tschanz, D. Finan, P. Iyer, A. Singh, T. Jacob, S. Jain, S. Venkataraman, Y. Hoskote, and N. Borkar, "An 80-tile 1.28TFLOPS network-on-chip in 60nm CMOS," in *Proc. Int. Solid-State Circuits Conf.*, Feb. 2007, pp. 98–99.

[9] M. Girkar and C. D. Polychronopoulos, "Automatic extraction of functional parallelism from ordinary programs," *IEEE Trans. Parallel & Distrib. Systems*, vol. 3, no. 2, pp. 166–178, Mar. 1992.

[10] C. Lee, M. Potkonjak, and W. H. Mangione-Smith, "MediaBench: A tool for evaluating and synthesizing multimedia and communications systems," in *Proc. Int. Symp. Microarchitecture*, Dec. 1997, pp. 330–337.

[11] *Aristotle manual*, Aristotle research group, http://www.cc.gatech.edu/aristotle/.

[12] D. A. Patterson and J. L. Hennessy, *Computer Architecture: A Quantitative Approach*. Morgan Kaufmann Publishers, San Mateo, CA, 1989.

[13] *Design Compiler*. Synopsys Inc. (http://www.synopsys.com).

[14] *CB-11 Cell Based IC Product Family*. NEC Electronics, Inc. (http://www.necel.com).

[15] *PowerTheater*, Sequence Design Inc., http://www.sequencedesign.com.

[16] K. Deb, A. Pratap, S. Agarwal, and T. Meyarivan, "A fast and elitist multi-objective genetic algorithm: NSGA-II," *IEEE Trans. Evolutionary Computation*, vol. 6, no. 2, pp. 182–197, Apr. 2002.

[17] K. Wakabayashi, "C-based synthesis experiences with a behavior synthesizer, "Cyber"," in *Proc. Design Automation & Test Europe Conf.*, Mar. 1999, pp. 390–393.

[18] R. P. Dick, D. L. Rhodes, and W. Wolf, "TGFF: Task graphs for free," in *Proc. Int. Symp. Hardware/Software Codesign*, Mar. 1998, pp. 97–101.

[19] Y.-K. Kwok and I. Ahmad, "Benchmarking and comparison of the task graph scheduling algorithms," *J. Parallel & Distributed Computing*, vol. 59, no. 3, pp. 381–422, Dec. 1999.

[20] M.-Y. Wu and D. D. Gajski, "Hypertool: A programming aid for message-passing systems," *IEEE Trans. Parallel & Distrib. Systems*, vol. 1, no. 3, pp. 330–343, July 1990.

[21] R. C. Gonzalez and R. E. Woods, *Digital Image Processing, 2nd edition*. Addison-Wesley Longman Publishing Co., Inc. Boston, MA, USA, 1992.

[22] W. Wolf, "A decade of hardware/software codesign," *Computer*, vol. 36, no. 4, pp. 38–43, Apr. 2003.

[23] G. De Micheli, R. Ernst, and W. Wolf, *Readings in Hardware/Software Co-design*. Morgan Kaufmann Publishers Inc., San Francisco, CA, 2001.

[24] P. Yu and T. Mitra, "Characterizing embedded applications for instruction-set extensible processors," in *Proc. Design Automation Conf.*, June 2004, pp. 723–728.

[25] K. Atasu, L. Pozzi, and P. Ienne, "Automatic application-specific instruction-set extensions under microarchitectural constraints," in *Proc. Design Automation Conf.*, June 2003, pp. 256–261.

[26] N. Clark, H. Zhong, W. Tang, and S. Mahlke, "Processor acceleration through automated instruction set customization," in *Proc. Int. Symp. Microarchitecture*, Dec. 2003, pp. 40–47.

[27] N. Cheung, S. Parameswaran, J. Henkel, and J. Chan, "MINCE: Matching instructions using combinational equivalence for extensible proces-

sors," in *Proc. Design Automation & Test Europe Conf.*, Feb. 2004, pp. 1020–1027.

[28] D. Goodwin and D. Petkov, "Automatic generation of application specific processors," in *Proc. Int. Conf. Compilers, Architecture, and Synthesis for Embedded Systems*, Oct. 2003, pp. 137–147.

[29] J. Cong, Y. Fan, G. Han, and Z. Zhang, "Application-specific instruction generation for configurable processor architectures," in *Proc. Symp. Field Programmable Gate Arrays*, Feb. 2004, pp. 183–189.

[30] J. Henkel and R. Ernst, "An approach to automated hardware/software partitioning using a flexible granularity that is driven by high-level estimation techniques," *IEEE Trans. VLSI Systems*, vol. 9, no. 2, pp. 273–289, Apr. 2001.

Chapter 5

Design and Run Time Code Compression
for Embedded Systems

Sri Parameswaran[1], Jörg Henkel[2], Andhi Janapsatya[1], Talal Bonny[2]
Aleksandar Ignjatovic[1]

[1]*School of Computer Science and Engineering*
The University of New South Wales
NSW 2033
Australia

[2]*Department of Computer Science*
Karlsruhe University
Zirkel 2, D-76131 Karlsruhe
Germany

Abstract   Compression has long been utilized in electronic systems to improve performance, reduce transmission costs, and to minimize code size. In this chapter we
show two separate techniques to compress instructions. The first technique compresses instruction traces, so that the compressed trace can be used to explore the
best cache configuration to be used in an embedded system. Trace compression
enables rapid cache space exploration. The second technique uses compressed
instruction in memory, to be expanded just before execution in the processor. This enables a smaller code footprint, and reduced power consumption.
This chapter explains the methods, and shows the benefits of two orthogonal
approaches to the design of an embedded system.

Keywords: Embedded Systems, HW-SW Codesign, Embedded Processors, Trace compression, Code Compression

J. Henkel and S. Parameswaran (eds.), Designing Embedded Processors – A Low Power Perspective,
97–128.

1.   Introduction

Compression has played a crucial part in making electronics ubiquitous. Be it the condensing of a billion transistors into a tiny chip or the contraction of a high definition picture into a small file, compression has enabled technology to be omnipresent. Compression reduces cost by occupying a smaller area, reduces power consumption by often having less data to process, and improves performance by transmitting and processing a fraction of the original data.

In contrast to systems mentioned above, this chapter discusses two methods to compress code. These two methods, one at design time and the other at run time, make systems based upon embedded processors more efficient. One of the major sources of power consumption, and therefore a prime contender for aggressive optimization, is the cache memory of an embedded system. Since embedded systems typically run a single application or a class of applications (unlike a general purpose processing system), it is possible to determine a priori the best cache configuration to suit. To decide upon this cache configuration, the designer typically takes a trace of the memory activity of the processor, and uses a trace driven cache simulator to find hit rates of differing configurations before choosing the best cache structure. Cache traces can be up to several hundred billions of lines, and trying to execute it several times through cache simulators (once for each configuration) is tedious. Thus, the first method shown here uses a compressed trace to perform cache simulation without uncompressing the trace, enabling the cache architecture exploration of an embedded system in a much shorter time than otherwise possible. The second method discussed in this chapter shows how instructions can be compressed at design time and how these instructions can be uncompressed at run time to execute correctly. Such a scheme allows a smaller memory to operate the system, reducing the footprint and the power consumption. Such a run time system requires a look-up table to evaluate the real code from the compressed code. It is imperative that the look-up table be as small as possible to achieve the best possible outcome in terms of area and power. Thus the second part of the chapter explores a novel method to make the look-up table smaller.

1.1   Design Time Compression

The availability of customizable cache memory in a system provides a designer the option of specifying an optimal cache memory configuration to match the application requirements in order to:

- minimize cache energy cost per access;

- minimize access time per cache access;

- minimize the number of cache misses and reduce the number of off-chip memory accesses; or,

- any combination of the above;

In a typical design process, cache simulation is performed for various cache configurations to determine the optimal cache parameters for the given program trace. However, a typical program trace can be several tens of gigabytes long. For a 1 GHz processor, a one second snapshot stored as a program trace might translate to a 10 GB file. This becomes even more critical, as tens of processors are included in a single embedded system. Although disk space may no longer be an issue in the near future with the emergence of large hard drives (up to 750 GB (Technology, 2006)), getting large amounts of data to and from the disk is still a time consuming process.

One method to alleviate the large cost of processing a program trace file is to compress the trace file. Compression methods have been proposed in the past to reduce the size of these program trace files. Although compression allows the reduction of the program trace file size, the required intermediate memory during decompression and program trace analysis is still large, requiring enormous numbers of reads and writes to disks.

In this chapter, a method is shown to allow cache simulation to be performed from a compressed program trace file such that only 'partial decompression' is necessary. To compress the program trace file, we developed a compression methodology that allows for random access decompression (Lekatsas et al., 2000). Random access decompression is a term used to describe a compression methodology that allows the decompression to start at any point in the compressed file. The random access decompression feature allows the described cache simulation methodology to operate on the compressed program trace file and provide the opportunity to parallelize the cache simulation methodology.

Compared to existing data compression tools (such as gzip (Ziv and Lempel, 1977)), the compression rate shown here is anywhere from two to ten times worse when compared to gzip. This is because the compression methodology is not designed to achieve maximal compression rate, instead it is designed for minimal processing cost. If a high compression rate is required, a post-compression step with LZ compression can be used to further improve the compression rate.

Besides cache analysis, there are many reasons for wanting to analyze a program trace, such as estimating the energy of a system, evaluating system performance, etc. Depending on the type of analysis to be performed, different compression algorithms can be developed to overcome the bottleneck posed by the large amount of time needed to read and write from disks.

1.2   Run Time Code Compression

Increasing software size requires larger memory and therefore may significantly increase the cost of an embedded system. The beginning of this trend has already been recognized in the early 1990s, where first approaches for code compression of embedded applications arose (Wolfe and Chanin, 1992). Code compression in an embedded system works as follows (see also Lekatsas and Wolf, 1999; Benini et al., 2001; Lekatsas et al., 2005): the compressed code, generated by compressing the object code using a code compression tool (at *design time*), is stored in the embedded system's instruction memory. At *run time*, the compressed instructions are decompressed and executed by the processor. Proposed compression schemes may be classified into two general groups: *statistical* and *dictionary* schemes (Bell et al., 1990). In statistical compression schemes, the frequency of instruction sequences is used to choose the size of the codewords that replace the original ones. Thereby, shorter codewords are used for the most frequent sequences of instructions, whereas longer codewords are replaced by less frequent sequences. In dictionary compression methods, entire sequences of common instructions are selected and replaced by a single new codeword which is then used as an index to the dictionary that contains the original sequence of instructions. In both cases, *Look-up Tables* are used to store the original instructions. The compressed instructions serve as indices to those tables. One of the major problems is that the tables can become large in size and thus they diminish the advantages that could be obtained by compressing the code. However, the entire research in the area has always focused on achieving better code compression without explicitly targeting the problem of large *Look-up Table* sizes. We reduce the size of the *Look-up Tables* by sorting the table's entries to decrease the number of bits toggle between every two sequential instructions (as we will show within this paper). Sorting the table's entries to find the optimal solution is an NP-Hard problem. For that we use the *Lin-Kernighan* heuristic algorithm (Helsgaun, 2000) which generates optimal or near-optimal solutions for the symmetric traveling salesman problem (*TSP*).

Interestingly, *our method is orthogonal to any kind of approach that uses a priori knowledge of the instruction set of a specific architecture*. Hence, all results we report can be further improved by *ISA-specific* compression approaches.

2.   Related Work

There are two types of compression algorithms: lossless and lossy. Lossless compression allows the exact original source to be reproduced from the compressed format. In comparison, lossy compression generally results in a higher compression ratio by removing some information from the data. Thus, it is not possible to recreate the original file. The work presented in this chapter uses lossless. The lossless schemes can be further divided into hardware-based techniques (Benini et al., 2002, Xu et al., 2004) and software-based techniques (Tuduce and Gross, 2005, Yang et al., 2005). The software based technique are beneficial in systems where the process of code decompression does not influence the performance due to system architecture application requirements.

An existing data compression method, the LZ77 introduced by Ziv and Lempel (1977), is a well-known compression method used in gzip. The algorithm reads a file as a stream and uses previous texts in the stream to encode the incoming texts in the stream. Ziv and Lempel (1978) presented another version of their algorithm, known as LZ78. In 1983, Welch (1984) introduces an improvement to the LZ78 to limit the growth of the dictionary file, known as the LZW algorithm. Pleszkun (1994) presented a two pass algorithm for compressing a program trace file. The first pass is a preprocessing step to identify the dynamic basic blocks, procedure calls, etc. The trace is encoded by specifying the basic block and its successor. Johnson and Ha (1994) and Johnson et al. (2001) presented an offset-encoding compression scheme known as PDATS. Each address reference in the trace encoded as its offset from the previous address reference. In addition, a run-length coding is used to encode sequences of addresses with the same offset. Their experimental results show an average compression ratio of seven for large traces and they achieved a speedup factor of 10 when executing Dinero with PDATS traces. Nevill-Manning and Witten (1997) introduced the hierarchical compression method known as 'SEQUITUR'. It constructs grammar based on the two rules: no pair of adjacent symbols appears more than once in the grammar, and every rule is used more than once. SEQUITUR can be applied to any type of information streams. The processing time of SEQUITUR to process one symbol is $O(\sqrt{n})$, where n is the number of input symbols encountered. Our work differs from SEQUITUR as we allow symbols to appear more than once in a single grammar; this is because the goal of our compression method is to allow minimal processing cost of the compressed program trace. Milenkovic and Milenkovic (2003) presented a compression methodology called 'Stream-Based Compression' (SBC). SBC performs compression by replacing each address stream by its index in the stream table. Address streams are identified according to basic blocks in the program. Burtscher (2004) introduced VPC3 which runs in a single pass in linear time over the trace data. The trace data in VPC3 contains many attributes in addition to the memory address. It encodes the trace data using value predictors. Luo and John (2004) presented a 'Locality Based Trace Compression' (LBTC). The methodology employs two techniques; the first technique is offset encoding of the memory references, and the second technique is to statically encode the attributes associated with a memory location through the assumption that most attributes are static and do not change frequently from one dynamic access to another. Their experimental results showed that they improve the compression rate by $2\times$ over the PDATS method. Zhang and Gupta (2004, 2005) presented a unified representation of profiles called 'Whole Execution Trace' (WET). WET is constructed by labeling a static program representation with profile information. Their experimental results showed that their method achieved an average compression ratio of 41.

For run time code compression there are several related approaches that use dictionary-based schemes. In Corliss et al. (2003) instruction operand parameters are deployed to catch a larger number of identical instruction sequences and replace them with codewords in the dictionary. In Yoshida et al. (1997), the authors developed a compression algorithm to unify the duplicated instructions of the embedded program and assign a compressed object code. Their technique typically needs a large external ROM. In Das et al. (2004, 2005) the authors developed a dictionary-based algorithm that utilizes unused encoding space in the *ISA* (Instruction Set Architecture) to encode codewords and addresses issues arising from variable-length instructions. Okuma et al. (1998), proposed an instruction encoding technique that re-encodes the immediate field of an instruction. Their technique requires complex decoder logic and furthermore restricts the maximum number of immediate values that are simultaneously allowed in an application. Lefurgy et al. (1997), suggested a method for compressing programs in embedded processors by analyzing a program and replacing *common sequences* of instructions with a single new instruction. The authors proposed LZW-based algorithms to compress branch blocks using a coding table. Wolfe and Chanin (1992) have designed a new RISC architecture called *CCRP* (Code Compressed RISC Processor) which has an instruction cache that is modified to run compressed programs. Since codewords are of variable length, a translation table is always needed to map uncompressed branch addresses to compressed ones. In Larin and Conte (1999), compiler-driven *Huffman-based* compression with a compressed instruction cache has been proposed for a VLIW architecture. An *Address Translation Table* (ATT) is generated by the compiler, containing the original and compressed addresses of each basic block and its size. In Game (2000), another example of *Huffman-based* code compression is deployed in real systems (called Code-Pack) for its embedded *PowerPC*. It resembles CCRP in that it is part of the memory system and a LAT-like device is needed to map between the native and compressed address spaces. Lekatsas and Wolf (1999), proposed a new code compression technique SAMC which is based on *Arithmetic Coding* and a *Markov model*. Common to all work in code compression is that *Look-up Tables* are deployed for decoding (they can come as LAT or ATT). In any case, these tables will take space in memory and may significantly impact the total (compressed code size plus table size) compression ratio.

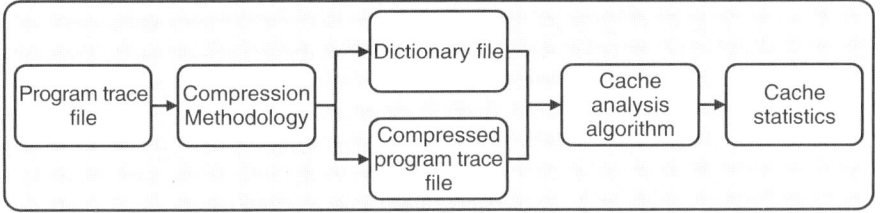

Fig. 5.1 Compression and cache analysis methodology

3. Design Time – Cache Trace Compression and Cache Size Prediction

3.1 Methodology

The methodology flow of the compression and cache analysis process are as shown in Figure 5.1. Input to the process is a program trace file. The compression process outputs a dictionary file and a compressed program trace file. Cache analysis uses the dictionary file and the compressed program trace file as its input, and it outputs the cache statistics. As mentioned, the cache analysis process is performed directly on the compressed trace, without decompression.

Format of the Trace.   We view the program trace as a bit-stream with no knowledge of the program functionality. We do not know the number of instructions in the program trace, nor how instructions interact with each other, and we have no knowledge of which instructions belong to which process or function. Nevertheless, our method can achieve high compression rates.

The program trace file contains a trace of memory references. Our trace is identical to the Dinero input format. An example of the program trace are shown in Figures 5.4a, b, and c. The numbers on the left column are the sequence number and the numbers on the right column are the memory references. We assume all memory addresses are cacheable, and no memory reference contain attributes such as data dependency etc. If required, a data dependencies graph can be built using the information available in the trace. At the current moment, our tool only looks at traces of instruction memory references.

3.2 Compression Algorithm

The compression algorithm is designed to compress the trace file but also allows the compressed trace to be analyzed without the need for decompression. The compression algorithm is designed as a random-access compression algorithm to allow analysis and decompression to start from anywhere in the compressed trace file.

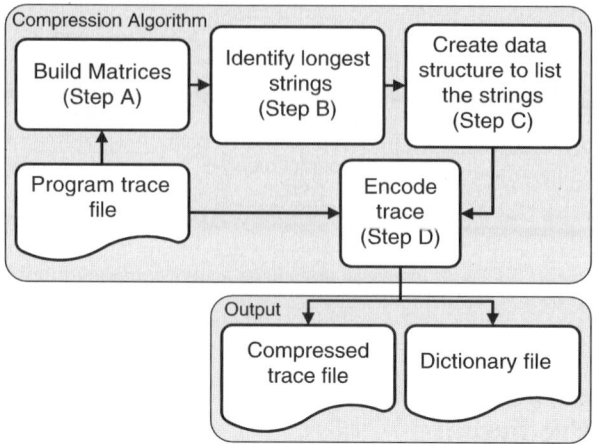

Fig. 5.2 Compression algorithm

Figure 5.2 shows the compression algorithm. For ease of explanation, the boxes shown in Figure 5.2 are labeled Step A, Step B, Step C, and Step D. In Step A and Step B, the trace is analyzed and the longest repetitive strings are identified. These repetitive strings are then stored within a forest data structure, this is Step C. Step D will use the forest data structure to encode strings of instructions as read from the program trace file.

Step A.   The purpose of Step A and Step B is to identify patterns that exist in a program trace file. In our algorithm, we identify two occurring patterns. The first pattern is the consecutive repetition of the same instructions string, for example 'abcabcabcabcabc'. The second pattern is the reoccurrence of a string repetitively but not consecutively, for example the string 'abc' in the text **'abcqweabc**rtyu**abc**io **abc**k**abc**fg' occurs repetitively but not consecutively and has a non-uniform stride between occurrences.

```
\\ Creating the matrix
\\ n = total number instructions in the window
\\ h = the height of the matrix, equal to the maximal distance to look for
\\      repeated instructions
for i = 1 to n
    for j = 1 to h
        if (W[i] == W[i + j])
            M[i, j] = 1;
        else
            M[i, j] = 0;
```

Fig. 5.3   Building matrix M

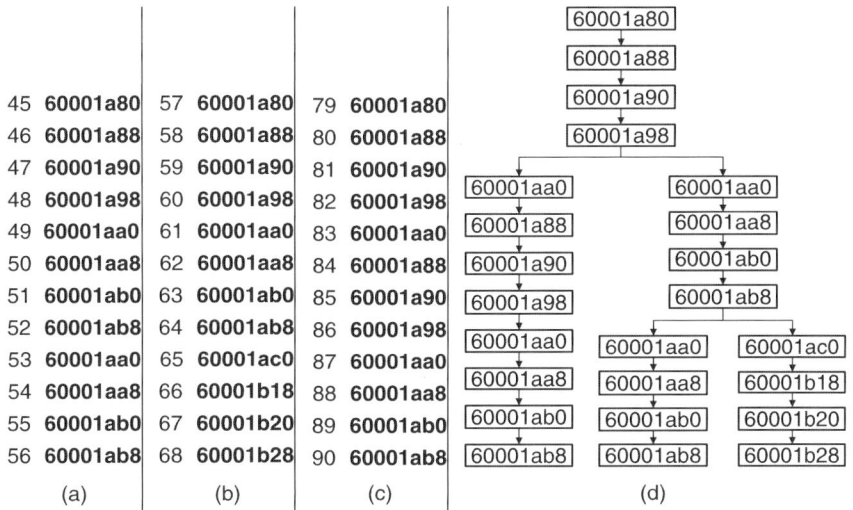

Fig. 5.4   Program trace and data structure examples

To identify the pattern within the instruction trace file, a matrix is built to assist the process. Figures 5.4a, b, and c show examples of a program trace. Figure 5.5 shows the matrix built from the program trace shown in Figure 5.4.

The numbers shown on the bottom of Figure 5.5 represent the instruction sequence, and the numbers on the left (first column) show the distance. Since a program trace can be very long, the matrix of width n covers only a substring of the trace consisting of n consecutive instructions. Even though this is a small substring of the whole trace, if n is reasonably large (in our implementation of the algorithm $n = 4000$), each loop in the program should be identifiable within a single window. To cover the whole program trace, a sliding window is used. To prevent some loops being lost in between windows boundary, the sliding window mechanism will advance the window by $n/2$ instructions at a time.

The matrix M of size $n \times h$ is built as follows. We take n consecutive instructions as one window $W(i)$, $1 \leq i \leq n$. The entry in the ith column and j^{th} row of our matrix M is equal to one if and only if $W(i) = W(i+j)$, $1 \leq i \leq n$, $1 \leq j \leq h$; otherwise it is equal to zero.

In the matrix shown in Figure 5.5, the grey shading highlighting the string of '1s' found in the matrix indicates the repeated occurrence of the same instruction. These strings are identified in Step A and analyzed in Step B (shown below) to ensure no repeated sequences are identified.

Figure 5.3 shows the algorithm for identifying the patterns in the program trace file by utilizing the matrix shown in Figure 5.5. We now construct another matrix S of the size $(n - h) \times h$, as shown in Figure 5.6. The entry $S[i, j]$ is

	45	46	47	48	49	50	51	52	53	54	55	56	57	58	59	60	61	62	65	64	65	66	67	68
+14	0	0	0	0	0	0	0	0	0	0	0	0	0	0	0	0	0	0	0	0	0	0	0	0
+13	0	0	0	0	0	0	0	0	0	0	0	0	0	0	0	0	0	0	0	0	0	0	0	0
+12	1	1	1	1	1	1	1	1	0	0	0	0	0	0	0	0	0	0	0	0	0	0	0	0
+11	0	0	0	0	0	0	0	0	0	0	0	0	0	0	0	0	0	0	0	0	0	0	0	0
+10	0	0	0	0	0	0	0	0	0	0	0	0	0	0	0	0	0	0	0	0	0	0	0	0
+9	0	0	0	0	0	0	0	0	0	0	0	0	0	0	0	0	0	0	0	0	0	0	0	0
+8	0	0	0	0	0	0	0	0	1	1	1	1	0	0	0	0	0	0	0	0	0	0	0	0
+7	0	0	0	0	0	0	0	0	0	0	0	0	0	0	0	0	0	0	0	0	0	0	0	0
+6	0	0	0	0	0	0	0	0	0	0	0	0	0	0	0	0	0	0	0	0	0	0	0	0
+5	0	0	0	0	0	0	0	0	0	0	0	0	0	0	0	0	0	0	0	0	0	0	0	0
+4	0	0	0	0	1	1	1	1	0	0	0	0	0	0	0	0	0	0	0	0	0	0	0	0
+3	0	0	0	0	0	0	0	0	0	0	0	0	0	0	0	0	0	0	0	0	0	0	0	0
+2	0	0	0	0	0	0	0	0	0	0	0	0	0	0	0	0	0	0	0	0	0	0	0	0
+1	0	0	0	0	0	0	0	0	0	0	0	0	0	0	0	0	0	0	0	0	0	0	0	0

Fig. 5.5   Matrix M

```
for j = 1 to h
   for i = 1 to n − h
S[i, j] = 0; \\ Creating the counting matrix S of size n − h × h
for j = 1 to h
   for i = n − h downto 1
      if (M[i, j] == 1)
         S[i, j] = S[i + 1, j] + 1;
      else
         S[i, j] = 0;
```

Fig. 5.6   Building matrix S

```
\\ Evaluating the gain of strings that are candidates for the dictionary
for i = 1 to n − h
   for j = 1 to h
      if (S[i, j] ≥ j)
         Gain(W[i..i + j]) = IntegerPart(S[i, j]/j) * (j − 1);
         j = h;
         i = i + IntegerPart(S[i, j]/j);
      else if (S[i, j] > 0)
         for k = 1 to h
            if S[i, k] ≥ S[i, j]
               p = p + 1;
         if (p * (S[i, j] − 1) > Gain(W[i..i + j]))
            Gain(W[i..i + j]) = p * (S[i, j] − 1);
         i = i + 1;
```

Fig. 5.7   Evaluating the gain of strings

equal to the number of consecutive '1s', in row j and columns k of matrix M such that $k \geq i$ (see Figure 5.8).

Step B.   We now identify strings that are potentially useful for compression as follows.

	45	46	47	48	49	50	51	52	53	54	55	56	57	58	59	60	61	62	65	64	65	66	67	68
+14	0	0	0	0	0	0	0	0	0	0	0	0	0	0	0	0	0	0	0	0	0	0	0	0
+13	0	0	0	0	0	0	0	0	0	0	0	0	0	0	0	0	0	0	0	0	0	0	0	0
+12	8	7	6	5	4	3	2	1	0	0	0	0	0	0	0	0	0	0	0	0	0	0	0	0
+11	0	0	0	0	0	0	0	0	0	0	0	0	0	0	0	0	0	0	0	0	0	0	0	0
+10	0	0	0	0	0	0	0	0	0	0	0	0	0	0	0	0	0	0	0	0	0	0	0	0
+9	0	0	0	0	0	0	0	0	0	0	0	0	0	0	0	0	0	0	0	0	0	0	0	0
+8	0	0	0	0	0	0	0	0	4	3	2	1	0	0	0	0	0	0	0	0	0	0	0	0
+7	0	0	0	0	0	0	0	0	0	0	0	0	0	0	0	0	0	0	0	0	0	0	0	0
+6	0	0	0	0	0	0	0	0	0	0	0	0	0	0	0	0	0	0	0	0	0	0	0	0
+5	0	0	0	0	0	0	0	0	0	0	0	0	0	0	0	0	0	0	0	0	0	0	0	0
+4	0	0	0	0	4	3	2	1	0	0	0	0	0	0	0	0	0	0	0	0	0	0	0	0
+3	0	0	0	0	0	0	0	0	0	0	0	0	0	0	0	0	0	0	0	0	0	0	0	0
+2	0	0	0	0	0	0	0	0	0	0	0	0	0	0	0	0	0	0	0	0	0	0	0	0
+1	0	0	0	0	0	0	0	0	0	0	0	0	0	0	0	0	0	0	0	0	0	0	0	0

Fig. 5.8 Matrix S

The heuristics behind the value of $\text{Gain}(W[i..i+j])$ of the string $W[i..i+j]$ is to assess how useful the string is for the purpose of compression. We want to use the strings that are both long and frequent, i.e., those that have high gain. In the algorithm shown in Figure 5.7, the first 'if' clause $(S[i,j] \geq j)$ detects strings that repeat consecutively, such as loops in programs; the second 'if' $(S[i,j] > 0)$ finds remaining strings that are executed frequently within the same window, but in a scattered manner. For each i the algorithm picks a string with the highest gain. The algorithm can be easily extended so that if there are more than one string with the highest gain, the longest string is picked.

Step C.   We now create a forest data structure to store the strings identified in Step B; we store the starting memory address of the sequences identified in Step B. Each tree within the forest is used to store the subsequent memory addresses in the strings that were identified. An example of the forest data structure is shown in Figure 5.4d. This tree corresponds to the program trace shown in Figure 5.4; the left branch corresponds to the trace shown in Figure 5.4c and the left branch of the right branch corresponds to the trace shown in Figure 5.4a, and the right part of the right branch corresponds to the trace shown in Figure 5.4b. Each node where a string terminates has a counter associated with it; this counter is incremented each time the corresponding string is found. Thus, the final value of the counter will reflect how many times the corresponding string has been found in Step B.

Step D.   In Step D, we encode the trace. The encoded trace file is a sequence of either addresses that are not initial elements of a string from the forest (i.e., not a root of any of the trees in the forest) or as a sequence of indices that point to the dictionary entries. Figure 5.9 shows the trace encoding procedure. We read the trace file matching read addresses with strings in the tree, until we

```
trace_index = 0;
dict_index = 1;
while trace_file != EOF
        max_string = null;
        addr = read(trace_file, trace_index);
        trace_index++;
        string = null;
        until found_in_the_tree(string + addr) == null
                string = string + addr;
                if gain(string) ≥ gain(max_string)
                   max_string = string;
                addr=read(trace_file, trace_index);
                trace_index++;

        if max_string == NULL
           compress = compress + addr;
        else
           if dict_index(max_string) == 0
              dict_index(max_string)=dict_index;
              dict_index++;
           compress = compress + index(max_string);
           dictionary = dictionary + (dict_index(max_string), max_string);
           trace_index = trace_index(max_string) + 1;
```

Fig. 5.9   Trace encoding procedure

cannot traverse down the tree. We look at all substrings of the traversed branch of the tree and pick the string with largest gain (if several strings have such gain, the longest string is picked).

Computational Complexity.   Building the matrix M for a window of size n is quadratic in n; building the counting matrix S is cubic in n. Note that this is an extremely pessimistic estimate and in practice the time complexity is essentially quadratic in n, because the matrix M is very sparse, and the largest sequence of consecutive ones to be counted is very small compared to the window size n. If the trace is of size t, then the number of windows to process is $2t/n$. Thus the total work prior to encoding is $O(t \cdot n)$ (expected), $O(t \cdot n^2)$ (unrealistic worst case). Finally, encoding is linear in t. Thus, as verified in practice, the expected run time is proportional to $t \cdot n$. Notice that the compression efficiency, for as long as reasonably feasible, is not very critical. This is because the compression is just a preprocessing step that is done only once for each program trace. Subsequently, a set of stored preprocessed (compressed) traces is reused many times for different statistical analyses of various applications with various inputs.

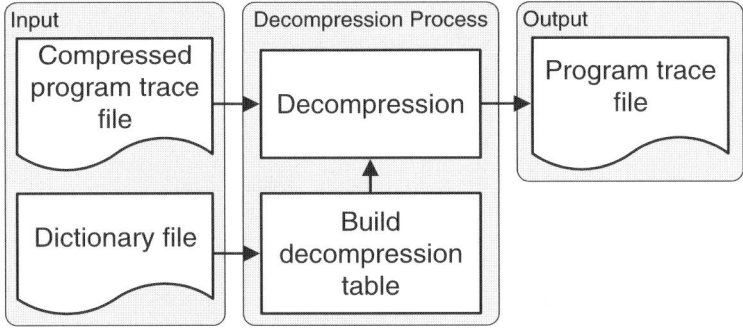

Fig. 5.10 Decompression process

Decompression for the Purpose of Verification. To verify and guarantee the correctness of the compression algorithm, a decompression algorithm was built to reproduce the trace file given the dictionary file and the compressed trace file. Note that in actual use, decompression is not needed; our analysis algorithm executes on the compressed trace.

The decompression algorithm is shown in Figure 5.10. The decompression algorithm takes the compressed trace file and the dictionary file as inputs, and outputs the trace file.

As shown in Figure 5.10, the algorithm reads the dictionary file to build a decompression table. The decompression step reads the list of indices stored in the compressed trace file and uses the information in the decompression table to produce the trace file.

3.3 Cache Simulation Algorithm

Our Cache simulation algorithm aims to calculate the number of cache misses that can occur given a cache configuration and a compressed program trace. Inputs to the cache simulation algorithm are the cache parameters, the dictionary file, and the compressed program trace file. Figure 5.11 shows the procedures of the cache simulation algorithm.

Each entry in the dictionary is an instruction string that represents the sequence of memory references. Figure 5.12 displays the cache simulation algorithm. The algorithm reads the compressed program trace file and determines the number of cache misses that occur.

The cache simulation algorithm starts by initializing an array to keep track of the content of each cache location. Size of the array is dependent on the cache parameters (i.e., cache associativity, cache line size, and cache size).

The dictionary file is then read and cache simulation (function sim_cache) is performed for each instruction string entry in the dictionary. Cache simula-

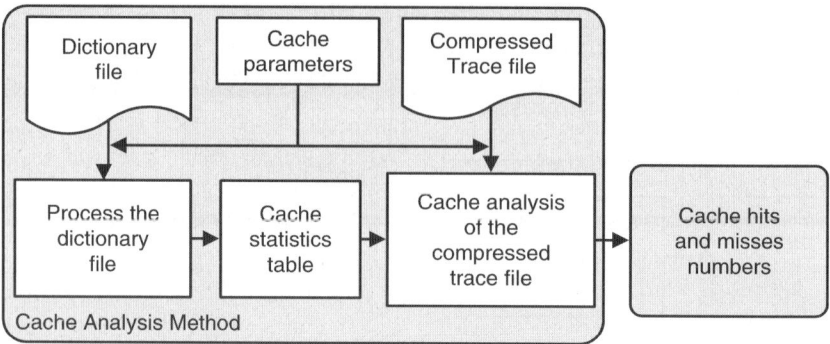

Fig. 5.11   Cache analysis methodology

tion on the instruction strings will calculate the cache compulsory misses for loading the particular string into the cache, the cache conflict misses for executing the string, and finally the resulting cache footprint after the execution of the instruction string.

The sim_cache O function simulates the instruction string as if the instructions were to be executed through the cache memory. After processing the entries in the dictionary, a cache statistics table is obtained. The table is indexed using the instruction string index and each table entry stores the following information:

- $cache\_compulsory\_content$, a list of instructions that occupy the cache before pre-execution of the instruction string

- $cache\_conflict$, the amount of cache conflict misses number due to the execution of the string; and,

- $cache\_footprint$, a list of instructions that occupy the cache after-execution of the instruction string;

The cache simulation algorithm then continues by reading the compressed program trace file. If a memory reference is read, it is compared with the existing entry or entries (depending on cache associativity parameters) in the same cache location and a cache hit or miss can then be determined and the cache content for that location is to be updated with the read memory reference. Otherwise, if a dictionary index is read from the compressed trace file, the index is used to access the entry in the cache statistics table using the function $update\_cache$. The following three steps are performed by the $update\_cache$ function:

1. the existing cache entry is compared with the entries in the $cache\_compulsory\_content$ list from the table to determine the number of cache hit or misses occurred;

```
\\ Initialize the cache address array
\\ M = total number of cache location
\\ M is equal to number of cache set for a direct-mapped cache
for a = 0 to M
    cache = 0;

\\ process the dictionary file
\\ N = total number on entries in the dictionary based on the dictionary index
for b = 1 to N
    sim_cache(b);

\\ start of cache analysis
\\ S = total number of entries in the compressed trace file
for c = 0 to S
    if (c == dictionary_index)
        update_cache(dict_data(c));
    else
        update_cache(c);
```

Fig. 5.12   Cache simulation algorithm

2. the $cache\_conflict$ value is added to the total cache miss number;

3. the content of the cache is updated with the resulting cache footprint, $cache\_footprint$ list from the table.

Performance improvement over existing cache analysis tools such as Dinero IV (Edler and Hill, 1999) is expected due to the smaller compressed trace file to process, compared to the existing trace file size, and due to pre-simulation of subsets of the trace when analyzing the dictionary entries. Results from the cache simulation is verified by comparing against the output from DineroIV (Edler and Hill, 1999). We found our cache simulation algorithm to be 100% accurate compared to DineroIV outputs (i.e., there are no errors in the output).

3.4   Experimental Setup and Results

Experiments were performed to gauge the compression ratio of the compression algorithm and to verify the cache simulation algorithm with an existing tool (Dinero IV (Edler and Hill, 1999)).

Benchmarks were taken from Mediabench (Lee et al., 1997). Column 1 in Table 5.1 shows the six benchmarks used in experiments. The rest of Table 5.1 displays the following information: column 2 shows the number of memory references in the trace, column 3 shows the size of the trace file in bytes, and column 4 shows the size of the gzip trace file in bytes. Table 5.2 shows the compression result and its comparison to gzip; column 1 shows the application name, column 2 shows the resulting compressed file size in bytes, column 3 shows the ratio of the compressed trace file compared to the original trace file (column 3 divided by column 3 of Table 5.1), column 4 shows the size of the compressed trace file with 'gzip', column 5 shows the ratio of Dinero IV total run time compared to the total run time of our cache simulation methodology, and column 6 shows the ratio of Dinero IV with inputs piped from a gzip trace file using zcat compared to the total run time of our simulation.

Program traces were generated using Instruction Set Simulator (ISS) for Tensilica Xtensa processor (Processor, 2006). The compression algorithm was implemented in C and compiled with gcc version 4.0.2; the program was executed in a dual processor Opteron 2.2 GHz machine.

Cache analysis is performed for various cache configurations. The cache parameters used were: cache size of 128 bytes to 16384 bytes, cache line size of 8 bytes to 256 bytes, and cache associativity from 1 to 32 (each parameter is incremented in powers of two; total cache configuration simulated is 206). Cache simulations were executed on the compressed file and the cache miss numbers output from the cache simulation was verified against Dinero IV to ensure correctness of the cache simulation output.

Figure 5.13 shows the comparison of the total execution time for simulating multiple cache configurations with Dinero IV, Dinero IV with gzip, and our simulation tool. Due to large differences between the total run time of the individual benchmarks, the bar graphs in Figure 5.13 are not on the same scale. Benchmarks 'jpegenc' and 'jpegdec' total run time bar is measured in minutes, while 'g721enc', 'g721dec', 'mpeg2enc', and 'mpeg2dec' run time are measured in hours. The ratio between the time taken by Dinero IV over our simulation methodology is shown in column 8 on Table 5.1.

Figure 5.14 shows a comparison of the run time of Dinero IV compared to our simulation methodology for simulating individual cache configurations of 'g721enc' program trace file. The left bars shown in Figure 5.14 indicate

Table 5.1 Benchmark list

Application	Trace size	File size (bytes)	gzip File size (bytes)
g721enc	108,635,308	977,717,772	12,878,673
g721dec	105,800,842	952,207,578	12,161,013
jpegenc	11,508,580	103,577,220	1,051,695
jpegdec	2,842,673	25,584,057	422,638
mpeg2enc	2,359,305,107	21,233,745,963	297,736,877
mpeg2dec	937,319,532	8,435,875,788	111,485,710

Table 5.2 Compression result

Application	Compressed file size (bytes)	Compression gzip ratio	Compressed file size (bytes)	Run time ratio	Run time ratio (gzip)
g721enc	38,049,493	25.7	2,743,007	6.41	6.35
g721dec	25,646,458	37.1	1,473,615	8.08	8.0
jpegenc	9,895,273	10.5	195,292	15.32	14.10
jpegdec	2,681,652	9.5	42,771	15	14.08
mpeg2enc	1,139,867,708	18.6	46,256,026	5.39	5.35
mpeg2dec	255,749,085	33.0	11,621,180	7.8	6.78

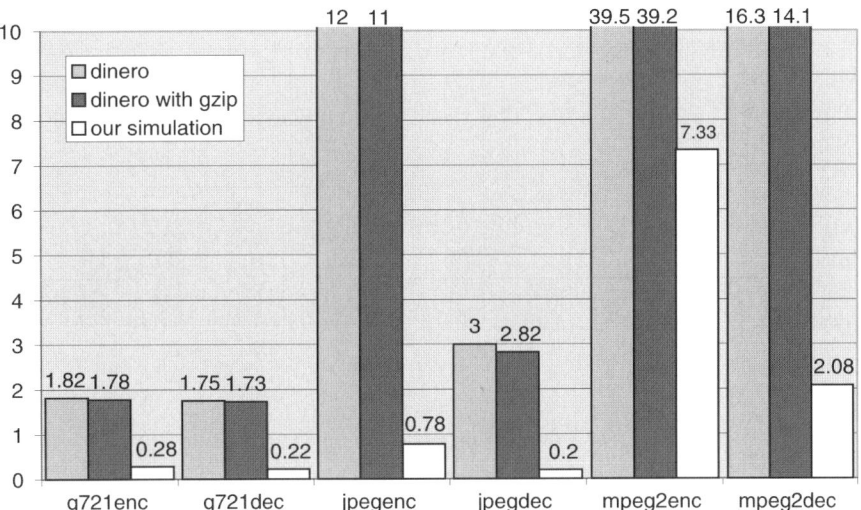

Fig. 5.13 Cache simulation total run time

the Dinero IV run time and the right bars indicate the run time of our simulation methodology, the y-axis shows the run time in seconds. Table 5.3 shows the cache configuration for each of the results shown in Figure 5.14. The bar comparisons in Figure 5.13 and 5.14 show that our simulation methodology is always faster when compared to the Dinero IV run time.

Column 6 in Table 5.1 shows that the compression algorithm can compress program traces by an average factor of 22.4. Results shown in column 8 of Table 5.1 show that by using our simulation methodology, it is possible to accurately perform faster cache simulations by an average factor of 9.67 when compared to Dinero IV, and average speedup factor of 9.10 when compared to Dinero IV with gzip.

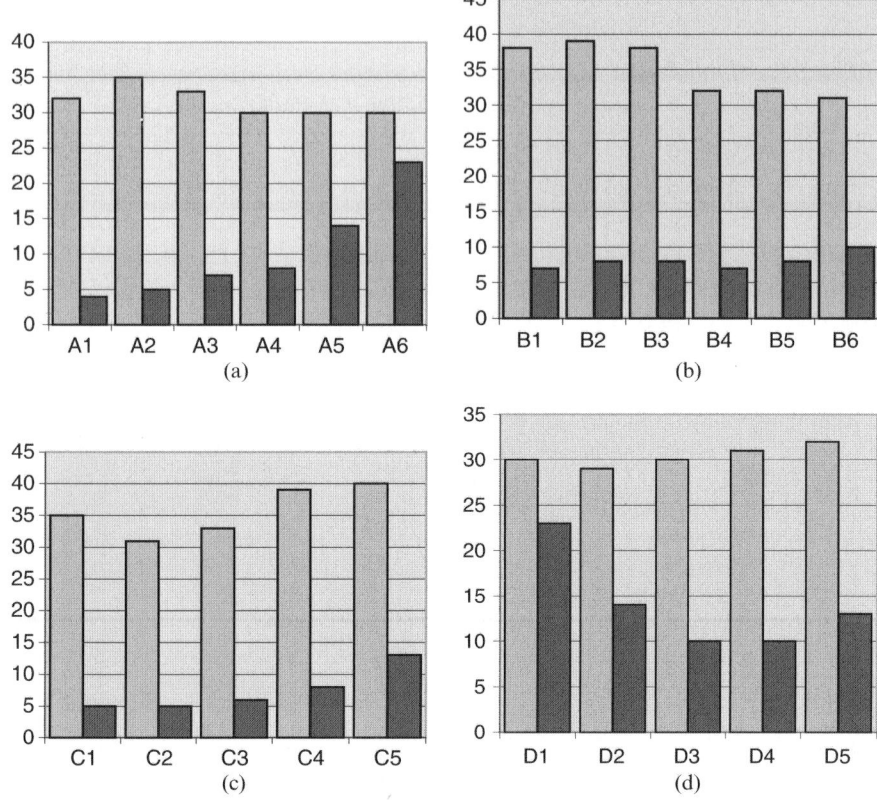

Fig. 5.14   Cache simulation run time comparison

4.   Run Time – Code Compression

4.1   Code Compression Technique

The process of compressing and decompressing code passes through *off-line* (i.e., design time) and *on-line* (i.e., run time) phases, respectively. In the off-line phase, the original code is compressed and decoding table is generated. During the on-line phase, the original instructions are generated from the compressed ones by using the decoding table. But the decoding table has a negative impact on the final compression ratio CR:

$$CR = \frac{size(compressed\_instructions+tables)}{size(original\_code)}$$

Hence, an efficient compression ratio can be accomplished by minimizing both the code itself and the tables. This is crucial since table sizes can reach more than 20% compared to the code size as we have found through a large set of

Table 5.3 Cache configuration

Cache config.	Cache parameters size, line size, assoc.	Cache config.	Cache parameters size, line size, assoc.
A1	512,8,1	B1	512,8,8
A2	1024,8,1	B2	1024,8,8
A3	2048,8,1	B3	2048,8,8
A4	4096,8,1	B4	4096,8,8
A5	8192,8,1	B5	8192,8,8
A6	16384,8,1	B6	16384,8,8
C1	1024,8,1	D1	16384,8,1
C2	1024,8,2	D2	16384,8,2
C3	1024,8,4	D3	16384,8,4
C4	1024,8,8	D4	16384,8,8
C5	1024,8,16	D5	16384,8,16

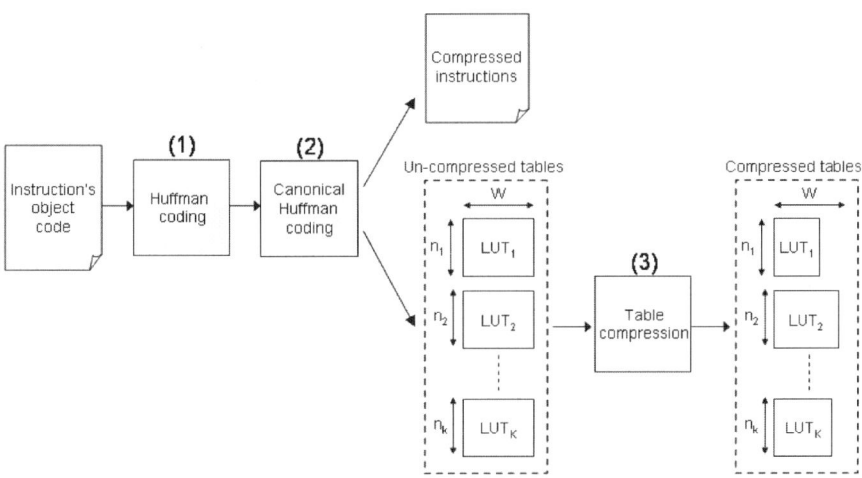

Fig. 5.15 Code compression steps

applications (see Figure 5.19). Therefore, our main contribution is a table compression scheme that we use along with a *Canonical Huffman Coding* scheme. In this scheme we conduct the following steps (see Figure 5.15):

1. The object code of the instructions are encoded with a variable-length code using *Huffman Coding*.
   Input: An object code for the instructions and their frequencies.
   Output: *Huffman* encoded instructions.

2. The *Huffman* encoded instructions are re-encoded again using *Canonical Huffman Coding*.
   Input: *Huffman* encoded instruction length and their length frequencies (number of instructions of each length).
   Output: *Canonical Huffman-encoded* instructions (*compressed instructions*). The unique original instructions are stored in different *Look-up Tables* depending on their compressed instruction length (one *Look-up Table* for each compressed instruction length).

3. The *Look-up Table* is compressed by sorting its entries using the *Lin-Kernighan* Algorithm (Helsgaun, 2000).
   Input: The *Look-up Table* which has a number of columns equal to instruction word length W.
   Output: The compressed *Look-up Table*.

Huffman Coding.   The first step in our compression scheme is to encode the instruction's object code using *Huffman Coding* (Bell et al., 1990). It is a well-known method based on probabilistic distribution. Most frequent input symbols are encoded with shortest codes and vice versa. The problem in *Huffman Coding* is the decoding because of the variable-length codes. This is a major problem when it comes to hardware implementation. To overcome this problem we re-encode the *Huffman* encoded instructions using *Canonical Huffman Coding*.

Canonical Huffman Coding.   *Canonical* codes are a sub-class of *Huffman* codes, that have a numerical sequence property, i.e., codewords with the same length are binary representations of consecutive integers (Nekritch, 2000). Using *Canonical Huffman* codes therefore allows for a space- and time-efficient decoding (Klein, 1997) because it requires less information to be

Table 5.4   Example for constructing Canonical Huffman codes

Symbols	Distribution	Huffman code	Canonical Huffman code
A	3/28	001	010
B	1/28	0000	0000
C	2/28	1001	0001
D	5/28	101	011
E	5/28	01	10
F	1/28	0001	0010
G	1/28	1000	0011
H	10/28	11	11

stored in encoded files. To re-encode the instructions using *Canonical Huffman Coding*, we need information about the length of *Huffman* compressed instructions and the number of compressed instructions for each length (Length Frequency). We can get these information from the previous step after encoding the instructions using *Huffman Coding*. Table 5.4 illustrates an example for constructing the *Canonical Huffman* code from a *Huffman* code for given symbol probabilities (assuming that every symbol denotes a unique object instruction code). In the first step, we compute the first codeword for each symbol length starting from the longest one (i.e., for length 4, 3, and 2). Then we create the remaining codewords for each length by adding '1' to the previous one.

The generated *Canonical Huffman* codes can be considered the compressed codes of the application's instructions. To use these codes as indices for our LUT, the indices need to have the same length. Hence, the codes with the same length are the indices to one specific LUT. We have as many LUTs as we have different code lengths. See the output of Step 2 in Figure 5.15.

Algorithm 5.1 DTM: Decoding Table Minimization

Number of table entries: n
Width of Table: W
Length of table index: L
/* **Initialization** */
0. Number of Toggles T=0, Column cost C=0
/* **Algorithm Start** */
1.   **for**(*each column i of W*) **do**
2.   **for** (*each entry j of n*) **do**
3.   **if** (*Toggle(j)* [$0 \rightarrow 1$ *or* $1 \rightarrow 0$] *is true*) **then**
4. $T(i)=T(i)+1$ *// Count the toggles*
5.   **end if**
6.   **end for**
7.   *cost of column i is* $C(i)=T(i)\times L$
/* *check if the column is compressible* */
8.   **if** (*n* > *C(i)*) **then**   *// Compress the column*
9.   *i. save the index at every toggle*
10.   *ii. table cost=table cost +C(i)*
11.   **else**
12.   *i. keep column(i)without compression*
13.   *ii. table cost=table cost + n*
14.   **end if**
15. **end for**

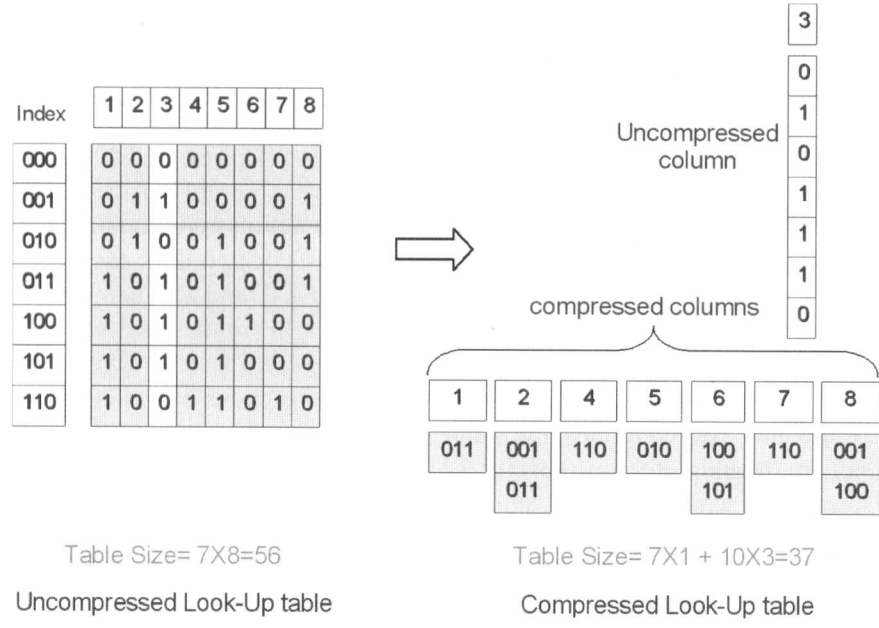

Fig. 5.16   Example for compressing table with 8-bit symbols

Look-up Table Compression.   Considering the fact that the *Look-up Table*
consists of a number of columns which is equal to the uncompressed instruction word length W (in bits), minimizing the LUT cost can be achieved by
reducing the height (i.e., number of rows) of its columns. The basic idea is
to re-sort the table in such a fashion that as many columns as possible can be
reduced in height. The basic principle of compressing the table is to minimize
the number of bit transitions per column (for example from top to bottom) and
then saving the indices only when a bit toggle occurs instead of saving the
complete column. Algorithm 5.1 explains our Decoding Table Minimization
scheme called DTM which is executed during the off-line phase. Starting from
the first column, the algorithm counts the number of bit toggles T occurring in
the same column and computes the cost of compressing this column. This cost
is the sum C of the table's indices where bit toggles occur in this column. If the
sum is less than the original column cost n, then the column is compressible,
otherwise the column is left without compression. The algorithm repeats the
previous steps for each table column.

Figure 5.16 shows an example for table compression. The size of the uncompressed table is 56. The number of unique instructions is $8 \Rightarrow$ index length $= 3$.
Hence, the column is compressible if it has a maximum of two transitions (because more than two transitions will need more than 7 bits which is the cost
of the uncompressed column). In that case, seven columns will be compressed

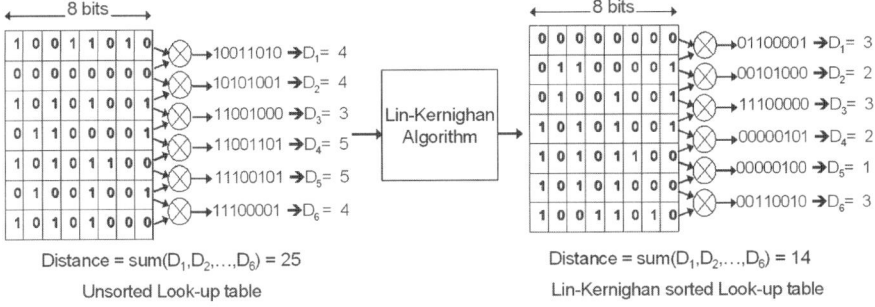

Fig. 5.17   Example for Lin-Kernighan sorted table with 8-bit symbols

and one column will be left without compression. The size of the table after compression is minimized to 37 bits.

Achieving a higher table compression ratio by compressing more table columns depends on how its entries are sorted. Finding the optimum solution of sorting the entries is NP-Hard. To sort the table's entries, we use the *Lin-Kernighan* heuristic algorithm (Helsgaun, 2000) which generates an optimal or near-optimal solution for the symmetric traveling salesman problem (TSP). TSP is a problem in discrete or combinatorial optimization to find (for a given number of symbols and costs of traveling from any one to others) the minimum sum of the distances between each two symbols covering all symbols. We use it to find the minimum sum of symbol distances between any two entries (see Algorithm 5.2). The distance between two entries is the Hamming distance which is the sum of bit toggles between these entries. In the hardware, this can be computed using XOR gates. Figure 5.17-left illustrates an example for a *Look-up Table* with 7 symbols and 8 bits each. The figure on the left shows unsorted symbols, the distance between every two consecutive symbols and the distance between the first and the last one. The sum of distances from the top of the table to the bottom is 25. Using *Lin-Kernighan* algorithm (Figure 5.17-right), the sum of the distances is decreased to 14. Now, if we compress the unsorted and the sorted tables using our DTM algorithm, we find that their costs are 53 bits and 36 bits, respectively.

4.2   Hardware Implementation

Hardware implementation is based on a scalable hardware prototyping board (Platinum) from *Pro-Design* (pro, 2006). To decode the encoded instructions, we use a pipelined parallel *Huffman* decoder (Rudberg and Wanhammar, 1997). We have developed the hardware to decode the compressed *Look-up Tables*. The decoder architecture is illustrated in Figure 5.18. It consists of two main parts: (1) *Length Decoder* that calculates the length of the *Canonical-*

Algorithm 5.2 Lin-Kernighan Sorting

Number of table's entries: n
Length of entry: W
Entries of table: $S_1, S_2, \ldots, S_n$
/* Initialization */
Distance =0
New_Distance =100000   *// Big number as initial value*
Min_Distance =1000   *// minimum distance defined*
/* Algorithm Start */
1. **call** sort($S_1, S_2, \ldots, S_n$) *// sort entries using Lin_Kernighan algorithm*
2. **for**(*all entries i =1* **to** *n-1*) **do**   *// Compute distance*
3. $A = S_i$ **XOR** S_{i+1}
4. $D_i = \sum_{j=1}^{W} A_j$ *//j is index for bit position in A*
5. $Distance = Distance + D_i$
6. **end for**
7. **if**($Distance < New\_Distance$) **do**
8. $New\_Distance = Distance$
9.   **if**($New\_Distance <= Min\_Distance$) **do**
10.   goto 19
11.   **else**
12.   goto 1   *// Repeat for better solution*
13.   **end if**
14. **else**   *// Couldn't find better solution*
15.   goto 1   *// Repeat to find better solution*
16. **end if**
17. **return**($S_1', S_2', \ldots, S_n'$)   *// Returns sorted entries*

---

encoded instructions; and, (2) *Instruction Decoder* that retrieves the original instruction from the compressed *Look-up Table*. The decoder operates as follows: the 32-bit shift register 1 receives 32-bit code which is some compressed instructions and shifts it continuously from left to right to the second shift register. The pipelining unit consists of a number of stages equal to the number of different code lengths k, every one has a comparator to compare the code received from the shift register 2 with the first code of one particular length. The pipelining unit represents the output as a separate signal for every code length. In every cycle, one code length is checked in the pipelining unit. The counter at the same time counts the number of cycles needed to detect the code length and activates the MUX and DE-MUX to pass the compressed instruction to the corresponding compressed *Look-up Table*. The compressed *Look-up Table* consists of compressed columns that are stored in FPGA Block RAM.

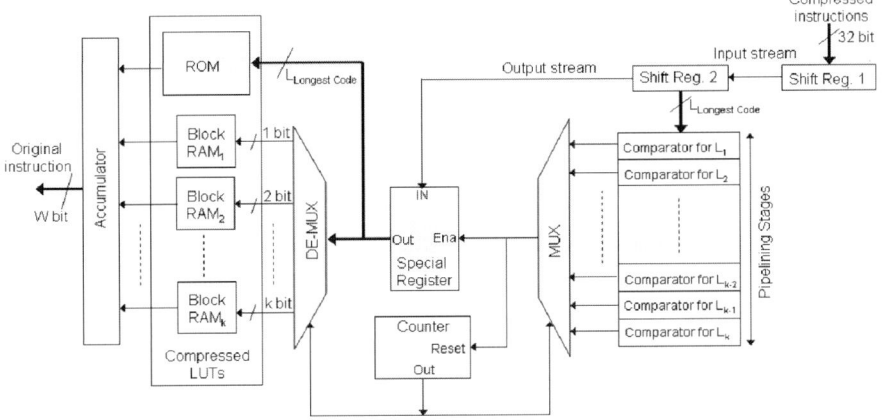

Fig. 5.18 Hardware Decoder

Uncompressed columns are stored in ROM. The compressed instruction is de-multiplexed to one of the Block RAMs depending on its length and compared with the column indices to find the position where the compressed instruction is larger than the respective indices. The original bits of uncompressed table columns are retrieved directly from the ROM. The accumulator concatenates the bits into the correct positions and generates the W bits decompressed instruction.

4.3 Experiments and Results

To show the efficiency of our method we have applied it on different bench-mark sizes serving as a representative set of applications. We have conducted evaluations deploying the ARM(SA-110), MIPS(4KC), and PowerPC (MPC85) architectures. The final results are presented in Tables 5.5, 5.6, and 5.7. They account for the overhead stemming from the *Look-up Tables* implemented as FPGA block RAM. To give an idea of how important table compression actually can be, we analyzed a large number of applications with a wide range of different sizes and found that the amount of unique instructions is approximately proportional to the application's size and can account for more than 20% of all instructions, even for very large applications (Figure 5.19). Hence, the LUT has a significant effect on the final compression ratio. Before we discuss the results, we explain some of the terms used in the tables as follows: we refer to *unique instruction size* as a measure (in #bytes) for the table cost as it is generated by the compression scheme, i.e., before our table compression scheme has been applied. The term *code size* denotes the compressed instruction cost (in #bytes). The term *table size* stands for the cost (in #bytes) of the Block RAMs and ROM hosting the compressed and uncompressed table columns, respectively. We use *total size* as the sum of *compressed instructions size* and *table size*. Furthermore, TCR is the table compression ra-

Table 5.5   Experimental results using ARM target architecture (size is in byte)

Bench-mark	Original code		Compressed code		
	Code size	Uniq. inst. size	Comp. inst. size	Table size /TCR	Total size /CR
Hanoi	16,200	9,336	5,161	5,217 55.88%	10,378 64.06%
Whetstone	31,264	15,716	10,404	8,500 54.08%	18,904 60.46%
JPEG	91,024	39,608	34,264	20,714 52.29%	54,978 60.39%
Lame	290,924	123,256	117,480	59,117 47.96%	176,597 60.70%
Blowfish	968,688	174,948	351,166	80,651 46.09%	431,817 44.57%
GSM	1,104,104	191,564	407,393	87,401 45.62%	494,794 44.81%

Table 5.6   Experimental results using MIPS target architecture (size is in bytes)

Bench-mark	Original code		Compressed code		
	Code size	Uniq. inst. size	Comp. inst. size	Table size /TCR	Total size /CR
Hanoi	17,832	7,156	4,276	4,203 58.74%	8,479 47.54%
Whetstone	30,856	10,668	8,161	6,204 58.16%	14,365 46.55%
JPEG	126,252	37,180	41,315	20,850 56.07%	62,165 49.23%
Lame	311,352	127,764	119,661	65,082 50.94%	184,743 59.33%
Blowfish	569,060	182,312	218,771	87,527 48.01%	306,298 53.82%
GSM	653,480	208,032	256,635	98,191 47.20%	354,826 54.29%

tio, i.e., the relationship of the table size after and before table compression. CR stands for the *final* compression ratio.

From the Tables 5.5, 5.6, and 5.7, we can observe the following: (1) *Look-up Tables* can be compressed up to 45% depending on the size of application. TCR becomes better if the number of the instructions increases, i.e., for big

Table 5.7   Experimental results using PowerPC target architecture (size is in bytes)

Bench-mark	Original code		Compressed code		
	Code size	Uniq. inst. size	Comp. inst. size	Table size /TCR	Total size /CR
Hanoi	16,404	8,836	5,077	5,154 58.33%	10,231 62.36%
Whetstone	28,192	12,676	8,967	7,326 57.80%	16,293 57.79%
JPEG	99,656	39,040	36,544	21,686 55.55%	58,230 58.43%
Lame	275,060	134,240	115,153	67,348 50.17%	182,501 66.34%
Blowfish	483,896	182,560	205,388	87,482 47.92%	292,870 60.52%
GSM	557,772	206,776	240,177	99,045 47.90%	339,222 60.81%

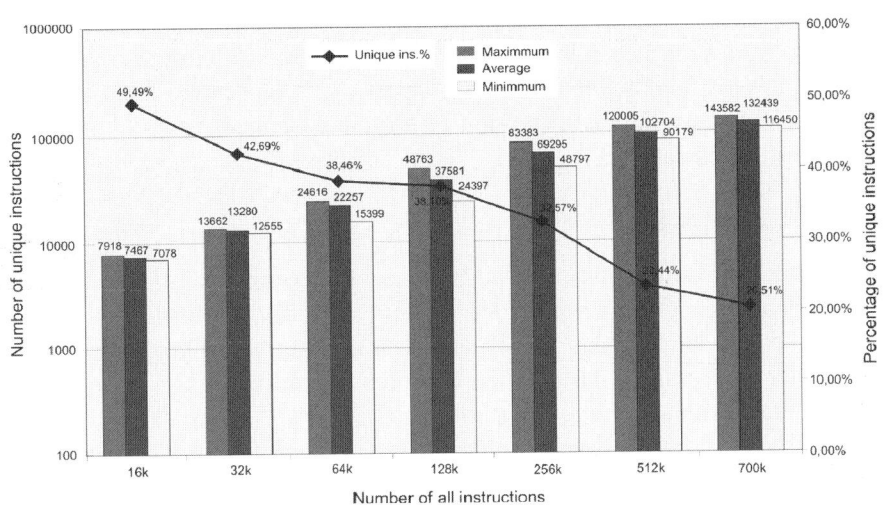

Fig. 5.19   Look-up Table size (in percentage of the whole application size) compared to the total number of unique instructions of an application. Each bar set represents a different application

applications because the number of unique instructions increases accordingly. Ex: *Hanoi* and GSM for all ISAs. This is obvious since good compression ratios are achieved through re-occurring patterns. However, since the compression scheme cannot work well in those cases by principle, the relevancy of

LUT compression increases since large numbers of unique instructions result in large LUTs. Hence, LUT compression is especially useful in those cases where the code is less compressible. It is therefore a powerful method in such cases where traditional code compression schemes do not perform well. (2) For the applications compiled for the ARM processor, CR is better if the application is big (like GSM) and a compression ratio of 44% can be achieved including the LUT overhead. But, for the applications compiled for the MIPS processor, CR is better for small applications (like *Hanoi*). This can be interpreted as follows: comparing the ratio of compressed instruction size to the code size for the smallest and biggest applications, i.e., Hanoi and GSM, we find that the ratio increased from 31% to 36% for the ARM processor and from 24% to 39% for the MIPS processor. This is because GSM has a large number of instructions for ARM compared to MIPS processors with pretty much the same number of unique instructions. That leads to the increase in number of repetition and improves the *Canonical Huffman* property, i.e., more repetition gives shorter code and better compressed instruction ratio. (3) No ISA-specific knowledge has been used to obtain these ratios. Therefore, our scheme is orthogonal to the respective processor architecture and hence, even higher compression ratios could be achieved if ISA-specific knowledge was applied. Finally we want to point out that the decompression hardware has been implemented using Xilinx ISE 6.3 for VirtexII. The average maximum frequency of the hardware is about 180 MHz. The number of slices needed for the decoder is about 850 slices. This is a fraction of a typical application as shown in the final compression ratios.

4.4   Conclusion

This chapter presented two methods based on compression for efficiently designing and executing embedded systems. Firstly, a novel method was presented for cache analysis, which analyzes a compressed program trace file without decompression. An average compression ratio of 22.4 is seen in the experimental results for large program trace files with size up to 21 GB. Cache simulation method is observed to have an average speedup of 9.67 over Dinero IV, and an average speedup factor of 9.10 is observed when compared to Dinero IV with gzip. Secondly, a approach for embedded system code compression was presented, based on *Canonical Huffman Coding*. The *Look-up Table* overhead is reduced using the *Lin-Kernighan* algorithm. These methods are orthogonal to any ISA-specific characteristic. Hence, these are the basis for even higher compression ratios when ISA-specific knowledge is considered. Without ISA-specific knowledge we have already achieved a compression ratio of up to 45% including the LUT overhead.

References

http://www.prodesigncad.com (2007).

Bell, Timothy C., Cleary, John G., and Witten, Ian H. (1990) Text compression. Englewood Cliffs. Prentice-Hall.

Benini, L., Bruni, D., Macii, A., and Macii, E. (2002) Hardware-assisted data compression for energy minimization in systems with embedded processors. In *DATE '02: Proceedings of the Conference on Design, Automation and Test in Europe*, page 449, Washington, DC, USA. IEEE Computer Society.

Benini, Luca, Macii, Alberto, and Nannarelli, Alberto (2001) Cached-code compression for energy minimization in embedded processors. In *ISLPED '01: Proceedings of the 2001 International Symposium on Low Power Electronics and Design*, pages 322–327, New York, NY, USA. ACM Press.

Burtscher, Martin (2004) Vpc3: A fast and effective trace-compression algorithm. In *SIGMETRICS '04/Performance '04: Proceedings of the Joint International Conference on Measurement and Modeling of Computer Systems*, pages 167–176, New York, NY, USA. ACM Press.

Corliss, Marc L., Lewis, E. Christopher, and Roth, Amir (2003) Dise: a programmable macro engine for customizing applications. In *ISCA '03: Proceedings of the 30th Annual International Symposium on Computer Architecture*, pages 362–373, New York, NY, USA. ACM Press.

Das, Dipankar, Kumar, Rajeev, and Chakrabarti, P.P. (2004) Code compression using unused encoding space for variable length instruction encodings. In *Proceedings of the 8th VLSI Design and Test Workshop (VDAT)* Mysore, India.

Das, Dipankar, Kumar, Rajeev, and Chakrabarti, P.P. (2005) Dictionary based code compression for variable length instruction encodings. In *VLSID '05: Proceedings of the 18th International Conference on VLSI Design held jointly with 4th International Conference on Embedded Systems Design (VLSID'05)*, pages 545–550, Washington, DC, USA. IEEE Computer Society.

Edler, Jan and Hill, Mark D. Dinero iv trace-driven uniprocessor cache simulator. In *http://www.cs.wisc.edu/˜markhill/DineroIV/*. 1999

Game, Mark B. (2000) Codepack: Code compression for powerpc processors. In *PowerPC Embedded Processor Solutions*.

Helsgaun, K. (2000) An effective implementation of the Lin-Kernighan traveling salesman heuristic. *European Journal of Operational Research*, 126(1):106–130.

Johnson, Eric E. and Ha, Jiheng (1994) PDATS: Lossless address space compression for reducing file size and access time. In *Proceedings of 1994 IEEE International Phoenix Conference on Computers and Communication*.

Johnson, Eric E., Ha, Jiheng, and Zaidi, M. Baqar (2001) Lossless trace compression. *IEEE Transactions on Computers*, 50(2):158–173.

Klein, Shmuel T. (1997) Space- and time-efficient decoding with canonical huffman trees. In *CPM '97: Proceedings of the 8th Annual Symposium on Combinatorial Pattern Matching*, pages 65–75, London, UK. Springer-Verlag.

Larin, Sergei Y. and Conte, Thomas M. (1999) Compiler-driven cached code compression schemes for embedded ilp processors. In *MICRO 32: Proceedings of the 32nd Annual ACM/IEEE International Symposium on Microarchitecture*, pages 82–92, Washington, DC, USA. IEEE Computer Society.

Lee, Chunho, Potkonjak, Miodrag, and Mangione-Smith, William H. (1997) Mediabench: A tool for evaluating and synthesizing multimedia and communicatons systems. In *MICRO 30: Proceedings of the 30th Annual ACM/IEEE International Symposium on Microarchitecture*, pages 330–335, Washington, DC, USA. IEEE Computer Society.

Lefurgy, Charles, Bird, Peter, Chen, I-Cheng, and Mudge, Trevor (1997) Improving code density using compression techniques. In *MICRO 30: Proceedings of the 30th Annual ACM/IEEE International Symposium on Microarchitecture*, pages 194–203, Washington, DC, USA. IEEE Computer Society.

Lekatsas, H. and Wolf, Wayne (1999) SAMC: A code compression algorithm for embedded processors. *IEEE Transactions on CAD*, 18(12):1689–1701.

Lekatsas, Haris, Henkel, Joerg, and Wolf, Wayne (2000) Code compression for low power embedded system design. In *DAC '00: Proceedings of the 37th Conference on Design Automation*, pages 294–299, New York, NY, USA. ACM Press.

Lekatsas, Haris, Henkel, Joerg, Jakkula, Venkata, and Chakradhar, Srimat (2005) A unified architecture for adaptive compression of data and code on embedded systems. In *VLSID '05: Proceedings of the 18th International Conference on VLSI Design held jointly with 4th International Conference on Embedded Systems Design (VLSID'05)*, pages 117–123, Washington, DC, USA. IEEE Computer Society.

Luo, Yue and John, Lizy Kurian (2004) Locality-based online trace compression. *IEEE Transactions on Computers*, 53(6):723–731.

Milenkovic, Aleksandar and Milenkovic, Milena (2003) Exploiting streams in instruction and data address trace compression. In *WWC-6 2003 IEEE International Workshop on Workload Characterization*, pages 99–107.

Nekritch, Yakov (2000) Decoding of canonical huffman codes with look-up tables. In *DCC '00: Proceedings of the Conference on Data Compression*, page 566, Washington, DC, USA. IEEE Computer Society.

Nevill-Manning, C. and Witten, I. (1997) Identifying hierarchical structure in sequences: A linear-time algorithm. In *Journal of Artificial Intelligence Research (JAIR)*, 7:67–82.

Okuma, T., Tomiyama, H., Inoue, A., Fajar, E., and Yasuura, H. (1998) Instruction encoding techniques for area minimization of instruction rom. In *ISSS '98: Proceedings of the 11th International Symposium on System Synthesis*, pages 125–130, Washington, DC, USA. IEEE Computer Society.

Pleszkun, Andrew R. (1994) Techniques for compressing program address traces. In *MICRO 27: Proceedings of the 27th Annual International Symposium on Microarchitecture*, pages 32–39, New York, NY, USA. ACM Press.

Processor, Xtensa. http://www.tensilica.com.

Rudberg, Mikael K. and Wanhammar, Lars (1997) High speed pipelined parallel huffman decoding. In *ISCAS '97: Proceedings of 1997 IEEE International Symposium on Circuits and Systems*, pages 2080–2083.

Technology, (2006) Seagate. http://www.seagate.com/docs/pdf/marketing/po_db35.pdf.

Tuduce, Irina C. and Gross, Thomas (2005) Adaptive main memory compression. pages 237–250.

Welch, Terry A. (1984) A technique for high-performance data compression. *IEEE Computer*, 17(6):8–19.

Wolfe, Andrew and Chanin, Alex (1992) Executing compressed programs on an embedded risc architecture. In *MICRO 25: Proceedings of the 25th Annual International Symposium on Microarchitecture*, pages 81–91, Los Alamitos, CA, USA. IEEE Computer Society Press.

Xu, X.H., Clarke, C.T., and Jones, S.R. (2004) High performance code compression architecture for the embedded arm/thumb processor. In *CF '04: Proceedings of the 1st Conference on Computing Frontiers*, pages 451–456, New York, NY, USA. ACM Press.

Yang, Lei, Dick, Robert P., Lekatsas, Haris, and Chakradhar, Srimat (2005) Crames: Compressed ram for embedded systems. In *CODES+ISSS '05: Proceedings of the 3rd IEEE/ACM/IFIP International Conference on Hardware/Software Codesign and System Synthesis*, pages 93–98.

Yoshida, Yukihiro, Song, Bao-Yu, Okuhata, Hiroyuki, Onoye, Takao, and Shirakawa, Isao (1997) An object code compression approach to embedded processors. In *ISLPED '97: Proceedings of the 1997 International Symposium on Low Power Electronics and Design*, pages 265–268, New York, NY, USA. ACM Press.

Zhang, Xiangyu and Gupta, Rajiv (2004) Whole execution traces. In *MICRO 37: Proceedings of the 37th Annual IEEE/ACM International Symposium on Microarchitecture*, pages 105–116, Washington, DC, USA. IEEE Computer Society.

Zhang, Xiangyu and Gupta, Rajiv (2005) Whole execution traces and their applications. *ACM Trans. Archit. Code Optim.*, 2(3):301–334.

Ziv, Jacob and Lempel, Abraham (1977) A universal algorithm for sequential data compression. *IEEE Transactions on Information Theory*, 23(3):337–343.

Ziv, Jacob and Lempel, Abraham (1978) Compression of individual sequences via variable-rate coding. *IEEE Transactions on Information Theory*, 24(5):530–536.

II

Embedded Memories

Chapter 6

Power Optimisation Strategies Targeting the Memory Subsystem

Preeti Ranjan Panda
Department of Computer Science and Engineering
Indian Institute of Technology Delhi
Hauz Khas
New Delhi 110016
India

Abstract   Power optimisations targeting the memory subsystem have received considerable attention in recent years because of the dominant role played by memory in the overall system power. The more complex the application, the greater the volume of instructions and data involved, and hence, the greater the significance of issues involving power-efficient storage and retrieval of these instructions and data. In this chapter we give a brief overview of how memory architecture and accesses affect system power dissipation, and some recent proposals on reducing memory-related power through diverse mechanisms: optimisations of the traditional cache memory system, architectural innovations targeting application specific designs, compiler optimisations, and other techniques.

Keywords: power optimisation; memory subsystem; cache memory; scratch pad memory.

1.   Introduction

The memory subsystem plays a dominant role in every type of modern electronic design, starting from general purpose microprocessors to customised application specific systems. As applications get more complex, the memory storage and retrieval plays a critical role in every design dimension: it dominates the area of the design since the increasing amounts of instructions and data lead to increasing memory requirements; with increasing memory size, the access times for memory are larger than typical computation times, leading to memory becoming a significant determinant of performance;

J. Henkel and S. Parameswaran (eds.), Designing Embedded Processors – A Low Power Perspective,
131–155.

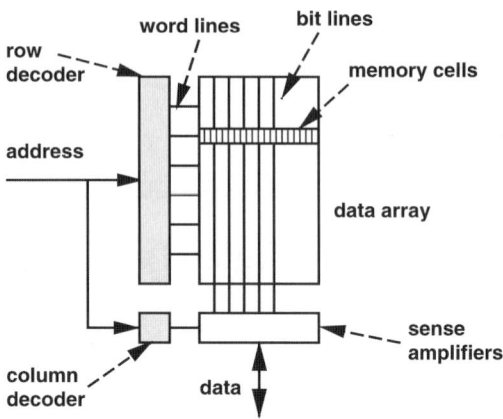

Fig. 6.1   Simplified view of typical memory structure

and larger memories also lead to high static and dynamic power dissipation (Panda et al., 1999b).

This section focusses on different aspects of memory-related optimisations targeting power efficiency. In this chapter, we give an overview of the different power optimisation opportunities that arise in the context of memory accesses. The next two chapters elaborate on two selected areas within this domain.

Figure 6.1 shows a simplified view of a typical memory structure. The core storage area is organised into rows and columns of memory cells. The address is split into two parts: a *row address* consisting of the higher order bits, and a *column address* consisting of the lower order bits. The row address is decoded in the *row decoder*, resulting in the activation of one of the *word lines*. This selects one row of cells in the memory, which causes transfer of data between the cell and the *bit-lines* – these lines run through all the rows of the memory. A *sense amplifier* detects the transitions on the bit-lines and transmits the result on the data bus. A *column decoder* selects the bits from the row of cells and transmits the addressed bits to the data bus.

Power dissipation during memory accesses can be attributed to three main components: (1) address decoders and word lines; (2) data array, sense amplifiers, and the bit-lines; and (3) the data and address buses leading to the memory. All three components are significant as each involves the driving of high capacitance wires that requires a considerable amount of energy: word lines, bit-lines, and data/address buses. Power optimisations for the memory subsystem indirectly target one of these components and can be classified into the following broad categories:

- *Power-efficient memory architectures* – novel architectural concepts that aid power reduction, both in traditional cache memory design and in other unconventional memory architectures such as scratch pad memory, banked memory, etc.

- *Compiler optimisations targeting memory power* – where code is generated for general-purpose processors explicitly targeting power reduction.

- *Application specific memory customisation* – where the memory system can be tailored for the particular application, leading to superior solutions than a standard memory hierarchy.

- *Other techniques: dynamic voltage scaling, compression, encoding, etc.* – these are known techniques from other domains that are also applicable to memory power reduction.

In the following sections, we study the above topics in greater detail.

2. Power-Efficient Memory Architectures

The memory subsystem in embedded processor based systems usually consist of cache memory, along with other memory modules possibly customised for the application. Because of the dominating role of instruction and data caches, new low power memory architectures have been in the area of improving traditional cache designs to make them power-efficient using a variety of techniques.

2.1 Partitioned Memory and Caches

Partitioning the memory structure into smaller memories (or banks) is one of the mechanisms used to reduce the effective length of the bit-lines driven during memory operations, thereby reducing power dissipation. In multibank memories (Figure 6.2), each bit-line is smaller and the sense amplifiers are replicated in each bank. Since only one of the banks would be activated during a memory operation, this would lead to reduced power consumption in the overall structure because smaller bit-lines are associated with a smaller capacitance, and hence, lower energy. Banked structures for caches were proposed in Ko et al. (1995) and Su and Despain (1995). In the banking architecture proposed in Ko et al. (1995), the address decoding circuit is selectable in the sense that the active decoding (and hence, any resulting switching of capacitive loads) takes place only for the bank currently addressed, and not for the other banks. The sense amplifier circuitry is also selectable. The bit-line segmentation technique (Ghose and Kamble, 1999), where a longer bit-line is divided into smaller segments so that each bit-line is effectively smaller and the sense amplifiers are also smaller, is conceptually a similar optimisation.

One variant proposed in Kim et al. (2001) is to make the smaller units complete caches as opposed to just banks. The added flexibility here is that the different caches need not be homogeneous. A prediction mechanism such as *most recently used* is employed to predict which sub-cache will be accessed next.

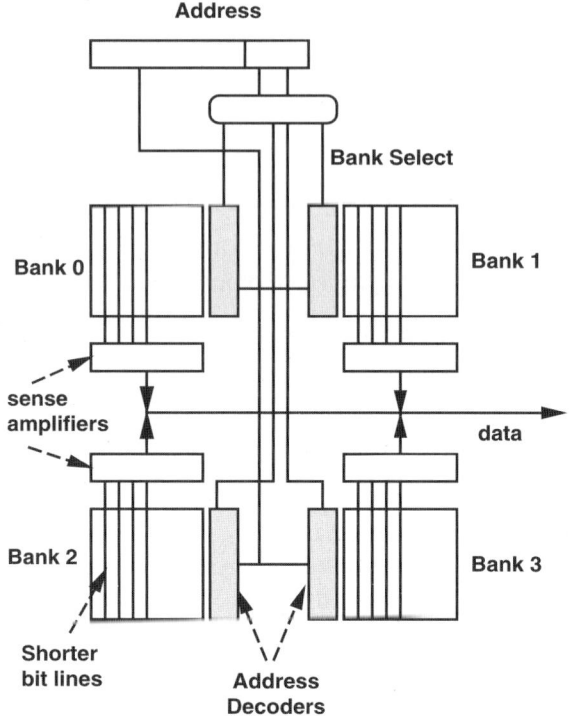

Fig. 6.2   Banking reduces bit-line capacitance

2.2   Augmenting with Additional Buffers/Caches

A large number of low power cache ideas have been centred around one central principle: add an extra cache or buffer, usually small in size, and design the system to fetch data directly from this buffer, thereby preventing an access to the L1 cache altogether. Since the buffer is relatively small, we can achieve significant power savings if we can ensure a high hit rate to the buffer.

The technique of *block buffering* (Su and Despain, 1995) stores the previously accessed cache line in a buffer. If the next access is from the same line, then the buffer is directly read and there is no need to access the core of the cache. This scheme successfully reduces power when there is a significant amount of spatial locality in memory references (Figure 6.3a). The idea of a block buffer can be extended to include more than one line instead of just the last accessed line. In (Ghose and Kamble, 1999), a fully associative buffer consisting of four lines is used for this purpose.

One simple strategy is to introduce another level of hierarchy before the L1 cache, placing a very small cache (called a *filter cache* in Kin et al. (1997)) between the processor and the regular cache. This causes a performance overhead

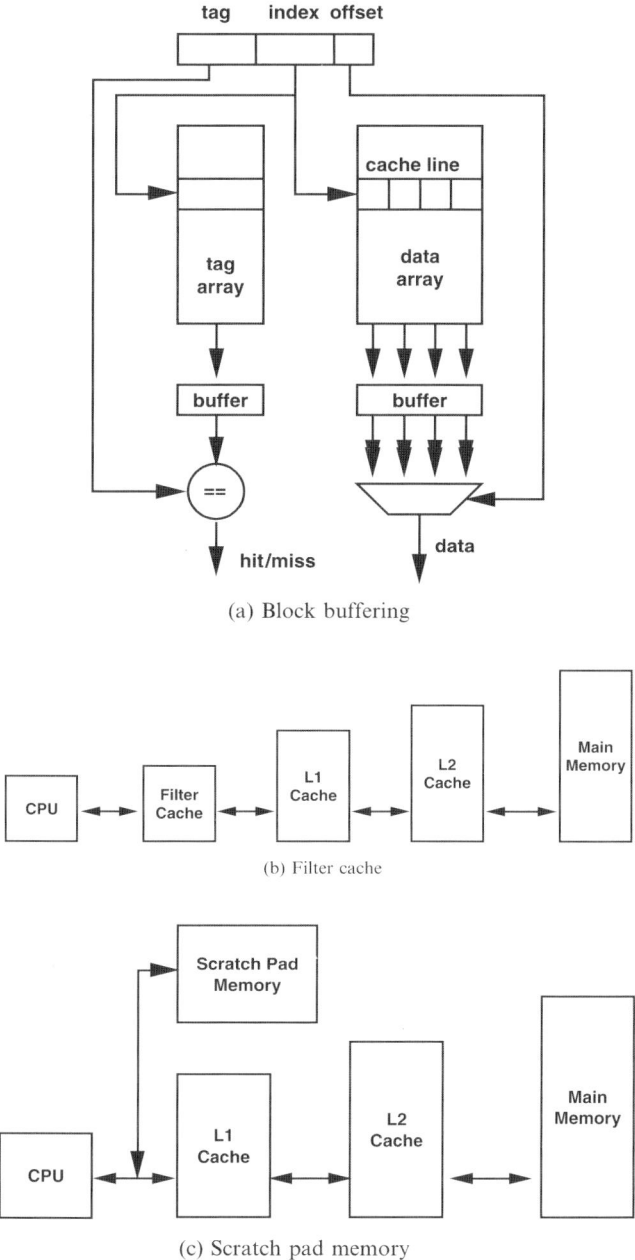

(a) Block buffering

(b) Filter cache

(c) Scratch pad memory

Fig. 6.3 Caches augmented by buffers

because the hit ratio of this small cache is bound to be lower, but it lowers total power dissipation because the hits to this small cache cause a much lower power dissipation. This is illustrated in Figure 6.3b. No overall modification is proposed to the overall cache hierarchy, except that the filter cache is unusually small compared to a regular L1 cache.

Data and instructions can also be statically assigned to an additional on-chip memory whose address space is disjoint from the cached part. In the *S-Cache* (Panwar and Rennels, 1995), basic blocks of instructions are statically assigned. The compiler is responsible for the layout of the instructions, based on profile data, etc. Similarly, in the *Scratch Pad Memory* (Panda et al., 2000), data is statically assigned by the compiler keeping in mind the data size and frequency of access (Figure 6.3c). Data in scratch pad memory is never evicted (unlike in data caches) – the compiler explicitly manages the space and "hits" are guaranteed. Compiler techniques to exploit this architectural enhancement are discussed in the next section.

Since it is well known that programs tend to spend a lot of time inside relatively small loop structures, researchers have investigated the use of specialised hardware structures to exploit this phenomenon. *Loop cache* (Bellas et al., 2000) is one such structure consisting of an augmentation to the normal cache hierarchy. Frequently executed basic blocks within loops are stored in the loop cache. The processor first accesses the loop cache for an instruction; if it is present, there is no need to access the normal cache hierarchy, else the instruction cache is accessed. The *Decoded Instruction Buffer* (Bajwa et al., 1997) is analogous to the loop cache idea, but here, the decoded instructions occurring in a loop are stored in the buffer, to prevent the power overhead associated with instruction decoding. The decoded instructions are written to the buffer in the first loop iterations; in subsequent iterations, they are read off the buffer instead of the L1 instruction cache. An *L0-cache* is proposed in Bellas et al. (1999) for storing basic blocks of code that are frequently accessed. A prediction mechanism and confidence estimator based on modifications to the processor's branch prediction unit is used to decide whether a branch target is selected for insertion into the L0 cache.

Some variants of the above fundamental ideas have been reported in recent research. A decoded filter cache is used in Tang et al. (2002), which is based on the filter cache idea except that it stores decoded instructions. One problem that arises in this context is that the lengths of the decoded instructions are widely varying, leading to waste of space if the width is set to the widest one. Thus, only a subset of instructions are considered cacheable. The hardware is accompanied by a prediction mechanism that selects between the decoded buffer, the line buffers discussed earlier, and the instruction cache. An extension to the filter cache and L0-cache strategy is the *HotSpot cache* (Yang and Lee, 2004), where the filter cache is placed parallel to the main cache hierarchy, and a dynamic mechanism is used to monitor the instruction access patterns and copy frequently executed basic blocks to the filter cache. Recognising that a significant fraction of data memory accesses during program execution consist of return address stores and callee saved registers during function calls, a separate data memory organised as a stack was proposed in Mamidipaka and Dutt (2003) so that these data are directly stored and retrieved from here rather from main memory through the cache system.

2.3   Reducing Tag Comparison Power

The line of research discussed above aims at achieving power efficiency by using an additional small buffer which causes a reduction in the number of accesses to the L1 cache. In this section we study power optimisation along a slightly orthogonal direction. For performance reasons, the tag array and the data array in the cache are accessed *simultaneously* so that by the time the address of the resident cache line is fetched from the tag array and compared with the required address to detect hit or miss, the corresponding data is already available for forwarding to the processor (Hennessy and Patterson, 1994). Thus, the data is fetched even before we know whether the access is a hit or a miss; on a miss, it is simply discarded. Since we expect most accesses to be hits, this parallel access strategy improves peformance significantly. In a set-associative cache, all the tag arrays and data arrays are accessed at once. While designed for optimal performance, this overall strategy results in waste of power. There is a significant scope here for trade-offs between performance and power.

The simplest power optimisation addressing the above issue is to sequentialise the accesses to the tag and data arrays – that is, to fetch from the data array only if the tag fetch indicates a cache hit. This prevents dynamic power dissipation incurred when data is fetched from the data array in spite of a cache miss (Hasegawa et al., 1995). Also, data only needs to be fetched from the way that matched, not from the other ways. The idea is illustrated in Figure 6.4a and b. Shaded blocks indicate data and tag arrays that are active in the respective cycles. This approach does compromise on the cache performance, though. Another simple idea is (in case of instruction cache) to retain the address of the last accessed cache line, and to fetch from the tag array only if the next instruction refers to a different line (Panwar and Rennels, 1995). If the reference is to the same line, then we are guaranteed a cache hit, and power is saved by preventing the redundant access of the tag array. Using a similar logic, we can assert that if there has been no cache miss since the last reference to a basic block of instructions, then there is no need to fetch the tag from the instruction cache in the next reference (since the previously fetched instruction has not had an opportunity to get evicted yet). The information about whether the target of a branch instruction exists in the cache is recorded in the Branch

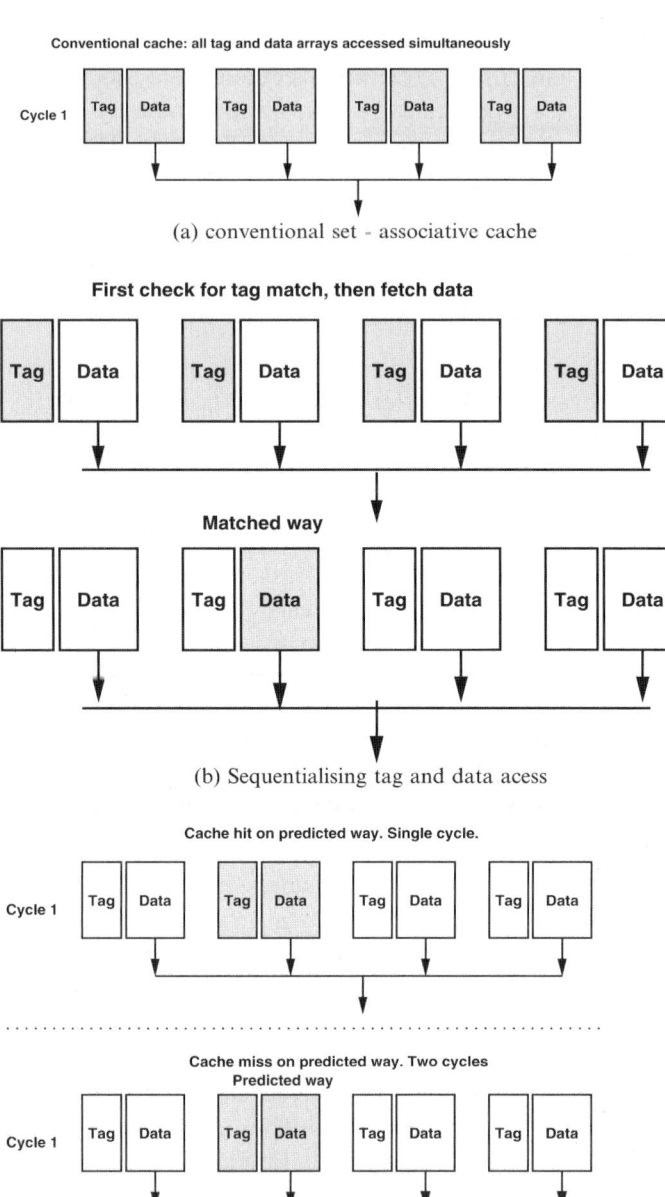

Fig. 6.4 Reducing tag comparisons (Inoue et al., 1999)

Target Buffer, a commonly used structure in modern processors. If the condition is satisfied, then the fetch from the tag array is disabled, saving memory activity in the process (Inoue et al., 2002).

The observation that, in set-associative caches, consecutive references tend to access data from the same way, can be exploited in a mechanism that predicts the way number for the next access to be the same as the previous one. On the next access, only the tag and data arrays of the predicted way are activated, resulting in a significant amount of dynamic power savings when the prediction is correct (Inoue et al., 1999). When the prediction turns out to be incorrect, the rest of the ways are fetched in the next cycle, which incurs a performance penalty. This is illustrated in Figure 6.4. An alternative method of incorporating way prediction is through the *location cache* (Min et al., 2004) – a proposal for the L2 cache. This is an extra cache that is used to indicate which way in the actual cache contains the data. A hit in the location cache indicates the way number, and hence, we only need to access the specific way in the L2 cache, thereby avoiding reading all the tag and data arrays of the set-associative cache. A miss in the location cache indicates that we do not have a prediction, and leads to a regular access from all the ways in the L2 cache.

A certain amount of flexibility can be built into set-associative caches to control accesses to the different ways – ways can be selectively enabled or disabled depending on the characteristics of an application. For example, in the L2 cache, we can reserve separate ways for instruction and separate ones for data so as to prevent conflicts. Also, for small programs where instruction cache conflicts are not expected, some of the ways assigned to instructions can be disabled to reduce activity in their tag and data arrays (Malik et al., 2000). In the *way-halting cache* (Zhang et al., 2004), some least significant tag bits from each way are stored in a separate array, which is first accessed and the corresponding bits of the address compared. If the comparison fails, then a matching failure is guaranteed for that way, and hence, the actual fetch of the tag and data arrays is prevented, saving power. Another way of building flexibility into the cache structure is to allow for different cache sets to have different effective associativities (Ishihara and Fallah, 2005). The disabling of the ways can be done in various ways – at cache design time by disconnecting the power supply of redundant ways, disconnecting the memory cells from bit-lines, etc. The finding of the optimum number of ways for each set can be done concurrently with code placement in instruction memory by the compiler.

2.4 Reducing Cache Leakage Power

The techniques discussed in previous sections target the dynamic power consumption in caches. As we mentioned earlier, the importance of static power has been growing in recent years, and with smaller geometries, the contribution of static power to the overall power consumption in a chip is expected to grow. Static power is dissipated as long as a voltage is supplied to the circuit, and can be eliminated by turning power supply off (in which case memory data is lost) or reduced by turning the voltage down (in which case data can be retained, but accessing the data requires us to raise the voltage again). A few strategies have been proposed to address the static power dissipation in caches.

An important observation regarding lifetime of cache contents is that the data tends to be accessed frequently in small intervals of time, and is essentially dead for large periods when it is never accessed. This leads to an interesting strategy – can we turn off power to a cache line if we predict that the line will not be accessed in the near future? The cache decay technique (Kaxiras et al., 2001) relies on this approach. A counter is maintained for each line; when it reaches a threshold value, the power to the line is turned off after updating the L2 cache with the new value if necessary. To keep the overhead of maintaining the counters low, there is only one global counter, but a two-bit derived counter is placed in each line to control the power supply. The threshold value of the counter is determined from the values of the static energy dissipated and the energy expended in re-fetching the line from the L2 cache.

An alternative to turning off the power to cache lines is to turn down the voltage so that data is retained, but cannot be accessed directly. In order to access the data, the line would have to be first switched to high voltage (causing a performance overhead). The power saved per line in this manner is smaller than that in the decay technique where the power supply to the line is turned off, but this may permit more lines to be moved into the *drowsy* state (Flautner et al., 2002). The idea is proposed to be used for all lines in the L2 cache, and some lines of the L1 cache. A simple strategy of putting the entire cache to drowsy mode once in a while works quite well, as opposed to introducing active circuitry to maintain counters on a per line basis. A variation on this theme is to use predictive schemes to selectively move lines to and from drowsy state. In Hu et al. (2003), information about cache lines with high temporal locality is maintained in the BTB, and these lines are kept active. When a cache line is accessed, the sequentially following line is moved to active state, anticipating its use. A very fine grain control can be exercised by permitting the compiler to insert instructions to turn individual cache lines off when its use is not anticipated for a long time (Zhang et al., 2002).

2.5   Other Cache Ideas

The observation that data values are frequently duplicated in typical caches has been used to design a non-redundant cache (Molina et al., 2003), where pointers to data are stored in the cache instead of the data itself. If the value is small enough (which does happen frequently in practice), then the space for pointer itself is used to store the value. This saves power as the pointer can use a small number of bits instead of the typical data word size, and hence, its access would consume less power than the conventional cache access.

Other types of correlation observed in data and addresses can also be exploited. One observation is that the source addresses of consecutive load instructions tend to have the same difference – a phenomenon that is observed in memory loads in loops. In correlating caches (Mallik et al., 2004), a load instruction leads to a prefetch of the correlated instruction also. A small correlating buffer acts as the primary cache, reducing access energy.

3. Compiler Optimisations Targeting Memory Power

Compiler optimisations targeting high performance generally also reduce power and energy indirectly. When the optimised performance results in lesser number of instructions executed, it also means a smaller number of accesses to instruction memory. Since memory energy is proportional to the number of accesses, this also reduces the total energy consumed. Along the same lines, optimisations that reduce the number of accesses to data memory also reduce the total memory energy. Thus, for example, all register allocation related optimisations, which increase the efficiency of register usage, are also favourable with respect to power, as it is more power-efficient to access registers instead of memory. This argument also generalises to all levels of the memory hierarchy. Performance optimisations that increase the hit ratio to the L1 cache are also power optimisations, since the L1 cache access dissipates lesser energy than an L2 cache access. The extent of performance improvement due to an optimisation may be different from the extent of power improvement. However, the optimisations are generally in the same direction, and if a memory related optimisation improves performance, then it also reduces power. However, interesting exceptions do exist – good examples being those that rely on speculative memory loads. In such cases, the access latency may be hidden by other CPU activity, but the associated energy dissipated cannot be undone. Such an optimisation improves performance, but reduces energy efficiency.

In this section we study some power optimisation techniques that are applicable at the compiler stage. Making the compiler (or synthesis tool, if custom hardware is generated for an application) explicitly aware of the performance/energy optimised features present in the memory subsystem increases the compilation time, but yields the power benefits without any run-time overhead and without the need of expensive hardware. While most standard compiler optimisations including constant folding and propagation, algebraic simplifications, copy propagation, common subexpression elimination, loop invariant code motion, loop transformations such as pipelining and interchange, etc. (Muchnick, 1997) are also relevant for power reduction, some others that increase the code size (such as loop unrolling and function inlining) need more careful attention. Optimisations such as unrolling and inlining increase the code size, thereby increasing the instruction memory size. Since larger memories are associated with increased access energy, these transformations may actually end up decreasing energy efficiency.

3.1   Data Layout

Decisions regarding organisation of data in memory frequently follow rules specific to a programming language. However, where safety can be guaranteed, an aggressive compiler can determine the data organisation in order to optimise performance or power. In an embedded systems, such a consideration is worth the extra time such analysis may involve.

This flexibility to decide the relative order of program variables can be exploited to improve performance and reduce power by clustering variables into cache-line-sized blocks. If the same set of variables are frequently accessed within a small time-span, their addresses can be clustered so as to map into the same cache line, resulting in temporal locality during execution. This avoids cache possible conflicts. A similar strategy can be applied to array variables. Based on the access patterns during loops and the sizes of the arrays and caches, the possibility of conflicts between the respective array elements can be minimised by adjusting the start positions of the arrays in memory (Panda and Dutt, 1999). The arrays can also be interleaved to bring about a similar effect (Panda and Dutt, 1999; Kulkarni et al., 2000). A generalisation of array interleaving occurs when we have an array of structures/records. Since the access patterns to different fields in the structures can be different (e.g., in one loop, only one of two fields of a structure may be accessed), it may be worthwhile to re-organise the arrays so that the fields are grouped according to the way they are accessed in the program loops, not according to the user declaration (Panda et al., 2001).

Data layout decisions are also applicable in the context of DRAM based designs. Here, instead of a cache line, the cluster size is that of a memory row, since accessing consecutive variables from the same row is more efficient than from different rows both in terms of performance as well as energy. This idea can be generalised to also take advantage of burst mode memory accesses, where it is more efficient to read a sequence of consecutive memory elements than to fetch them successively (Panda et al., 1998; Choi and Kim, 2003).

3.2   Instruction Layout

The layout of instructions in memory can also be controlled in order to improve instruction cache performance and power. While the sequence of instructions within a basic block usually needs to be maintained, the compiler has the flexibility of assigning the start addresses of basic blocks. This can be used to ensure the reduction of cache conflicts if we can predict the interference of instructions in the cache based on the relative positions, cache size, and frequency of execution. The central idea is that if two basic blocks are frequently accessed in an interleaved manner (this can happen if both are in the same loop), then it is a good idea to assign them to contiguous locations in memory. This minimises the possibility of cache conflicts since they are likely to map to different parts of the cache. If the two blocks together fit into the cache, then conflict-free access is guaranteed. Even otherwise, conflicts could be minimised. Early work in this area (McFarling, 1989) was extended in Tomiyama and Yasuura (1997) to include an interprocedural analysis. Traces of program execution are used to determine which basic blocks are executed frequently in succession. In Parameswaran and Henkel (2001), basic blocks are first assigned cache regions so as to ensure that the more frequently executed ones are displaced minimum number of times. The mapping from cache locations to memory locations is carried out in a second phase.

Both data and instruction layout strategies usually incur some memory area overhead – optimising the cache locations of instructions and data implies exact control over the main memory locations, and may lead to unoccupied memory locations. This overhead is generally not large enough to affect the other design metrics such as cost and chip area significantly.

3.3   Scratch Pad Memory Utilisation

A significant role can be played by the compiler when the architecture contains structures scratch pad memory, as these memories are directly managed by the compiler. Compile-time analysis is involved in determining which data and instructions should be mapped to the scratch pad memory. The analysis for mapping data to scratch pad memory presented in Panda et al. (2000) considers the size of the data, the frequency of access, the array lifetimes, and the possibility of cache conflicts between different arrays in the mapping decision. A problem of this nature maps approximately to the well known *Knapsack Problem* in computer science (Garey and Johnson, 1979), where we are given a knapsack of a fixed size, and can fill it with objects of different sizes and profit values. The objective is to select a subset of the objects that fit into the knapsack while maximising the profit. In the scratch pad case, the memory size is the equivalent of the knapsack size, object size corresponds to the array size, and the profit can be measured in terms of the number of accesses to the variables mapped into the scratch pad. This is because access to the scratch pad involves less energy dissipation than a cache access, as the scratch pad is a simple SRAM with no additional tag-related circuitry characterising the cache, and hence, no associated dynamic power dissipation. In terms of performance, the guaranteed "hit" to the scratch pad ensures no cache-miss related delays.

The formulation above can be extended in several directions. In Kandemir et al. (2001), the authors make the data mapping decisions dynamic. Different blocks of data, particularly in the context of the *loop tiling* compiler optimisation, are mapped to the scratch pad memory at different time instances. Since these blocks can even be fragments of an array, this may overcome the difficulty of array size preventing the compiler's ability to map an array to the scratch pad. A similar argument as above can also be made in the context of instructions. Frequently executed instructions or basic blocks of instructions can also be mapped to the scratch pad so as to prevent the energy and performance-related consequences of being evicted from the instruction cache. Power is saved both on account of the elimination of tag storage and access, and both performance and energy improves because of reduced cache misses. In fact, a unified formulation can use the same scratch pad memory to map either instructions or data (Steinke et al., 2002; Janapsatya et al., 2004). In a multiprocessor environment, we can associate a scratch pad memory with each individual processor, and the set of all scratch pads can be connected into a virtually shared scratch pad memory (VS-SPM) (Kandemir et al., 2002), where access to local or remote scratch pads can be fast for all processors, but accesses to other memory can be more expensive. If the scratch pad is to be dynamically managed, an explicit transfer protocol has to be designed for transferring data between it and the main memory. (Francesco et al., 2004) uses a DMA engine to perform these transfers, and presents a framework with an application programming interface to explicitly account for the latency of these transfers.

3.4   Memory Bank Utilisation

The presence of multiple memory banks creates interesting optimisation opportunities for the compiler. Traditionally a few DSP processors used a dual-bank on-chip memory architecture, but in modern systems, banking is used in various contexts for various objectives. In synchronous DRAMs (SDRAMs), banking is used to improve performance by keeping multiple data buffers from different banks ready for data access. In application specific systems, dividing a monolithic memory into several banks leads to considerable performance improvement and power savings. The performance improvement comes from the ability to simultaneously access multiple data words, while the power savings arise from smaller addressing circuitry, word lines, and bit-lines, as observed earlier.

The essential problem for the compiler is to assign data to memory banks in order to minimise certain objective functions. In terms of performance, we would like to be able to simultaneously access data in different banks so that computation time decreases, assuming multiple datapath resources are available. In terms of power dissipation, there is the possibility of moving specific banks to *sleep mode* when there are periods of zero activity. (Lyuh and Kim, 2004) addresses the problem of allocating data to banks and scheduling memory accesses in order to improve energy efficiency in the memory banks. A specific instance of the bank allocation problem is discussed in more detail in Chapter 8.

4. Application Specific Memory Customisation

One of the most important characteristics of an embedded system is that the hardware architecture can be customised for the specific application or set of applications that the system will be used on. This customisation can help improve area, performance, and power efficiency of the implementation for the given application. Even when the overall platform may be fixed, different subsystems can be independently tailored to suit the requirements of the application. The memory subsystem is a fertile ground for such customisation. Cache memory can be organised in many different ways by varying a large number of parameters: cache size, line size, associativity, write policy, replacement policy, etc. Since the number of possibile configurations is huge, an explicitly simulation-based selection strategy is too time-consuming. A large body of research work has focussed on making the memory exploration automatic or semi-automatic.

Finding the ideal cache size for an application involves important trade-offs from the energy perspective. On one hand, larger caches reduce cache misses and hence decrease energy. On the other hand, the per-access energy increases. Therefore, the trade-offs need to be carefully evaluated.

A framework for estimating the effect of different memory parameters such as cache line size, cache size, and scratch pad memory size was presented in Panda et al. (1999a). The estimation process first performs a data partitioning between scratch pad memory and cache, and for data mapped into cache, considers the effect of different line sizes on the array access patterns in loop bodies. The work in Li and Henkel (1998) takes into account the effect of different software optimisations on the instruction cache. Estimates are used for the effect of transformations on system energy using an energy model of caches. The energy impact of the tiling optimisation has been experimentally evaluated in Shiue and Chakrabarti (1999), with the overall conclusion that the optimal configurations are typically different for performance and energy metrics. The need for a whole-program analysis is also emphasised, since the best configuration for the entire program is different from those identified for code kernels within the program. The relationship between cache parameters and the exernal interface parameters such as bus width is studied in Givargis et al. (1999), where the role of the power dissipation due to the CPU–cache bus in newer technologies is emphasised.

Analogous to data, customisation of the memory hierarchy for instructions can also be similarly performed. The loop cache is to instructions what scratch pad memory is to data. A study of loop cache customisation is covered in Cotterel and Vahid (2002). The customisation is based on an estimation of the energy saved by using different loop cache sizes, using information such as the loop sizes, execution frequencies, and whether they fit into the loop cache.

The column caching technique (Chiou et al., 2000) involves a hardware design to dynamically change the associativity of the cache, leading to the possibility to effectively turn a part of the cache into a scratch pad. A programmable bit vector determines which cache ways are restricted in this manner, so that the cache replacement algorithm selectively replaces lines only from the unrestricted ways. In the method suggested by Petrov and Orailoglu (2001), the number of bits used for tag comparisons in the data cache is programmable. It may happen that during typical array accesses in loops, only a small number of least significant bits in the tag may differ – on an average, the number of tag bits needed may be small. Reducing the number of tag bits used correspondingly reduces the energy dissipation during fetches from the data cache tag memory. Based on this observation, an encoding scheme was also proposed in Petrov and Orailoglu (2002) for effective compression the tag bits, with the objective of reducing tag memory power dissipation. In Grun et al. (2001), the authors classify program variables into different types of locality – spatial and temporal, and using this information, map them to be accessed through different (spatial and temporal) cache structures.

Customising the banking structure of data memories can lead to significant improvements in performance and power. One can take advantage of different effective word lengths of variables used in typical programs to have different banks consisting of different widths, thereby reducing power when accessing data from banks with smaller widths (Cao et al., 2002). The power implications of splitting the memory into several different physical modules was addressed in Benini et al. (2000). Frequently accessed address ranges are targeted for smaller memories, and less frequently accessed ranges for larger memories.

In typical cache organisations, the lower order address bits are used to index into the cache. However, in an application specific environment, we have the flexibility to use a different mapping function if it can be shown to reduce cache misses. A trace of memory addresses accessed during program execution can be used to determine such an application specific mapping function in Givargis (2003).

5.   Other Techniques: Dynamic Voltage Scaling, Compression, Encoding, etc.

In this section we review a few other power optimisation techniques targeting the memory system.

5.1   Dynamic Voltage Scaling

Dynamic voltage scaling is an important new strategy that aims at obtaining power reduction through modifying the voltage supply. Since the power dissipation is a quadratic function of the voltage, this can help significantly improve power. The basic idea here is to retain high voltage for those operations that are present in the performance-critical paths, but reduce the voltage for those that are not performance-critical, thereby reducing the overall system power without a negative effect on performance. In Cho and Chang (2004), it is observed that the scaling is very different for the CPU/logic and memory subsystems, and hence, frequency and voltage scaling must take into account the relative effects on both the CPU and memory, in order to achieve system-wide power reduction.

5.2   Power Management in DRAM

Modern DRAMs provide explicit mechanisms to transfer them to low power state, where power dissipation is reduced. When multiple DRAM modules exist, one of the simplest power management strategies is to move the DRAM to low power state when it is unused. The DRAM controller takes the decision to move the DRAM to and from the low power state. There is usually a performance penalty for change of power states; specifically, a number of cycles are wasted when moving a DRAM unit from low power mode to active mode. While one may consider several sophisicated techniques to predict when to change state, the study in Fan et al. (2001) indicates that the relatively simple strategy of always transferring the state to low power after a burst access gives the best results. DRAM power management functions can be performed by many different entities: DRAM controller can use prediction techniques; the compiler can analyse the source code and insert explicit instructions to change DRAM power state; and the operating system's scheduler can also monitor activity on the DRAM and make the management decisions. Since the operating system is able to simultaneously monitor activities of different processes, it may discover patterns not expected by the compiler (Delaluz et al., 2002). The proposal in Huang et al. (2005) is to merge DRAM accesses with short idle periods between them into clusters so as to artificially create longer idle periods; consequently, the memory can be switched to low power mode for longer periods or can be switched to a lower power mode than previously possible.

For synchronous DRAM systems, Joo et al. (2002) analysed the power dissipation in different components and built a power model. Among the optimisations identified was the observation that for reducing the address bus energy, it is not sufficient to only reduce bit transitions as static power is dissipated when the bus is driven. Hence, acceptable alternatives are the tristating of the address buses when it is idle, and setting it to all 1s.

5.3   Encoding

The topic of encoding address and data buses connecting to memory in order to reduce bus switching power has been investigated in detail over the last decade. Since address and data buses may be long, and hence, high capacitance, switching activity on the buses can consume a significant amount of energy.

An initial proposal to reduce bus switching was to add an extra bit to the bus, which would indicate to the receiver/decoder whether or not the word should be bitwise inverted. The choice would be made depending on which option reduced the hamming distance between the current and previous words (Stan and Burleson, 1995).

One way to reduce address bus transitions is to modify the layout of data in memory, when the flexibility exists. For array data in programs, the choice exists between row-major, column-major, and various types of block or tile-baseds storage (Panda and Dutt, 1999). The typical situation of different arrays being accessed in loops leads to different effective working zones during program execution. The proposal in Musoll et al. (1997) was to send the start addresses of the zones once, and in subsequent accesses, to only send the offset within the zone. In the Beach solution (Benini et al., 1998), mechanisms are proposed to exploit blocks of addresses that occur repeatedly in address sequences, particularly to instruction memory. The T0 code (Benini and Micheli, 2000) adds an extra line to the bus, which is used to indicate whether the next address is consecutive and is generated by incrementing the previous one. This scheme may lead to near zero transitions in the steady state when there is a sequence of consecutive addresses; this happens often during data array accesses in loops.

A practical implementation of encoding schemes has to take into account that the bus may be time-multiplexed between instruction and data. Also, sequential addresses on instruction or data buses may not be perfectly sequential, but due to the word size being some multiple of a byte (say 4 bytes per instruction or data word) in a byte-addressable memory, the least significant bits may need to be omitted from the encoding scheme.

For a more detailed survey of low power bus encoding techniques, the reader is referred to Benini and Micheli (2000).

5.4   Compression

Various strategies for compression of data and instructions in memory and cache have been studied in the context of reducing memory traffic and consequently improving performance. The reduction in memory traffic also indirectly reduces power, although we need to take into account the power overhead of the decompression circuit. One way to directly effect instruction compression is to use a small instruction set with narrow instructions, as in the 16-bit Thumb instruction set used in the ARM7TDMI processor (Segars et al., 1995). The narrow instructions increase the memory bandwidth and reduce power dissipation due to instruction fetch.

The instruction compression proposal in Yoshida et al. (1997) is based on the observation that in typical programs, only a small fraction of the complete instruction set of the processor is actually utilised. Even here, there is a significant degree of repetition of the same instructions. These frequently occurring instructions could be compressed into smaller bit patterns. The decompression to generate the original instructions is done through accessing a table indexed by the compressed pattern. A variant proposed by Benini et al. (1999) is to only compress those instructions that are known to be frequently used. This overcomes one problem with the previous approach – if the actual set of used instructions is large, then it leads to an increase in the number of bits to represent the set, a very large decompression table, and the associated power overheads in the decompression. In Chandar et al. (1999), the instruction encoding for compression is permitted to be reconfigurable – the actual codes for individual instructions are modified during different phases of program execution, since frequency of instruction execution may vary in different program phases.

Architecturally, there are two major options of the logical placement of the decompression unit: (1) between the cache and the main memory and (2) between the CPU and the cache. When the decompression unit is between the cache and the main memory, the cache stores the decompressed instructions, and the decompressor is activated only on a cache miss – the associated energy overhead is small as it is not active during cache hits. When the decompression unit is between the CPU and the cache, the effective compression is better as the compressed instructions are stored in the cache, but there is the energy overhead from the decompression unit being activated on every instruction fetch. Lekatsas et al. (2002) and Benini et al. (2001) propose architectures where the decompression unit resides between the CPU and the cache, while in Benini et al. (2002) the unit resides between the cache and main memory. Macii et al. (2003) targets the data cache, and places the compression and decompression units between the data cache and main memory. The idea of using compression can also be applied to data stored in scratch pad memory so that more data could be placed in a given scratch pad space (Ozturk et al., 2004).

Summary

In this chapter we discussed some of the major ideas proposed in recent years on the topic of power optimisations in the memory subsystem. In embedded systems, where the designer has a considerable amount of flexibility to tailor the architectural details to an application, the memory architecture can be significantly optimised to reduce power. Consequently, many different memory organisations have been proposed for instruction and data memory in embedded systems. A correspondingly strong compiler support is needed to suitably utilise the innovative architectures. It also helps to have a powerful archicture exploration and customisation framework in the loop, since it is unlikely that there is a single optimal architecture for a given application – many different candidate architectures need to be evaluated.

In the next two chapters, two of the topics are discussed in greater detail. Chapter 7 elaborates on the topic of assigning data values in programs to specific layers in the memory hierarchy, while Chapter 8 considers the memory banking idea in depth.

References

Bajwa, R.S., Hiraki, M., Kojima, H., Gorny, D.J.., Nitta, K., Shridhar, A., Seki, K., and Sasaki, K. (1997) Instruction buffering to reduce power in processors for signal processing. *IEEE Transactions on VLSI Systems*, 5(4):417–424.

Bellas, N., Hajj, I., and Polychronopoulos, C. (1999) Using dynamic cache management techniques to reduce energy in a high-performance processor. In *International Symposium on Low Power Electronics and Design*, pages 64–69, San Diego, USA.

Bellas, N., Hajj, I.N., Polychronopoulos, C.D., and Stamoulis, G. (2000) Architectural and compiler techniques for energy reduction in high-performance microprocessors. *IEEE Transactions on VLSI Systems*, 8(3): 317–326.

Benini, L. and Micheli, G.De (2000) System level power optimization: Techniques and tools. *ACM Transactions on Design Automation of Electronic Systems*, 5(2):115–192.

Benini, L., de Micheli, G., Macii, E., Poncino, M., and Quer, S. (1998) Power optimization of core-based systems by address bus encoding. *IEEE Transactions on VLSI Systems*, 6(4):554–562.

Benini, L., Macii, A., Macii, E., and Poncino, M. (1999) Selective instruction compression for memory energy reduction in embedded systems. In *International Symposium on Low Power Electronics and Design*, pages 206–211, San Diego, USA.

Benini, L., Macii, A., and Poncino, M. (2000) A recursive algorithm for low-power memory partitioning. In *International Symopsium on Low Power Electronics and Design*, Rapallo, Italy.

Benini, L., Macii, A., and Nannarelli, A. (2001) Cached-code compression for energy minimization in embedded processors. In *International symposium on Low Power Electronics and Design*, pages 322–327, Huntington Beach, USA.

Benini, L., Bruni, D., Macii, A., and Macii, E. (2002) Hardware-assisted data compression for energy minimization in systems with embedded processors. In *Design Automation and Test in Europe*, pages 449–453, Paris, France.

Cao, Y., Tomiyama, H., Okuma, T., and Yasuura, H. (2002) Data memory design considering effective bitwidth for low-energy embedded systems. In *International Symposium on System Synthesis*, pages 201–206, Kyoto, Japan.

Chandar, S.G., Mehendale, M., and Govindarajan, R. (1999) Area and power reduction of embedded dsp systems using instruction compression and re-configurable encoding. In *Proceedings of the IEEE/ACM International Conference on Computer Aided Design*, pages 631–634, San Jose, CA.

Chiou, D., Jain, P., Rudolph, L., and Devadas, S. (2000) Application-specific memory management for embedded systems using software-controlled caches. In *Design Automation Conference*, pages 416–419, Los Angeles, USA.

Cho, Y. and Chang, N. (2004) Memory-aware energy-optimal frequency assignment for dynamic supply voltage scaling. In *International Symposium on Low Power Electronics and Design*, pages 387–392, Newport Beach, USA.

Choi, Y. and Kim, T. (2003) Memory layout techniques for variables utilizing efficient dram access modes in embedded system design. In *Design Automation Conference*, pages 881–886, Anaheim, USA.

Cotterel, S. and Vahid, F. (2002) Synthesis of customized loop caches for core based embedded systems. In *Proceedings of the IEEE International Conference on Computer Aided Design*, pages 665–662, San Jose, USA.

Delaluz, V., Sivasubramaniam, A., Kandemir, M., Vijaykrishnan, N., and Irwin, M.J. (2002) Scheduler-based dram energy management. In *Design Automation Conference*, New Orleans, USA.

Fan, X., Ellis, C., and Lebeck, A. (2001) Memory controller policies for dram power management. In *International Symposium on Low Power Electronics and Design*, pages 129–134, Huntington Beach, USA.

Flautner, K., Kim, N.S., Martin, S., Blaauw, D., and Mudge, T. (2002) Drowsy caches: Simple techniques for reducing leakage power. In *International Symposium on Computer Architecture*, pages 240–251, Anchorage, USA.

Francesco, P., Marchal, P., Atienza, D., Benini, L., Catthoor, F., and Mendias, J.M. (2004) An integrated hardware/software approach for run-time scratch-pad management. In *Design Automation Conference*, pages 238–243, San Diego, USA.

Garey, M.R. and Johnson, D.S. (1979) *Computers and Intractibility – A Guide to the Theory of NP-Completeness*. W.H. Freeman.

Ghose, K. and Kamble, M.B. (1999) Reducing power in superscalar processor caches using subbanking, multiple line buffers and bit-line segmentation.

In *International Symposium on Low Power Electronics and Design*, pages 70–75, San Diego, USA.

Givargis, T. (2003) Improved indexing for cache miss reduction in embedded systems. In *Design Automation Conference*, pages 875–880, Anaheim, USA.

Givargis, T., Henkel, J., and Vahid, F. (1999) Interface and cache power exploration for core-based embedded system design. In *International Conference on Computer Aided Design*, pages 270–273, San Jose, USA.

Grun, P., Dutt, N., and Nicolau, A. (2001) Access pattern based local memory customization for low power embedded systems. In *Design Automation and Test in Europe*, pages 778–784, Munich, Germany.

Hasegawa, A., Kawasaki, I., Yamada, K., Yoshioka, S., Kawasaki, S., and Biswas, P. (1995) SH3: High code density, low power. *IEEE Micro*, 15(6):11–19.

Hennessy, J.L. and Patterson, D.A. (1994) *Computer Architecture – A Quantitative Approach*. Morgan Kaufman, San Francisco, CA.

Hu, J.S., Nadgir, A., Vijaykrishnan, N., Irwin, M.J., and Kandemir, M. (2003) Exploiting program hotspots and code sequentiality for instruction cache leakage management. In *International Symposium on Low Power Electronics and Design*, pages 402–407, Seoul, Korea.

Huang, H., Shin, K.G., Lefurgy, C., and Keller, T. (2005) Improving energy efficiency by making dram less randomly accessed. In *International Symposium on Low Power Electronics and Design*, pages 393–398, San Diego, USA.

Inoue, K., Ishihara, T., and Murakami, K. (1999) Way-predicting set-associative cache for high performance and low energy consumption. In *International Symposium on Low Power Electronics and Design*, pages 273–275, San Diego, USA.

Inoue, K., Moshnyaga, V.G., and Murakami, K. (2002) A history-based I-cache for low-energy multimedia applications. In *International Symposium on Low Power Electronics and Design*, pages 148–153, Monterey, USA.

Ishihara, T. and Fallah, F. (2005) A non-uniform cache architecture for low power system design. In *International Symposium on Low Power Electronics and Design*, pages 363–368, San Diego, USA.

Janapsatya, A., Parameswaran, S., and Ignjatovic, A. (2004) Hardware/software managed scratchpad memory for embedded systems. In *Proceedings of the IEEE/ACM International Conference on Computer Aided Design*, pages 370–377, San Jose, USA.

Joo, Y., Choi, Y., Shim, H., Lee, H.G., Kim, K., and Chang, N. (2002) Energy exploration and reduction of sdram memory systems. In *Design Automation Conference*, pages 892–897, New Orleans, USA.

Kandemir, M., Ramanujam, J., Irwin, M.J., Vijaykrishnan, N., Kadayif, I., and Parikh, A. (2001) Dynamic management of scratch-pad memory space. In *ACM/IEEE Design Automation Conference*, pages 690–695, Los Vegas, USA.

Kandemir, M., Ramanujam, J., and Choudhary, A. (2002) Exploiting shared scratch pad memory space in embedded multiprocessor systems. In *ACM/IEEE Design Automation Conference*, pages 219–224, New Orleans, USA.

Kaxiras, S., Hu, Z., and Martonosi, M. (2001) Cache decay: Exploiting generational behavior to reduce cache leakage power. In *International Symposium on Computer Architecture*, pages 240–251, Goteberg, Sweden.

Kim, S., Vijaykrishnan, N., Kandemir, M., Sivasubramaniam, A., Irwin, M.J., and Geethanjali, E. (2001) Power-aware partitioned cache architectures. In *International Symposium on Low Power Electronics and Design*, pages 64–67, Huntington Beach, USA.

Kin, J., Gupta, M., and Mangione-Smith, W.H. (1997) The filter cache: an energy efficient memory structure. In *International Symposium on Microarchitecture*, pages 184–193, Research Triangle Park, USA.

Ko, U., Balsara, P.T., and Nanda, A.K. (1995) Energy optimization of multi-level processor cache architectures. In *International Symposium on Low Power Design*, pages 45–49, New York, USA.

Kulkarni, C., Catthoor, F., and Man, H.De (2000) Advanced data layout organization for multi-media applications. In *Proceedings Workshop on Parallel and Distributed Computing in Image Processing, Video Processing, and Multimedia (PDIVM'2000)*, Cancun, Mexico.

Lekatsas, H., Henkel, J., and Wolf, W. (2002) Code compression for low power embedded system design. In *Design Automation Conference*, pages 294–299, Los Angeles, USA.

Li, Y. and Henkel, J. (1998) A framework for estimating and minimizing energy dissipation of embedded hw/sw systems. In *Design Automation Conference*, pages 188–193, San Francisco, USA.

Lyuh, C.-G. and Kim, T. (2004) Memory access scheduling and binding considering energy minimization in multi-bank memory systems. In *Design Automation Conference*, pages 81–86, San Diego, USA.

Macii, A., Macii, E., Crudo, F., and Zafalon, R. (2003) A new algorithm for energy-driven data compression in vliw embedded processors. In *Design Automation and Test in Europe*, pages 1024–1029, Munich, Germany.

Mallik, A., Wildrick, M.C., and Memik, G. (2004) Design and implementation of correlating caches. In *International Symposium on Low Power Electronics and Design*, pages 58–61, Newport Beach, USA.

Malik, A., Moyer, B., and Cermak, D. (2000) A low power unified cache architecture providing power and performance flexibility. In *International Symposium on Low Power Electronics and Design*, pages 241–243, Rapallo, Italy.

Mamidipaka, M. and Dutt, N. (2003) On-chip stack based memory organization for low power embedded architectures. In *Design Automation and Test in Europe*, pages 1082–1089, Munich, Germany.

McFarling, S. (1989) Program optimization for instruction caches. In *Third International Conference on Architectural Support for Programming Languages and Operating Systems*, pages 183–191, Boston, MA.

Min, R., Jone, W.-B., and Hu, Y. (2004) Location cache: A low-power l2 cache system. In *International Symposium on Low Power Electronics and Design*, pages 120–125, Newport Beach, USA.

Molina, C., Aliagas, C., Garcia, M., Gonzalez, A., and Tubella, J. (2003) Non redundant data cache. In *International Symposium on Low Power Electronics and Design*, pages 274–277, Seoul, Korea.

Muchnick, S. (1997) *Advanced Compiler Design and Implementation*. Morgan Kaufman, San Francisco, CA.

Musoll, E., Lang, T., and Cortadella, J. (1997) Exploiting the locality of memory references to reduce the address bus energy. In *International Symposium on Low Power Electronics and Design*, pages 202–207, Monterey, CA.

Ozturk, O., Kandemir, M., Demirkiran, I., Chen, G., and Irwin, M.J. (2004) Data compression for improving spm behavior. In *Design Automation Conference*, pages 401–406, San Diego, USA.

Panda, P.R. and Dutt, N.D. (1999) Low-power memory mapping through reducing address bus activity. *IEEE Transactions on VLSI Systems*, 7(3): 309–320.

Panda, P.R., Dutt, N.D., and Nicolau, A. (1998) Incorporating DRAM access modes into high-level synthesis. *IEEE Transactions on Computer Aided Design*, 17(2):96–109.

Panda, P.R., Dutt, N.D., and Nicolau, A. (1999a) Local memory exploration and optimization in embedded systems. *IEEE Transactions on Computer Aided Design*, 18(1):3–13.

Panda, P.R., Dutt, N.D., and Nicolau, A. (1999b) *Memory Issues in Embedded Systems-On-Chip: Optimizations and Exploration*. Kluwer Academic Publishers, Norwell, MA.

Panda, P.R., Dutt, N.D., and Nicolau, A. (2000) On-chip vs. off-chip memory: The data partitioning problem in embedded processor-based systems. *ACM Transactions on Design Automation of Electronic Systems*, 5(3):682–704.

Panda, P.R., Semeria, L., and de Micheli, G. (2001) Cache-efficient memory layout of aggregate data structures. In *International Symposium on System Synthesis*, Montreal, Canada.

Panwar, R. and Rennels, D. (1995) Reducing the frequency of tag compares for low power i-cache design. In *International Symposium on Low Power Design*, pages 57–62, New York, USA.

Parameswaran, S. and Henkel, J. (2001) I-copes: Fast instruction code placement for embedded systems to improve performance and energy efficiency. In *Proceedings of the IEEE/ACM International Conference on Computer Aided Design*, pages 635–641.

Petrov, P. and Orailoglu, A. (2001) Data cache energy minimization through programmable tag size matching to the applications. In *International Symposium on System Synthesis*, pages 113–117, Montreal, Canada.

Petrov, P. and Orailoglu, A. (2002) Low-power data memory communication for application-specific embedded processors. In *International Symposium on System Synthesis*, pages 219–224, Kyoto, Japan.

Segars, S., Clarke, K., and Goudge, L. (1995) Embedded control problems, thumb, and the arm7tdmi. *IEEE Micro*, 15(5):20–30.

Shiue, W.-T. and Chakrabarti, C. (1999) Memory exploration for low power embedded systems. In *Design Automation Conference*, pages 140–145.

Stan, M.R. and Burleson, W.P. (1995) Bus-invert coding for low power I/O. *IEEE Transactions on VLSI Systems*, 3(1):49–58.

Steinke, S., Wehmeyer, L., Lee, B., and Marwedel, P. (2002) Assigning program and data objects to scratchpad for energy reduction. In *Design Automation and Test in Europe*, pages 409–417, Paris, France.

Su, C.-L. and Despain, A.M. (1995) Cache design trade-offs for power and performance optimization: A case study. In *International Symposium on Low Power Design*, pages 63–68, New York, NY.

Tang, W., Gupta, R., and Nicolau, A. (2002) Power savings in embedded processors through decode filter cache. In *Design Automation and Test in Europe*, pages 443–448, Paris, France.

Tomiyama, H. and Yasuura, H. (1997) Code placement techniques for cache miss rate reduction. *ACM Transactions on Design Automation of Electronic Systems*, 2(4):410–429.

Yang, C.-L. and Lee, C.-H. (2004) Hotspot cache: Joint temporal and spatial locality exploitation for I-cache energy reduction. In *International Symposium on Low Power Electronics and Design*, pages 114–119, Newport Beach, USA.

Yoshida, Y., Song, B.-Y., Okuhata, H., Onoye, T., and Shirakawa, I. (1997) An object code compression approach to embedded processors. In *International Symposium on Low Power Electronics and Design*, pages 265–268, Monterey, USA.

Zhang, C., Vahid, F., Yang, J., and Najjar, W. (2004) A way-halting cache for low-energy high-performance systems. In *International Symposium on Low Power Electronics and Design*, pages 126–131, Newport Beach, USA.

Zhang, W., Hu, J.S., Degalahal, V., Kandemir, M., Vijaykrishnan, N., and Irwin, M.J. (2002) Compiler directed instruction cache leakage optimization. In *International Symposium on Microarchitecture*, pages 208–218, Istanbul, Turkey.

Chapter 7

Layer Assignment Techniques for Low Energy in Multi-Layered Memory Organizations

Erik Brockmeyer[1], Bart Durinck[2], Henk Corporaal[3], and Francky Catthoor[4]

[1] *IMEC vzw*
Belgium

[2] *IMEC vzw*
Belgium

[3] *Technische University Eindhoven*
The Netherlands

[4] *IMEC vzw*
Belgium
Also professor at the Katholieke University Leuven
Belgium

Abstract   Nearly all platforms use a multi-layer memory hierarchy to bridge the enormous latency gap between the large off-chip memories and local register files. However, most of previous work on HW or SW controlled techniques for layer assignment have been mainly focused on performance. As a result, the intermediate layers have been assigned too large sizes leading to energy inefficiency. In this chapter we present a technique that takes advantage of both the temporal locality and limited lifetime of the arrays of the application for trading performance and energy consumption under layer size constraints. These trade-off points are the so-called *Pareto points*, which represent solutions which are not only the optimal points in energy or time, but also intermediate points in a way that it is not possible to gain energy without loosing time or vice versa. A prototype tool has been developed and tested using two real life applications of industrial relevance. Following this approach we have been able to half the energy consumed by the data memory hierarchy for each of our drivers.

Keywords: memory hierarchy; low power; scratch pad memory management.

J. Henkel and S. Parameswaran (eds.), Designing Embedded Processors – A Low Power Perspective,
157–190.
© 2007 *Springer.*

1.   Introduction

In general purpose computing memory hierarchies are the standard technique to bridge the growing gap in performance between the processor and the main memory. The evolution of the access time for bulk DRAM memory has not kept up with the clock speed of the processor (7%/year versus 60%/year). The GHz race of the late 1990s and early 2000s has resulted in processors needing hundreds to thousands of clock cycles for a random off-chip DRAM access.

This performance hurting effect is mitigated by introducing several (up to three) layers of on-chip SRAM based cache memories that exploit the spatial and temporal locality of the memory accesses found in all applications in less or greater extent. Dedicated hardware resources on the processor chip try to store copies of the most used data as close as possible to the processing unit, in smaller layers of memory, costing less time to access.

The existence of the cache architecture is largely abstracted away from the view of the programmer who can write portable and reusable source code in a flat memory space heaven. Only the designers of the chip and the cache architecture have to go through the hell of finding a one-size-fits-all solution with a good performance versus a low cost in chip area, design complexity and power budget.

In 2004 the limits of what is feasible are reached because of several reasons. One is that the marginal gain is low: doubling the size from 1 to 2 MB of cache memory results in more than half of the chip area being used for cache (chip area translates directly to manufacturing cost), but only gives a 0%–20% performance improvement. Another is that the required power is reaching the limit of what can be cooled by air flow (around 100 W).

In embedded systems and especially the booming market of battery powered consumer devices, the power and energy limitation are more severe because users expect to be able to use their pda, smart phone, digital camera, music player, ... for hours, days and even weeks without recharging. The battery peak power is maximum a few W and the total energy budget a few tens of Wh. This budget is for the whole system, where parts like screen, sound, wireless interface, ... take a big piece leaving even less for the digital processing.

There is also the trend of more and more complex tasks like high-quality sound, video and 3D graphics (multi-media), resulting in higher demands on the digital processing. Dedicated hardware often does not provide the required flexibility which is especially important due to the scaled technology costs. So as a result large volume platforms with instruction-set processors (ISPs) and embedded software are becoming more and more established. Different processor low-power micro-architectures address different needs: RISC based architectures (ARM, MIPS) are often chosen for control tasks, where digital signal processors (DSPs) are more suited for bulk data processing. Hybrid solutions are popular, combining different processing units (RISC + DSP), with a micro-architecture that has a hybrid feature set: RISC processors with extended media instructions (XScale) or DSPs with an architecture optimized for control tasks (BlackFin).

However diverse the processors used in embedded systems, they all need significant amounts of (off-chip) memory to store the (multi-media) data. The typical clock speeds are in a few hundreds of MHz, with the 1 GHz mark reached by the high-end TI TMS320C64 DSP in 2004. The gap between processor and memory is also widening, but the hardware resources available to tackle this problem are far more constrained than in general purpose computing. Up to two levels of on-chip cache can be present in embedded processors, but they come at a cost. Especially energy consumption is identified as the main bottleneck, where more than 50% of the energy of the processor is going to the on-chip caches, and that is even without the energy going to off-chip SDRAM. Figure 7.1 illustrates this point for a TI TMS320C64 DSP [Texas Instruments, 2003]. Also power analysis for a RISC processor like the ARM9 shows similar results [SimpleScalar, 2002].

Quite some embedded processors have on-chip scratchpad memories, which can be used to bridge the memory-processor gap in a more efficient way. The control of data placement and replacement, which is a fixed hardware policy in case of a cache, is placed in the hands of the software developer. Using application knowledge, a custom policy can be implemented in software, putting often used data in scratchpad memory close to the processor and using software-controlled transfers to and from further layers to maximize the use of the closest layers (both faster and more energy-efficient).

This can result in significant gains in performance and energy, because of two reasons: knowledge on the future can be exploited and a scratchpad

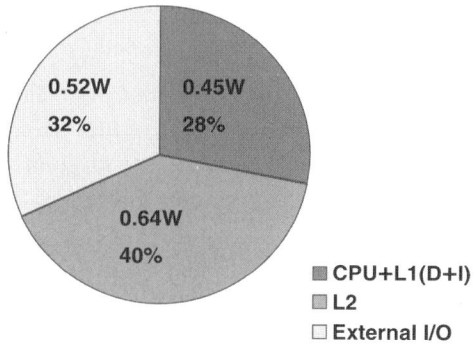

Fig. 7.1   Power consumption for a 130 nm TI TMS320C64 DSP

memory of the same size as a cache is faster and cheaper in energy. Actually the size of the scratchpad will be smaller than the cache to fit the same data, resulting in smaller memories with less static energy.

Scratchpad memory exploitation comes at the cost of the time and effort of the software developer and it breaks the programming abstraction of a flat memory space, hurting software platform independence and reuse. For embedded application this effort can be acceptable, because only a limited amount of software kernels dominate the execution. Still, finding optimal data assignments to (scratchpad) memories in terms of performance and energy is not a trivial task as the search space of the possibilities of which data to (re)place, where and at what time, is huge.

This software controlled placement of data in a multi-layer memory hierarchy is the problem that the proposed memory hierarchy layer assignment (MHLA) techniques tackles at design-time. The software developer is assisted by design-time software tools. Application source code is automatically analyzed and information on the spacial and temporal locality of the data is derived (data reuse analysis). Using cost models for performance and energy of the processor, the memories and the data transfers a high-level estimation of execution time and energy consumption can be made automatically for different possible solutions. An optimal assignment of data to layers in terms of time or energy can be found in the big search space using greedy heuristics.

These techniques are also applicable on platforms with caches. The analysis of the locality can provide hints for the application data layout to conform with the cache structure. Also on some architectures explicit hints like prefetch instructions can be passed to the hardware. In addition scratchpad memories can be combined with caches, providing a hybrid solution: the MHLA techniques are applied to exploit the scratchpad, but parts of the software that are not analyzable or where the effort compared to the gain is too high, are handled by a cache. This shows that the application domain is bigger than scratchpad based memory organizations alone.

2.   Basic Problem Definition

Existing platforms nearly always have more than one layer in their memory subsystem. These layers are inserted to bridge the enormous performance, latency and energy consumption gap between the large off-chip memories and the processor. Memory hierarchy layers can contain normal (software controlled) memories or caches.

An application has to be mapped efficiently on this memory hierarchy. Often this requires that smaller copies are made from larger data arrays which can be stored in the smaller layers (Achteren et al., 2002; Issenin et al., 2004). Those copies must be selected such that they minimize the miss cost of all the layers globally. Any transfer of data from a higher layer to the current one is considered to be a miss for the current layer. This happens most efficiently under software control because a compiler can take a global view. In the case of local memories, copy operations should be explicitly present in the code. However, in the case of hardware controlled caches, the cache controller will make the copies of signals at the moment they are accessed (and the copy is not present in the cache yet). So the code must be written such that the controller is forced to make the right decision (Kulkarni, 2001; van Meeuwen, 2002).

Memory hierarchy layer assignment (MHLA) will take advantage of temporal locality and limited lifetime of the arrays in order to optimize the energy consumption and performance within the resource constraints while taking into account the copy overhead. In current designs, the intermediate layers are not used efficiently and can be made factors smaller consuming less energy while maintaining an equally small miss rate and meeting the performance requirements. Infact, a trade-off exists between speed and energy consumption.

A power, area and performance trade-off exists because some copies are good for energy while others are good for performance and only a limited set can be selected within the given memory resources. Therefore a fast high-level power, time and size estimation must be made to fit as many as possible data objects into the local layers that optimize these criteria. These costs are also the real criteria in which the designer is interested.

We will not indicate the basic steps and illustrate them on a small example.

2.1   Data Reuse Analysis

By exploiting data reuse, a part of an array is copied from one layer to the lower layer from where it is read multiple times. As a result, energy can be saved since most accesses take place on the smaller copy and not on the large more energy consuming original array.

Many different opportunities exist for making a data reuse copy. These are called copy candidates (CCs). Only when it has been decided to instantiate a CC we call it a copy. A relation exists between the size of a CC and the number of transfers from the higher layer, typically called misses (see Figure 7.2). This figure shows a loop nest with one reference to an array A with size 250. The array has 10,000 accesses. Several CCs for array A are possible. For example we could add a copy A'' of size 10 which is made in front of the k-loop by adding the statement "for $(z = 0; z < 10; z++)$ $A2[z] = A[j*10+z];$". This copy[1] statement is executed 100 times, resulting in 1000 misses to the A array. This CC point is shown in Figure 7.2b. Note that the good spatial locality in this example does not influence the amount of misses to the next level.

---

[1] Though the copy candidates will be instantiated as arrays in the application we reserve the name array for the "original array".

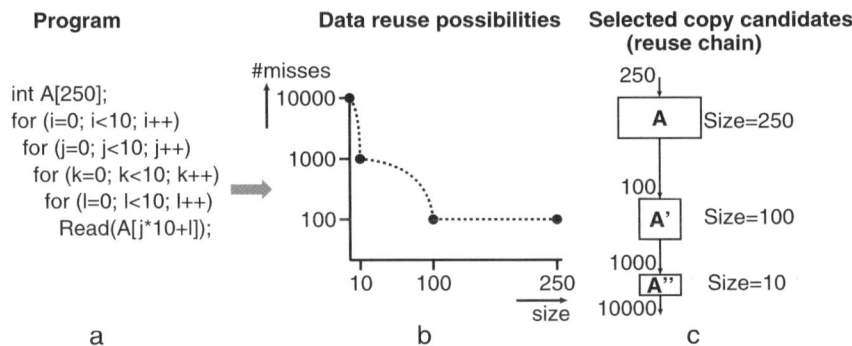

Fig. 7.2   Data reuse information

In theory any CC size ranging from one element to the full array size is a potential candidate. However, in practice only a limited set of CCs leads to efficient solutions. Obviously, a larger CC can retain the data longer and can therefore avoid more misses. The relation between misses and copy size for all possible CCs are shown in Figure 7.2b. The most promising CC sizes and miss counts are kept and put into a data reuse chain as shown in Figure 7.2c. These are exactly those that have a relation to the loop bounds. This data reuse chain is completed with the 250 writes to the array. The above example considers only a single array with one reference. In practice multiple arrays exist, each with one or more references. To each read reference corresponds a reuse chain. These chains are combined in a reuse tree. For example, the upper left of Figure 7.3 shows two data reuse trees for the two arrays (A and B from which array A has two references). Indeed, the second reference of A has no promising CC. More details on identification of data reuse chains and trees can be found in Catthoor et al. (1998), Achteren et al. (2002), Kandemir and Choudhary (2002), Issenin et al. (2004).

2.2   Memory Hierarchy Assignment

In the next step, CCs and arrays are mapped to a data memory hierarchy. We consider a generic target platform. It contains multiple memory layers L_i, where each layer contains multiple memory modules. All memory modules within a layer are of the same type but can have different sizes and number of ports. Typical types are software controlled SRAM or DRAM (typically called scratchpad memories), off-chip (S)DRAM and caches.

Now we are capable of defining the MHLA problem: MHLA = selecting a set of CCs and assigning them together with the arrays to the memory layers of

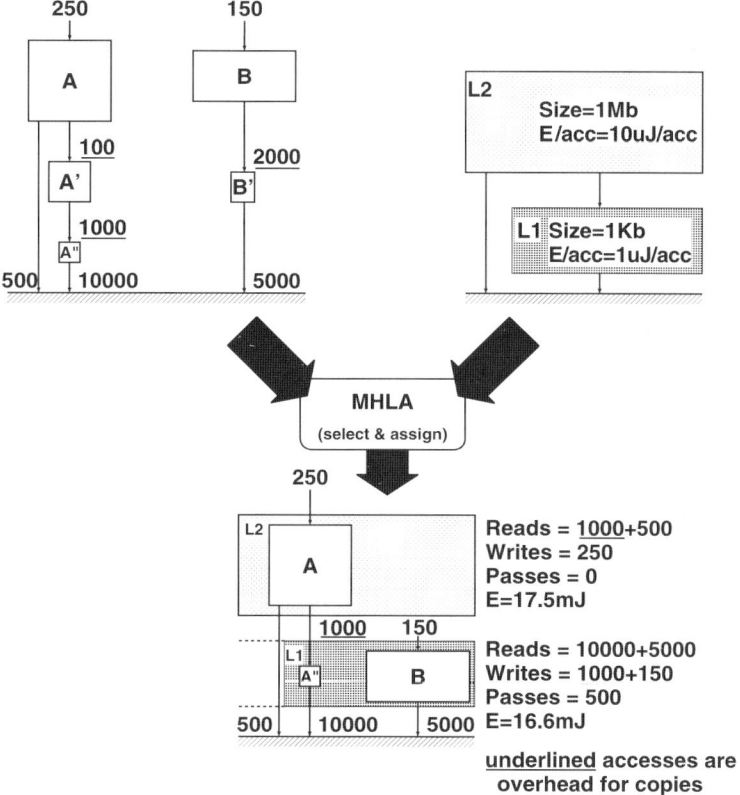

Fig. 7.3 MHLA problem definition

the platform, to find optimal energy consumption and execution time or trade-off points between them.

MHLA determines the energy consumption of a mapping by calculating the activity of the individual partitions and using the energy consumption per access. This cost is a function of size and other memory parameters and is modeled in a memory library. The MHLA process and its mapping result are depicted in Figure 7.3. MHLA has selected the A'' and B to be stored in L1 and the A array in L2. As a result 250 writes occur on L2 for A, 500 reads for the first access and 1,000 misses for A''. The L1 layer has 150 writes for the B array, 1,000 writes due to the misses of A'' and 15,000 reads for both A'' and B. Note that the 500 accesses of the first A reference do not affect the activity of the L1 for this architecture. This may not be the case for hardware controlled caches. Because all accesses have to pass through the cache when no bypass is foreseen. Also note that for caches no explicit copies are introduced in the code. However, the cache controller can be enforced to make the desired copy by a proper memory layout (Kulkarni, 2001).

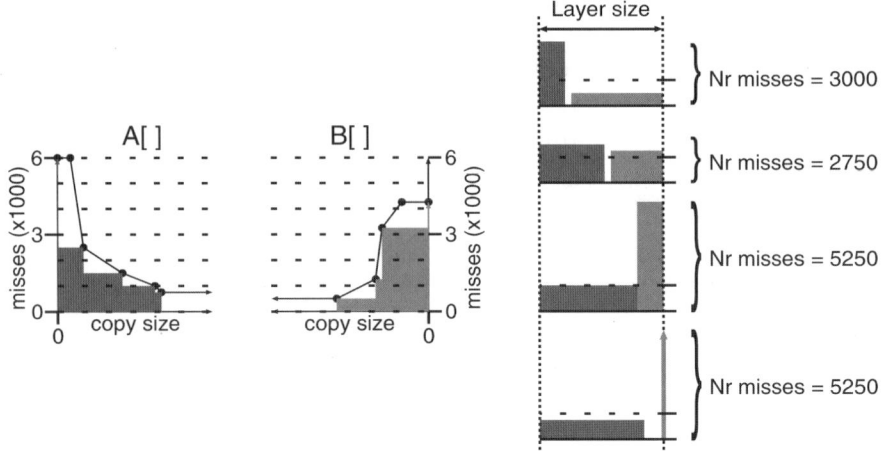

Fig. 7.4 Trading of the copy candidate size

The mapping of arrays and CCs must be performed globally in order mini-mize the energy consumption. The size of one copy must be traded for the size of another copy because layer size is constraint or must be kept low for energy reduction. A simple illustrative example of the data reuse trade-off having mul-tiple references is given in Figure 7.4. The two left hand side curves show the miss count versus copy size trade-off for two access references in an appli-cation. The access to array A has a maximum of 6000 misses which can be reduced below 1000 misses when the largest interesting CC is selected. Sim-ilarly, the array B has about 4000 misses without CC and 500 for the largest CC. The number of misses is minimal when selecting the largest CC for both. However, this assignment leads to an infeasible implementation as the total size does not obey the layer size constraint given at the right hand side of Figure 7.4. The column at the right hand side of this figure shows the feasible combination of CCs in this layer. The upper solution combines the largest CC of B and the largest feasible CC of A. The total number of 3000 cache misses is the sum of the individual CCs. While constructing the other solutions, some space of the B copy candidate is traded for the A copy candidate. As a result the number of misses for A reduces and the number of misses of B increases. In this simple example it is easy to determine the optimal solution having 2750 misses. In general however realistic applications have too many possibilities to still find the best solutions manually. Certainly when trading the selection and assign-ment of CCs together with the assignment of arrays to a layer. This is needed in order to avoid unnecessary copying from one layer to the next. Additionally the problem must be considered over the memory hierarchy globally because misses of one layer can be traded for misses on another layer. So an automated technique and tool is needed.

2.3 Power, Area and Time Trade-off

In the same way as in the previous subsection less misses to a bigger memory layer lead to less energy, less misses also lead to less execution time. This is because smaller memories require less energy per access as well as less time per access.

There still is a subtle difference in the energy/size and energy/time trade-off, because execution time can be gained by fetching the data early enough so that the processor does not have to stall when it is really needed. So copies without reuse can be used to win execution time at the cost of memory space. This concept is called prefetching, it does not reduce the number of misses as such, so no energy is won as a direct consequence.

In case when fetching data from SDRAM there is another issue which has to be taken into account for execution time: SDRAM page accesses. SDRAM has the concept of an open page, which contains the current row out of the 2D memory cell matrix. Contemporary SDRAMs have page sizes of 512 bytes ranging to a few KBs. Consecutive accesses to the same page are a lot faster than accesses that need to open a new page. They can also be pipelined, where the request of the next element overlaps with returning the value of the current element.

On some architectures the decoupling of data transfers from processing is done by a Direct Memory Access (DMA) controller, making prefetching possible and allowing for the grouping of the accesses to use the SDRAM page in an efficient way (see Figure 7.5).

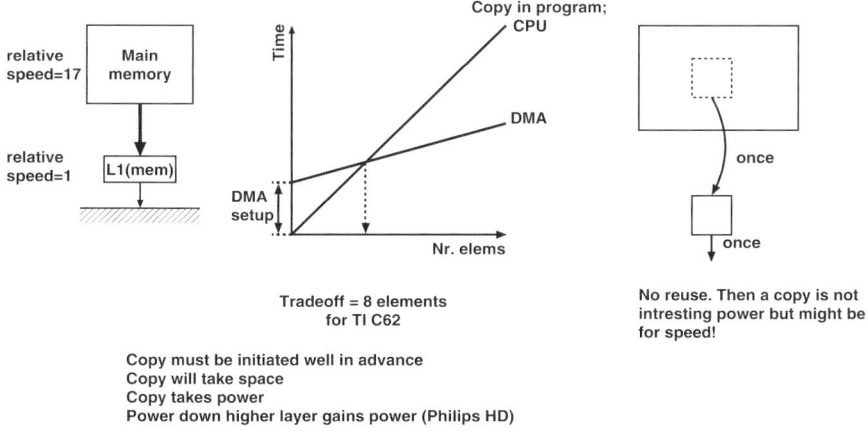

Fig. 7.5 MHLA time trade-offs

All this can cause a trade-off in the assignment of CCs. A CC with a *shape* "going with the direction" of the SDRAM page will result in faster transfer. Other CCs going against the page can be factors slower.

Because there is both a trade-off of size versus energy and size versus time, for a fixed size, there is a trade-off between time and energy. Prefetching and SDRAM page effects have a different magnitude of effect on time and energy, so that some CCs have more effect per size for the energy or time. This means that the assignment optimal for time is not the same as the assignment optimal for energy.

Figure 7.5 illustrates that if a copy exploits no reuse and does not save energy through reducing the number of accesses to a further memory,[2] it can still save time by prefetching the data well in advance. Because of the extra size needed to store the prefetched data, there will be an associated energy cost.

2.4   Overall Methodology for MHLA

The information flow in the implementation of the MHLA tool is depicted in Figure 7.6. At the basis is a "C in C out" flow, where the source code coming from previous steps in the DTSE steps goes in, and transformed C goes out. The primary result of MHLA is an assignment of arrays and CCs to the layers of memory. The transformed C output implements the selected copies and their transfers.

Data reuse analysis can be performed analytically on the source code. It requires information about certain aspects of the code: loops, iterator bounds,

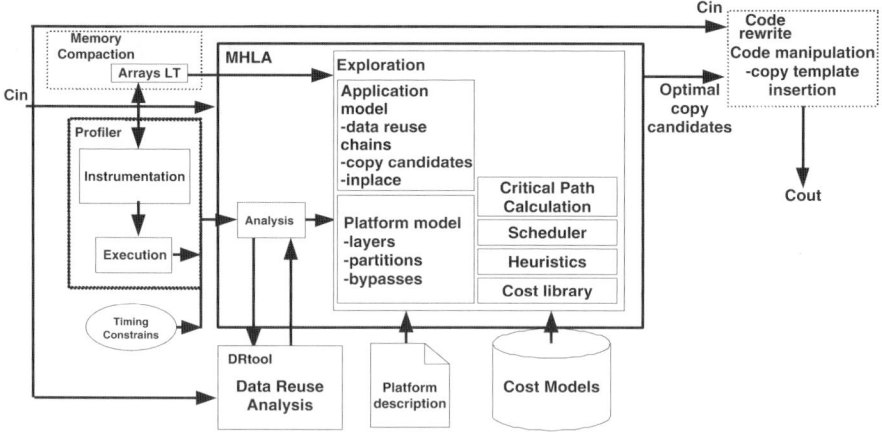

Fig. 7.6   MHLA flow overview

---

[2] As a secondary effect energy will be won, because a faster execution will save on static energy.

dimensions of arrays, array index expressions. The analysis will result in an analytical representation of the reuse possibilities: the copy candidates.

Additionally MHLA gets profiling information on the program: execution counts of basic blocks of code, number of accesses to arrays, ... combined with the analytical reuse information, number of accesses and misses of copies and arrays can be derived. Note that this hybrid analytical versus dynamic profiling is a hybrid technique. A subset of the application can be extracted and a powerful analytical approach can be performed on it, whereas the other parts of the application can still be characterized using the profiling information. Combining the information is even more powerful.

In the exploration loop, different array and CC assignments to memory layers are tried out. For this it needs a description of the platform and more specifically about the memory hierarchy: the size and interconnection of the layers. A certain assignment results in a cost estimate (where cost is energy consumption or execution time), based on given cost models. These energy and time estimates are part of the result of an MHLA exploration; it is not only the assignment and C out.

Constraints can also be imposed on the assignment: e.g. certain arrays can be disallowed to be put in on-chip memory because of I/O requirements. MHLA can take this pre-assignment information as input.

The exploration models more aspects of the application: it also takes into account the limited lifetime of arrays and CCs and tries to estimate the inplace opportunities (a.k.a. "location reuse", it is an estimate of the data layout that ATOMIUM/MEMORY COMPACTION [The Atomium club, 2004 performs]. See Section 4 for details.

Because the assignment search space can easily grow too big for realistically sized applications to explore exhaustively, the exploration needs a heuristic steering to find optimal solutions faster. The heuristics is tuned to find first energy optimal, execution time optimal or Pareto points between them.

So the result of MHLA is more than one solution, possibly multiple Pareto optimal, representing different points in the energy/time trade-off space.

3. Data Reuse Analysis

An introduction to the concepts of data reuse is given earlier in Section 2.1 and illustrated by Figure 7.2. This section will go deeper into the concepts that are needed to understand the MHLA technique, without going into the mathematical details that are available in Durinck et al. (2005).

It is important to note that the copy candidates can be derived from the given information in the source code, one does not need to run the program to see which data is reused. This is a requirement if code implementing the copies has to be generated. The array dimensions, the loop structure and bounds and the access index expression fully determine the access pattern, and it hence can be analytically described. The model focuses on array accesses with affine index expressions in term of iterators.

Data reuse analysis reasons on a *geometrical* level on the data and its possible reuse. It does not consider individual array elements, but looks at geometrical shapes covered by a loop iteration and the change to the next iteration. Our DR implementation uses *bounding boxes* as it basic shapes, which simplifies analysis to upper and lower bound analysis in each dimension of the array. For 1D arrays the CCs are *lines*, for 2D arrays the CCs are *rectangles*, for 3D arrays 3D hypercubes. The hypercube approximation will not always result in the theoretical smallest sized CCs, there might be holes or corners in the bounding box that do not have reuse or are even not used at all. But the regularity and simplicity of making and accessing these nicely shared copies will in most practical cases outweigh the "waste of space".

3.1 Copy Candidates

We start with a small motion estimation (ME) example (see Figure 7.7) that is easy to understand, and is capable to show the basic features of DR. A motion estimation (ME) is based on the correlation between successive frames. Therefore, the motion estimation searches for the blocks that have the biggest similarity. A block in the current frame is matched with a block in the previous frame and a motion vector (MV) points to this matching block. The ME example used here works on blocks of size 16×16 and a search windows of ± 7 pixels.

Figure 7.7 gives conceptual information on the results of the DR step for the frame_prev: copy candidates are in a hierarchy corresponding to the loop nesting level, forming per array a tree with the part of the program tree that has accesses to this array. This tree is called the *reuse tree*, with at the root the array and the leafs are the accesses. The first child of the root corresponds to the outermost loop in the hierarchy: the y-loop (note the dotted arrow connects them). Going down the copy candidate tree corresponding to deeper nested loops, note that the sizes get smaller going towards the access, whereas the number of misses gets bigger.

3.2 Block Transfers

A *block transfer* (BT) is a concept related to copy candidates and it is important to understand this relation. Whereas a copy stands for storage, the block transfer represents data movement. A copy candidate has one or more block transfers which fill the copy candidate with data. For instance the prev_x CC of Figure 7.7 has two BTs. The initial block transfer (BTinit) copies the data

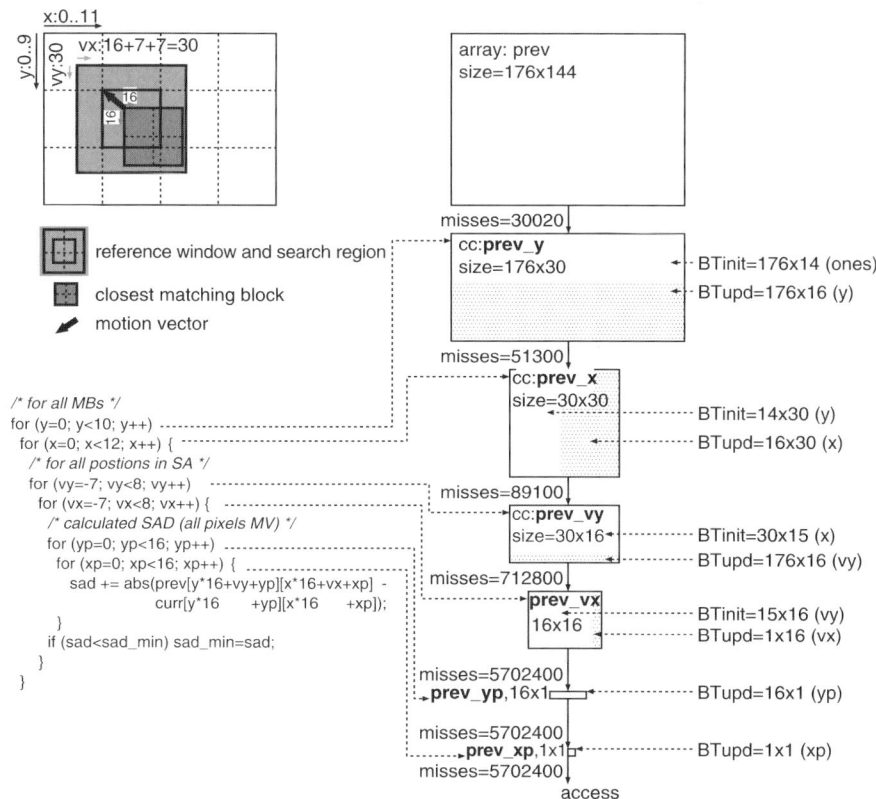

Fig. 7.7 Straightforward ME kernel

(of size 14×30) just before entering the x-loop and the second block transfer (BTupd of size 16×30) is executed every x-iteration to update part of the CC.

A graphical view of this process is shown in the Figure 7.8. For iteration $x = 0$, the blue 30×30 block is the reference window, for the next iteration $x = 1$ the red 30×30 block is the reference window. The overlap is the purple 14×30 rectangle. An initial block transfer before the x-loop will copy the overlapping data, that is the data being reused between the first and the second iteration of x. An update block transfer will copy a 16×30 block every iteration of x, every iteration 30 pixels more to the right. A 14×30 block from the previous iteration will remain, exploiting the reuse between the previous and current iteration. On a platform with DMA capabilities, it is likely that a block transfer will be mapped to a DMA transfer.

The miss counts can be calculated from these BTs. The BTinit is responsible for 3780 misses because it has a size of 420 and is in the y-loop (and thus executed 9 $\times$). Additionally, the BTupd has 47,520 misses because it has a size of

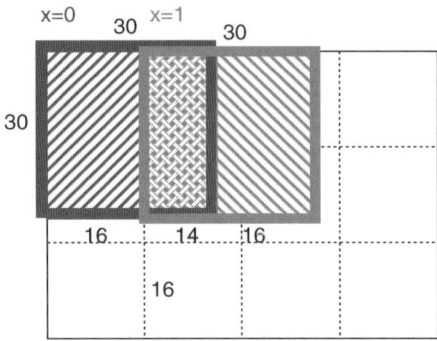

Fig. 7.8   Block Transfers for motion estimation copy at x level

480 and is in the x-loop and thus executed 99 times. As a whole this is 51,300 misses. This number is annotated to the arrow entering the CC in Figure 7.7.

3.3   Non-Carried Copy Candidates

Because of the concept of BTs that fill the copy, additional copies can be identified at each loop level. For instance, there is another possibility to exploit reuse at the level of the x-loop: the prev_x copy candidate, for which, in every iteration of x, the 30×30 block is simply copied in. This alternative does not exploit the reuse carried from one iteration of x to the next (therefore the copy is called for non-carried), but does capture the reuse at deeper loop levels. The size is the same and the number of misses is higher (89,100) than the carried CC. It might seem that the non-carried CC can never become interesting. However, the non-carried copy implies one BT only, which may be more efficient if starting a BT has significant overhead. Additionally, the lifetime of the non-carried copy is smaller. Suppose the y-loop has more sub-loops than only the x-loop (like in Figure 7.9). In that case choosing the non-carried copy with the same size is not alive during these other sub-loops. Hence additional data of other copies can be assigned to this layer. MHLA can explore these trade-offs.

Figure 7.10 shows the detailed DR-tool output. Now, at each loop level two different copies are identified, the first loop is loop carried while the other is not. As a result, 12 copies are identified instead of the 6 in Figure 7.7. The root of the tree (the array in red) is at the top of the picture, the leaf (the access in green) is at the bottom. In-between are the copy candidates (CCs). The naming convention for the copy candidates (e.g. prev_frame_x_e0) is as follows: it starts with the name of the corresponding array, it is followed by the iterator name of the loop level and the letters "nc" are added if the copy is non-carried and it ends with a suffix to make it a unique identifier. Let us focus on the copy candidate selected by MHLA earlier: frame_prev_x_e0. It is a 30×30 block (900

```
.................for (y=0; y<9; y++)           /* for all MBs */
 ..........for (x=0; x<11; x++) {
             for (vy=-8; vy<8; vy++)          /* for all positions in SA */
              for (vx=-8; vx<8; vx++) {
               for (yp=0; yp<16; yp++)        /* calculate SAD (all pixels MB) */
                for (xp=0; xp<16; xp++) {
                  sad += abs(prev[y*16+vy+yp][x*16+vx+xp]-
                        curr[y*16  +yp][x*16  +xp]);
                }
              if (sad<sad_min) sad_min=sad;
 ...............}
             for (i=0; i<10; i++)
               {
                 // another piece of code.
                 // here we can reuse the locations of frame_prev_x_nc_e0
                 // but we canNOT reuse the location of frame_prev_x_e0
               }
 .....................}
```

Fig. 7.9 Copies have different lifetimes that affects inplace oportunities that influence the size

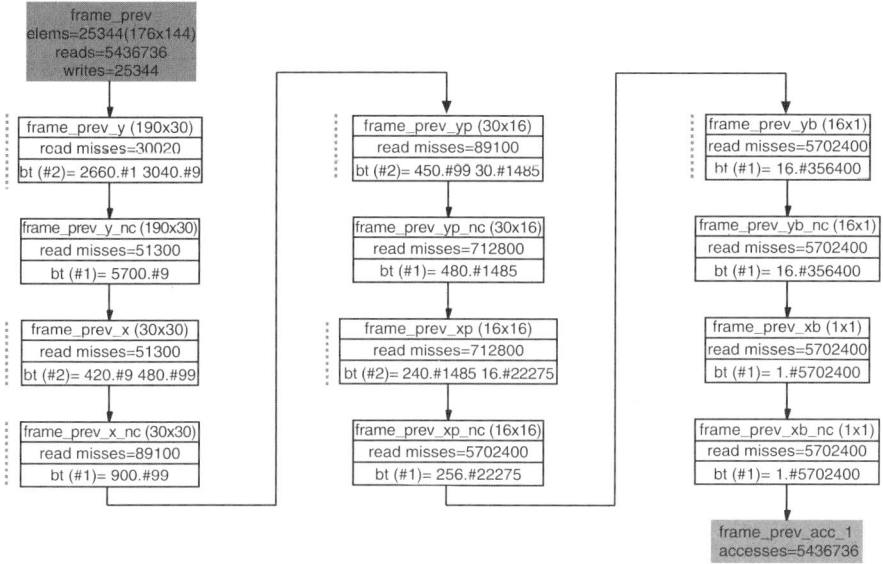

Fig. 7.10 Data reuse tree for straight forward ME kernel

elements, and also 900 bytes) with 51,300 misses. The last line contains some information on the *block transfers* (BTs) corresponding to this copy candidate. The 'bt(#2): 420.#9 490.#99' line says there are two block transfers: one of 420 (14 × 30) elements executed 9 times (the number of y-iterations) and the second of 480 (16 × 30) elements executed 99 times (the number of x-iterations).

3.4   Branches in Reuse Tree

On the loop levels where different branches come together you have a lot
of possibilities to exploit inter-access reuse. Let us consider the example in
Figure 7.11. At the x-loop level two branches come together from the two
nested sub-loops as both access the same array frame. For the individual
accesses, two copies can be found of size 256. Unfortunately, the two accesses
do not completely over and no common copy of size 256 exist. However, the
very same data is accessed by the second sub-loop three iterations of the x-
loop later. This data can be reused when a copy of size 64×16 is introduced
(prev_x_both). Obviously, this copy is significantly bigger than the two separately. This example was for two accesses, of course, many accesses can branch
together at a certain level. In principle any combination (in the mathematical
sense) of these copies makes sense. Depending on the geometry of the overlap,
DR analysis will build up a hierarchy of these copy candidates. Many alternatives can be discarded because they are worse in terms of size and misses
compared to another copy. Still, many copies can potentially be interesting
and cannot be pruned. This analysis has the potential of producing CCs that
exclude each other; assigning one means another cannot be assigned. These
are the so-called *XOR copy candidates*. In principle the data of a copy can be

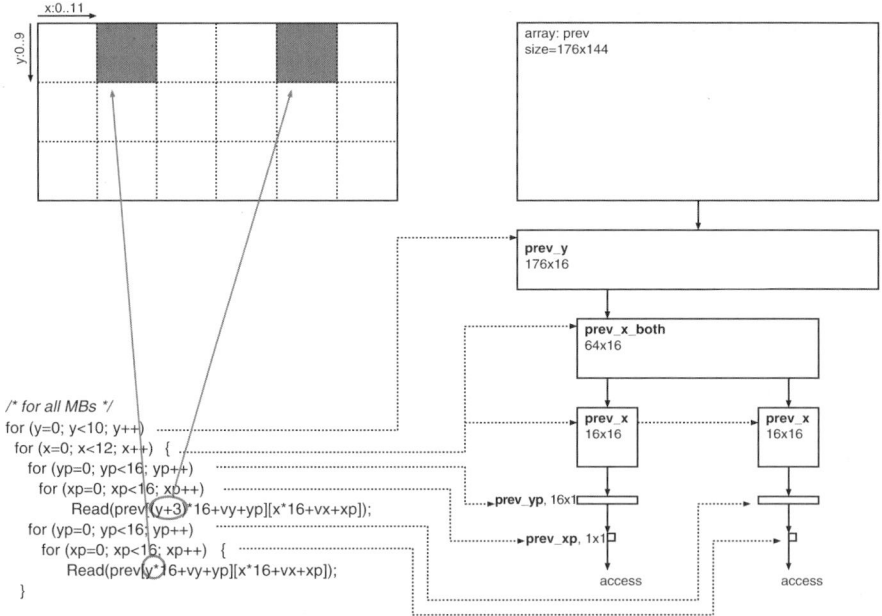

Fig. 7.11   Example with multiple opportunities for copies at the x-loop level because there are
multiple references

combined out of multiple parent copies. But this very much complicates the code and is therefore not desired.

3.5 Write Accesses in the Reuse Tree

Although the discussion so far focused on read accesses because they will have a higher reuse factor than write accesses, it does make sense to take the latter into account. Not that it is likely to find write reuse (it normally does not make sense to overwrite a value with another before it has ever been used by a read), but supporting writes means that a mix of reads and writes can be better analyzed and exploited. For instance, a write can be made to a local copy, and then read back from this copy, before the data is written back to the array. Also write transfers which do not exploit reuse, but group accesses can be interesting for execution time in case a DMA controller is present (see Section 2).

Supporting write accesses in the reuse tree requires some extra care in order to guarantee consistency. Data reuse analysis follows a rather simple approach: a write block transfer is always "mirrored" by a read block transfer reading in the block: this ensures that if not all data in the block is written to (holes), the values of the original array are preserved (because they were read upfront). As a result a copy can have up to four block transfers: BTinitRead, BTupdRead, BTinitWrite and BTupdWrite.

4. High-Level Estimations

The designer is interested in energy, area and timing rather than data reuse analysis. Therefore we have developed high-level estimators to evaluate these criteria given a particular copy selection and assignment.

4.1 Energy Estimation

Based on the platform information we can use a memory energy model to get the energy needed for an access in a certain memory. An MHLA assignment of arrays and copies to layers, the number of accesses to each layer can be easily calculated. Multiplying these numbers with the energy per access (based on commercial SRAM vendor data sheets), and adding it up over the different layers, will give a dynamic energy estimate.

Knowing which (parts of) memories are active for how long (see time estimation in the next subsection), all information is available to calculate the static energy consumption based on the high-level memory model.

4.2 Size Estimation

The exploitation of limited lifetime allows to have smaller layers or to store more data in an equal sized layer. Both can have a huge impact on the (energy)

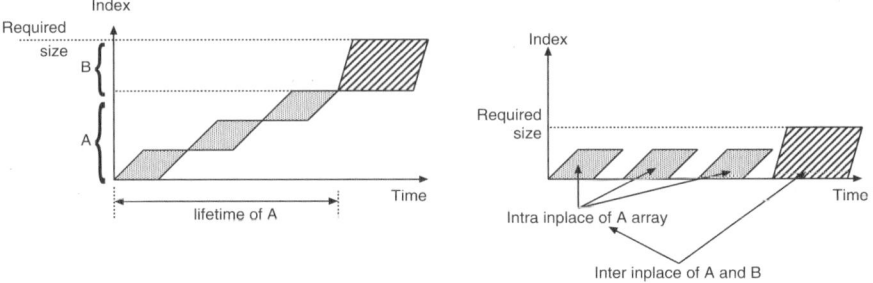

Fig. 7.12   Inplace concept explanation

performance. Especially, the short lifetime of the CCs should be considered carefully. Also, it can be expected that a technique without inplace estimation is useless for larger applications. All arrays will have relatively a smaller lifetime when the application size increases.

Figure 7.12 explains how we can exploit limited lifetimes and reuse locations. The left part shows how elements of A and B are used in time. The shaded areas indicate which elements are used and for how long. Clearly, the declared size of array A can be reduced by a factor three by reusing the same locations. This is called intra-inplace (Greef, 1998). Furthermore the lifetime of the elements of array B do not overlap with A. Therefore B can also reuse the locations of A. This is called inter-inplace (Greef, 1998). The result of both inplace opportunities is shown on the right of Figure 7.12.

We have implemented a low complexity inter-inplace estimation. The storage size estimation is based on the simultaneous alive data in the most inner loops of the application. As a result we only have to update the storage size of those inner loops that span the lifetime of the CC. Typically, most copy candidates have a short lifetime and span only a few inner loops.

The lifetime and size of both the arrays and copies must be known. In fact, we need to know the effective size and not the declared size. To this end we have coupled to the ATOMIUM/MEMORY COMPACTION (MC) tool Catthoor et al. (1998). MC provides the lifetime and intra-inplace size of all arrays. For the copy candidates the lifetime information can be determined at data reuse analysis time: it will be dependent of on which loop level the CC resides, which array accesses read from it and whether the reuse is loop-carried or not. The lifetime of a CC is simply the body of the corresponding loop level. Also the CC size is known from the data reuse and cannot be further reduced by intra-inplace.

This gives us all the needed information for a high-level estimation model illustrated in Figure 7.13. We only need to look at the innermost loops, because other loops will always have a subset of the data alive at the innermost places.

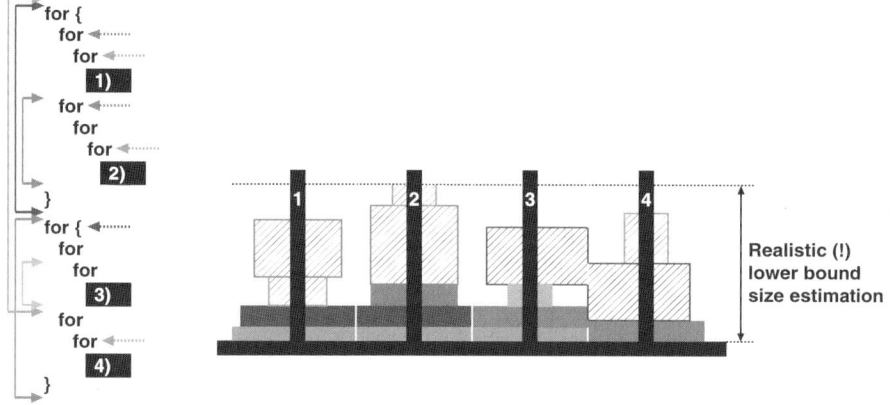

Fig. 7.13 The towers of inplace

So the innermost loops represent the places where the size constraint is to be met. We can represent the innermost loops as the poles around which the towers of inplace grow. As arrays and CCs are added by the incremental assignment, their size is added to the poles of the innermost loops spanned by the lifetime.

This estimation greatly reduces the complexity, instead of having to investigate the footprint at all the (loop) blocks in the program, only the innermost loops need to be considered. Also the estimation is close to what is really feasible, as is illustrated in Figure 7.14. In fact, the estimation gives a realistic underbound for the inplace step that is performed later in the data layout step of the DTSE script. Note that for the subsequent MC step the copies have become arrays and that they can be treated the same way as the "original" arrays for the actual data layout.

4.3   Time Estimation

Much in the same way as energy estimation, we can count the number of accesses to the different layers and multiply them by the number of cycles that an access cost.

There is one additional issue though estimating the time needed to make the copy. We can make a significant error if we do not take into account that copying blocks of data can be done more efficiently using DMA. Also, when copying from SDRAM it is important to take the SDRAM page into account. So we can use a high-level DMA time model.

For a DMA transfer, five components were identified. The five identified components of a DMA transfer are all expressed in processor clock cycles in Table 7.1.

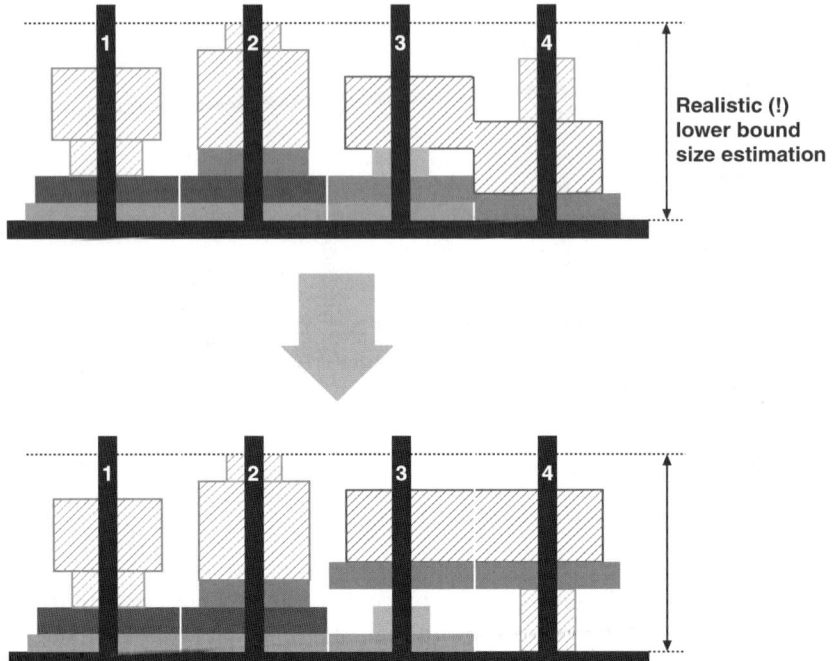

Realistic (!)
lower bound
size estimation

Fig. 7.14 The towers of inplace provide a lower bound estimate

Table 7.1 Main components of block transfer timing

Description	Duration
Processor overhead cycles	1
DMA controller overhead cycles	13
Cycles to copy a single data element	2
Line gap cycles	16
Line setup cycles	6

The number of processor overhead cycles is the number of cycles it takes for the processor to send the block transfer message to the DMA controller. This operation is considered to take only processor time, and not any time of the DMA controller. Next is a number of cycles the DMA controller needs to process the message internally. The third item in Table 7.1 represents the single element copy duration. The DMA controller copies data stored in arrays which consists of several lines. To switch from one line to another, an extra amount of time is accounted for (the line gap time).[3] The last row in the table is the access of the first line, as it differs from any other due to an extra setup time.

---

[3] A line gap actually comes from the SDRAM page miss penalty.

Table 7.2 DMA timing examples

Array	Number of elements	Line gap	Size dependent timing	Total time
11	1	0	$(1 \cdot 2) + (0 \cdot 16) = 2$	22
25	10	4	$(10 \cdot 2) + (4 \cdot 16) = 84$	104
52	10	1	$(10 \cdot 2) + (1 \cdot 16) = 36$	56

In the last column in Table 7.1, the numbers from the data sheets for a typical DMA controller are depicted. The first two elements and the last element are independent of the size of the block transfer and form an offset to the size dependent numbers.

To illustrate the impact of the DMA timing, Table 7.2 is composed. In this table, the block transfer times for three arrays (sized 11, 25 and 52) are precomputed. The first column lists the size of the arrays copied. The second column shows the total number of elements for that particular block array. Next, the number of line gaps in the block transfer is listed, and in the forth column the variable part of the timings is computed for all cases. The last column lists the total transfer time, which equals the previous column plus a fixed offset. This offset equals $1 + 13 + 6 = 20$ (from the last column in Table 7.1). This shows that not only the size of a transfer, but also the shape matters. Related work to this is the block-based data layout for SDRAM proposed in kim et al. (2003).

For the 11 array, the size equals $1 \cdot 1 = 1$, as shown in the second column. This block transfer does not involve any line gaps, and the size dependent timing part equals one time the per-element transfer time, and zero times line-gap time, which equals 2 in the forth column. To compute the total transfer time, the fixed offset of 20 is added to this in the last column. In a similar way the results for the 25 and 52 arrays are computed. Both of them consist of 10 elements, but due to the difference in the number of line gaps, copying a 25 array takes almost twice as long as copying a 52 array. But even the slowest of the two is much faster than ten individual element transfers, as that would take $22 \cdot 10 = 220$ clock cycles.

Although exploiting data reuse opportunities provides great benefits, performance can be further improved by scheduling some block transfers (BTs) earlier, which we call time extensions (TEs). TEs enable us to increase the performance of a system, with minimal cost. They enable the selective prefetching of copy candidates from off-chip memory layers to on-chip memory layers, exploiting the lifetime information of copies. Indeed, usually CCs consume temporary on-chip memory space, for the part of the code that they are active. We will demonstrate the importance of the TE with a simple example (see Figure 7.2).

```
for(i=0;i<3;i++)
  for(j=0;j<10;j++)
    read(A)
    read(B)

  for(i=0;i<6;i++)
    for(j=0;j<10;j++)
      read(B)

for(i=0;i<3;i++)
  for(j=0;j<10;j++)
    read(A)

for(i4=0;i4<10;i++)
  for(j4=0;j4<8;j++)
    read(C)
```

Fig. 7.15   Copy candidates for arrays A, B, C have limited lifetimes

For the example in a sample toy code given in Figure 7.15, we find out that a copy candidate (A′) for an array A has a lifetime of three loops (loops 1, 2, 3), a copy candidate (B′) for an array B has a lifetime of two loops (loops 1, 2) while a copy candidate (C′) for an array C has a lifetime of 1 loop (loop 4).

In the towers of inplace model (Figure 7.16), we place the lifetime information of every CC for every loop and examine whether on-chip memory size is exceeded or not.

The copy candidates of our toy code (Figure 7.2) are modeled as boxes in the inplace tower model (Figure 7.16a). In our technique, we try to schedule the BT earlier (prefetch), thus we can hide the time required for the copy to be made.

We model this by extending one loop at a time, covering every copy candidate and then checking the constraints. Copy candidates A′ and B′ cannot be extended over an earlier loop, because there are no loops before to accommodate the lifetime. On the other hand, copy candidate C′ (located at loop 4) can be extended because there are loops before it (loops 1, 2, 3). First, CC C′ is extended one loop before, at loop three (Figure 7.16b). On-chip memory size required by C′ is added to the existing on-chip memory size of loop three (visualized as a new box on top of the A′ box). The new total on-chip memory required by loop three is under the constraint of on-chip memory size, thus this extension is valid. If we extend one loop further (loop two), we reach an invalid state (Figure 7.16c), because we have a size violation. It has to be noted though,

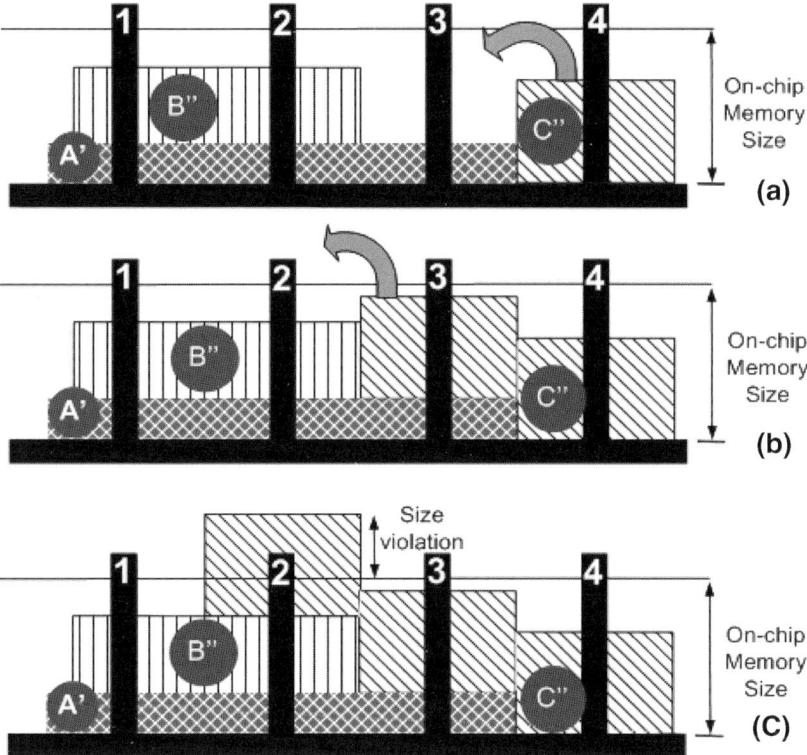

Fig. 7.16 InPlace towers with time extensions

that if the cycles of copying elements to C' can be hidden by extension to loop three, then no further time extension is required.

Simple time extensions exploit the freedom to move the issue of block transfer before earlier sub-loops than the sub-loop with the accesses to the copy.

```
for y
  for x1
    B[]

  BT_ISSUE_AND_SYNC(A_x2)
  for x2
    A_x2[]
```

becomes:

```
for y
  BT_ISSUE(A_x2)
  for x1
    B[]
```

```
BT_SYNC(A_x2)
for x2
  A_x2[]
```

For the duration of the $x1$-loop, the block transfer will be in parallel with processing.

If no earlier sub-loops are available, or if not all block transfer duration is put in parallel yet, *pipelined time extensions* exploit the freedom to move the issue of block transfer before sub-loops in an earlier iteration of the loop.

```
for y=0..W
  BT_ISSUE_AND_SYNC(B_x1)
  for x1
    B_x1[]
  for x2
    A[]
```

becomes:

```
BT_ISSUE(B_x1)
for y=0..W
  BT_SYNC(B x1)
  for x1
    B_x1[]
  if (y < W)
    BT_ISSUE(B_x1)
  for x2
    A[]
    }
```

Except for the first block transfer, the block transfer will be in parallel with processing of the $x2$-loop.

If not all block transfer duration can be hidden by a simple or pipelined time extension or if no sub-loops are present (fully nested case), a *parallel pipelined time extension* can put the transfer for the n-th iteration in parallel with the processing of the $n-1$-th iteration.

```
for y=0..W
  BT_ISSUE_AND_SYNC(A_x)
  for x
    A_x[]
```

becomes:

```
BT_ISSUE(A_x[0])
for y=0..W
  BT_SYNC(A_x[y%2])
  if (y<W)
```

```
  BT_ISSUE(A_x[(y+1)%2])
for x
  A_x[]
```

Except for the first block transfer, the block transfer will be in parallel with processing of the x-loop. Note that this overlaps the lifetimes of the data of the block transfers of two successive iterations and MHLA must check whether that size is available.

When scheduling the block transfers earlier, the freedom can be limited by several type of dependencies. The block transfer "inherits" the dependencies from the access it is for: e.g. if an access depends on a value written in an earlier statement, the block transfer will also depend on this statement and cannot be scheduled before it.

Note that memory dependencies[4] with writes will not exist within the loop body because the logic behind the additional CCs by selection of accesses will not allow an overlapping write not to be included (see Section 3). This ensures that the write is also going to the copy.

When scheduling block transfers over an iteration boundary, we do have to take into account the memory dependencies. Using scalar-level analysis will find dependencies between all references to an array, even if the data elements they actually access are not dependent, because they do not overlap. Scalar-level dependency analysis is not good enough for the memory dependencies, because the freedom will be too limited. Array-level dependency analysis taking into account the accessed part of the array is needed to find more freedom. Note that this extra freedom might exactly be the goal of loop and data flow transformations in earlier steps of the DTSE methodology, so it is not acceptable that MHLA would not be able to analyze and exploit this freedom.

Array-level dependency analysis is a matter of looking at the overlap in the accesses' extents in the previous and next iteration, which can easily be done in the geometrical view of the data reuse analysis. Additional dependency information on each CC expresses which accesses are dependent, and the sub-loops which contain these accesses are the limits of the block transfer scheduling freedom.

Other dependencies than the memory dependencies, reference (data dependent array index) and control (conditions), can be traced by the scalar factored use-def (FUD) analysis kernels of the ATOMIUM tool (also used in the ATOMIUM/MEMORY ARCHITECT tool). Scalar-level analysis is sufficient for this type of dependencies because data dependent array indices and control expression will mostly be based on scalar values and no freedom is lost.

---

[4]We are using the term memory dependencies here to refer to the types of dependencies that can exist between read and write access couples, they group flow (read after write), output (write after write) and anti- (write after read) dependencies. These terms come from compiler research terminology.

5.   Exploration Methodology for MHLA Search Space

An efficient mapping of arrays and CCs to partitions must be found within reasonable time. The exploration is performed in two phases. The first phase assigns all the arrays to the partitions. Indeed, all arrays have to be assigned to guarantee functionality. In the second phase, a selection of CCs is mapped to the remaining partition space for each valid array assignment. In principle we can fully search all possible mappings of arrays and CCs. This may take several hours though. Therefore we have implemented a steering heuristic and have optimized the implementation by the so-called incremental assignment.

5.1   Steering Heuristic

The arrays having the highest access over size ratio are assigned to the cheapest possible partition first. Intuitively this makes sense, as then the cheapest partitions will get most accesses. A similar heuristic is used for the CCs. However, for CCs we do not know the number of accesses before all possible CCs are mapped. For example, in Figure 7.2 the number of reads to A' is 1,000 or 10,000 depending on whether A'' is mapped or not. Therefore we use the highest ratio of reduction in misses over the additional CC size based on the current situation. Instead of only adding copies it is possible that a copy replaces its assigned children. In this case the additional size will the smaller than the CC size which can be more beneficial. So it can be that first A'' is selected because it has a higher ratio, but in a later iteration of the algorithm it is replaced by A' because there is still space available.

5.2   Incremental Assignment

The MHLA exploration works in an incremental way for exploration efficiency. From a given, current assignment a new one is constructed; see Figure 7.17. The left hand side shows an example of the current assignment where the array A is assigned to L3, A'' to L0 and the copy candidate A' unassigned. First a new CC is selected. In the example, A' is the only CC remaining and is selected. This CC is assigned iteratively to those memory partitions that are in between the layers to which the higher and lower CC are assigned. In Figure 7.17, A' can only be assigned to partitions in L1 and L2. Other partitions do not have to be searched for. Independent of the partition assignment of A', the misses 1000 of A'' do not take place anymore on L3 but are replaced by the 100 misses of A'. If A' is assigned to L1, the 100 misses of A' and the 1000 misses of the next lower assigned CC (A'') are added to L1. Furthermore the size impact of the assignment to L1 must be evaluated. The most simple size estimation adds the CC size to its assigned partition. This will lead however to a very poor layer usage. An improved estimation based on limited lifetime is explained in Section 4. The *changes* in size, accesses and the effects on the

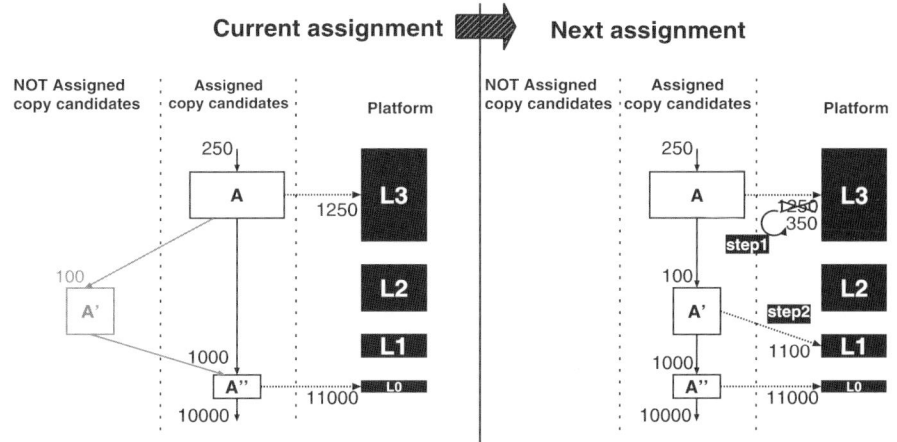

Fig. 7.17 The next assignment is incrementally calculated

total energy consumption can be calculated without knowing all details of the already assigned CCs.

6. Case Studies

Two realistic demonstrators having different characteristics are selected to present the impact of MHLA. The first demonstrator from the video compression fields works mainly on two dimensional arrays and has a lot of data reuse. The second demonstrator is a wireless receiver with limited reuse and only single dimensional arrays. It is worth noting that both algorithms span several pages of complex C code.

Initially we assume the architecture to have two layers. The used partition energy model is based on a real memory model and is displayed by the solid line in Figure 7.19. Relative energy figures are sufficient for the tool explore on. All the memory model numbers are relative to the fixed size off-chip memory of 1MB. The largest on-chip memory of 16KB is a factor 3 less energy consuming than off-chip memory. This conservative factor is realistic and surely not in favor of the method. The energy model is slightly super logarithmic so a memory which is $256 \times$ larger consumes $8.6 \times$ more energy per access. This same energy model is used for L0 and L1 in both drivers. The energy consumption is computed by multiplying by the memory activity. The memory activity is obtained by executing an instrumentation version of the source code.

6.1 QSDPCM

The Quadtree Structured Difference Pulse Code Modulation (QSDPCM) algorithm is an inter-frame compression technique for video images. It involves a hierarchical motion estimation step, and a quadtree based encoding of the motion compensated frame-to-frame difference signal (Strobach, 1988).

A global vie w of the QSDPCM main signals and their reuse is given in Figure 7.18. Many data reuse opportunities exists for the QSDPCM application as can be seen from the many data reuse chains. Different runs of the MHLA tool explore the L1 size. Figure 7.19 shows the energy contribution of the L1

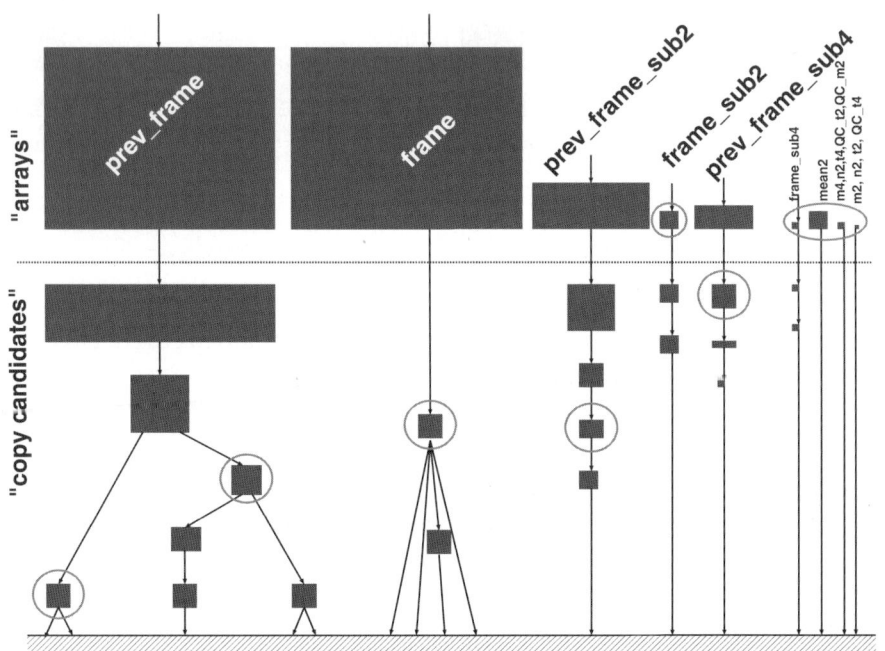

Fig. 7.18   Assignment to memory hierachy

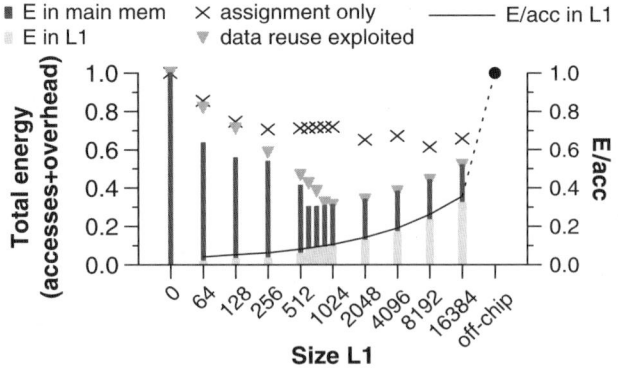

Fig. 7.19   Energy for varying L1 size

(bottom bar) and main memory (top bar). When increasing the layer size, the energy goes down because fewer accesses occur on the more energy consuming main memory. The L1 miss rate reduction does not decrease much further for a L1 size larger than 640. Therefore the L1 energy per access increase penalty is not compensated by the lower amount of misses. Hence the overall energy consumption increases. The optimal assignment corresponding to this optimum is given in Figure 7.18. The circled arrays and CCs are stored in the L1. The other arrays are stored in the main memory. The not circled CCs are not selected. Inserting an extra smaller third layer did not significantly reduce the energy, as the L1 in the two-layer optimum architecture is already small. When we compare the presented technique to an array assignment technique (crosses) we gain a factor 2 in energy. When switching of the inplace estimation (triangles) we require a L1 of 1K instead of 640 elements to reach the minimum energy. Moreover, it consumes 3.2% more energy.

6.2 DAB Wireless Receiver

Digital Radio is a typical application for portable usage. Therefore low energy issues are very important to extend the battery lifetime. The DAB channel coding standard involves several complex data-dominant algorithms.

Similar to the previous driver, a L1 size exploration is performed and shown in the top curve of Figure 7.20. The minimum energy is consumed for a L1 size of 8K elements. The relatively large decrease in energy while increasing the L1 size from 0 to 64 reveals that it might be interesting to insert an L0 of size 64. Therefore we repeat the experiments with an additional layer of size 64 and 128. The additional energy reduction is 15% for both layer sizes at the optimal point. Deeper investigation showed that the additional gain was obtained in the FFT function. Up to 25% of the L1 accesses were removed due to this additional layer.

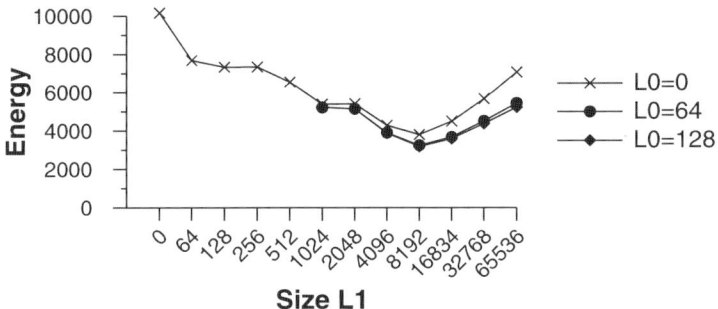

Fig. 7.20 Architecture exploration for DAB

Table 7.3   DAB results on TriMedia simulator

	#reg	#ld/st	pred #ld/st	Data cache misses	Cycles
Standard register usage	6	–	23,532	–	–
Mild register usage	22	18,408	17,152	2,895	91,552
Aggressive register usage	70	11,837	11,232	763	47,341

The most energy-efficient assignment has been used to map the DAB application to the TriMedia processor. This processor is selected because it has a data memory hierarchy that matches the optimal architecture. The processor has an L0 layer of 128 registers, 16KB cache and 8Mb of SDRAM. Importantly, the exploitation of the L0 register file has to be carefully evaluated. On one hand, the number of data load stores will reduce because more data remains in the L1. On the other hand, the higher register pressure might counteract this gain as register spilling is required to schedule all instructions. Also the required unrolling, in order to keep the data in the register file, needs more instruction cache space. The trade-off between the load stores, spilling and instruction cache is given in Table 7.3. The native TriMedia simulator is used for the evaluation of three differently transformed implementations having a more or less aggressive L0 register usage (second column). The large reduction of 34% in memory accesses has of course a large impact on data memory consumption and performance. Also very important to note, the prediction of the MHLA is very close to the actual number of accesses made by the compiled code. The small difference between the MHLA predicted activity and simulation results are largely explained by the few register spillings and some low level compiler details.

6.3   Execution Time Measurements

The tool exploration time is an important factor next to the quality of the results. Table 7.4 shows the number of explored assignments before finding the optimal solution in comparison to the total number in the huge exploration space. The last column clearly shows that the chosen heuristic allows to find the best solutions within reasonable time. Interesting to note is that the current implementation makes about 20,000 evaluations per second on a Pentium-IV.

7.   Related Work

Optimizing the memory hierarchy for performance is a well-explored topic (Anderson et al., 1995; Ancourt, 1997; Kampe and Dahlgren, 2000; Lim and Yew, 2000). Also some relatively early data memory hierarchy related papers

Table 7.4 Tool performance on QSDPCM

Size L1	nr. explored assignments	nr. valid assignments	Optimum in # iterations
64	46,951	4767	3,333
128	351,819	45,976	34,027
256	256,3636	540,651	1,514
512	934,606	6,175,295	9,279
1,024	20,711,631	12,356,321	1,077
2,048	25,552,456	23,460,160	786
4,096	28,311,552	28,311,552	12

address the energy related issues Benini et al. (2000) but they are follow ups of our early work Catthoor et al. (1998) and they do not yet implement all the steps and techniques that have been addressed in this chapter.

The characterization of Grun et al. (2001) clusters the data sets into different types. These different clusters have certain memory type preferences and are assigned accordingly. The mapping is suboptimal (especially for the regular accesses) because it is based on some average characteristics and does not allow and accurate prediction. Afterward, the performance is measured by simulation. Also Panda et al. (1998) have made a distinction between caches and scratchpad memories. However, no real layer assignment was made. The technique presented in Steinke et al. (2002) assign data (and instructions) to the scratchpad. However, no consideration is made to benefit from reuse and inplace opportunities.

The closest work to ours is presented in Kandemir and Choudhary (2002). It analyses and exploits the temporal locality by inserting local copies. Their layer assignment builds a separate hierarchy per loop nests and then combines them into a single hierarchy. However, a global view of the assignment and lifetime of the arrays and copies is required for real life applications having imperfect nested loops. In the end, no accurate estimation for size, time and energy is made. For instance, no overhead estimation is made which makes it impossible to trade-off copies versus arrays in a certain layer. Similarly, the work published in Masselos (2001) lacks of a global application view.

Access trace based analysis techniques like presented Benini et al. (2002) have limited optimization capabilities. The quality of the analysis depends on the preceding compilation step. For instance, from access trace point of view all elements of an array are accessed equally while a small data reuse copy could be present. Also, two data objects that are at the location cannot be distiguished. As a result, the exploration space cannot be searched properly.

To our knowledge no one has combined the opportunities for data reuse and inplace in an automatable technique for real life application that is not based on simulation. Most related work does rely fully on an access trace. This does not allow to explore properly and generate correct code in general. Our approach finds all the Pareto optimal by adding the relevant estimations (power, area and time).

8. Conclusion and Future Work

This chapter has presented the first automated technique to perform layer assignment taking data reuse and inplace into account. A fast exploration technique and heuristic is proposed. This technique is implemented in a prototype tool that has allowed us to do exploration on real life demonstrators of industrial relevance. The energy is more than halved by exploiting limited lifetime and reuse in arrays.

The intention is to extend the technique such that it can handle more memory types. The currently supported types are both software and hardware controlled memories. Currently we are adding cycle budget estimation such that it trades energy for performance. Limited data dependent conditions and data dependent addressing is supported. Further research is required to remove more dynamic and data dependent limitations.

References

Achteren, T. Van, R. Lauwereins, and F. Catthoor (2002). Systematic data reuse exploration techniques for non-homogeneous access patterns. In *Proceeding's of the 5th ACM/IEEE Design and Test in Europe Conference (DATE)*, pages 428–435, Paris, France.

C. Ancourt, et al. (1997). Automatic data mapping of signal processing applications. In *Proceedings of International Conference on Application Specific Array Processors*, pages 350–362, Zurich, Switzerland.

J. Anderson, S. Amarasinghe, and M. Lam (1995). Data and computation transformations for multiprocessors. In *5th ACM SIGPLAN Symposium on Principles and Practice of Parallel Programming*, pages 39–50.

L. Benini, A. Bogliolo, and Micheli, G. De (2000). A survey of design techniques for system-level dynamic power management. *IEEE Transactions on VLSI Systems*, pages 299–316.

L. Benini, L. Macchiarulo, A. Macii, and M. Poncino (2002). Layout-driven memory synthesis for embedded system-on-chip. *IEEE Transactions on VLSI*, pages 96–105.

Catthoor, Francky, Wuytack, Sven, De Greef, Eddy, Balasa, Florin, Nachtergaele, Lode, and Vandecappelle, Arnout (1998). *Custom Memory Management Methodology, Exploration of Memory Organization for Embedded Multimedia System Design*. Kluwer Academic Publ., Boston, MA.

B. Durinck, Brockmeyer, Erik, Vandercappelle, Arnout, Greef, Eddy De, vander Vegt, Jeroen, and Dasygenis, Minas (2005). Extensions for data reuse and single processor memory hierarchy layer assignment. Project Deliverable M4.DDT-2, IMEC vzw.

Greef, E. De (1998). *Storage Size Reduction for Multimedia Applications*. Doctoral Dissertation, ESAT/KUL, Belgium.

P. Grun, N. Dutt, and A. Nicolau (2001). Apex: access pattern based memory architecture exploration. In *The 14th International Symposium on System Synthesis*, pages 25–32, Montreal, Canada.

Issenin, Ilya, Brockmeyer, Erik, Miranda, Miguel, and Dutt, Nikil (2004). Data reuse analysis technique for software-controlled memory hierarchies. In *7th ACM/IEEE Design and Test in Europe Conference*, pages 202–207, Paris, France.

M. Kampe and F. Dahlgren (2000). Exploration of spatial locality on emerging applications and the consequences for cache performance. In *Proceedings of the International Parallel and Distribution Processing Symposium (IPDPS)*, pages 163–170, Cancun, Mexico.

Kandemir, Mahmut and Choudhary, A. (2002). Compiler-directed scratch pad memory hierarchy design and management. In *Proceedings of the 39th ACM/IEEE Design Automation Conference*, pages 690–695, Las Vegas, NV.

H.S. Kim, N. Vijaykrishnan, M. Kandemir, E. Brockmeyer, F. Catthoor, M.J. Irwin, Estimating Influence of Data Layout Optimizations on SDRAM Energy Consumption, Proc. of ISLPED03. PP. 40–43, Aug 2003.

C. Kulkarni (2001). *Cache Optimization for Multimedia Applications*. Doctoral dissertation, ESAT/KUL, Belgium.

H-B. Lim and P-C. Yew (2000). Efficient integration of compiler-directed cache coherence and data prefetching. In *Proceedings of the International Parallel and Distributed Processing Symposium (IPDPS)*, pages 331–339, Cancun, Mexico.

K. Masselos, et al. (2001). Memory hierarchy layer assignment for data re-use exploitation in multimedia algorithms realized on predefined processor architectures. In *The 8th IEEE International Conference on Electronics, Circuits and Systems (ICECS)*, pages 285–288.

P.R. Panda, N.D. Dutt, and A. Nicolau (1998). Data cache sizing for embedded processor applications. In *Proceedings of the 1st ACM/IEEE Design and Test in Europe Conference (DATE)*, pages 925–926, Paris, France.

SimpleScalar (2002). The simplescalar-arm power modeling project; poweranalyzer for pocket computers. WWW page. `http://www.eecs.umich.edu/ panalyzer`.

Steinke, Stefan, Wehmeyer, Lars, Lee, Bo-Sik, and Marwedel, Peter (2002). Assigning program and data objects to scratchpad for energy reduction. In *Proceedings of the 5th ACM/IEEE Design and Test in Europe Conference*, pages 409–415, Paris, France.

Strobach, P. (1988). QSDPCM – a new technique in scene adaptive coding. In *Proceedings of the 4th European Signal Processing Conference*, pages 1141–1144, Grenoble, France. Elsevier Publ., Amsterdam.

The Atomium club (2004). Atomium/mc 2.2.1 user's manual. Technical Report Version 1.2, IMEC vzw.

Texas Instruments (2003). Application report tms320c6414/15/16 power consumption summary. Technical reference spra811c, Texas Instruments.

van Meeuwen, Tycho (2002). Data-cache conflict-miss reduction by high-level data-layout transformations. Master's thesis report, Technische Universiteit Eindhoven, The Netherlands.

Chapter 8

Memory Bank Locality and Its Usage in Reducing Energy Consumption

Mahmut Kandemir
Pennsylvania State University

Abstract   Bank locality can be defined, in the context of a multi-bank memory system, as localizing the number of load/store accesses to a small set of memory banks at a given time. An optimizing compiler can modify a given input code to improve its bank locality. There are several practical advantages of enhancing bank locality, the most important of which is reduced memory energy consumption. Recent trends indicate that energy consumption is fast becoming a first-order design parameter as processor-based systems continue to become more complex and multi-functional. Off-chip memory energy consumption in particular can be a limiting factor in many embedded system designs. This paper presents a novel compiler-based strategy for maximizing the benefits of low-power operating modes available in some recent DRAM-based multi-bank memory systems. In this strategy, the compiler uses linear algebra to represent and optimize bank locality in a mathematical framework. We discuss that exploiting bank locality can be cast as loop (iteration space) and array layout (data space) transformations. We also present experimental data showing the effectiveness of our optimization strategy. Our results show that exploiting bank locality can result in large energy savings.

Keywords: memory banking; energy optimization; DRAM

1.   Introduction and Motivation

Compared with application-specific custom circuitry, many processor-based systems can exhibit a factor of 100–1000 worse energy-delay product. Consequently, many research groups are trying to reduce this gap by re-examining the hardware–software interface, and designing hardware and software level solutions for reducing energy consumption. The role of power and energy

J. Henkel and S. Parameswaran (eds.), Designing Embedded Processors – A Low Power Perspective,
191–216.

consumption in real-time and embedded systems is well recognized and becoming an increasingly important area of research.

In many system designs, large off-chip main memories can be a major energy consumer (Catthoor et al., 1998; Farkas et al., 2000). There are two main reasons for high dynamic energy consumption of off-chip data memories. First, off-chip data memories present a large capacitive load to the rest of the system. Second, most data-intensive embedded applications exhibit poor data reuse (cache locality), resulting in frequent visits to the main memory. Consequently, hardware and software techniques that target at reducing memory energy are critical in many embedded designs.

One approach to attack this problem from the hardware perspective is to design a banked memory architecture, where the main memory is partitioned into multiple banks (Ram, 1999). Since this effectively reduces the size of the memory accessed during a reference, it also helps to reduce per access dynamic energy consumption. In addition, many new banked memory architectures provide multiple low-power operating modes. Typically, an idle bank (i.e., a bank not used by the current computation) can be placed into a low-power operating mode, increasing energy savings further. An effective use of such low-power modes is critical for obtaining the best energy behavior, and has received some attention recently (Lebeck et al., 2000; Delaluz et al., 2001; Kandemir et al., 2001; Fan et al., 2002).

In this paper, we attack this important problem using a mathematical framework embedded into an optimizing compiler. Specifically, we present a compiler-based strategy for maximizing the benefits of low-power operating modes available in DRAM architectures. In this strategy, an optimizing compiler employs loop and data transformations to increase bank inter-access times, thereby allowing a memory bank to remain in the idle state for longer periods of time. This is achieved by clustering accesses with temporal affinity in a small set of banks. In most cases, this has a great impact on memory energy consumption as a longer bank inter-access time means a more aggressive (i.e., more energy saving) low-power mode. In other words, unlike most existing compiler frameworks that use loop/data transformations for either cache locality and/or loop-level parallelism (Wolf and Lam, 1991; Li 1993; Wolfe, 1996), we use these transformations for enhancing bank locality to improve energy behavior of banked memory architectures.

To test the effectiveness of the proposed compiler-based energy optimization strategy, we implemented it in a source-to-source translator (Amarasinghe et al., 1996) and evaluated its performance using five array-intensive embedded applications. Our experimental results indicate that the proposed strategy is very successful in practice and improves bank locality significantly. As a result, it reduces off-chip memory energy consumption substantially over a strategy that uses low-power operating modes without our optimization. Our results also demonstrate that these energy improvements are consistent across different systems/software parameters (e.g., the number of banks, array layouts). In addition, the impact of our approach on execution cycles was found to be negligible. Based on these results, we strongly encourage compiler writers for embedded systems that employ banked memories to focus their loop/data optimizations on memory energy.

The rest of this paper is organized as follows. Section 2 describes low-power operating modes and makes an informal introduction to memory bank locality. Section 3 gives formal definition of array mappings to virtual banks. Section 4 presents intra-reference and inter-reference constraints for bank locality. Section 5 demonstrates how loop (iteration space) transformations can be used for improving bank locality. Section 6 discusses how data (memory layout) transformations can be used for implementing array decompositions across banks. Section 7 describes global optimizations, and virtual-to-physical bank mappings are discussed in Section 8. Section 9 presents our experimental results, and Section 10 concludes the paper by summarizing our major contributions and gives a brief introduction to our future work.

2.   Banked Memory Architecture and Low-Power Operating Modes

We focus on an RDRAM-like off-chip memory architecture (Ram, 1999), where off-chip memory is partitioned into several (equally-sized) banks. We also assume that each bank can be placed into a low-power operating mode (independently) when it is not in active use. In a low-power operating mode, a bank typically consumes much less energy than it would consume when it is in active (i.e., fully-operational) mode. However, when a bank in low-power mode is accessed, it takes some amount of time until it is available to service the request. This time delay is called re-synchronization penalty (or re-synchronization cost), and is the major factor that prevents us from using the most aggressive (i.e., the most energy-saving) low-power mode as soon as the bank becomes idle. Typically, given a set of low-power operating modes, there is a trade-off between energy saving and re-synchronization latency. In other words, the most energy-saving low-power mode has also the highest re-synchronization penalty of all. Therefore, one should be careful in selecting the suitable low-power mode when a memory bank becomes idle. Figure 8.1 lists energy consumptions and re-synchronization costs for the operating modes used in this study. The energy values shown in this figure have been obtained from the measured current values associated with memory banks documented in memory data sheets (for 3.3V, 2.5 nsec cycle time, 8MB memory) (Ram, 1999). The re-synchronization latencies have also been obtained from the same data sheets. Based on the trends gleaned from data sheets, the energy values are increased by 30% when bank size is doubled. Note, however, that our approach is general enough in that it can accommodate

	Energy Consumption (nJ)	Re-synchronization Cost (cycles)
active	3.570	0
napping	0.320	30
power-down	0.005	9,000

Fig. 8.1   Bank operating modes used in this work

any number of low-power operating modes with different energy consumptions and re-synchronization costs.

Prior research shows that compiler-based (Delaluz et al., 2001; Kandemir et al., 2001), OS-based (Lebeck et al., 2000), and pure hardware-based schemes (Delaluz et al., 2001) can be adopted to decide the most suitable low-power mode to use when a memory bank is detected to be idle. Since in this work we focus on array-intensive embedded applications, we opted to use a compiler-based approach, where an optimizing compiler (taking into account loop access patterns and array-to-bank mappings – that is, the layout of data in banked memory) decides which operating mode to use. Note that (where applicable) compiler-based strategy has an important advantage over OS-based and hardware-based techniques. The compiler-based strategy (unlike OS or pure hardware-based strategies) does not rely on history information; that is, the compiler can predict (quite accurately for array-based embedded codes) the future access patterns (and also, the future idle times), and select the most appropriate operating mode to use when idleness is predicted. In addition, the compiler can also predict when an idle bank will be requested, and can pre-activate the bank in question in an attempt to eliminate re-synchronization latency. The details of the compiler-based low-power mode detection strategy are beyond the scope of this paper and can be found in Delaluz et al. (2001) and Kandemir et al. (2001). Kandemir et al. (2002) evaluates the impact of classical loop optimizations on energy consumption of banked memories. Delaluz and Kandemir (2004) study the impact of array regrouping, a data optimization, on energy-efficiency of banked memory systems. In contrast, the work presented in this paper is oriented toward increasing the benefits of low-power modes by loop transformations and data distributions across memory banks. Farrahi et al. (1998) show how a sleep mode can be exploited for memory partitions. Sudarsanam and Malik (2000) and Saghir et al. (1996) discuss techniques for exploiting dual banks for ASIPs and DSPs, respectively. Panda (1999) addresses the problem of incorporating the application-specific customization of memory bank configuration into behavioral synthesis. In comparison, we study how compiler-directed data decomposition and loop optimizations can improve energy behavior of a multi-banked system.

While a compiler-based low-power mode management strategy can lead to large energy savings in memory, it is possible to increase these savings even further by increasing the bank inter-access times (i.e., the duration between two successive accesses to a given memory bank). One way of achieving this is to modify the program access pattern such that when a bank is accessed, it is used as much as possible before moving to another bank. Programs that exhibit such bank access patterns are said to have bank locality. We demonstrate in this paper that, for array-intensive embedded applications, loop (iteration space) transformations can be used for improving bank locality.

It should be stressed that all the results presented in this paper (whether optimized for bank locality or not) have been obtained using the same compiler-based low-power mode detection/activation strategy. That is, the benefits associated with the optimized codes come solely from our optimizations, not from a specific low-power mode selection policy. We also need to mention that in this work we assume that no virtual memory support exists in the system under consideration. Consequently, the compiler can directly work with physical addresses; that is, it can layout data in physical memory and place banks into low-power modes based on the information it collected during program analysis. Note that there exist many embedded systems that work without a virtual memory support (Hwu, 1997). Work is in progress to extend the techniques discussed in this paper to environments with virtual memory (by enlisting help from OS).

3.   Affine Mappings of Arrays to Banks

We assume an affine program, where both array references and loop bounds are affine functions of enclosing loop indices and loop-invariant unknowns. Many embedded image/video-processing codes fit into this description (Catthoor et al., 1998). We represent each iteration of a given n-dimensional loop nest (i.e., a nest that contains n loops) using an n-entry vector (called iteration vector), where the ith element from top refers to the value of the ith loop index. Then, an iteration space, which is a collection of loop iterations, defines a polytope in an n-dimensional space. In a similar way, we can view data space (i.e., the set of all possible array indices) of an m-dimensional array as a rectilinear polytope, where each point (called index vector) corresponds to an array element. An affine mapping from an iteration space to an array space corresponds to an array reference, and can be represented by the affine function $\phi(I) = \theta(I) + \delta =$ where I is the iteration vector (n-entry), θ is an m-by-n linear access matrix and δ is an offset (displacement) vector. For example, a given array reference $U[i+1][j-2]$ in a two-dimensional nest (with loops i and j) has an identity access matrix and a displacement vector of $[1 \ -2]^{\text{T}}$. It is to be noted that the iteration vector here is $[i \ j]^{\text{T}}$.

Based on these definitions, we can define an affine mapping of array elements to memory banks using an affine function $\Omega(d) = \Pi d + \pi$, where d is an array index vector, Π is a b-by-m linear mapping matrix (called bank decomposition matrix or array decomposition matrix), and π is a b-entry offset (displacement) vector. In this paper, we refer to Ω as the bank decomposition function. Based on this, the array-to-bank assignment problem can be cast as one of determining an affine mapping Ω for each array involved so as the bank locality is improved as much as possible.

An important issue here is to understand what an affine bank decomposition function represents. A bank decomposition function Ω indicates how each array element is mapped to a virtual bank architecture. A virtual bank architecture is a very fine-granular banked memory architecture. Each bank in this architecture is represented using a b-entry vector (in other words, the bank space is b-dimensional). A bank decomposition function maps an array element into a virtual bank from the virtual bank architecture (also called virtual bank space). Two array elements are stored in the same virtual bank if and only if the corresponding index vectors result in the same output (i.e., the same virtual bank) when given to the bank decomposition function. In mathematical terms, let d_1 and d_2 refer to (the index vectors of) two array elements. They map to the same virtual bank (i.e., they are co-located) if and only if:

$$\Omega(d_1) = \Omega(d_2)$$

Assuming, $\Omega(d) = \Pi d + \pi$, the above condition implies that $\Pi d_1 + \pi = \Pi d_2 + \pi$, which in turn indicates that $\Pi(d_1 - d_2) = 0$; that is $(d_1 - d_2) \in Ker\{\Pi\}$. In other words, these two array elements will be mapped to the same virtual bank if and only if the difference between them is in the kernel set (null set) of the bank decomposition matrix.[1] As an example, if $\Pi = [1 \ 0]$, then $d_1 = [3 \ 5]^T$ and $d_2 = [3 \ 7]^T$ will be mapped onto the same virtual bank. However, $d_1 = [3 \ 5]^T$ and $d_2 = [4 \ 5]^T$ will not be mapped to the same bank under the same Π. A trivial bank decomposition matrix is a zero matrix that maps all elements to the same bank. However, such a decomposition matrix does not exploit any bank structure, and, consequently, is hardly useful for our objective (which is optimizing memory energy consumption).

Before going into more technical discussion about bank decompositions and loop and data transformations, let us briefly mention three important issues. First, if two array elements of an array are accessed at similar times (during loop execution) – e.g., during the same or successive loop iterations – it is important that they map on the same bank. This is because (as discussed earlier) such a mapping allows us to take advantage of low-power operating modes better. As we will show in this paper (Section 5), this can be achieved, to a large extent, by employing loop transformations. Second, co-locating some array elements means (in some cases) that we need to separate some array elements (in physical memory) that are consecutive in logical memory (i.e., from the

---

[1] A vector is said to be in the kernel set of a matrix if its dot product with every row of that matrix is 0.

viewpoint of programmer/compiler). This can create problems in addressing such array elements. We show in Section 6 that data (array layout) transformations can be used to ensure proper addressing of array elements. Third, we intentionally keep our bank decomposition matrices very general. That is, we can represent any matrix, meaning that we can perform mappings to multi-dimensional bank spaces (if desired). This flexibility allows us to explore the impact of compiler-level loop and data transformations on a wide variety of banked memory architectures. It also promotes modularity within the compiler in the following sense. When we determine loop and data transformations and bank decomposition functions, we can work with a multi-dimensional decomposition matrix (if we want to). Later, when we are to map our array data to a given banked architecture, we can use a folding function that eliminates some virtual banks (i.e., fold them into physical banks) in a systematic way. In other words, a folding function maps a virtual bank architecture (space) to a physical bank architecture (space). In doing so, some array elements that are not co-located in the virtual space can be co-located in the physical space. Section 8 discusses our folding functions in more detail.

4. Constraints for Bank Locality

Given a pair of references ($\varphi_1(I) = \theta_1 I + \delta_1$ and $\varphi_2(I) = \theta_2 I + \delta_2$) to the same array, in order to ensure accesses with spatial locality to the memory banks (i.e., bank locality), we need two types of constraints: *intra-reference constrains and inter-reference constraints*. An intra-reference constraint originates from a single reference and indicates the condition that the reference in question should access the same memory bank in two consecutive loop iterations. For example, assuming zero displacement vectors (π), in order for φ_1 to exhibit bank locality,

$$\Pi(\theta_1 I + \delta_1) = \Pi(\theta_1 I^+ + \delta_1)$$

should be satisfied. In this expression, I^+ denotes the loop iteration that immediately follows I. For simplicity, we assume that when we move from I to I^+, no loop boundary (upper bound) is crossed. From this constraint, we can obtain $\Pi\theta_1(I^+ - I) = 0$, which means that

$$\Pi\rho_1 = 0 \Rightarrow \Pi \in Ker\{\rho_1\},$$

where ρ_1 is the last column of θ_1. Note that a similar condition must be satisfied for the second reference as well (i.e., $\Pi \in Ker\{\rho_2\}$). An inter-reference constraint, on the other hand, implies that the data elements accessed by a given iteration through two references should be on the same bank. In mathematical terms, one should have

$$\Pi(\theta_1 I + \delta_1) = \Pi(\theta_2 I + \delta_2).$$

From this constraint, we can derive that $\Pi(\theta_1 - \theta_2)I = \Pi(\delta_1 - \delta_2)$. In this case, if $\theta_1 = \theta_2$ then we have

$$\Pi(\delta_1 - \delta_2) = 0 \Rightarrow \Pi \in Ker\{\delta_1 - \delta_2\}.$$

On the other hand, if $\theta_1 = \theta_2$ does not hold true, it is not possible to simplify the expression above further.

It should be noted that even if we assume that $\theta_1 = \theta_2$, requiring that $\Pi \in Ker\{\rho_1\}$ and $\Pi \in Ker\{\rho_2\}$ $\Pi \in Ker\{\delta_1 - \delta_2\}$ are satisfied at the same time implies that ρ_1, ρ_2, and $\delta_1 - \delta_2$ should be linearly dependent to each other, which may or may not be true. Of course, since in general $\theta_1 = \theta_2$ may not be hold at all, finding a suitable Π that satisfies both intra-reference constraint and inter-reference constraint becomes much harder. In the next section, we show how loop transformations can help in finding a suitable bank assignment matrix such that both intra-reference and inter-reference bank locality constraints are satisfied.

5.   Loop Transformations for Bank Locality

Previous research used many loop transformations for improving cache locality and parallelism (e.g., see Wolfe, 1996 and the references therein). In this section, we demonstrate how loop transformations can also be used for enhancing bank locality (and thus for increasing bank inter-access times). The idea behind this approach is to select a loop transformation matrix such that the chances for satisfying both intra-reference and inter-reference constraints (after the transformation) are increased.

A linear loop transformation, also called iteration space mapping, for a given n-dimensional nest (i.e., a nest that contains n loops) can be represented using an n-by-n linear matrix Λ. This linear matrix maps each iteration vector in the original nest to a new vector in the transformed nest (Wolfe, 1996). For example, if $I = [i \ j]^T$ is the original iteration vector, the loop transformation matrix

$$\Lambda = \begin{bmatrix} 0 & 1 \\ 1 & 1 \end{bmatrix}$$

results in the new (transformed) iteration vector $I' = [i' \ j']^T = \Lambda[i \ j]^T = [j \ i]^T$. In other words, such a loop transformation maps an original loop iteration $[i \ j]^T$ to a new loop iteration $[j \ i]^T$. The issues related to code generation after a loop transformation (i.e., re-writing loop body as well as re-computing loop bounds) have been studied extensively by prior research and can be found elsewhere (Li 1993; Wolfe, 1996).

Let us first focus on a single nest (which might contain multiple loops) and a single array. In Section 7, we generalize our approach to multiple arrays and multiple nests. Suppose that we apply a loop transformation represented by Λ

to a nest that contains two references φ_1 and φ_2 to the same array. Let us use I' to denote a loop iteration after the transformation (i.e., $I' = \Lambda I$ and $I = \Lambda^{-1} I'$). We can now write the intra-reference constraint as

$$\Pi(\theta_1 \Lambda^{-1} I' + \delta_1) = \Pi(\theta_1 \Lambda^{-1} I'^+ + \delta_1).$$

Consequently, after the transformation, we obtain $\Pi(\theta_1 \Lambda^{-1}(I'^+ - I') = 0$, which means that $\Pi \theta_1 q = 0 \Rightarrow \Pi \in Ker\{\theta_1 q\}$, where q is the last column of the inverse of the loop transformation matrix (i.e., the last column of Λ^{-1}). What this expression means is that by changing q (i.e., by selecting different q columns), we can obtain different bank decomposition matrices.

We can also re-write the inter-reference constraints after the loop transformation. Specifically, assuming $I' = \Lambda I$, we have

$$\Pi(\theta_1 \Lambda^{-1} I' + \delta_1) = \Pi(\theta_2 \Lambda^{-1} I'^+ + \delta_2).$$

From this last constraint, we can derive that $\Pi(\theta_1 - \theta_2)\Lambda^{-1} I' = \Pi(\delta_2 - \delta_1)$. So, if $\theta_1 = \theta_2$ then we have $\Pi(\delta_2 - \delta_1) = 0 \Rightarrow \Pi \in Ker\{\delta_2 - \delta_1\}$. However, now even if $\theta_1 = \theta_2$ does not hold, if we can select a Λ^{-1} from the kernel set of $(\theta_1 - \theta_2)$, we make sure that we can satisfy the inter-reference constraint by ensuring $\Pi \in Ker\{\delta_2 - \delta_1\}$. In other words, by selecting a suitable loop transformation matrix, we can increase the chances that our selection of bank decomposition matrix is constrained only by δ_1 and δ_2. Consequently, our approach to determining a loop transformation matrix (Λ) and a bank decomposition matrix (Π) for satisfying both intra-reference and inter-reference constraints for a given loop nest performs the following two tasks:

- Select a loop transformation matrix Λ from the kernel set of $(\theta_1 - \theta_2)$.

- Select a bank transformation matrix Π that satisfies $\Pi \in Ker\{\theta_1 q\}, \Pi \in Ker\{\theta_2 q\}$, and $\Pi \in Ker\{\delta_2 - \delta_1\}$.

The most important point to note here is that the chances of satisfying the second constraint here are higher than the case without loop transformations. This is because, in this case, we might be able to select a suitable q such that $\{\theta_1 q\}$, $\{\theta_2 q\}$, and $\{\delta_2 - \delta_1\}$ become linear-dependent on each other. In the following discussion, we study this issue in more detail.

We first consider a special case where $\theta_1 = \theta_2$. This case frequently occurs in many array-intensive applications from embedded image/video processing (especially in those that perform stencil-types of computations) (Catthoor et al., 1998). In this case, we are left with two constraints that need to be satisfied: $\Pi \theta_1 q = 0$ and $\Pi(\delta_2 - \delta_1) = 0$. Therefore, one can first determine Π from the second equation and then use it in solving the first equation for q. After determining q, it can be completed to a full Λ using, for example, the dependence-aware matrix completion strategy by Li (1993). This discussion clearly indicates that when $\theta_1 = \theta_2$ holds, the loop transformation enables us satisfy both intra-reference and inter-reference constraints.

If $\theta_1 = \theta_2$ does not hold true, this means that we need to satisfy these three constraints: $\Pi\theta_1 q = 0$, $\Pi\theta_2 q = 0$, and $\Pi(\theta_1 - \theta_2)\Lambda^{-1}I' = \Pi(\delta_2 - \delta_1)$. In this case, we first try to find a Λ^{-1} such that $(\theta_1 - \theta_2)\Lambda^{-1} = 0$. If such a Λ^{-1} can be found, then we try to select a bank decomposition matrix Π such that, these three constraints are satisfied: $\Pi\theta_1 q = 0$, $\Pi\theta_2 q = 0$, and $\Pi(\delta_2 - \delta_1) = 0$. If, on the other hand, we cannot find a Λ^{-1} that satisfies $(\theta_1 - \theta_2)\Lambda^{-1} = 0$, then there is no point in trying to satisfy $\Pi(\delta_2 - \delta_1) = 0$, and, so, we try to select a Π such that only $\Pi\theta_1 q = 0$ and $\Pi\theta_2 q = 0$ are satisfied. This means that if we cannot find a Λ^{-1} that satisfies $(\theta_1 - \theta_2)\Lambda^{-1} = 0$, we sacrifice the inter-reference locality (i.e., we do not try to satisfy the inter-reference constraint).

Example: Consider the following loop nest that contains two loops and two references to the same two-dimensional array (U):

```
for i = 1, N
    for j = 1, N − 1
        ...U[i][j]...U[i − 1][j + 1]...
```

Let us first assume that we will not use any loop transformation, and check whether it is possible to come up with a data distribution (decomposition) strategy that satisfies both intra-reference and inter-reference bank locality constraints. In this example, we have $\theta_1 = \theta_2$ and $\rho_1 = \rho_2 = [0 \ 1]^T$. In addition, $\delta_1 = [0 \ 0]^T$ and $\delta_2 = [-1 \ 1]^T$. Since we need to come up with a Π to satisfy both $\Pi\rho_1 = 0$ and $\Pi(\delta_1 - \delta_2) = 0$, we should have both

$$\Pi \begin{bmatrix} 0 \\ 1 \end{bmatrix} = 0 \quad \text{and} \quad \Pi \begin{bmatrix} -1 \\ 1 \end{bmatrix} = 0$$

at the same time. Since it is not possible to select a non-zero Π to satisfy both of these equations (since δ_1 and δ_2 are linearly independent), we can conclude that it is not possible to exploit intra-reference and inter-reference bank locality at the same time.

On the other hand, if we use a loop transformation Π, we need to satisfy $\Pi\theta_1 q = 0$ and $\Pi(\delta_1 - \delta_2) = 0$. From the second constraint, we can determine $\Pi = [1 \ 1]$. By substituting this in the first constraint, we have

$$[1 \ 1] \begin{bmatrix} 1 & 0 \\ 0 & 1 \end{bmatrix} q = 0,$$

which means $q = [1 \ -1]^T$. By completing this to a full matrix, we obtain

$$\Lambda^{-1} = \begin{bmatrix} 1 & 1 \\ 0 & -1 \end{bmatrix}.$$

Consequently, assuming that there are no data dependences in the loop nest, the transformed array references are $U[i' + j'][-j']$ and $U[i' + j' - 1][1 - j']$. Now, let us check the intra-reference and inter-reference bank locality constraints. Since

$$[1\ 1] \begin{bmatrix} i' + j' \\ -j' \end{bmatrix} = [1\ 1] \begin{bmatrix} i' + j' - 1 \\ 1 - j' \end{bmatrix} = i',$$

the inter-reference constraint is satisfied. Similarly, since

$$[1\ 1] \begin{bmatrix} i' + (j' + 1) \\ -(j' + 1) \end{bmatrix} [1\ 1] \begin{bmatrix} i' + j' \\ -j' \end{bmatrix} = i'$$

and

$$[1\ 1] \begin{bmatrix} i' + (j' + 1) - 1 \\ 1 - j' \end{bmatrix} = [1\ 1] \begin{bmatrix} i' + (j' + 1) - 1 \\ 1 - (j' + 1) \end{bmatrix} = i',$$

the intra-reference constraints (one per array reference) are also satisfied. Therefore, we can conclude that using loop transformation (Λ) can enable the compiler to exploit memory bank locality fully (which was not possible without loop transformation).

The results presented above can be extended to cases where we have more than two references to the same array in a given loop nest. Specifically, let us assume that we have υ references to the same array, namely, $\varphi_1, \varphi_2, ..., \varphi_\upsilon$. In this case, if $\theta_1 = \theta_2 = \cdots = \theta_\upsilon$ holds, we need to satisfy the following constraints: $\Pi\theta_1 q = 0$ and $\Pi(\delta_j - \delta_i) = 0$, where $1 \leq i < j \leq \upsilon$. This second expression represents a group of constraints (in fact, each pair of references contributes one constraint to the system of constraints). In this case, we first try to find a bank decomposition matrix from $\Pi(\delta_j - \delta_i) = 0$, and then, we determine a suitable q from $\Pi\theta_1 q = 0$. If, however, the said equality does not hold, we first try to find a Λ^{-1} such that the equality $(\theta_i - \theta_j)\Lambda^{-1} = 0$ will be true. If such a Λ^{-1} can be found, then we try to select a Π such that, the following constraints are satisfied: $\Pi\theta_k q = 0$ and $\Pi(\delta_j - \delta_i) = 0$, where $1 \leq k \leq \upsilon$ and $1 \leq i < j \leq \upsilon$. If, on the other hand, we cannot find a Λ^{-1} that satisfies $(\theta_i - \theta_j)\Lambda^{-1} = 0$, then it is not meaningful to satisfy $\Pi(\delta_j - \delta_i) = 0$, and we simply try to select a Π such that only the constraints $\Pi\theta_k q = 0$ are satisfied.

It should be noted that during this solution process we might not be able to satisfy all constraints. That is, we may not be able to select Π and Λ^{-1} such that all our constraints are satisfied (especially when we have too many references to the array). In such cases, to find a solution, we need to drop some constraints from consideration. It should be observed that our constraints come from either a single reference (e.g., $\Pi\theta_1 q = 0$ or from two references (e.g., $\Pi(\delta_j - \delta_i) = 0$). Therefore, we can select the constraint(s) to be dropped by deciding relative importance for array references with respect to each other. For example, a reference that is accessed conditionally (i.e., due to an enclosing if-statement) would be less important than a reference that will definitely be accessed (unconditionally) once the loop is entered. In this case, it makes sense to drop the former reference from the consideration (instead of the latter) if required. More specifically, we can associate *weights* with references, and in deciding which constraint(s) to drop, we can use these weights. Our current implementation uses the number of compiler-estimated accesses to a reference as its weight. In other words, a more frequently accessed reference will have a larger weight than a less frequently accessed reference. Based on this, the weight of an intra-reference constraint is set to the weight of the reference it contains; and the weight of an inter-reference constraint is set to sum of the weights of the references involved. As an example, suppose that within a nested loop we have three references: $U[i][j]$, $U[i-1][j+1]$, and $U[i-2][j+2]$. Assume further that the first reference occurs three times, the second one two times, and third one only once. In this case, if the compiler cannot satisfy all the bank locality constraints, it drops those belonging to the reference $U[i-2][j+2]$, and tries to satisfy the remaining ones. If it is still not successful, it drops the ones belonging to $U[i-1][j+1]$, and tries to satisfy the remaining ones and so on. This process continues until the compiler satisfies the remaining constraints, or it cannot satisfy any of them (in which case we do not optimize the array in question).

6.   Implementing Array Decompositions

The discussion so far helps us determine suitable loop transformations and bank decompositions for enhancing bank locality. However, it does not tell us anything about how such bank decompositions can actually be implemented. In this section, we demonstrate how data transformations can be used for implementing such decompositions. In particular, we study answers to the questions of the type "what does it mean to distribute a row-major array across memory banks in a column-by-column fashion?"

We recall from data transformation theory (O'Boyle and Knijnenburg, 1998) that applying a data transformation represented by a linear matrix M to an array transforms a reference $\theta I + \delta$ (to that array) to $M(\theta I + \delta)$. In fact, using such linear transformation matrices has been the most convenient way for implementing memory layout transformations in several high-level languages such as C and Fortran (Leung and Zahorjan, 1995). For example, using an inverse identity matrix as M converts the memory layout of a multi-dimensional array from column-major to row-major, or vice versa. We refer the reader to O'Boyle and Knijnenburg (1998) and Leung and Zahorjan (1995) for a comprehensive discussion of data transformation theory and low-level code generation issues after a data transformation. While the main goal of prior work on data transformations has been enhancing data locality in cache memory based architectures, the main objective in our work is to implement array decompositions across banks of a multi-bank architecture.

In our context, however, we use data transformations for a different purpose. Suppose, for example, that our approach discussed above resulted in a bank decomposition matrix $\Pi = [1 \ -1]^T$. What this means is that two array elements such as d1 and d2 are to be mapped into the same memory bank if and only if

$$\Omega(d_1) = \Omega(d_2) \Rightarrow [1 \ -1]d_1 = [1 \ -1]d_2.$$

In other words, two array elements such as $[5 \ 3]^T$ and $[6 \ 4]^T$ need to be co-located. But, we know that C language stores arrays using a row-major layout; that is, $[5 \ 3]^T$ and $[6 \ 4]^T$ are not consecutive in memory (as far as the compiler's view of layout is concerned). In order to reconcile our array decomposition with row-major storage, we need to reshuffle array elements (in memory) such that these two elements are consecutive in memory. This can be done by selecting a suitable data transformation matrix and applying it to the array in question. Our data transformation matrix should be such that after the transformation array elements such as $[5 \ 3]^T$ and $[6 \ 4]^T$ should be stored one after another.

In mathematical terms, let us assume that Π is the desired decomposition, and d1 and d2 are the array elements that need to be brought together, the selected data transformation matrix M should satisfy the following equality (focusing on a one-dimensional bank layout and a two-dimensional array): $\Pi M = [1 \ 0]$. This is because, under the mentioned assumptions $[1 \ 0]$ represents the distribution of array elements across virtual banks in a row-by-row fashion. By post-multiplying Π by M, we make sure that the elements that are required to be co-located by the decomposition matrix should be treated as if they are consecutive in memory.

7. Global Optimizations

So far, we have discussed how loop transformations can be used for improving memory bank locality, and how data transformations can be used for implementing the desired decomposition of arrays across memory banks. However, our discussion focused on a single nest (which might contain multiple loops) and a single array. In reality, most large-scale embedded applications have multiple nests, each manipulating a subset of the arrays declared in the code. In this section, we extend our approach to multiple arrays and multiple nests.

7.1 Single Nest, Multiple Arrays

Let us first focus on a case where we have a single nest (which may contain multiple loops) that manipulates multiple arrays. For clarity of presentation, we will focus on a special case with two arrays only; however, the results we obtain can easily be generalized to larger number of arrays as well.

Let Π_u and Π_v be the bank decomposition matrices for arrays U and V, respectively. Also, let $\theta_1 I + \delta_1$ and $\theta_2 I + \delta_2$ (resp. $\theta_3 I + \delta_3$ and $\theta_4 I + \delta_4$) be two references to array U (resp. V) in the nest. Then, assuming a loop transformation matrix Λ, for the best bank locality, the following constraints should be satisfied:

- $\Pi_u \theta_1 q = 0$

- $\Pi_u \theta_2 q = 0$

- $\Pi_u (\theta_1 - \theta_2) \Lambda^{-1} I' = \Pi_u (\delta_2 - \delta_1)$

- $\Pi_v \theta_3 q = 0$

- $\Pi_v \theta_4 q = 0$

- $\Pi_v (\theta_3 - \theta_4) \Lambda^{-1} I' = \Pi_v (\delta_4 - \delta_3)$

As before, in these expressions, q denotes the last column of the inverse of loop transformation matrix, and I' is the transformed iteration vector.

Let us first focus on the case where $\theta_1 = \theta_2$ and $\theta_3 = \theta_4$. In this case, we find a Π_u from $\Pi_u(\delta_2 - \delta_1) = 0$, and a Π_v from $\Pi_v(\delta_4 - \delta_3) = 0$. Then, using these Π_u and Π_v, we determine a suitable q by simultaneously solving $\Pi_u \theta_1 q = 0$ and $\Pi_v \theta_3 q = 0$. Of course, when such a q is not feasible, we need to drop of one or more equations. The concept of reference weights discussed earlier (in Section 5) can be used for choosing the victim.

If $\theta_1 = \theta_2$ and $\theta_3 = \theta_4$ do not hold, then, we proceed as follows. First, we check whether it is possible to find a Λ^{-1} such that both $(\theta_1 - \theta_2)\Lambda^{-1} = 0$ and $(\theta_3 - \theta_4)\Lambda^{-1} = 0$ will be satisfied. If it is, then, we can solve $\Pi_u \theta_1 q = 0$, $\Pi_u \theta_1 q = 0$, and $\Pi_u(\delta_2 - \delta_1) = 0$ for Π_u; and similarly, we can solve $\Pi_v \theta_3 q = 0$, $\Pi_v \theta_4 q = 0$, and $\Pi_v(\delta_4 - \delta_3) = 0$ for Π_v. As before, a conflict resolution strategy is used when necessary. If, on the other hand, it is not possible to find a Λ^{-1} such that both $(\theta_1 - \theta_2)\Lambda^{-1} = 0$ and $(\theta_3 - \theta_4)\Lambda^{-1} = 0$ are satisfied, then the third and sixth constraints given above can be dropped from consideration (i.e., inter-reference locality is sacrificed). Consequently, we first select a q, and then using $\Pi_u \theta_1 q = 0$, $\Pi_u \theta_1 q = 0$, $\Pi_v \theta_3 q = 0$, and $\Pi_v \theta_4 q = 0$, we determine Π_u and Π_v. Note that the case where $\theta_1 = \theta_2$ holds but $\theta_3 = \theta_4$ does not (or vice versa) can be handled by modifying the strategy explained above slightly. Details are omitted due to lack of space.

7.2 Multiple Nest, Single Array

We now concentrate our attention on another interesting case, where we have multiple nests manipulating one array. Again, for ease of presentation, we focus on a special case where we have only two nests. It should be straightforward to extend these results to multiple nests case. Assuming that Λ_1 and Λ_2

are the loop transformation matrices (to be determined) for the first and second nest, and that q_1 and q_2 are the last columns of Λ_1^{-1} and Λ_2^{-1}, respectively, we need to satisfy the following constraints:

- $\Pi\theta_1 q = 0$

- $\Pi\theta_2 q = 0$

- $\Pi(\theta_1 - \theta_2)\Lambda^{-1}I' = \Pi(\delta_2 - \delta_1)$

- $\Pi\theta_3 q = 0$

- $\Pi\theta_4 q = 0$

- $\Pi(\theta_3 - \theta_4)\Lambda^{-1}J' = \Pi(\delta_4 - \delta_3)$

In these expressions, I' (which is $\Lambda_1 I$) and J' (which is $\Lambda_2 J$) are the iteration vectors for the first and second nests, respectively, after the loop transformations. Also, $\theta_1 I + \delta_1$ and $\theta_2 I + \delta_2$ (resp. $\theta_3 J + \delta_3$ and $\theta_4 J + \delta_4$) are the references to the array in the first (resp. second) nest.

As before, let us first consider the case where $\theta_1 = \theta_2$ and $\theta_3 = \theta_4$. In this case, we can find a Π by solving $\Pi(\delta_2 - \delta_1) = 0$ and $\Pi(\delta_4 - \delta_3) = 0$ together. Then, using this Π, one can determine a q_1 by solving $\Pi\theta_1 q_1 = 0$, and a q_2 by solving $\Pi\theta_3 q_2 = 0$.

If the above equalities do not hold true, however, we can proceed as follows. First, we check whether it is possible to find a Λ_1^{-1} such that $(\theta_1 - \theta_2)\Lambda_1^{-1} = 0$ and a Λ_2^{-1} such that $(\theta_3 - \theta_4)\Lambda_2^{-1} = 0$ will be satisfied. If such Λ_1 and Λ_2 are possible, then, we can solve $\Pi\theta_1 q_1 = 0$, $\Pi\theta_2 q_1 = 0$, $\Pi\theta_3 q_2 = 0$, $\Pi\theta_4 q_2 = 0$ $\Pi(\delta_2 - \delta_1) = 0$, and $\Pi(\delta_4 - \delta_3) = 0$ for Π. If necessary, a conflict resolution strategy can be used to drop some constraint(s). If, however, it is not possible to find such Λ_1^{-1} and Λ_2^{-1} such that both $(\theta_1 - \theta_2)\Lambda_1^{-1} = 0$ and $(\theta_3 - \theta_4)\Lambda_2^{-1} = 0$ are satisfied, then the third and sixth constraints above can be dropped from consideration. In this case, we can select an arbitrary Π, and then using $\Pi\theta_1 q_1 = 0$, $\Pi\theta_2 q_1 = 0$, $\Pi\theta_3 q_3 = 0$, and $\Pi\theta_4 q_3 = 0$, determine q_1 and q_2.

Example: As an example, let us assume that a program has two separate nests that access a two-dimensional array U. Assume further that the references in the first nest are $U[i][j]$ and $U[i+1][j-1]$, and those in the second nest are $U[j][3i]$ and $U[j+2][3i-2]$. First, from $\Pi(\delta_2 - \delta_1) = 0$ and $\Pi(\delta_4 - \delta_3) = 0$, we need to determine a Π. In this case, we find that $\Pi = [1\ 1]$. Then, using this decomposition matrix, we need to determine a q_1 (for the first nest) from $\Pi\theta_1 q_1 = 0$, and a q_2 (for the second nest) from $\Pi\theta_3 q_3 = 0$. In this example, we find that $q_1 = [1\ -1]^T$ and $q_2 = [1\ -3]^T$. After that, using Li's method (Li 1993), these vectors (last columns) can be completed to Λ_1^{-1} and Λ_2^{-1}.

7.3   Multiple Nests, Multiple Arrays

This is the most general case where multiple nests operate on multiple arrays. A solution to this case can be developed by combining the solutions presented in the last two sections (single nest, multiple arrays) and (multiple nests, single array). Therefore, instead of giving the details (of the solution procedure) here, we just present the overall solution in an algorithmic form in Figure 8.2.

In this figure, $\theta_{i,j,k}$ represents the access matrix to the kth reference to array j in nest i. $\delta_{i,j,k}$ is the corresponding offset vector. Π_j, Λ_i, and q_i denote, respectively, the bank decomposition matrix for array j, the loop transformation matrix for nest i, and the last column of Λ_i-1. Also, F, G, and $H(i,j)$ denote, respectively, the number of nests, the number of arrays, and the number of references to array j in nest i. In lines 2 and 3, we build intra-reference and inter-reference constraints (for all arrays, references, and nests in the code). The variable trace indicates whether inter-reference constraints should be taken into account. If the loop in line 4 determines that for some arrays, inter-reference constraints can be solved, it determines the bank decomposition matrices taking into account those constraints as well (in addition to intra-reference constraints). For the arrays that it is not possible, only intra-reference constraints are accounted for. In lines 4, 6, and 7, we use the weight concept defined earlier for resolving conflicts (i.e., for dropping constraints). Finally, some remarks on line 7 follow. In order to solve such a system, we use an incremental approach. For example, we can first fix a Π_j and determine (using this Π_j) a q_i. This q_i in turn can be used to find a $\Pi_{j'}$ (where $j' \neq j$), which can be used, in turn, for determining a $q_{i'}$ (where $i' \neq i$), and so on.

1. $track \leftarrow false$
2. Build $\Pi_j\theta_{i,j,k}q_i = 0$
        for $1 \leq i \leq F$, $1 \leq j \leq G$, and $1 \leq k \leq H(i,j)$
3. Build $Pi_j(\theta_{i,j,k} - \theta_{i,j,k'}\Lambda^{-1}I' = \Pi_j(\delta_{i,j,k} - \delta_{i,j,k'})$
        for $1 \leq i \leq F$, $1 \leq j \leq G$, $1 \leq k,k' \leq H(i,j)$, and $k \neq k'$
4. For each i and j:
        If $(\theta_{i,j,k} = \theta_{i,j,k'})$ for any $k \neq k'$ then
            $track \leftarrow true$
            Solve $(\Pi_j$ from $\Pi_j(\delta_{i,j,k} - \delta_{i,j,k'}) = 0$ trying to satisfy
            as many such k and k' as possible
5. If $track = true$ for all j such that $\theta_{i,j,k} = \theta_{i,j,k'}$ does not hold for any i, k, k'
        select arbitrary Π_j
6. If $track = true$ true, then
        determine q_i from $\Pi_j\theta_{i,j,k}q_i = 0$
7. If track is not true, determine Π_j and q_i from $\Pi_j\theta_{i,j,k}q_i = 0$
8. Complete q_i columns to Λ_i^{-1} matrices taking care of data dependences

Fig. 8.2   Compiler algorithm for bank locality

7.4   Discussion

It should be stressed that so far our presentation has centered on intra-reference and inter-reference constraints. Both of these constraints are in fact what we can call intra-array constraints. That is, they are defined with respect to a given array. One can also define inter-array constraints. The idea would be clustering array elements (that belong to different arrays) that are accessed by the same loop iteration ((Delaluz et al., 2001) tries to achieve this by employing clustering heuristics). The displacement vectors (that we have not exploit so far) in bank decomposition function can be used for this purpose (along with bank decomposition matrices). Handling inter-array constraints can enable aggressive bank locality optimizations such as array interleaving and might result in a very complex set of equations (to be satisfied). Due to lack of space, we postpone the detailed discussion of inter-array constraints to a future study. However, it should be mentioned that intra-array constraints and inter-array constraints are complementary; that is, an optimizing compiler should try to satisfy both of them to the extent possible.

8.   Folding Functions

A folding function specifies a mapping of virtual banks to physical banks (i.e., the banks that actually exist in the system). In doing so, it can reduce the dimensionality and/or extents (dimension sizes) of the virtual bank space. A folding function Δ is specified by giving a decomposition style for each dimension of the virtual bank space along with the physical bank size in each dimension (of the physical bank space). For example, a folding function such as

$$\Delta : (b_1, b_2, b_3) \rightarrow (b_1/N, *, *)$$

indicates that a three-dimensional virtual bank space is mapped to a one-dimensional physical bank space. The $*$ notation here indicates that the corresponding virtual bank dimension is not distributed across physical banks. This indicates, for the example above, that the second and the third dimensions are not distributed; instead, they are folded. The notation b_1/N (where/denotes integer division) in the first dimension, on the other hand, reveals that this dimension (of the virtual bank space) is distributed across N physical banks. So, assuming N is 8, under such a folding function, the virtual bank $(16, i, j)$ is mapped to physical bank 2 for all values i and j can take. This folding function definition is quite general and encompasses very different types of virtual-to-physical bank mappings. For example, it can accommodate functions such as

$$\Delta : (b_1, b_2, b_3) \rightarrow (b_1/N, b_2/M, *),$$

which indicates that a three-dimensional virtual bank space is mapped to a two-dimensional physical bank space that contains $N \times M$ banks (this might be

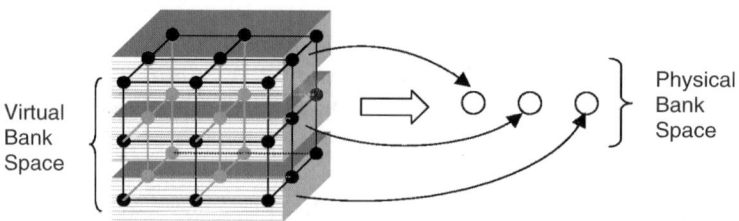

Fig. 8.3   An example folding function

useful, for example, for some SDRAMs, where memory banks actually form a two-dimensional grid). It should also be noted that when multiple virtual banks are mapped to the same physical bank, the loop iterations that access those virtual banks are localized (i.e., they exhibit bank locality as they now – after folding – access the same physical bank). Therefore, selection of a suitable folding function can be very important. In this paper, we set the dimensionality of the virtual bank space (for a given application) to the number of dimensions of the array (in that application) with the largest number of dimensions. Also, we restrict ourselves to one-dimensional physical bank spaces. Consequently, we use folding functions of the following type:

$$\Delta : (b_1, b_2, ..., b_r) \rightarrow (b_1/N, *, ..., *).$$

Such a folding function tries to take advantage of spatial locality between neighboring banks. In the future, we will study the impact of other types of folding functions on memory energy behavior. Figure 8.3 shows an example folding, where a three-dimensional virtual bank space is mapped to a one-dimensional physical bank space. We see that each two-dimensional slice of the virtual bank space is mapped to a single node of the physical bank space.

9.   Experiments

Our objective in this section is to demonstrate how successful our approach is in practice. In Section 9.1, we introduce our benchmarks and experimental setup. In Section 9.2, we give detailed experimental results.

9.1   Benchmark Codes and Experimental Framework

Figure 8.4 lists the benchmark codes used in this study and their important characteristics. Our applications are array-intensive codes that are typical in many embedded environments that target image and video processing. Bss is a blind signal separation and deconvolution application. Flt is a particle filter algorithm for wireless communication. Spe is combined spectral envelope normalization and subtraction algorithm. Seg is an image segregation code based

Benchmark Name	Number of Lines	Input Size	Base Energy	Base Time
Bss	486	44MB	274.3 mJ	8.11 nsec
Flt	324	62MB	188.1 mJ	5.20 nsec
Spe	156	61MB	108.1 mJ	2.14 nsec
Seg	144	58MB	106.4 mJ	2.08 nsec
Vid	1618	62MB	342.7 mJ	9.65 nsec

Fig. 8.4   Benchmark codes used in our experiments and baseline values

on pitch tracking and amplitude modulation. Finally, Vid is a video-enhancing algorithm that modifies transient locations. The reason that we selected these applications for our study is that they represent real-life embedded codes and are available to us. The third column in Figure 8.4 gives the total size of the data manipulated by each application. The last two columns, on the other hand, give the memory energy consumptions and execution times of our applications when a 8 × 8MB (total 64MB without any data cache) memory configuration is employed and when our low-power operating modes are used (whenever the banks are idle). In other words, even our base configuration takes advantage of low-power operating modes to the extent possible, and the benefits reported in Section 9.2 come solely from our optimization. The energy numbers reported in the next subsection are the values normalized with respect to the values given in the fourth column of Figure 8.4. Unless stated otherwise, 8 × 8MB (i.e., 8 memory banks, each has a capacity of 8MB) is our default bank configuration. Note that if a bank is not accessed during the entire execution of an application, the said bank is never activated for both the original and the optimized code versions.

Our loop and data transformations are implemented using SUIF from Stanford University (Amarasinghe et al., 1996). SUIF has independently developed compilation passes that work together by using a common intermediate format (IF) to represent programs. A typical compilation framework based on SUIF includes the following components: front end, data dependence analysis, and several optimization modules. Our framework is implemented as a separate optimization module within SUIF. We also use a powerful back-end compiler (when converting the C code to the executable) that performs instruction scheduling and graph coloring based global register allocation.

All energy numbers presented in this paper have been obtained using a custom memory energy simulator. This simulator takes as input a C program and a banked memory description (i.e., the number and sizes of memory banks as well as available low-power operating modes with their energy-saving factors and re-synchronization penalties). As output, it gives the energy consumption in memory banks along with a detailed bank inter-access time profiles. By giving the original and the optimized programs to this simulator as input, one can measure the impact of loop and data optimizations on memory system energy of a banked architecture.

9.2   Results

We start by presenting the percentage energy improvements when our optimization strategy is applied to the codes in our experimental suite. The results illustrated in Figure 8.5 (the first bar for each application) indicate that our strategy improves the energy consumption of all benchmark codes in our experimental suite. The average percentage memory energy improvement (across all benchmarks) is 19.1%. Note that our approach (in addition to determining a loop transformation for each nest in the code) also determines a suitable data transformation for each array (to implement array decompositions as explained in Section 6). These data transformations are dependent on the initial layout of data (row-major in our case); however, the resulting memory layouts are independent of the initial (default) memory layout. That is, independent of the initial layout of the arrays, our strategy determines the most suitable memory layouts for them (depending on the loop transformations selected for the nests) from the viewpoint of bank locality. However, the base energy numbers given in Figure 8.4 are dependent on the initial layout. Therefore, it is important to see whether our optimizations are still effective (in improving memory energy behavior) when the original data layout is highly optimized. To analyze this issue, we also measured the percentage energy savings over an energy optimization strategy that does not apply any loop transformations but selects the most suitable bank layouts for each array in the code being optimized (this version is from Delaluz et al. (2001)). The results given in Figure 8.5 (the second bar for each application) clearly show that our approach is still able to improve

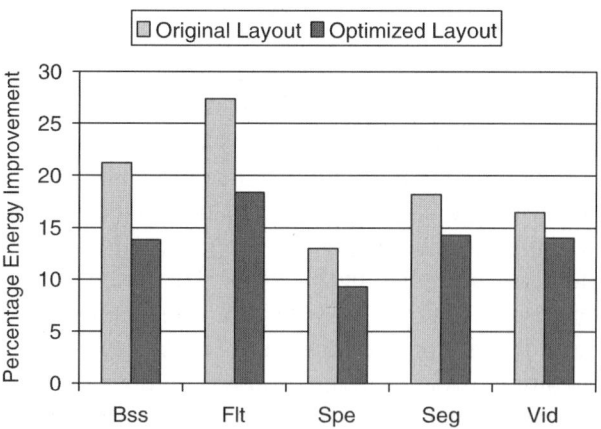

Fig. 8.5   Percentage energy savings over original (unoptimized) and optimized layouts

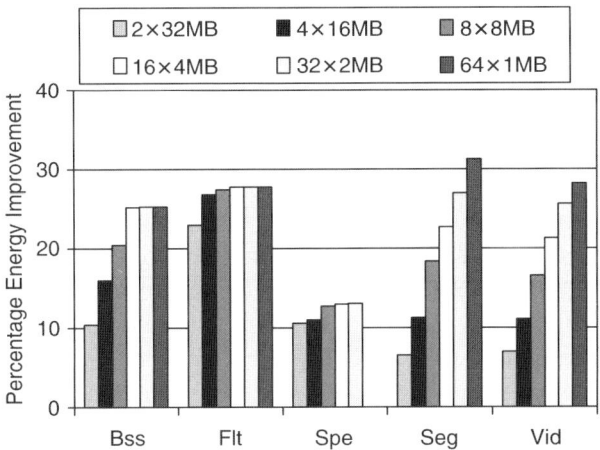

Fig. 8.6   Percentage energy savings with different bank configurations (over original layouts)

memory energy consumption for all benchmarks in our suite, achieving an average memory energy improvement of 13.8%. It should also be observed that
the energy benefits due to our optimizations vary depending on the application
under consideration. Specifically, we obtain the best results with Flt. This is
due to two main reasons. First, there are not many data dependences in this application; consequently, the compiler finds a legal (semantic-preserving) loop
transformation most of the time. Second, there are not many array references
per loop, and so, the compiler does not need to drop some (bank locality) constraints from consideration frequently. In fact, in compiling this application,
the compiler dropped only five constraints (the average value across all benchmarks was thirteen). In comparison, our savings with Spe are relatively low
due to large number of data dependences in the code (i.e., although most of the
time the compiler was able to find a loop transformation matrix Λ, the resulting
transformation was found to be illegal – not semantic-preserving).

In order to evaluate the impact of our approach with different bank configurations, we performed another set of experiments. More specifically, keeping
the total memory size fixed at 64MB, we conducted experiments with 2, 4, 8,
16, 32, and 64 banks. The energy results are given in Figure 8.6 (again as percentage improvements). One can observe from these results that working with
larger number of banks (i.e., with smaller bank sizes) in general increases the
benefits due to our optimization strategy. This is because smaller bank sizes
give our strategy more opportunities for energy-managing even smaller portions of main memory. Such a finer-grain management, in turn, increases the
energy benefits. However, one can also see from the same graph that, in two
applications (Bss and Flt), increasing the number of memory banks beyond 16

does not bring further energy benefits. This is because in these two applications increasing the number of banks beyond 16, say to 32, increases the average number of banks accessed per loop iteration. This in turn offsets the potential benefits of working with a larger number of banks. We faced the same problem with the remaining three applications only when we went beyond 64 banks (the results of which are not shown here).

So far, our experiments involved no cache memory. Including a data cache in the architecture can impact memory system energy consumption in at least two ways. First, the cache filters out some memory references that would otherwise go to off-chip memory. Consequently, we can expect an increase in bank inter-access times. Second, the cache reduces the overall energy consumption due to memory system as typically cache accesses consume much less energy than off-chip memory accesses. To calculate the data cache energy consumption, our simulator employs the cache energy model proposed in Kamble and Ghose (1997). Figure 8.7 gives the percentage memory system energy improvements with different on-chip data cache configurations over both original layouts and optimized layouts. Note that memory system energy with the cache versions includes the energy consumption in the cache as well as that in the off-chip memory. All cache configurations used are two-way set-associative with a block size of 32 bytes. One can clearly see from Figure 8.7 that, while increasing the cache size reduces the benefits of our optimization strategy, we still achieve reasonable energy savings. Therefore, even in a cache-based environment, applying bank-aware loop and data transformations can be useful. One should observe, however, that if the compiler can have an accurate

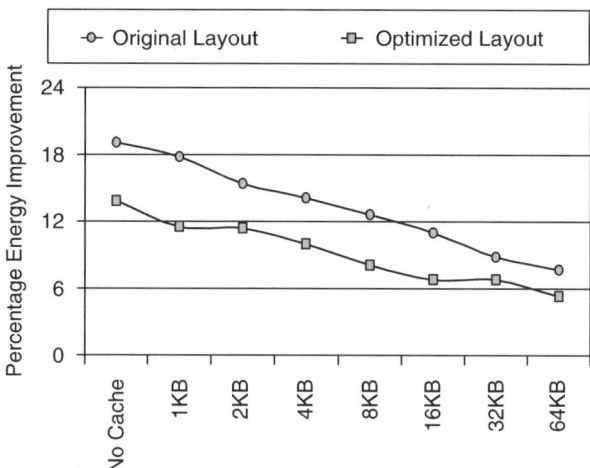

Fig. 8.7   Percentage energy savings with different cache configurations (averaged over all benchmarks)

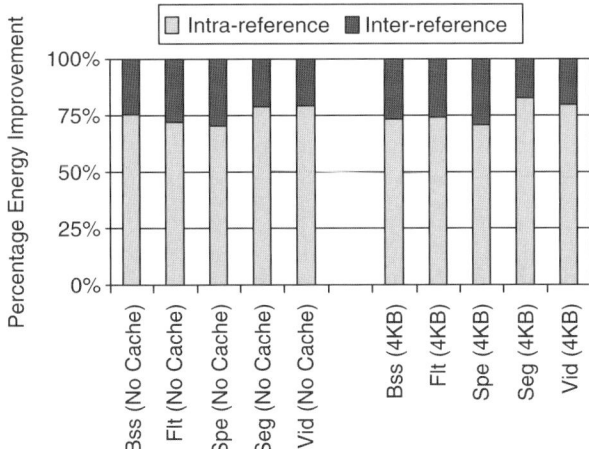

Fig. 8.8 Contributions of intra-reference and inter-reference constraints to memory energy savings (averaged over all benchmarks)

hit/miss prediction strategy, it could potentially do a much better job in managing memory banks. Implementing such a strategy is in our future agenda.

The energy benefits shown in this have been obtained by trying to satisfy both intra-reference and inter-reference constraints. We also wanted to see how these constraints contribute to these energy savings. In Figure 8.8, we illustrate how energy benefits are broken down between those coming from satisfying inter-reference constraints and those coming from satisfying intra-reference constraints (for both a cacheless memory system and for a memory system with 4KB data cache – both with our default bank configuration). These results reveal that most of the energy savings are coming from satisfying intra-reference constraints. There are two main reasons for this. First, the algorithm given Figure 8.2 does not try to satisfy inter-reference constraints if $\theta_{i,j,k} = \theta_{i,j,k'}$ checks fail. Second, in many cases, satisfying intra-reference constraints brings more benefits as such constraints capture access behavior across loop iterations. This is in contrast to inter-reference constraints whose existence is limited by the number of references to the same array in the loop body. Nevertheless, we still observe that the contribution of satisfying inter-reference constraints to overall memory energy savings is around 25% on the average, which indicates that it is important to take care of them as well.

Finally, we studied the performance impact of power-mode control with/without our loop and data transformations on performance. Since in this work we focus on array-intensive embedded applications (with compile-time analyzable data access patterns) and adopted a compiler-based operating mode control strategy, the compiler can also predict future bank re-activations and perform ahead-of-time bank re-activation (called pre-activation). We refer the reader to (Delaluz et al., 2001) for a discussion of memory bank pre-activation and its implementation details. Consequently, the performance penalty of power mode control itself was negligible (less than 2% on the average). When our optimizations are applied, we did not observe any significant performance variation in the cacheless memory architecture. In a cache-based memory architecture, on the other hand, we observed nearly 11% performance improvement on the average (mainly coming from the improved cache access patterns).

10.   Concluding Remarks and Future Work

Energy consumption is becoming increasingly important for both embedded and high-performance computing systems. The dynamic energy spent in off-chip DRAMs can be a significant portion of the overall energy budget. In this paper, we have presented a compiler-based optimization framework for array-intensive embedded applications to increase energy savings that can be obtained from low-power operating modes. Our experiments with a prototype implementation and five array-intensive embedded applications show significant savings in memory system energy.

Our future work consists of two different directions:

- First, we would like to study the impact of data optimizations on the effectiveness of our optimization strategy. While in this work we employed data transformations for implementing array decompositions, they can also be used for improving bank locality. This can be particularly useful in applications such as Vid, where we have lots of data dependences preventing loop transformations from modifying the access pattern to enhance bank locality. It is also possible that combining loop and data transformations under an integrated framework can further increase energy savings.

- Second, we would like to apply our optimization to an on-chip multiprocessor architecture. In particular, we believe that studying how bank localities of individual processors interact with each other can reveal important data, which can be used to reduce memory energy consumption in such architectures.

Acknowledgment

The author would like to thank U. Sezer of University of Wisconsin-Madison, and G. Chen, H. Saputra, and M. J. Irwin of Pennsylvania State University for their comments on earlier versions of this paper. This work is supported in part by NSF Career Award #0093082 and a fund from GSRC.

References

Ram (1999) 128/144-mbit direct RDRAM data sheet, rambus inc.

Amarasinghe, S.P., Anderson, J.M., Wilson, C.S., Liao, S.-W., Murphy, B.R., French, R.S., Lam, M.S., and Hall, M.W. (1996) Multiprocessors from a software perspective. *IEEE Micro*, pp. 52–61.

Catthoor, F., Wuytack, S., Greef, E.D., Balasa, F., Nachtergaele, L., and Vandecappelle, A. (1998) *Custom Memory Management Methodology – Exploration of Memory Organization for Embedded Multimedia System Design*. Kluwer Academic Publishers.

Delaluz, V. and Kandemir, M. (2004) Array regrouping and its use in compiling data-intensive embedded applications. *IEEE Transactions on Computers*. Vol. 53, No.1.

Delaluz, V., Kandemir, M., Vijaykrishnan, N., Sivasubramaniam, A., and Irwin, M.J. (2001) Dram energy management using software and hardware directed power mode control. In *Proceedings of the 7th Int'l Symposium on High Performance Computer Architecture*.

Fan, X., Ellis, C.S., and Lebeck, A.R. (2002) Modeling of dram power control policies using deterministic and stochastic petri nets. In *Proceedings of Workshop on Power-Aware Computer Systems*.

Farkas, K.I., Flinn, J., Back, G., Grunwald, D., and Anderson, J.-A.M. (2000) Quantifying the energy consumption of a pocket computer and a java virtual machine. In *Proceedings of SIGMETRICS*, pp. 252–263.

Farrahi, A., Tellez, G., and Sarrafzadeh, M. (1998) Exploiting sleep mode for memory partitions and other applications. *VLSI Design*, 7(3):271–287.

Hwu, W-M.W. (1997) Embedded microprocessor comparison. http://www.crhc. uiuc.edu/IMPACT/ece412/public_html/Notes/412_lec1/ppframe.htm.

Kamble, M.B. and Ghose, K. (1997) Analytical energy dissipation models for low power caches. In *Proceedings of International Symposium on Low-Power Electronics and Design*.

Kandemir, M., Kolcu, I., and Kadayif, I. (2002) Influence of loop optimizations on energy consumption of multi-bank memory systems. In *Proceedings of the International Conference on Compiler Construction*, Grenoble, France.

Kandemir, M., Sezer, U., and Delaluz, V. (2001) Improving memory energy using access pattern classification. In *Proceedings of the International Conference on Computer Aided Design*, San Jose, CA.

Lebeck, A.R., Fan, X., Zeng, H., and Ellis, C.S. (2000) Power-aware page allocation. In *Proceedings of 9th International Conference on Architectural Support for Programming Languages and Operating Systems*.

Leung, S.T. and Zahorjan, J. (1995) Optimizing data locality by array restructuring. Technical Report TR 95-09-01, Department of Computer Science and Engineering, University of Washington.

Li, W. (1993) *Compiling for NUMA Parallel Machines*. PhD thesis, Computer Science Department, Cornell University, Ithaca, NY.

O'Boyle, M. and Knijnenburg, P. (1998) Integrating loop and data transformations for global optimization. In *Proceedings of the International Conference on Parallel Architectures and Compilation Techniques*, Paris, France.

Panda, P.R. (1999) Memory bank customization and assignment in behavioral synthesis. In *Proceedings of ICCAD*.

Saghir, M.A.R., Chow, P., and Lee, C.G. (1996) Exploiting dual data-memory banks in digital signal processors. In *Proceedings of the International conference on Architectural Support for Programming Languages and Operating Systems*, pages 234–243, Cambridge, MA.

Sudarsanam, A. and Malik, S. (2000) Simultaneous reference allocation in code generation for dual data memory bank ASIPs. *ACM Transactions on Design Automation of Electronic Systems*, 5:242–264.

Wolf, M. and Lam, M. (1991) A data locality optimizing algorithm. In *Proceedings of the ACM Conference on Programming Language Design and Implementation*, pp. 30–44.

Wolfe, M. (1996) *High Performance Compilers for Parallel Computing*. Addison-Wesley Publishing Company.

III

Dynamic Voltage and Frequency Scaling

Chapter 9

Fundamentals of Power-Aware Scheduling

Xiaobo Sharon Hu[1] and Gang Quan[2]

[1]*Department of Computer Science and Engineering*
University of Notre Dame, Notre Dame, IN 46556, U.S.A.

[2]*Department of Computer Science and Engineering*
University of South Carolina, Columbia, SC 29208, U.S.A.

Abstract   Power-aware scheduling plays a key role in curtailing the energy consumption in real-time embedded systems. Since there is a vast variance in the composition and functionality of real-time embedded systems, different power-aware scheduling techniques are naturally needed. However, certain fundamental principles are applicable to all such systems. This chapter provides an overview of the basics in power and performance tradeoff and in real-time system scheduling. It also discusses the benefit of power-aware scheduling via a simple example. A categorization of different power-aware scheduling techniques are presented at the end.

Keywords: Power-aware, real-time scheduling, RTOS, dynamic voltage/frequency scaling, dynamic power management

1.   Introduction

Advances in both hardware and software technology have enabled system designers to develop large, complex embedded systems. As more powerful and sophisticated components being incorporated into embedded systems, many such systems consume a large amount of power. Though advances in low-power design techniques for embedded processors provide a variety of alternatives to mitigate power increase, it has been widely recognized that

J. Henkel and S. Parameswaran (eds.), Designing Embedded Processors – A Low Power Perspective,
219–229.

decisions made in the early design phases, particularly at the system level, are of critical importance in keeping power/energy demands in check.

Power-manageable hardware resources (such as processors and memory components) are finding their way into an ever increasing number of embedded systems. Runtime power reduction mechanisms employed by such power-manageable hardware resources can be categorized into two main types: (i) shutting down the processing unit when it is idle, commonly known as dynamic power management (DPM), and (ii) scaling the processor's supply voltages and working frequencies, usually referred as dynamic voltage/frequency scaling (DVFS). Using either of the power-saving techniques inevitably degrades system computing performance. As most embedded systems must satisfy certain timing requirements, the decreased computing capability threatens to cause these timing requirements to be violated and to make the systems unacceptable.

Embedded systems are usually designed to provide high peak performance when needed. Opportunities arise naturally to reduce power consumption by dynamically modulating performance according to workload variation. A power-aware system is one capable of providing the *right* power at the *right* place and at the *right* time. Power-aware scheduling thus aims at maximally exploiting the benefit of power-manageable resources through judiciously switching among different power-saving modes.

This and the following three chapters focus on embedded systems having real-time requirements. Real-time embedded systems (RTES) are prevalent in many real-world applications such as communication devices, transportation machines, entertainment appliances, and medical instruments. A distinctive feature of these systems is that they must produce results not only correctly but also *timely*. In the rest of the chapter, we first review pertinent properties of a CMOS circuit, the basic building block in today's microprocessors. Then, the fundamentals of real-time scheduling are summarized. We end the chapter by a motivational example to illustrate the impact of power-aware scheduling.

2.   Power and Performance Tradeoff in CMOS Circuits

In a CMOS circuit, power consumption includes two components [36]: dynamic and static power. Dynamic power consumption consists of switching power for charging and discharging the load capacitance, and short-circuit power due to the non-zero rising and falling time of the input and output signals. In general, dynamic power (P_{dyn}) can be represented as

$$P_{dyn} = \alpha C_L V^2 f, \qquad (9.1)$$

where α is the switching activity, C_L is the load capacitance, V is the supply voltage, and f is the system clock frequency. Typically, switching power dominates the dynamic component of the total power consumption. The static power consumption is primarily due to leakage currents, and the leakage power (P_{leak}) can be expressed as

$$P_{leak} = I_{leak}V, \tag{9.2}$$

where I_{leak} is the leakage current which consists of both the subthreshold current and the reverse bias junction current in the CMOS circuit. Leakage current increases rapidly with the scaling of the devices. It becomes particularly significant with the reduction of the threshold voltage.

When the processor is in the active mode, lowering supply voltage is one of the most effective ways to reduce switching power/energy consumption. According to Equations 9.1 and 9.2, reducing the supply voltage helps to reduce both the dynamic and leakage power consumption [27]. However, reducing the supply voltage increases the circuit delay (τ), which is given by

$$\tau \propto \frac{V}{(V - V_{th})^2}, \tag{9.3}$$

where V_{th} is the threshold voltage. Since a processor's maximal working frequency is inversely proportional to the circuit delay, reducing the supply voltage will increase the execution time of a software task, which may potentially violate the timing requirement of some tasks.

When the processor is idle, a major portion of the power consumption comes from the leakage. One simple technique to reduce the leakage power consumption is to switch a processor into a low-leakage mode when it is idle and switch it back to the normal execution mode when execution is required. Shutting down the processor can be considered as such a low-leakage mode. It has been reported that the power dissipation when the processor is idle can be on the order of 10^3 of that when the processor is shut down [9]. However, energy and timing overhead is unavoidable when shutting down and later waking up the processor. Other low-leakage modes may also be available on different processors. Depending on the actual techniques used for switching between different leakage modes, the timing overhead can be anywhere from tens of μs to tens of ms [35]. Again, such timing overhead may introduce unacceptable performance degradation.

The underlying principle of power-aware scheduling techniques is simply to exploit the slack times during software execution to accommodate the negative impacts on performance when applying DVFS and DPM. The challenge is to make appropriate decisions on processor speeds to guarantee the timing requirements while also considering the timing requirements of all the tasks in a system.

3.   Basics in Real-Time Scheduling

RTES are often modeled by a set of tasks, either dependent or independent, to be executed on one or more processors. The tasks, $\{\tau_1, \tau_2, \ldots, \tau_n\}$, can be periodic or aperiodic, and associated with the following parameters:

1. Period T_i: the time between the initiations of two consecutive requests (or jobs) of τ_i (and T_i is not a constant for aperiodic tasks)

2. Deadline D_i: the maximum time allowed from the initiation of a request of τ_i to the completion of the same request

3. Release time R_i: the initiation time of the first request of τ_i

4. Execution cycles C_i: the worst case clock cycles required to complete τ_i without any interruption. Note that the actual execution time of a task depends on the processor used and the processor speed selected

Scheduling determines when each request of a task is to be executed by the processor. One common strategy is to assign priorities to either jobs (requests of tasks) or tasks such that the job with the highest priority among all released jobs will be selected for execution. Depending on whether the arrival of a higher priority job can or cannot preempt the execution of the current job, schedulers are divided into preemptive ones and non-preemptive ones. The priority assignment can be either dynamic (i.e., the priority of a task can change from one request to another) or static (i.e., the priority of a task is fixed). The best known dynamic priority assignment is the earliest deadline first (EDF) algorithm and the best known static priority assignment is the rate-monotonic scheduling (RMS) [13]. Other scheduling policies may also be adopted by certain types of real-time systems, e.g., the cyclic approach and (weighted) round-robin approach [14]. In the cyclic approach, a complete schedule is derived off-line and tasks are dispatched online by following the schedule in a repeated manner. The (weighted) round-robin approach is often used for scheduling time-shared applications, where each ready task is executed for a fixed amount of time (called *time slice*). If more than one task is ready at a given time, they are selected based on the First-In-First-Out (FIFO) order. The selection of time slice size is a delicate balance between schedulability and scheduling overhead. The round-robin approach is not favored for systems with stringent timing requirements.

When an RTES employs multiple processors, tasks need to be first mapped to different processors and then scheduled on each processor. Though mapping can be done either dynamically (online) or statically (off-line), static mapping is often used for RTES in order to ensure a predictable system behavior. In general, multiprocessor mapping and scheduling are NP-hard problems [4]. Including power consumption adds more difficulty to the problem.

4. Impacts of Power-Aware Scheduling

To see how power-aware scheduling techniques can save energy, we study the example illustrated in Figure 9.1. Figure 9.1a shows a periodic task set consisting of two tasks (τ_1 and τ_2) scheduled according to the RMS, i.e., τ_1 has a higher fixed priority than τ_2. One strategy to save the energy consumption is to apply DPM, that is, to run the tasks with the highest processor speed, i.e., $S = 1$, and shut down the processor when idle as shown in Figure 9.1b. Assuming that power $P = V^3$ and processor speed $S = \frac{1}{V}$, the energy consumption within one least common multiple (LCM) of the periods, i.e., $[0, 120]$, is 70. Since the power consumption is a convex function of the supply voltage,

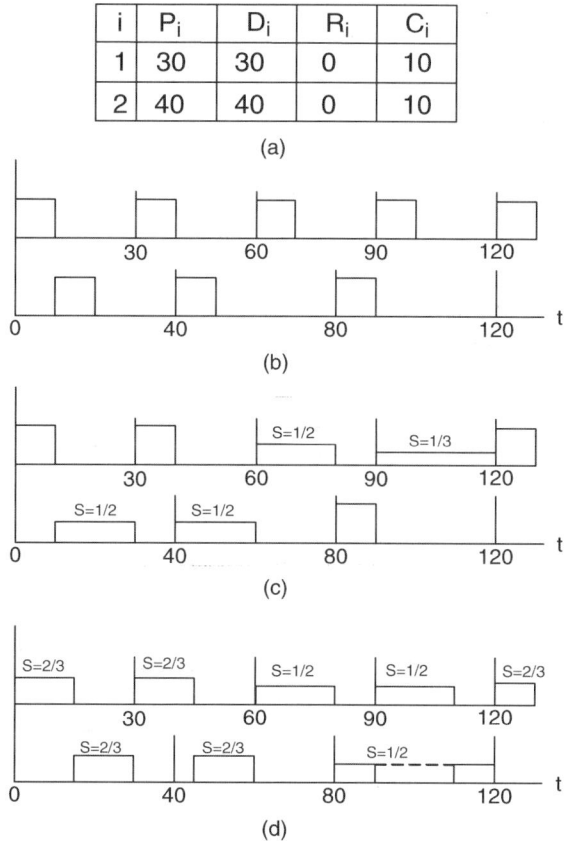

Fig. 9.1 (a) A real-time system with two periodic tasks. (b) τ_1 and τ_2 are executed at the maximal processor speed. (c) When there is only one task instance in the ready queue, reduce the processor speed and execute till its deadline or the release time of next task instance. (d) Vary the processor speed more aggressively to achieve higher energy saving. The default processor speed is 1.

reducing the supply voltage can be more effective in conserving the energy consumption. To reduce the supply voltage (and thus the processor speed) and satisfy the tasks' deadlines in the meantime, one simple strategy (as shown in Figure 9.1c) is to reduce the processor speed when there is only one task instance in the ready queue, and extend the execution of this task instance to its deadline or the release time of next task instance. In this case, the energy consumption within interval $[0, 120]$ becomes 38.6, or around 55% of that by shutting down the processor when it is idle. To achieve even more energy saving, we can be more aggressive in reducing the processor speed. For example, by carefully adjusting the processor speed as shown in Figure 9.1d, the energy consumption is reduced to as low as 25.3, while all the task deadlines are guaranteed. This example demonstrates that exploiting the characteristics of power-manageable resources and judiciously making decisions in real-time scheduling can substantially reduce energy consumption.

In the rest of the three chapters, three example DVFS techniques are introduced. The first two chapters consider priority-based (EDF and RMS), preemptive task systems executed on a single processor. Such systems can be found in a wide range of applications. Chapter 2 focuses on off-line DVFS techniques while Chapter 3 examines online DVFS techniques. The last chapter presents an off-line DVFS technique for tasks executed on a multiple processor system scheduled by a cyclic, non-preemptive policy.

5.   Further Reading

Extensive power-aware scheduling techniques have been proposed to satisfy a variety of timing constraints and minimize the power/energy consumption. Rooted in traditional real-time scheduling [33], the power-aware scheduling techniques can be classified along different dimensions. In what follows, we briefly survey these techniques from three perspectives, i.e., the real-time application characteristics, the scheduling approach, and the system architecture.

5.1   Application Characteristics

Power-aware scheduling techniques can be grouped by the nature of real-time applications. Real-time systems in general can be categorized as the *hard* or *soft* [8]. Power-aware scheduling techniques for hard real-time systems, such as those in [3, 5, 32, 38], strive to optimize the power/energy consumption under the assumption that all tasks must meet their deadlines. These techniques can further differ from each other by other characteristics such as the priority assignment, arrival periodicity, and granularity. For example, power-aware scheduling techniques are proposed to deal with fixed-priority systems [5, 25, 34] or EDF-based systems [10, 38, 44]. Some techniques target real-time application with periodic tasks, non-periodic tasks [22], or a mixture of two [31]. And some approaches also take the task granularity into considerations. For example, instead of assigning processor speed at the task level (so called the *inter-task* scheduling), Shin *et al.* proposed to assign different processor speeds based on different execution path inside a task (so called the *intra-task* scheduling [32]).

Not all real-time system are hard real-time systems. Other real-time systems, i.e., the *soft* real-time systems, can tolerate occasional deadline misses, which are dictated by different Quality of Service (QoS) requirements. The power-aware scheduling techniques for soft real-time system can be classified into two categories: the *best-effort* and the *guarantee* techniques. The *best-effort techniques* intend to maximize the QoS and minimize power/energy consumption, but with no assurance of achieving either of them. Examples of such approaches can be found in [2, 7, 17, 21, 26, 28]. The *guarantee approaches*, on the other hand, optimize energy usage while guaranteeing the specified QoS. A majority of the current guarantee approaches in power-aware scheduling are concerned with *statistical guarantee* (e.g., [8, 23, 39, 40]). However, using the statistical guarantee model such as the overall deadline miss rate may be problematic for some real-time systems (e.g., those in multimedia applications) that require the interval between subsequent deadline misses be kept small in order to restore the missing information satisfactorily. Niu *et al.* proposed a technique [20] that can provide a more deterministic guarantee for such applications.

5.2   Scheduling Decisions

Depending on how the scheduling decisions are made, power-aware scheduling techniques can be classified as either *off-line* and *online*. Off-line techniques (e.g., [18, 24, 25, 38]) are usually applied during design time, such as in the compilation and synthesis process, while online techniques (e.g., [3, 10, 11]) make scheduling decisions during runtime. The off-line scheduling techniques take advantage of the timing specifications known *a priori* and therefore can be highly predictable and also incur low scheduling overhead during runtime. One common practice in developing off-line schedules is to assume the worst case scenario, e.g., assuming that each task will take the worst case execution time. The problem, however, is that the worst case execution time can be much larger than the average case execution time, which could lead to rather pessimistic scheduling decisions and reduce severely power/energy savings. To tackle this problem, some approaches [5, 43] incorporate other known specifications, such as the probabilistic distribution of the execution time, into the scheduling making process. To further improve the energy efficiency, the off-line scheduling techniques can be combined with the appropriate online scheduling techniques resulting in some hybrid approaches, e.g., [12].

The online scheduling techniques are more flexible and adaptive to the run-time environment by using the runtime information to guide the scheduling decisions. As most embedded systems are highly dynamic in nature, the online approaches can lead to significant energy savings if used appropriately. For example, Kim *et al.* [10, 11] and Aydin *et al.* [3] proposed to dynamically reclaim the system resources if a task finishes earlier than that in the worst case scenarios. Zhu *et al.* [44] presented a technique to dynamically adjust processor speeds via feedback control. Since these scheduling techniques are applied online, the computation complexity must be kept very low. Otherwise, the techniques may be infeasible or severely compromise their power/energy efficiency.

5.3   Architectures

From embedded system architecture perspective, power-aware scheduling techniques can be categorized as uniprocessor or multiprocessor scheduling. A large body of power-aware scheduling techniques have targeted at uniprocessor systems. As IC technology and real-time application complexity continue to grow, we have seen increasing research interests in developing power-aware techniques for multiprocessor platform. Different from uniprocessor scheduling techniques, multiprocessor scheduling techniques need to decide not only when but also where a task needs to be executed. Therefore it becomes more difficult to find the optimal solution. The majority of the multiprocessor power-aware scheduling techniques have been focused on distributed systems with precedence relationships among tasks (e.g., [6, 15, 16, 30, 42]). There are also some techniques introduced for distributed system with independent tasks (e.g., [1]).

Besides the number of processing elements, other architectural issues are also dealt with in the current power-aware scheduling research. For example, Yan *et al.* [37] proposed to exploit the DVS capability and the adaptive body biasing (ABB) control—enabled by recent IC technology advances to address the increasing concerns of the leakage power consumption in the IC circuit— to reduce the overall power consumption. Other implementation issues such as discrete levels of supply voltage and the transition overhead existing in the practical commercial processors are also studied [19, 29, 41].

References

[1] T.A. AlEnawy and H. Aydin. Energy-aware task allocation for rate monotonic scheduling. *RTAS*, pages 213–223, 2005.

[2] H. Aydin, R. Melhem, D. Mosse, and P. Alvarez. Determining optimal processor speeds for periodic real-time tasks with different power characteristics. *ECRTS*, pages 225–232, June 2001.

[3] H. Aydin, R. Melhem, D. Mosse, and P. Alvarez. Dynamic and aggressive scheduling techniques for power aware real-time systems. *RTSS*, pages 95–105, 2001.

[4] M. Garey and D. Johnson. *Computers and Intractability: A Guide to the Theory of NP-Completeness*. FreeMan, San Francisco, CA, 1979.

[5] F. Gruian. Hard real-time scheduling for low energy using stochastic data and dvs processors. *ISLPED*, 46-51, 2001.

[6] F. Gruian and K. Kuchcinski. Lens: Task scheduling for low-energy systems using variable supply voltage processors. *ASPDAC*, pages 449–455, 2001.

[7] D. Grunwald, P. Levis, K.I. Farkas, C.B. Morrey III, and M. Neufeld. Policies for dynamic clock scheduling. *Proceedings of OSDI*, pages 73–86, 2000.

[8] S. Hua, G. Qu, and S. Bhattacharyya. Energy reduction techniques for multimedia applications with tolerance to deadline misses. *DAC*, pages 131–136, 2003.

[9] Intel. *PXA250 and PXA210 Applications Processors Design Guide*. Intel, 2002.

[10] W. Kim, J. Kim, and S.L. Min. A dynamic voltage scaling algorithm for dynamic-priority hard real-time systems using slack analysis. *DATE*, pages 788–794, 2002.

[11] W. Kim, J. Kim, and S.L. Min. Dynamic voltage scaling algorithm for fixed-priority real-time systems using work-demand analysis. *ISLPED*, pages 396–401, 2003.

[12] L. Leung, C.-Y. Tsui, and X. Hu. Joint dynamic voltage scaling and adpative body biasing for heterogeneous distributed real-time embedded systems. *DATE*, pages 634–639, 2005.

[13] C.L. Liu and J.W. Layland. Scheduling algorithms for multiprogramming in a hard real-time environment. *Journal of the ACM*, 17(2):46–61, 1973.

[14] J. Liu. *Real-Time Systems*. Prentice Hall, Upper Saddle River, NJ, USA, 2000.

[15] J. Liu, P.H. Chou, N. Bagherzadeh, and F.J. Kurdahi. Power-aware scheduling under timing constraints for mission-critical embedded systems. *DAC*, pages 840–845, 2001.

[16] J. Luo and N. Jha. Power-conscious joint scheduling of periodic task graphs and aperiodic tasks in distributed real-time embedded systems. *ICCAD*, pages 357–364, 2000.

[17] T. Ma and K. Shin. A user-customizable energy-adaptive combined static/dynamicscheduler for mobile applications. *RTSS*, pages 227–238, 2000.

[18] A. Manzak and C. Chakrabarti. Variable voltage task scheduling algorithms for minimizing energy. *ISPLED*, pages 279–282, 2001.

[19] B. Mochocki, X. Hu, and G. Quan. A realistic variable voltage scheduling model for real-time applications. *ICCAD*, pages 726–731, 2002.

[20] L. Niu and G. Quan. A hybrid static/dynamic dvs scheduling for real-time systems with (m, k)-guarantee. *RTSS*, pages 356–365, 2005.

[21] T. Pering, T. Burd, and R. Brodersen. The simulation and evaluation of dynamic voltage scaling algorithms. *ISLPED*, pages 76–81, 1998.

[22] A. Qadi, S. Goddard, and S. Farritor. A dynamic voltage scaling algorithm for sporadic tasks. *RTSS*, pages 52–62, 2003.

[23] Q. Qiu, Q. Wu, and M. Pedram. Dynamic power management in a mobile multimedia system with guaranteed Quality-of-Service. *DAC*, pages 834–839, 2001.

[24] G. Quan and X. Hu. Minimum energy fixed-priority scheduling for variable voltage processors. *DATE*, pages 782–787, 2002.

[25] G. Quan and X.S. Hu. Energy efficient fixed-priority scheduling for real-time systems on voltage variable processors. *DAC*, pages 828–833, 2001.

[26] K.F.S. Reinhardt and T. Mudge. Automatic performance-setting for dynamic voltage scaling. *MOBICOM*, pages 260–271, 2001.

[27] K. Roy, S. Mukhopadhyay, and H. Mahmoodi-Meimand.Leakage current mechanisms and leakage reduction techniques in deep-submicrometer cmos circuits. *Preceedings of IEEE*, 91(2):305–327, Feb 2003.

[28] C. Rusu, R. Melhem, and D. Mosse. Maximizing the system value while satisfying time and energy constraints. *IBM Journal on Res. & Dev.*, 47(5/6):689–701, Sep/Nov 2003.

[29] S. Saewong and R. Rajkumar. Practical voltage-scaling for fixed priority rt-system. *RTAS*, pages 106–114, 2003.

[30] M.T. Schmitz, B.M. Al-Hashimi, and P. Eles. Energy-efficient mapping and scheduling for dvs enabled distributed embedded systems. *DATE*, pages 321–330, 2002.

[31] D. Shin and J. Kim. Dynamic voltage scaling of periodic and aperiodic tasks in priority-driven systems. *ASPDAC*, pages 635–658, 2004.

[32] D. Shin, J. Kim, and S. Lee. Intra-task voltage scheduling for low-energy hard real-time applications. *IEEE Design and Test of Computers*, 18(2), March–April 2001.

[33] K.G. Shin and P. Ramanathan. Real-time computing: a new discipline of computer science and engineering. *Proceedings of IEEE*, 82(1):6–24, January 1994.

[34] Y. Shin and K. Choi. Power conscious fixed priority scheduling for hard real-time systems. *DAC*, pages 134–139, 1999.

[35] J. Tschanz, S. Narendra, Y. Ye, B. Bloechel, S. Borkar, and V. De. Dynamic sleep transistor and body bias for active leakage power control of microprocessors. *IEEE Journal of solid-State Circuits*, 38(11):1838–1845, 2003.

[36] N. Weste and D. Harris. *COMS VLSI Design: A Circuits And Systems Perspective (3rd)*. Addison Wesley, Boston, MA, 2005.

[37] L. Yan, J. Luo, and N.K. Jha. Joint dynamic voltage scaling and adpative body biasing for heterogeneous distributed real-time embedded systems. *IEEE Trans. on Computer-Aided Design*, 24(7), 2005.

[38] F. Yao, A. Demers, and S. Shenker. A scheduling model for reduced cpu energy. *FOCS*, pages 374–382, 1995.

[39] W. Yuan and K. Nahrstedt. Energy-efficient soft real-time cpu scheduing for mobile multimedia systems. In *SOSP'2003*, pages 149–163, 2003.

[40] W. Yuan and K. Nahrstedt. Recalendar: calendaring and scheduling applications with cpu and energy resource guarantees for mobile devices. In *PerCom*, pages 425–432, 2003.

[41] B. Zhai, D. Blaauw, D. Sylvester, and K. Flautner. Theoretical and practical limits of dynamic voltage scaling. *DAC*, pages 868–873, 2004.

[42] Y. Zhang, X. Hu, and D. Chen. Task scheduling and voltage selection for energy minimization. *DAC*, pages 183–188, 2002.

[43] Y. Zhang, Z. Lu, J. Lach, K. Skadron, and M.R. Stan. Optimal procrastinating voltage scheduling for hard real-time systems. *DAC*, pages 905–908, 2005.

[44] Y. Zhu and F. Mueller. Feedback edf scheduling exploiting dynamic voltage scaling. *RTAS*, pages 84–93, 2004.

Chapter 10

Static DVFS Scheduling

Gang Quan[1] and Xiaobo Sharon Hu[2]

[1]Department of Computer Science and Engineering
University of South Carolina, Columbia, SC 29208, U.S.A.

[2]Department of Computer Science and Engineering
University of Notre Dame, Notre Dame, IN 46556, U.S.A.

Abstract   DVFS processors, if used properly, can dramatically reduce the energy consumption of real-time systems employing such processors. In this chapter, two
static or off-line, voltage/frequency selection techniques are presented to maximally exploit the energy-saving benefit provided by DVFS processors. The first
technique targets a popular dynamic-priority task scheduling algorithm, i.e., the
Earliest Deadline First algorithm, while the second is applicable to any fixed-
priority task scheduling algorithm. Other related work is reviewed at the end of
the chapter.

Keywords: Real-time scheduling, dynamic voltage/frequency scaling, earliest deadline first
(EDF), rate montonic scheduling (RMS), off-line scheduling

1.   Introduction

It has been observed that for many real-time applications, the computation
requirements can be confined to a sequence of periodic or non-periodic tasks
with timing specifications known in advance. A static scheduler takes advantage of this fact and determines a suitable schedule off-line. A static schedule
can greatly reduce the extra demands on computing resources; more importantly, it is easier to guarantee in advance that the timing requirements can
be met.

J. Henkel and S. Parameswaran (eds.), Designing Embedded Processors – A Low Power Perspective,
231–242.

Consider a real-time system consists of consists of N independent jobs, $\mathcal{J} = \{J_1, J_2, \ldots, J_N\}$. These jobs can be regarded as sporadic jobs or instances of periodic tasks. Each job J_n is specified with the following parameters:

- r_n: the time at which job J_n is ready to be executed, referred to as *release time*.

- d_n: the time by which J_n must be completed, referred to as *deadline*.

- c_n: the maximum number of CPU cycles needed to complete job J_n without any interruption, referred to as *workload*.

Given a set of real-time jobs and a processor with DVFS capability, different processor speeds can be set at different times. A fundamental question is how to determine processing speeds during the entire job set execution interval, i.e., the *voltage schedule*, with which the lowest amount of energy is consumed and the jobs are all completed at or before their deadlines. For simplicity, we make the assumptions that the supply voltage of the processor can be updated continuously and instantaneously, and the leakage power consumption are ignored in this chapter. Under such assumptions, the following theorem helps constrain the search space for the optimal voltage schedule.

Theorem 10.1. *An optimal voltage schedule for a job set $\mathcal{J}$ must be defined on a set of time intervals in which the processor maintains a constant speed, and each of these intervals must start and end at either the release times or deadlines of the jobs.*

From Theorem 10.1, the processor speed needs to be updated only at the release time or the deadline of a job. We call such a point as the *scheduling point*. For the rest of this chapter, when we refer to a time t, we always mean a *scheduling point*.

The priority-driven preemptive scheduler is one of the most commonly used schedulers for many RTES [8]. In the priority-driven preemptive scheduling, different priority assignment schemes can exhibit drastically different characteristics and scheduling behaviors. In what follows, we study the static DVFS scheduling under two well-known priority assignment schemes: the EDF scheme and the fixed-priority (FP) scheme.

2.   EDF Scheduling

The EDF algorithm assigns priorities to individual jobs according to their absolute deadlines. It is optimal in the sense that the processor can be potentially fully utilized. In this section, we present a DVFS technique that can minimize energy when scheduling a set of jobs based on EDF [13]. Before we present the algorithm in detail, we first introduce several concepts.

Definition 10.1. *The **intensity** over the time interval* $[t_a, t_b]$*, denoted by* $I(t_a, t_b)$*, is defined as*

$$I(t_a, t_b) = \frac{\sum_i c_i}{t_b - t_a}, \tag{10.1}$$

where c_i *is the execution cycle for* J_i *such that* $[r_i, d_i] \subseteq [t_a, t_b]$*. The time interval* $[t_s, t_f]$ *with the largest intensity is called the* critical interval.

From Definition 10.1, it is not difficult to see that $I(t_a, t_b)$ sets the lower bound on the constant speed within interval $[t_a, t_b]$. Further, the following theorem shows that a critical interval and its intensity will determine a segment of the optimal schedule.

Theorem 10.2. *[13] Let* $[t_s, t_f]$ *be the critical interval for the job set* $\mathcal{J}$*. Then, an optimal voltage scheduler for* $\mathcal{J}$ *must have processor run at speed* $I(t_s, t_f)$ *for the entire interval of* $[t_s, t_f]$*, and* $I(t_s, t_f)$ *is the maximal speed in the schedule.*

Theorem 10.2 immediately lead to the following optimal DVFS algorithm that computes a sequence of critical interval iteratively.

We use an example to explain this algorithm. Consider the five jobs shown in Figure 10.1. (An up (down) arrow represents the release time (deadline) of a job, respectively.) Among all the intervals with different pairs of scheduling points, we can easily identify the first critical interval as [6,14] with intensity of 5/8. Then Algorithm 10.1 removes the interval together with the jobs in it, i.e., J_2 and J_5. At the same time, the release times and deadlines for the rest of the jobs are adjusted correspondingly and the resultant job set is shown in Figure 10.1b. The critical interval [4,14] with intensity of 1/2 is identified again from the job set in Figure 10.1b. This iterative procedure continues until all the jobs are removed from the original job sets. The final voltage schedule is shown in Figure 10.1c, which is constructed from the sequence of identified critical intervals and their corresponding intensities.

The complexity of Algorithm 10.1 is $O(N^3)$. With more sophistic data structure and implementation, the complexity can be reduced to as low as $O(N log^2 N)$ [13].

3. Fixed-Priority Scheduling

The fixed-priority (FP) scheduling policy is another popular scheduling scheme used in many real-time systems. The FP policy has the advantages of low overhead, ease of implementation, and high predictability [8]. Without loss of generality, we assume that for any two jobs, i.e., J_i and J_k in $\mathcal{J}$, J_i has a higher priority than J_k if $i < k$. While the optimal voltage schedule that consumes the least dynamic energy for a job set scheduled with EDF can be obtained in

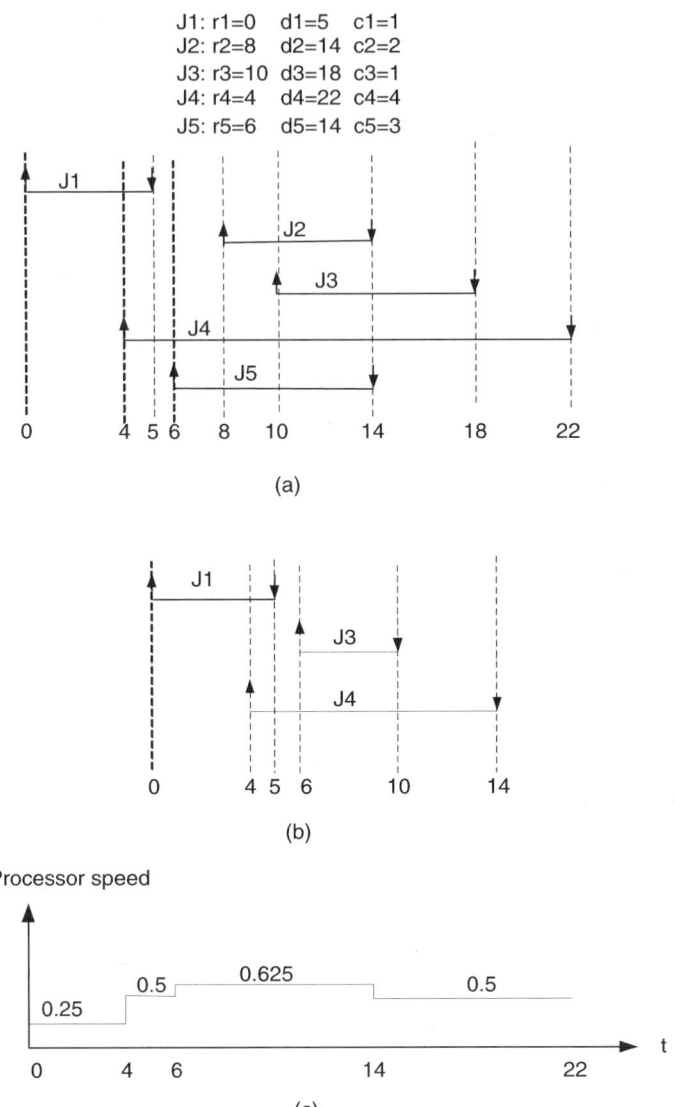

Fig. 10.1   (a) A real-time system with five jobs to be scheduled by EDF. (b) The job set after removing the first critical interval. (c) The final voltage schedule.

polynomial time, the problem becomes NP-hard when the jobs are scheduled according to the FP scheme [14].

In what follows, we introduce a heuristic technique that can find a voltage schedule with close to optimal energy saving in polynomial time. Furthermore, the technique can be readily extended to take into account practical factors such

---

Algorithm 10.1 Optimal voltage schedule for job set scheduled by EDF [13]

1: **LPEDF**$(\mathcal{J}, S_{opt})$
2: **Input:** A real-time job set $\mathcal{J} = \{J_1, ..., J_N\}$
3: **Output:** The optimal voltage schedule S_{opt}.
4: **while** $\mathcal{J}$ is not empty **do**
5:   Identify the critical interval $I^* = [t_s, t_f]$, such that $\overline{S}(t_s, t_f) = \frac{\sum_{J_i} C_i}{t_f - t_s}$ is
     the maximum, for any $J_i, [r_i, d_i] \subseteq [t_s, t_f]$.
6:   Add I^* in S_{opt};
7:   **for** any job $J_i \in \mathcal{J}$ **do**
8:     **if** $[r_i, d_i] \subseteq I^*$ **then**
9:       Remove J_i from $\mathcal{J}$;
10:     **end if**
11:     **if** $r_i \in I^*$ **then**
12:       $r_i = t_s$
13:     **else if** $r_i > t_f$ **then**
14:       $r_i = r_i - |I^*|$;
15:     **end if**
16:     **if** $d_i \in I^*$ **then**
17:       $d_i = t_s$
18:     **else if** $d_i > t_f$ **then**
19:       $d_i = d_i - |I^*|$;
20:     **end if**
21:   **end for**
22: **end while**

---

as transition overhead and leakage power. The technique consists of two steps: the first step computes the minimum constant speed that each job requires to complete by its deadline; the second step, based on the outcomes of the first one, constructs the energy-efficient voltage schedule. For the complete proofs and other technical details of this technique, reader can refer to [11].

3.1 Determining the Minimum Constant Speed for Each Job

The minimum constant speed for a job is the lowest processor speed that if it is applied during the whole execution interval, the job can meet its deadline. When deriving the minimum speed of J_n, only the higher priority jobs whose execution may interfere with J_n's execution need to be taken into considerations. The following definitions limit the number of jobs to be considered.

Definition 10.2. *A scheduling point t is called a J_n-scheduling point if $t = r_i, 1 \leq i \leq n$ or $t = d_n$. A J_n-scheduling point t is called* the earliest scheduling

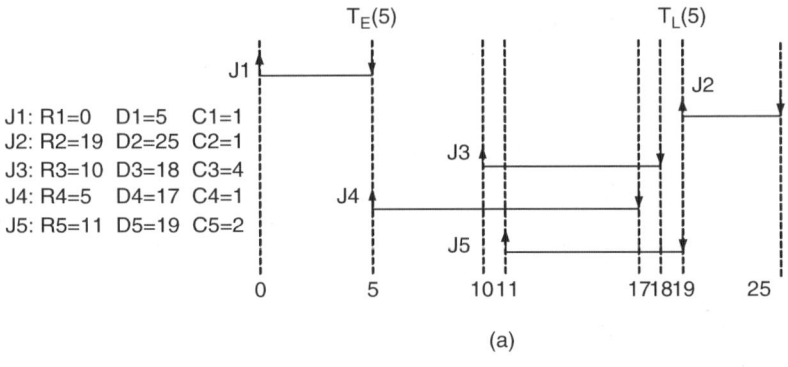

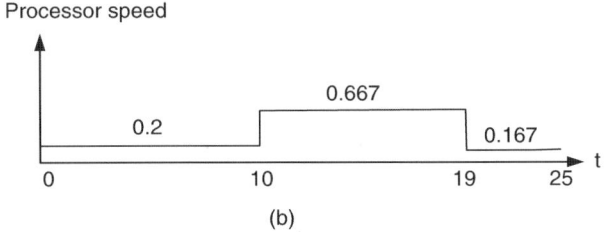

(a)

Fig. 10.2   (a) A real-time system with five jobs to be scheduled with FP scheme. $T_E(5)$ and $T_L(5)$ are the earliest and latest scheduling point of J_5, respectively. (b) The final voltage schedule.

point *of* J_n *and denoted by* $T_E(n)$ *if it is the largest* J_n*-scheduling point in* $[0, r_n]$ *that satisfies*

$$t \geq d_i \quad if \, t > r_i, \;\; 1 \leq i \leq n.$$

The latest time of J_n *by which* J_n *must be completed is called* the latest scheduling point *of* J_n *and is denoted by* $T_L(n)$.

For the system shown in Figure 10.2a, $T_E(5) = 5$ and $T_L(5) = 19$. It is not difficult to see that any higher priority jobs released prior to $T_E(n)$ or after $T_L(n)$ do not have any impact on the speed needed to complete J_n provided that these jobs are finished by their deadlines. Thus, when computing the minimum constant speed, we need only focus on jobs released within $[T_E(n), T_L(n)]$.

The computation of minimum constant speed depends on the length of interval under consideration as well as the workload demand within this interval. We introduce the definition of J_n-intensity to capture the concept of necessary constant speed for job J_n.

Definition 10.3. *Let* t_a, t_b *be two* J_n*-scheduling points,* J_n*-intensity* in the *interval* $[t_a, t_b]$, *denoted by* $I_n(t_a, t_b)$, *is defined to be*

$$I_n(t_a, t_b) = \frac{\sum_{i=1}^{n} \delta(J_i) * c_i}{t_b - t_a}, \tag{10.2}$$

where

$$\delta(J_i) = \begin{cases} 1 & t_a \leq r_i < t_b \\ 0 & otherwise \end{cases} \qquad (10.3)$$

In Figure 10.2a, for instance, we have $J_5(5, 19) = 1/2$. When a constant speed is applied during an interval, it is expected there is no idle time left. Otherwise, the constant speed may not be minimal. The following definition captures the idle time related concepts.

Definition 10.4. (Job Busy Interval) *Interval $[t_a, t_b]$ is a J_n-busy interval, if the following conditions are satisfied:*

- t_a, t_b *are J_n scheduling points.*

- $T_E(n) \leq t_a \leq r_n < t_b \leq T_L(n)$.

- *If speed $I_n(t_a, t_b)$ is applied within $[t_a, t_b]$, the processor is kept busy in $[t_a, t_b]$ by executing jobs with priorities higher or equal to that of J_n.*

Note that, in Figure 10.2a, intervals such as [11,17] and [10,19] are J_5-busy intervals, but interval [5,19] is not a J_5-busy interval since the processor will be idle from 7 to 10 if the speed $I_5[5,19] = 7/14$ is applied within this interval. To determine if an interval is a busy interval, we have the following lemma.

Lemma 10.1. *An interval $[t_a, t_b]$ is a J_n-busy interval if and only if*

$$I_n(t_a, t) \geq I_n(t_a, t_b) \qquad (10.4)$$

for every J_n-scheduling point $t \in (t_a, t_b]$.

For job J_n, it is not difficult to see that there may exist a number of J_n-busy intervals. The largest one among them, called the *essential interval*, has some very interesting properties for the computation of the minimal constant speed. Specifically,

Definition 10.5. *A J_n-busy interval $[t_s, t_f]$ is called the J_n-essential interval if for any J_n-busy interval $[t_a, t_b]$, we have*

$$t_s \leq t_a \quad and \quad t_b \leq t_f. \qquad (10.5)$$

The J_n-essential interval possesses the following important properties.

Theorem 10.3. *The J_n-essential interval, $[t_s, t_f]$, and the corresponding J_n-intensity, $I_n(t_s, t_f)$, satisfy*

$$I_n(t_s, t_f) = \min_{[t_a, t_b]} I_n(t_a, t_b), \qquad (10.6)$$

where $[t_a, t_b]$ is any J_n-busy interval within $[T_E(n), T_L(n)]$. Furthermore, if $I_n(t_s, t_f)$ is adopted as the processor speed during $[t_s, t_f]$, J_n is completed by its deadline.

According to Theorem 10.3, $I_n(t_s, t_f)$ is the valid minimum constant speed. Thus, determining the minimum constant speed for each job now becomes determining the essential interval and corresponding intensity associated with each job. Such an algorithm can be simply implemented based on Definition 10.5. In the following, we present another algorithm, Algorithm 10.2, which also follows the basic principle laid down in Definition 10.5 but employs a little different search mechanism, and on average takes less time than a straightforward implementation of Definition 10.5.

For ease of understanding, we use the example in Figure 10.2a and compute the essential interval for J_5 with Algorithm 10.2. An interval is initiated to be $[r_5, r_5]$ and will grow increasingly in length with the iteration of the *while* loop in the algorithm. Algorithms 10.1 first searches for the scheduling point t_1 such that for any other scheduling point t located on the right hand side of the interval we have $I_5(r_5, t_1) \leq I_5(r_5, t)$ (If a tie occurs, take the right most one.) In this example, $t_1 = 19$. Then the algorithm searches for scheduling point t_2 from those scheduling points on the left hand side of the interval such that $I_5(t_2, t_1) \geq I_5(t, t_1)$ (If a tie occurs, take the left most one.) We have $t_2 = 10$ in this example. The algorithm continues in this fashion until the interval stops

---

Algorithm 10.2 Algorithm for constructing the essential interval for a single job

---

1: **Input:** Job set $\mathcal{J} = \{J_1, ..., J_N\}$, and $T_E(n)$ and $T_L(n)$ for job J_n
2: **Output:** J_n-essential interval $[t_s, t_f]$ and S_n
3: $t_a' = t_b' = r_n$;
4: $t_a = T_E(n)$;
5: $t_b = T_L(n)$;
6: **while** $t_a \neq t_a'$ or $t_b' \neq t_b$ **do**
7:     $t_a = t_a'$;
8:     $t_b = t_b'$;
9:     **for** every J_n-scheduling point in $[t_b, T_L(n)]$ **do**
10:         Find t_b' such that $I(t_a, t_b')$ is the minimum;
11:     **end for**
12:     **for** every J_n-scheduling point in $[T_E(n), t_a]$ **do**
13:         Find t_a' such that $I(t_a', t_b')$ is the maximum;
14:     **end for**
15: **end while**
16: $t_s = t_a, t_f = t_b, S_n = I(t_a, t_b)$;

---

growing any longer, and the resultant interval, i.e., [10,19], is the essential interval for J_5.

3.2 Determining the Global Voltage Schedule

Algorithm 10.2 helps to identify the minimum constant speed of each job, but our goal is to find the voltage schedule within the whole job execution interval. To find the global voltage schedule, we borrow the concept of the *critical interval* from the DVFS EDF scheduling as discussed in the previous section.

Definition 10.6. *For job set J scheduled according to the FP scheme, the essential interval* $[t_s, t_f]$ *with the largest* J_n-*intensity is called the* critical interval.

We will then iteratively find the critical intervals and construct the global voltage schedule as that in Algorithm 10.1. The following theorem (Theorem 10.4) ensures that the result voltage schedule is a feasible, low-energy voltage schedule.

Theorem 10.4. *Given a job set J scheduled with FP scheme, let* $[t_s^k, t_f^k]$ *and* $\overline{S}(t_s^k, t_f^k)$ *for* $1 \leq k \leq K$ *be the critical intervals and corresponding speeds. Then*

- *Every job in J is guaranteed to be completed by its deadline if* $\overline{S}(t_s^k, t_f^k)$ *is used in the corresponding interval* $[t_s^k, t_f^k]$.

- $\overline{S}(t_s^1, t_f^1) \geq \overline{S}(t_s^2, t_f^2) \geq \cdots \geq \overline{S}(t_s^K, t_f^K)$.

From Theorem 10.4, the set of critical intervals and their associated speeds form a valid voltage schedule. Also, the speed for each critical interval is the lowest constant speed possible for that interval. Furthermore, the speed for the first critical interval can be used as the overall minimum constant speed for the entire job set J. This is an important parameter when designing systems with no sophisticated power management hardware but only simple on/off modes or where large voltage transition overhead is a concern

For the example shown in Figure 10.2a, the first critical interval can be readily identified as [10,19] with intensity as 5/9. Along with the removal of this interval, (i) job J_3 and J_5 are removed, (ii) the deadline of J_4 is changed to 10, and (iii) J_2 is "shifted" left correspondingly, i.e., with its updated release time and deadline as 11 and 16, respectively. This procedure continues until all the jobs are removed and we obtain a feasible and low-power schedule as shown in Figure 10.2b.

4.   Related Work

This chapter discusses several off-line uniprocessor scheduling techniques that exploit the DVS capability of modern processors in real-time scheduling for saving energy. These techniques take the worst case scenarios under considerations, which could be pessimistic for some real-time embedded systems. The online techniques, which are more flexible and adaptive, will be discussed in the next chapter. While an online approach can efficiently exploit runtime variations, the energy efficient scheduling approach usually need to incorporate more sophisticated off-line analysis results since many real-time embedded systems have highly deterministic timing specifications, and the energy consumption for such systems can be greatly reduced by aggressively taking advantage of these information when applying the DVS techniques. Furthermore, off-line approaches can afford to employ more advanced optimization algorithms without being constrained much by the time and energy consumption due to carrying out the approaches themselves.

Besides the techniques introduced in this chapter, other off-line techniques are also reported in the literature, e.g., [1, 3, 4, 5]. In [4], the lower power scheduling problem is formulated as an integer linear programming problem, and the system consists of a set of tasks with same arrive times and deadlines but different context switching activities. Aydin *et al.* proposed an optimal voltage schedule for periodic task sets with different power consumption characteristics in [2]. An off-line scheduling heuristic for non-preemptive hard real-time tasks is discussed in [3]. The techniques presented in [2, 3] target jobs scheduled according to the EDF priority assignment.

A number of papers have been published on off-line FP DVS scheduling techniques. In [14], Yun and Kim formally proved that the problem of finding the optimal voltage schedule for an FP real-time job set is NP-hard. Shin *et al.* [12] proposed to determine the lowest maximum processor speed based on the worst case response time analysis [6]. This approach is suitable for the periodic tasks having the same starting time or they can still be very pessimistic otherwise. Moreover, it is not difficult to see that the processor speeds can be further reduced for the jobs not having the worst case response time. In [9], Manzak and Chakrabarti proposed to use the Lagrange multiplier method to determine the processor speed for a periodic task set scheduled with RMS. The energy consumption is minimized under the constraint that the total utilization is a constant no bigger than the well-known utilization bound [7]. In [10], Quan and Hu proposed an approach to find the optimal voltage schedule for this problem. Unfortunately, this approach suffers prohibitively high computation cost, i.e., the computation complexity may increase exponentially with the number of jobs in the worst case, and therefore can only be applied to real-time systems with a small number of real-time jobs.

References

[1] H. Aydin, R. Melhem, D. Mosse, and P. Alvarez. Determining optimal processor speeds for periodic real-time tasks with different power characteristics. *ECRTS*, pages 225–232, June 2001.

[2] H. Aydin, R. Melhem, D. Mosse, and P. Alvarez. Dynamic and aggressive scheduling techniques for power aware real-time systems. *RTSS*, pages 95–105, 2001.

[3] I. Hong, D. Kirovski, G. Qu, M. Potkonjak, and M.B. Srivastava. Power optimization of variable voltage core-based systems. *Proceedings of DAC*, pages 176–181, 1998.

[4] T. Ishihara and H. Yasuura. Voltage scheduling problem for dynamically variable voltage processors. *ISLPED*, pages 197–202, August 1998.

[5] R. Jejurikar and R. Gupta. Energy-aware task scheduling with task synchronization for embedded real-time systems. *IEEE Trans. on Computer Aided Design of Integrated Circuits and Systems*, 25(6):1024–1037, June 2006.

[6] J. Lehoczky, L. Sha, and Y. Ding. The rate monotonic scheduling algorithm: Exact characterization and average case behavior. *RTSS*, pages 166–171, 1989.

[7] C.L. Liu and J.W. Layland. Scheduling algorithms for multiprogramming in a hard real-time environment. *Journal of the ACM*, 17(2):46–61, 1973.

[8] J. Liu. *Real-Time Systems*. Prentice Hall, Upper Saddle River,, NJ, USA, 2000.

[9] A. Makzak and C. Chakrabarti. Variable voltage task scheduling algorithms for minimizing energy/power. *IEEE Transactions on VLSI*, 11(2):270–276, April 2003.

[10] G. Quan and X. Hu. Minimum energy fixed-priority scheduling for variable voltage processors. *DATE*, pages 782–787, 2002.

[11] G. Quan and X. Hu. Energy efficient scheduling for real-time systems on variable voltage processors. *accepted by ACM Trans. on Embedded System Design*, 2006.

[12] Y. Shin, K. Choi, and T. Sakurai. Power optimization of real-time embedded systems on variable speed processors. *ICCAD*, pages 365–368, 2000.

[13] F. Yao, A. Demers, and S. Shenker. A scheduling model for reduced cpu energy. *FOCS*, pages 374–382, 1995.

[14] H.-S. Yun and J. Kim. On energy optimal voltage scheduling for fixed-prioirty hard real-time systems. *ACM Transactions on Embedded Computing Systems*, vol 2:393–430, 2003.

Chapter 11

Dynamic DVFS Scheduling

Padmanabhan S. Pillai[1] and Kang G. Shin[2]

[1] *Intel Research Pittsburgh*
Pittsburgh, PA 15213, USA

[2] *Electrical Engineering and Computer Science Department*
University of Michigan
Ann Arbor, MI 48105, USA

Abstract   As discussed in the previous chapter, offline analysis can be used to generate a schedule of DVFS state changes to minimize energy consumption, while ensuring sufficient processing cycles are available for all tasks to meet their deadlines, even under worst-case computation requirements. However, invocations of real-time tasks typically use less than their specified worst-case computation requirements, presenting an opportunity for further energy conservation. This chapter outlines three online, dynamic techniques to more aggressively scale back processing frequency and voltage to conserve energy when task computation cycles vary, yet continue to provide timeliness guarantees for worst-case execution time scenarios.

Keywords: EDF; RM; dynamic DVFS; online algorithm; schedulability

1.   Introduction

Techniques in the previous chapter show how DVFS can be used in conjunction with real-time scheduling: using offline analysis, DVFS settings can be determined to ensure that the task set remains schedulable. These offline, or static, DVFS techniques can indeed reduce power dissipation of real-time systems. However, even greater energy savings can be realized. In particular, tasks typically require fewer processing cycles than indicated by their specified worst-case execution times (WCETs). As the actual execution times cannot be

J. Henkel and S. Parameswaran (eds), Designing Embedded Processors – A Low Power Perspective,
243–258.

predicted in advance, the static techniques must always use the worst-case values. Online, or dynamic, techniques of selecting DVFS settings can take into account actual execution-time information that is only available at runtime to further reduce voltage, frequency, and energy consumption, while maintaining deadline guarantees. This chapter presents three techniques to dynamically select DVFS settings in real-time systems.

The techniques outlined in this chapter assume the classic model of a real-time system, in which a set of n tasks are executed periodically. Each task, T_i, has an associated period, P_i, and a worst-case computation time, C_i. The task is *released* (put in a runnable state) periodically once every P_i time units (actual units can be seconds, or processor cycles, or any other meaningful quanta). The task must complete its execution by its deadline. As in classical treatments of the real-time scheduling problem [11], the relative deadline is assumed to be equal to period, i.e., a task must complete execution before the next release of the task. Although not explicit in the model, aperiodic and sporadic tasks can be handled by a periodic or deferred server [9].

A real-time scheduler allocates processor time to tasks such that all tasks can complete their jobs/invocations before their deadlines, given that no task uses more than its specified WCET, and that the task set as a whole fits the schedulability conditions of the scheduling algorithm, which are often expressed in the form of schedulability tests. The two most-studied real-time schedulers, Rate Monotonic (RM) and Earliest-Deadline-First (EDF) schedulers [6, 7, 10, 11, 14], have been shown to be optimal in terms of task schedulability under the canonical real-time task model outlined above, assuming negligible preemption overheads. RM is a fixed-priority, preemptive scheduler that gives highest priority to tasks with the shortest periods, while EDF is a dynamic-priority, preemptive scheduler that gives the highest priority to the task with the most imminent deadline. The mechanisms presented in this chapter extend RM and EDF scheduling to make use of DVFS, while keeping the schedulability conditions unchanged, i.e., any task set that is schedulable under RM or EDF will remain so under the DVFS extensions presented here. In this treatment, the overheads of changing DVFS settings are assumed to be negligible.

The following section examines schedulability tests for EDF and RM scheduling. Following this, the bulk of this chapter describes and compares three dynamic DVFS selection algorithms. Finally, some related work and suggested reading on dynamic selection mechanisms for DVFS are presented.

2.   Schedulability Constraints for EDF and RM

This section introduces the task set schedulability tests for EDF and RM scheduling, and presents a very simple mechanism for statically selecting a voltage and frequency scaling based on these. The dynamic techniques presented later will build upon this concept. The naive static mechanism essentially selects the lowest frequency for which the task set can continue to pass the schedulability test. Although not as effective or aggressive as the techniques presented in the previous chapter, this static technique can serve as the simplest energy-conserving baseline against which the dynamic techniques can be compared.

In applying a DVFS setting, observe that scaling the operating frequency by a factor α ($0 < \alpha \leq 1$) effectively results in the worst-case computation time needed by a task being scaled by a factor $1/\alpha$, while the desired period (and deadline) remains unaffected. Taking the well-known schedulability tests for EDF and RM schedulers from real-time systems literature, and by using the scaled values for worst-case computation needs of the tasks, one can test a task set for schedulability at a particular frequency. The necessary and sufficient schedulability test for a task set under ideal EDF scheduling requires that the sum of the worst-case *utilizations* (computation time divided by period) be less than one, i.e., $C_1/P_1 + \cdots + C_n/P_n \leq 1$ [11]. Using the scaled computation time values, one can obtain the EDF schedulability test with frequency scaling factor α:

$$C_1/P_1 + \cdots + C_n/P_n \leq \alpha$$

Similarly, starting with the sufficient (but not necessary) condition for schedulability under RM scheduling [6] and using the scaled WCETs results in the test for a scaled frequency (see Figure 11.1). The operating frequency selected is the lowest one for which the modified schedulability test succeeds. The voltage, of course, is changed to match the operating frequency. Figure 11.1 summarizes the static voltage scaling for EDF and RM scheduling, where there are m operating frequencies $f_1, \ldots, f_m$ such that $f_1 < f_2 < \cdots < f_m$.

Figure 11.2 illustrates these mechanisms, showing sample worst-case execution traces under statically-scaled EDF and RM scheduling. The example uses the task set in Table 11.1, which indicates each task's period and worst-case computation time, and assumes that three normalized, discrete frequencies are available (0.5, 0.75, and 1.0). The figure also illustrates the difference between EDF and RM (i.e., deadline vs. rate for priority), and shows that statically-scaled RM cannot reduce frequency (and therefore reduce voltage and conserve energy) as aggressively as the EDF version.

3. Cycle-Conserving, Real-time DVFS

Although real-time task sets are specified with worst-case computation requirements, real-time tasks typically use much less than the worst case on most invocations. This presents an opportunity to further reduce energy consumption by reducing operating frequency and voltage when an invoca-

EDF_test (α):
    if ($C_1/P_1 + \cdots + C_n/P_n \leq \alpha$) return true;
    else return false;

RM_test (α):
    if ($\forall T_i \in \{T_1, \ldots, T_n | P_1 \leq \cdots \leq P_n\}$
      $\lceil P_i/P_1 \rceil * C_1 + \cdots + \lceil P_i/P_i \rceil * C_i \leq \alpha * P_i$)
    return true;
    else return false;

select_frequency:
    use lowest frequency $f_i \in \{f_1, \ldots, f_m | f_1 < \cdots < f_m\}$
    such that RM_test(f_i/f_m) or EDF_test(f_i/f_m) is true.

Fig. 11.1   Schedulability tests for EDF and RM scheduling, assuming a specified frequency scaling factor, and a simple static DVFS mechanism based on these

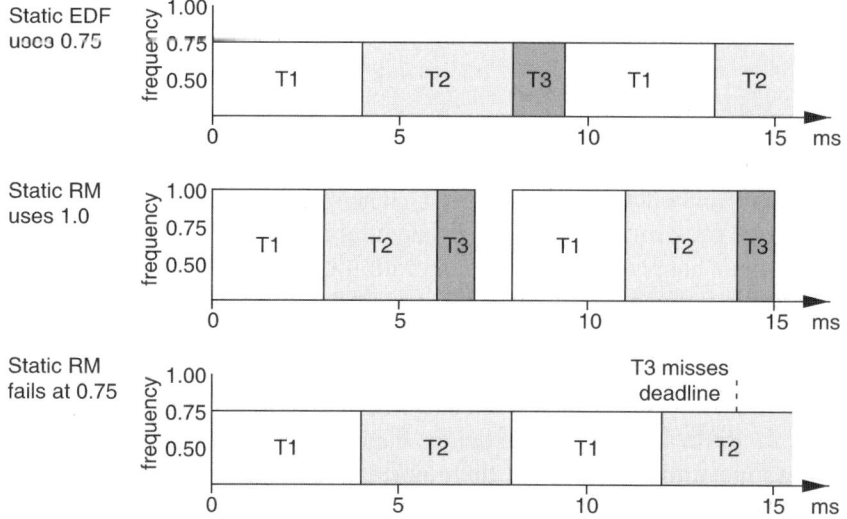

Fig. 11.2   Example of static frequency scaling based on schedulability tests

tion uses less than its WCET, and increasing frequency as needed to meet worst-case needs. In particular, when a task is released for execution, the actual computation that will be performed in the particular invocation is unknown, and DVFS settings can be set conservatively to account for WCET. When the task completes, the actual processor cycles used can be compared to the worst-case specification. Any unused cycles that were allotted to the task

Table 11.1 Example task set, where computing times are specified at the maximum processor frequency

Task	Computing time	Period
1	3 ms	8 ms
2	3 ms	10 ms
3	1 ms	14 ms

would be wasted, idling the processor. Instead of idling for extra processor cycles, dynamic DVFS selection algorithms can avoid wasting cycles (hence "cycle-conserving") by reducing the operating frequency. This is akin to slack-time stealing scheduling [8], except surplus time is used to run other remaining tasks at a lower CPU frequency rather than accomplish more work. These algorithms are tightly-coupled with the operating system's task management services, since the dynamic selection of DVFS settings must be coupled to task release and completion times. The main challenge in designing such algorithms is to ensure that deadline guarantees are not violated when the operating frequencies are reduced.

3.1   Cycle-Conserving EDF

As discussed above, EDF scheduling has a very simple task set schedulability test: as long as the sum of the worst-case task utilizations is less than α, the task set is schedulable when the operating frequency is set to the maximum frequency scaled by factor α. If the actual execution time of a task, T_i, during a particular invocation were known *a priori*, then one could treat this as the WCET for the task during its current invocation. Using this value to compute a reduced utilization and set frequency will result in a task set that is schedulable, at least until T_i's deadline. The intuition behind this is the number of processing cycles available to other tasks between T_i's release and deadline is unchanged from the case where WCET is assumed and is actually needed. Furthermore, if the system is operated at frequencies greater than what is required by the actual execution time, then this simply provides additional computation cycles and schedulability of the task set is not affected.

Based on these observations, a simple dynamic DVFS algorithm can set frequency using task utilization based on WCET at task release time, and then reduce frequency when the task completes, recomputing utilization using actual cycles of computation. This reduced value is used until the task is released again for its next invocation. Figure 11.3 illustrates this technique, using the same task set and available frequencies used with the earlier example, but using actual execution times from Table 11.2. At each scheduling point (task release or completion), the utilization is recomputed using the actual time

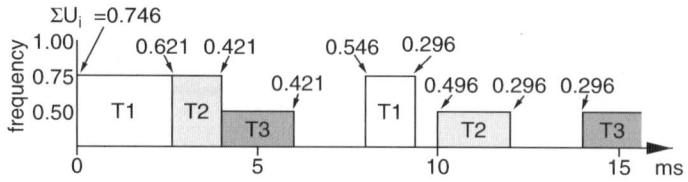

Fig. 11.3   Example of cycle-conserving EDF

Table 11.2   Actual computation requirements of the example task set (assuming execution at max. frequency)

Task	Invocation 1	Invocation 2
1	2 ms	1 ms
2	1 ms	1 ms
3	1 ms	1 ms

select_frequency():
    use lowest freq. $f_i \in \{f_1, \ldots, f_m | f_1 < \cdots < f_m\}$
    such that $U_1 + \cdots + U_n \leq f_i / f_m$

upon task_release(T_i):
    set U_i to C_i / P_i;
    select_frequency();

upon task_completion(T_i):
    set U_i to cc_i / P_i;
        /* cc_i is the actual cycles used this invocation */
    select_frequency();

Fig. 11.4   Cycle-conserving DVFS for EDF schedulers

for completed tasks and the specified worst case for the others, and the frequency is set accordingly. The numerical values in the figure show the total task utilizations computed using the information available at each point.

A direct implementation of this algorithm is given in Figure 11.4. Suppose a task T_i completes its current invocation after using cc_i cycles, a value typically much less than its worst-case computation time C_i. Reducing the task's utilization from U_i to cc_i / P_i reduces the computed total task set utilization, and may potentially allow a reduced frequency setting. Similarly, at a subsequent task release, U_i is restored to the worst-case C_i / P_i value, potentially requiring an increase in operating frequency. At first glance, this transient reduction in frequency does not appear to significantly reduce energy expenditure. However, due to arbitrary overlap of task execution and having multiple tasks simultaneously in the reduced-utilization state, the total savings can be significant.

3.2 Cycle-Conserving RM

The schedulability-test-based approach can also be applied to a cycle-conserving DVFS algorithm for RM scheduling. However, since the RM schedulability test is significantly more complex ($O(n^2)$, where n is the number of tasks to be scheduled), a different approach is warranted. First, observe that even assuming tasks always require their worst-case computation times, the statically-scaled RM mechanism discussed earlier can meet all real-time deadline guarantees. Any DVFS mechanism that uses RM scheduling while ensuring that all tasks make equal or better progress than in the worst case under the statically-scaled RM algorithm will ensure that all deadlines are met. Furthermore, by avoiding getting ahead of the worst-case execution pattern, the DVFS mechanisms can apply any surplus cycles gained by tasks completing early to safely reduce operating voltage and frequency.

Using the same example task set and execution times as before, Figure 11.5 illustrates how this can be accomplished. Initially, the algorithm starts with the worst-case schedule based on static-scaling (a), which, for this example, uses the maximum CPU frequency. To keep things simple, the algorithm does not look beyond the next deadline in the system. It then tries to spread out the work that should be accomplished before this deadline over the entire interval from the current time to the deadline (b). This provides a minimum operating frequency value, but since the frequency settings are discrete, this must be rounded up to the closest available setting, frequency $= 1.0$. After executing T_1, it repeats this computation of spreading out the remaining work over the remaining time until the next deadline (c). This results in a lower operating frequency since T_1 completed earlier than its worst-case specified computing time. Repeating this at each scheduling point results in the final execution trace (f).

Although conceptually simple, the actual algorithm (Figure 11.6) for this is somewhat complex due to the number of counters that must be maintained. This algorithm needs to keep track of the worst-case remaining cycles of computation, c_left_i, for each task, T_i. When task T_i is released, c_left_i is set to C_i. The algorithm then determines the progress that the statically-scaled RM mechanism would make, assuming WCETs, by the closest deadline for *any* task in the system. It computes s_j and s_m, the number of cycles to this next deadline, assuming operation at the statically-scaled and the maximum frequencies, respectively. The s_j cycles are allocated to the tasks according to RM priority order, with each task T_i receiving an allocation

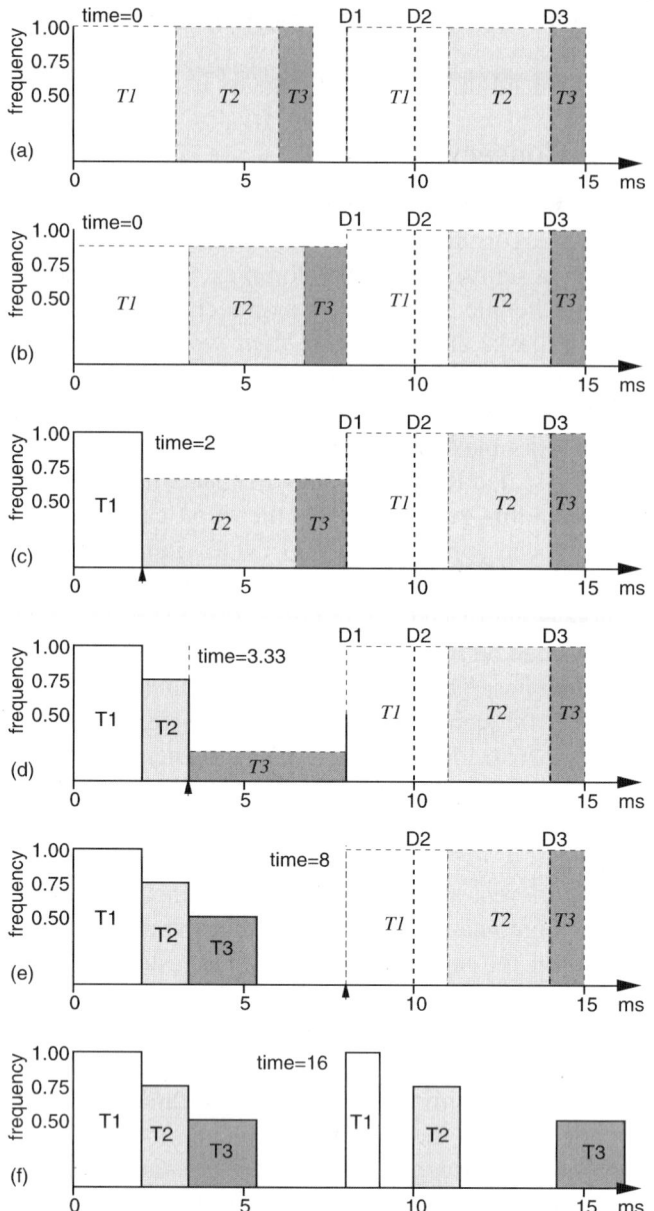

Fig. 11.5 Example of cycle-conserving RM: (a) initially use statically-scaled, worst-case RM
schedule as target; (b) determine minimum frequency so as to complete the same
work by D1; rounding up to the closest discrete setting requires frequency 1.0;
(c) after T1 completes (early), recompute the required frequency as 0.75; (d) once
T2 completes, a very low frequency (0.5) suffices to complete the remaining work
by D1; (e) T1 is re-released, and now, try to match the work that should be done by
D2; (f) execution trace through time 16 ms.

assume f_j is frequency set by static-scaling algorithm

select_frequency():
    set s_m = max_cycles_until_next_deadline();
    use lowest freq. $f_i \in \{f_1, \ldots, f_m | f_1 < \cdots < f_m\}$
    such that $(d_1 + \cdots + d_n)/s_m \le f_i/f_m$

upon task_release(T_i):
    set $c\_left_i = C_i$;
    set s_m = max_cycles_until_next_deadline();
    set $s_j = s_m * f_j/f_m$;
    allocate_cycles (s_j);
    select_frequency();

upon task_completion(T_i):
    set $c\_left_i = 0$;
    set $d_i = 0$;
    select_frequency();

during task_execution(T_i):
    decrement c_left_i and d_i;

allocate_cycles(k):
    for $i = 1$ to n, $T_i \in \{T_1, \ldots, T_n | P_1 \le \cdots \le P_n\}$
                      /* tasks sorted by period */
        if ($c\_left_i < k$)
            set $d_i = c\_left_i$;
            set $k = k - c\_left_i$;
        else
            set $d_i = k$;
            set $k = 0$;

Fig. 11.6   Cycle-conserving DVFS for RM schedulers

$d_i \le c\_left_i$ corresponding to the number of cycles that it would execute under the statically-scaled RM scenario over this interval. As long as for each task, T_i, the system runs T_i for at least d_i cycles, or T_i completes, by the next task deadline, execution for all tasks is keeping pace with the worst-case statically-scaled scenario. To ensure sufficient number of cycles are available, the ratio of the sum of d_i values to s_m is used to select the operating frequency and voltage. As tasks execute, their c_left and d values are decremented. When a task T_i

completes, c_left_i and d_i are both set to 0, and the frequency may be reduced. This algorithm ensures that all task invocations that would have completed by any given task deadline in the worst-case statically-scaled RM schedule would also be completed before the same deadline under dynamic DVFS settings, hence ensuring all deadlines are equally met.

These algorithms dynamically adjust frequency and voltage, reacting to the actual computational requirements of the real-time tasks. They illustrate two very different techniques for ensuring correctness, one through schedulability tests, and the other by controlling execution timing to keep pace with a different algorithm known to meet all deadlines. A great variety of such reactive, dynamic DVFS algorithms are possible. Instead of focusing on additional reactive techniques, the next section introduces a proactive DVFS mechanism that is qualitatively different from the algorithms presented thus far.

4.   Look-ahead DVFS

The final (and most aggressive) DVFS algorithm in this chapter attempts to achieve even better energy savings using a look-ahead technique to determine future computation need and defer task execution. The cycle-conserving approaches discussed above assume the worst case initially, execute at a high frequency until some tasks complete, and only then reduce operating frequency and voltage. In contrast, the look-ahead scheme tries to defer as much work as possible, and sets the operating frequency to meet the minimum work that must be done now to ensure all future deadlines are met. Of course, this may require high operating frequencies later in order to complete all of the deferred work in time. On the other hand, if tasks tend to use much less than their worst-case computing time allocations, the peak execution rates for deferred work may never be needed, and this heuristic will allow the system to continue operating at a low frequency and voltage while completing all tasks by their deadlines.

Continuing with the same example used earlier, Figure 11.7 illustrates the operation of the look-ahead RT-DVS EDF algorithm. The goal is to defer work beyond the earliest deadline in the system (D_1) so that a low frequency can be selected for immediate operation. The algorithm allocates time in the schedule for the worst-case execution of each task, in reverse EDF order, starting with the task with the latest deadline, T_3. T_3's work is spread out evenly (for simplicity) between D_1 and its own deadline, D_3, subject to a constraint reserving capacity for future invocations of the other tasks (a). This step is repeated for T_2, which cannot entirely fit between D_1 and D_2 after allocating T_3 and reserving capacity for future invocations of T_1. Additional work for T_2 and all of T_1 are allotted before D_1 (b). Note that more of T_2 could be deferred beyond D_1, some of T_3's allocation is deferred until after D_2, but, for simplicity, this is not considered. The total work allocated before D_1 is used to determine

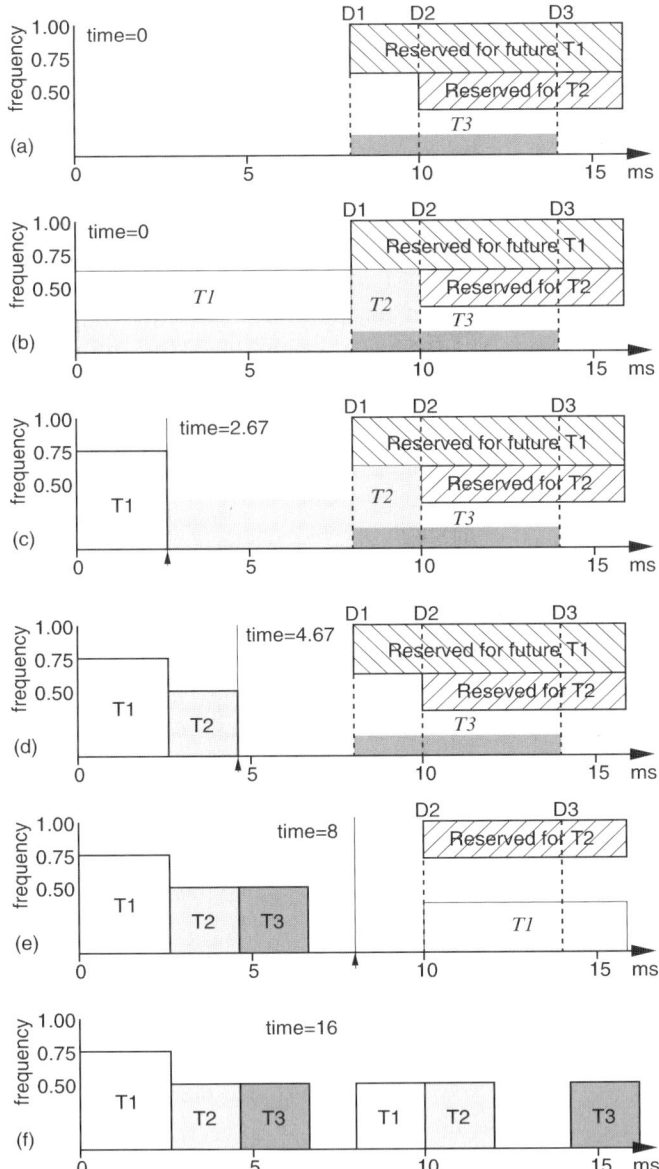

Fig. 11.7   Example of look-ahead EDF: (a) at time 0, plan to defer T3's execution until after
D1 (but by its deadline D3, and likewise, try to fit T2 between D1 and D2; (b) T1
and the portion of T2 that did not fit must execute before D1, requiring use of frequency 0.75; (c) after T1 completes, repeat calculations to find the new frequency
setting, 0.5; (d) repeating the calculation after T2 completes indicates that we do not
need to execute anything by D1, but EDF is work-conserving, so T3 executes at the
minimum frequency; (e) this occurs again when T1's next invocation is released;
(f) execution trace through time 16 ms.

the operating frequency. Once T_1 has completed, using less than its specified worst-case execution cycles, this process is repeated and finds a lower operating frequency (c). Repeating this method of greedily deferring work beyond the next deadline in the system ultimately results in the execution trace shown in (f).

The actual algorithm for look-ahead RT-DVS with EDF scheduling is shown in Figure 11.8. As with the cycle-conserving algorithm for RM, this algorithm keeps track of the worst-case remaining computation c_left_i for the current invocation of each task, T_i. This is set to C_i on task release, decremented as the task executes, and set to 0 on completion. The major step in this algorithm is the deferral function. This function computes s, the minimum number of cycles that must be executed before the next deadline in the system, D_n, in order to meet all future deadlines, assuming greedy deferral of work until after

select_frequency(x):
    use lowest freq. $f_i \in \{f_1, \ldots, f_m | f_1 < \cdots < f_m\}$
    such that $x \leq f_i / f_m$

upon task_release(T_i):
    set $c\_left_i = C_i$;
    defer();

upon task_completion(T_i):
    set $c\_left_i = 0$;
    defer();

during task_execution(T_i):
    decrement c_left_i;

defer():
    set $U = C_1/P_1 + \cdots + C_n/P_n$;
    set $s = 0$;
    for $i = 1$ to n, $T_i \in \{T_1, \ldots, T_n | D_1 \geq \cdots \geq D_n\}$
                /* Note: reverse EDF order of tasks */
        set $U = U - C_i/P_i$;
        set $x = \max(0, c\_left_i - (1 - U)(D_i - D_n))$;
        set $U = U + (c\_left_i - x)/(D_i - D_n)$;
        set $s = s + x$;
    select_frequency $(s/(D_n - \text{current\_time}))$;

Fig. 11.8   Look-ahead DVFS for EDF schedulers

D_n. The operating frequency is set just fast enough to execute s cycles over this interval until D_1. To calculate s, the running tasks are considered in reverse EDF order (i.e., latest deadline first). Assuming worst-case utilization by tasks with earlier deadlines (effectively reserving time for their future invocations), the deferral function calculates the minimum number of cycles, x, that the task must execute before the closest deadline, D_n, in order for it to complete by its own deadline. The cumulative utilization U is adjusted to reflect the actual utilization of the task for the time after D_n. This calculation is repeated for all of the tasks, using assumed worst-case utilization values for earlier-deadline tasks and the computed values for the later-deadline ones. s is simply the sum of the x values calculated for all of the tasks, and therefore reflects an estimate of the total number of cycles that must execute by D_n in order for all tasks to meet their deadlines.

Although this algorithm very aggressively reduces processor frequency and voltage, it is still conservative, and generally overestimates the cycles of computation needed before the next deadline in order to ensure all tasks meet their deadlines. In fact, this heuristic may occasionally request a frequency greater than the maximum operating speed. In such cases, it is safe to simply run at the maximum speed, since the heuristic always reserves capacity for all future task invocations, and the task set is schedulable under EDF.

5.   Evaluating Energy Performance of DVFS Algorithms

All of the DVFS algorithms presented in this chapter should be fairly easy to incorporate into a real-time operating system, and do not require significant processing costs. The dynamic schemes all require $O(n)$ computation (assuming the scheduler provides an EDF sorted task list), and should not require significant processing over the basic scheduler. The most significant overheads may come from the hardware voltage switching times. However, all of these algorithms require no more than two switches per task per invocation, so these overheads can easily be accounted for, and added to, the worst-case task computation times.

The processor energy consumption when employing these algorithms will vary significantly based on workload. One can analytically compute the energy dissipation given an accurate execution trace, e.g., from simulation or dumped from an emulator, and a model of the energy-per-cycle cost for different operating voltages. Based on the series of examples presented above, Table 11.3 shows the normalized energy dissipated in the example task set (Table 11.1) for the first 16 ms, using the actual execution times from Table 11.2. This assumes that a simple $E \propto V^2$ model suffices, the 0.5, 0.75, and 1.0 frequency settings need 3, 4, and 5 volts, respectively, and that idle cycles consume no energy. For this simple example, significant energy can be

Table 11.3   Normalized energy consumption for the example traces

DVFS method	Energy used
none (plain EDF)	1.0
statically-scaled RM	1.0
statically-scaled EDF	0.64
cycle-conserving EDF	0.52
cycle-conserving RM	0.71
look-ahead EDF	0.44

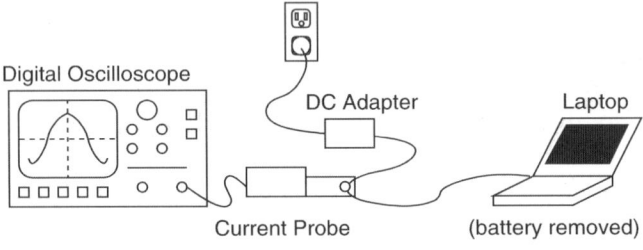

Fig. 11.9   Power measurement on laptop implementation

saved, and great differences do exist between the various algorithms. Such analysis can be readily incorporated into a simulation framework to report processor energy consumption for various DVFS techniques.

Given a platform that supports DVFS, one can also implement the algorithms and directly measure their energy performance. Figure 11.9 illustrates an experimental setup for measuring energy dissipation. A DVFS-capable laptop computer operates with the battery removed on external power. A digital oscilloscope records high resolution, high frequency measurements of current and voltage over long intervals. The product of these measures integrated over time gives the energy consumption of the computer. Most modern digital oscilloscopes are capable of performing long interval capture and the computation steps directly, without external data recording or post-processing. Due to the high sampling rate of the captured trace, this technique can resolve short duration phenomena, often down to tens of microseconds, limited by the frequency response of the current probe and the voltage regulator inside the laptop. One limitation is that this measured value is of the total system energy consumption, and not just the processor consumption. So care must be taken when comparing numbers or evaluating energy models to consider the high power consumption of display back lighting and hard disks that are not affected by DVFS techniques. Several papers, including [4, 13] present further evaluation of the DVFS algorithms presented in this chapter through simulation and measurements.

6. Related Readings

The dynamic DVFS techniques described in this chapter are primarily derived from [13]. This, and [3, 5, 15] are the earliest works to consider online techniques to improve energy performance when real-time tasks use less than their specified WCETs. Reference [5] uses an online scheduler to further improve an offline WCET-based schedule when tasks terminate early. Reference [15] uses a computationally intensive mechanisms to determine the best feasible schedule for DVFS. Reference [3] combines offline analysis with an online, stochastic mechanism that uses the probability distribution of task computation times.

In more recent works, reference [1] presents aggressive techniques to dynamically reclaim unused processing time to further reduce frequency. Several works [2, 12, 16] develop feedback-based mechanisms and control-theoretic formulations to predict task execution time and adjust DVFS aggressiveness accordingly. Reference [4] presents a comparison of several DVFS techniques.

References

[1] AYDIN, H., MELHEM, R., MOSSE, D., AND MEJIA-ALVAREZ, P. Power-aware scheduling for periodic real-time tasks. *IEEE Transactions on Computing 53*, 5 (2004), 584–600.

[2] DUDANI, A., MUELLER, F., AND ZHU, Y. Energy-conserving feedback EDF scheduling for embedded systems with real-time constraints. In *ACM SIGPLAN Joint Conference Languages, Compilers, and Tools for Embedded Systems (LCTES'02) and Software and Compilers for Embedded Systems (SCOPES'02)* (June 2002), pp. 213–222.

[3] GRUIAN, F. Hard real-time scheduling for low energy using stochastic data and DVS processors. In *Proceedings of the International Symposium on Low-Power Electronics and Design ISLPED'01* (Huntington Beach, CA, Aug. 2001).

[4] KIM, W., SHIN, D., YUN, H.-S., KIM, J., AND MIN, S.L. Performance comparison of dynamic voltage scaling algorithms for hard real-time systems. In *Proceedings of the 8th IEEE Real-Time and Embedded Technology and Applications Symposium (RTAS'02)* (2002).

[5] KRISHNA, C.M., AND LEE, Y.-H. Voltage-clock-scaling techniques for low power in hard real-time systems. In *Proceedings of the IEEE Real-Time Technology and Applications Symposium* (Washington, DC, May 2000), pp. 156–165.

[6] KRISHNA, C.M., AND SHIN, K.G. *Real-Time Systems*. McGraw-Hill, 1997.

[7] LEHOCZKY, J., SHA, L., AND DING, Y. The rate monotonic scheduling algorithm: exact characterization and average case behavior. In *Proceedings of the IEEE Real-Time Systems Symposium* (1989), pp. 166–171.

[8] LEHOCZKY, J., AND THUEL, S. Algorithms for scheduling hard aperiodic tasks in fixed-priority systems using slack stealing. In *Proceedings of the IEEE Real-Time Systems Symposium* (1994).

[9] LEHOCZKY, J.P., SHA, L., AND STROSNIDER, J.K. Enhanced aperiodic responsiveness in hard real-time environments. In *Proceedings of the 8th IEEE Real-Time Systems Symposium* (Los Alamitos, CA, Dec. 1987), pp. 261–270.

[10] LEUNG, J. Y.-T., AND WHITEHEAD, J. On the complexity of fixed-priority scheduling of periodic, real-time tasks. *Performance Evaluation* 2, 4 (1982), 237–250.

[11] LIU, C.L., AND LAYLAND, J.W. Scheduling algorithms for multiprogramming in a hard real-time environment. *Journal of the ACM 20*, 1 (1973), 46–61.

[12] LU, Z., HEIN, J., HUMPHREY, M., STAN, M., LACH, J., AND SKADRON, K. Control-theoretic dynamic frequency and voltage scaling for multimedia workloads. In *CASES '02: Proceedings of the 2002 International Conference on Compilers, Architecture, and Synthesis for Embedded Systems* (2002), pp. 156–163.

[13] PILLAI, P., AND SHIN, K. G. Real-time dynamic voltage scaling for low-power embedded operating systems. In *Proceedings of the 18th ACM Symposium on Operating Systems Principles* (Banff, Alberta, CA, Oct. 2001), pp. 89–102.

[14] STANKOVIC, J., ET AL. *Deadline Scheduling for Real-Time Systems.* Kluwer Academic Publishers, 1998.

[15] SWAMINATHAN, V., AND CHAKRABARTY, K. Real-time task scheduling for energy-aware embedded systems. In *Proceedings of the IEEE Real-Time Systems Symp. (Work-in-Progress Session)* (Orlando, FL, Nov. 2000).

[16] VARMA, A., GANESH, B., SEN, M., CHOUDHARY, S. R., SRINIVASAN, L., AND JACOB, B. A control-theoretic approach to dynamic voltage scaling. In *Proceedings of International Conference on Compilers, Architectures, and Synthesis for Embedded Systems (CASES 2003)* (Oct. 2003), pp. 255–266.

[17] ZHU, Y., AND MUELLER, F. Feedback EDF scheduling exploiting dynamic voltage scaling. In *10th IEEE Real-Time and Embedded Technology and Applications Symposium (RTAS'04)* (2004).

Chapter 12

Voltage Selection for Time-Constrained Multiprocessor Systems

Alexandru Andrei[1], Petru Eles[1], Zebo Peng[1], Marcus Schmitz[2],
and Bashir M. Al-Hashimi[2]

[1]*Department of Computer and Information Science*
Linköping University
SE-58 183 Linköping
Sweden

[2]*Department of Electronics and Computer Science*
University of Southampton
Southampton, SO17 1BJ
United Kingdom

Abstract   Dynamic voltage selection and adaptive body biasing have been shown to reduce
dynamic and leakage power consumption effectively. In this chapter we present
an energy optimization approach for time constrained applications implemented
on multiprocessor systems. We start by introducing a genetic algorithm that
performs the mapping and scheduling of the application on the target hardware
architecture. Then, we discuss in detail several voltage selection algorithms,
explicitly taking into account the transition overheads implied by changing voltage levels.

   We investigate the continuous voltage selection as well as its discrete counterpart, and we prove strong NP-hardness in the discrete case. The energy savings
achievable by the approach proposed in this chapter applied on a real-life application (GSM voice codec), are used for illustrating its effectiveness.

Keywords: multiprocessor systems, time constrained applications, voltage selection,
adaptive body biasing, transition overheads

J. Henkel and S. Parameswaran (eds.), Designing Embedded Processors – A Low Power Perspective,
259–284.

1.   Introduction

This chapter presents voltage selection algorithms for a set of tasks, possibly with dependencies, which are executed on multiprocessor systems under real-time constraints. While dynamic voltage selection (DVS) aims to reduce the dynamic power consumption by scaling down operational frequency and circuit supply voltage V_{dd}, adaptive body-biasing (ABB) is effective in reducing the leakage power by scaling down frequency and increasing the threshold voltage V_{th} through body-biasing. The trend in deep-submicron CMOS technology to statically reduce the supply voltage levels and consequently the threshold voltages (in order to maintain peak performance) is resulting in the fact that a substantial portion of the overall power dissipation will be due to leakage currents (Borkar, 1999; Kim and Roy, 2002). This makes the adaptive body-biasing approach and its combination with dynamic supply voltage selection important for energy-efficient designs in the foreseeable future.

In summary, this chapter presents the following issues:

(a) An algorithm for mapping and scheduling tasks with dependencies on a multiprocessor system.

(b) Supply voltage and body-bias voltage selection at the system-level, where several tasks with dependencies execute a time-constrained application on a multiprocessor system.

(c) Four different voltage selection schemes are formulated as nonlinear programming (NLP) and mixed integer linear programming (MILP) problems which can be solved optimally. The formulations are equally applicable to single and multiprocessor systems.

(d) A proof that discrete voltage selection with and without the consideration of transition overheads in terms of energy and time is strongly NP-hard, while the continuous voltage selection cases can be solved in polynomial time (with an arbitrary given approximation $\epsilon > 0$).

(e) Since voltage selection for components that operate with discrete voltages is proofed to be NP-hard, this chapter introduces a simple yet effective heuristic based on the NLP formulation for the continuous voltage selection problem.

The remainder of the chapter is organized as follows: preliminaries regarding the system specification, the processor power and delay models are given in Sections 2 and 3. A short overview on an algorithm for mapping and scheduling tasks with dependencies on a multiprocessor system is given in Section 4. This is followed by a motivational example in Section 5. The four investigated processor voltage selection problems are formulated in Section 6. Continuous and discrete voltage selection problems are discussed in Sections 7 and

8, respectively. Some experimental results are presented in Section 9, related work is presented in Section 10 followed by a short summary in Section 11.

2. System and Application Model

In this chapter we consider embedded systems which are realized as hetero-geneous distributed architectures. Such architectures consist of several differ-ent processing elements (PEs), such as programmable microprocessors, ASIPs, FPGAs, and ASICs, some of which feature DVS and ABB capability. These computational components communicate via an infrastructure of communica-tion links (CLs), like buses and point-to-point connections. We define $\mathcal{P}$ and $\mathcal{L}$ to be the sets of all processing elements and all links, respectively. An example architecture is shown in Figure 12.1a. The functionality of applications is cap-tured by task graphs $G(\Pi, \Gamma)$. Nodes $\tau \in \Pi$ in these directed acyclic graphs represent computational tasks, while edges $\gamma \in \Gamma$ indicate data dependencies between these tasks (communications). Tasks τ_i require in the worst case NC_i clock cycles to be executed, depending on the PE to which they are mapped. Further, tasks are annotated with deadlines dl that have to be met at run-time.

If two dependent tasks are assigned to different PEs, p_x and p_y with $x \neq y$, then the communication takes place over a CL, involving a certain amount of time and power.

After task mapping and scheduling has been performed, it is known where and in which order tasks and communications take place. Figure 12.1a shows an example task graph that has been mapped onto an architecture and Figure 12.1b depicts a possible execution order.

To tie the execution order into the application model, we perform the follow-ing transformation on the original task graph. First, all communications that take place over communication links are captured by communication tasks, as indicated by squares in Figure 12.1c. For instance, communication γ_{1-2} is replaced by task τ_6 and the edges connecting τ_6 to τ_1 and τ_2 are introduced. $\mathcal{K}$ defines the set of all such communication tasks. Furthermore, we denote

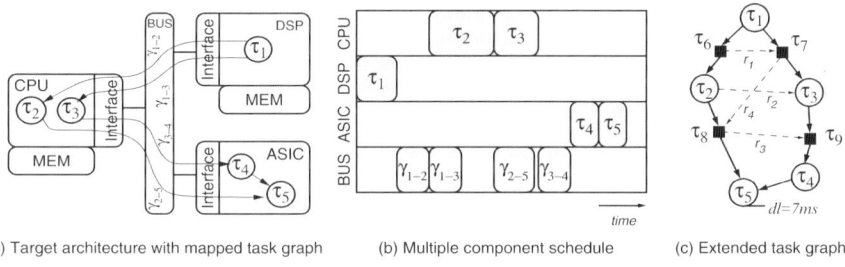

(a) Target architecture with mapped task graph   (b) Multiple component schedule   (c) Extended task graph

Fig. 12.1 System and application model

with $\mathcal{T} = \Pi \cup \mathcal{K}$ the set of all computations and communications. Second, on top of the precedence relations given by data dependencies between tasks, we introduce additional precedence relations $r \in \mathcal{R}$, generated as result of scheduling tasks mapped to the same PE and communications mapped on the same CL. In Figure 12.1c the dependencies $\mathcal{R}$ are represented as dotted edges. We define the set of all edges as $\mathcal{E} = \mathcal{C} \cup \mathcal{R}$. We construct the mapped and scheduled task graph $G(\mathcal{T}, \mathcal{E})$. Further, we define the set $\mathcal{E}^{\bullet} \subseteq \mathcal{E}$ of edges, as follows: an edge $(i,j) \in \mathcal{E}^{\bullet}$ if it connects task τ_i with its immediate successor τ_j (according to the schedule), where τ_i and τ_j are mapped on the same PE or CL.

3.   Processor Power and Delay Models

Digital CMOS circuitry has two major sources of power dissipation: (a) dynamic power P_{dyn}, which is dissipated whenever active computations are carried out (switching of logic states) and (b) leakage power P_{leak} which is consumed whenever the circuit is powered, even if no computations are performed. The dynamic power is expressed by Chandrakasan and Brodersen (1995) and Martin et al. (2002),

$$P_{dyn} = C_{eff} \cdot f \cdot V_{dd}^2 \tag{12.1}$$

where C_{eff}, f, and V_{dd} denote the effective charged capacitance, operational frequency, and circuit supply voltage, respectively. A similar equation for the dynamic power has been introduced in Equation 9.1 from Chapter 9. Although, until recently, the dynamic power dissipation had been dominating, the trend to reduce the overall circuit supply voltage and consequently threshold voltage is raising concerns about the leakage currents. For near future technology ($<$65 nm) it is expected that leakage will account for more than 50% of the total power. The leakage power is given by Martin et al. (2002),

$$P_{leak} = L_g \cdot V_{dd} \cdot K_3 \cdot e^{K_4 \cdot V_{dd}} \cdot e^{K_5 \cdot V_{bs}} + |V_{bs}| \cdot I_{Ju} \tag{12.2}$$

where V_{bs} is the body-bias voltage and I_{Ju} represents the body junction leakage current (constant for a given technology). The fitting parameters K_3, K_4, and K_5 denote circuit technology dependent constants and L_g reflects the number of gates. For clarity reasons we maintain the same indices as used in Martin et al. (2002), where also actual values for these constants are given. Please note that the leakage power is stronger influenced by V_{bs} than by V_{dd}, due to the fact that the constant K_5 is larger than the constant K_4 (e.g., for the Crusoe processor described in Martin et al. (2002), $K_5 = 4.19$ while $K_4 = 1.83$).

Nevertheless, scaling the supply and the body-bias voltage, in order to reduce the power consumption, has a side-effect on the circuit delay d and hence the operational frequency (Chandrakasan and Brodersen, 1995;

Martin et al., 2002):

$$f = \frac{1}{d} = \frac{((1 + K_1) \cdot V_{dd} + K_2 \cdot V_{bs} - V_{th1})^\alpha}{K_6 \cdot L_d \cdot V_{dd}} \tag{12.3}$$

where α reflects the velocity saturation imposed by the used technology (common values $1.4 \leq \alpha \leq 2$), L_d is the logic depth, and K_1, K_2, K_6 and V_{th1} are circuit dependent constants.

Another important issue, which often is overlooked in voltage scaling approaches, is the consideration of transition overheads, i.e., each time the processor's supply voltage and body-bias voltage are altered, the change requires a certain amount of extra energy and time. These energy $\epsilon_{k,j}$ and delay $\delta_{k,j}$ overheads, when switching from V_{dd_k} to V_{dd_j} and from V_{bs_k} to V_{bs_j}, are given by Martin et al. (2002),

$$\epsilon_{k,j} = C_r \cdot |V_{dd_k} - V_{dd_j}|^2 + C_s \cdot |V_{bs_k} - V_{bs_j}|^2 \tag{12.4}$$

$$\delta_{k,j} = \max(p_{Vdd} \cdot |V_{dd_k} - V_{dd_j}|, p_{Vbs} \cdot |V_{bs_k} - V_{bs_j}|) \tag{12.5}$$

where C_r denotes power rail capacitance, and C_s the total substrate and well capacitance. Since transition times for V_{dd} and V_{bs} are different, the two constants p_{Vdd} and p_{Vbs} are used to calculate both time overheads independently. Considering that supply and body-bias voltage can be scaled in parallel, the transition overhead $\delta_{k,j}$ depends on the maximum time required to reach the new voltage levels.

Voltage scaling is only rewarding if the energy saved through optimized voltages is not outdone by the transition overheads. Furthermore, it is obvious that disregarding transition time overhead can seriously affect the schedulablity of real-time systems.

In the following, we assume that the processors can operate in several execution modes. An execution mode m_z is characterized by a pair of supply and body-bias voltages: $m_z = (V_{dd_z}, V_{bs_z})$. As a result, an execution mode has an associated frequency and power consumption (dynamic and leakage) that can be calculated using Equation 12.3 and respectively Equations 12.1 and 12.2. Upon a mode change, the corresponding delay and energy penalties are computed using Equations 12.5 and 12.4.

Tasks that are mapped on different processors communicate over one or more shared buses. In this chapter, we assume that the buses are not voltage scalable and thus working at a given frequency. Each communication task has a fixed execution time and energy consumption depending proportionally on the amount of communication. For simplicity of the explanations, we will not differentiate between computation and communication tasks.

4.   Optimization of Mapping and Schedule for Voltage Selection

The voltage selection techniques presented in this chapter are applied to applications that have been statically mapped and scheduled beforehand. Nevertheless, the efficiency of the voltage selection depends significantly on the available slack times within the system schedule. For this reason a global optimization flow that incorporates mapping and scheduling should be tightly integrated with voltage selection.

This section briefly introduces an algorithm for the mapping and scheduling of a given application on a target system composed of voltage scalable processors. The overall flow is outlined in Figure 12.2. As input, we consider a set of tasks with precedence constraints $\mathcal{T} = \{\tau_i\}$ and a hardware architecture consisting of a set $\mathcal{P} = \{p_k\}$ of processors. For each task τ_i, the deadline dl_i is given. For each processor $p \in \mathcal{P}$, the worst case number of clock cycles to be executed NC_i^p and the switched capacitance $C_{eff_i}^p$ are also given. Each processor can vary its supply voltage V_{dd} and body-bias voltage V_{bs}. The power dissipation (leakage and dynamic) and the cycle time (processor speed) depend on the selected voltage pair (mode). Tasks are executed cycle by cycle, and each cycle can potentially execute at a different voltage pair, i.e., at a different speed. The goal is to find a mapping, schedule, and voltage pair assignment for each task such that the individual task deadlines are met and the total energy consumption is minimal. As indicated in Figure 12.2, the optimization flow is split in two parts:

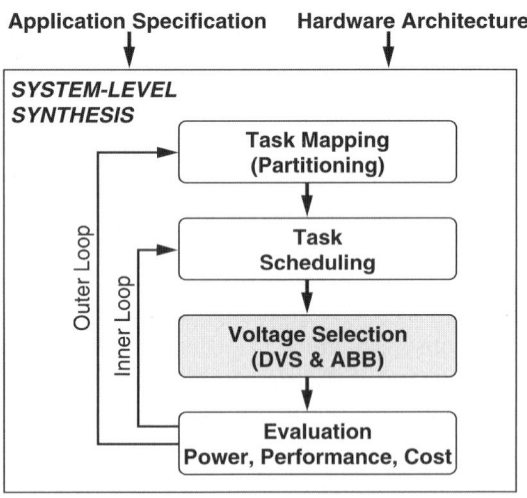

Fig. 12.2   Optimization flow for energy efficiency

- Task mapping optimization

- Schedule optimization

4.1   Genetic Task Mapping Algorithm

The task mapping step determines the allocation of each task on a particular processor. In the case of a heterogeneous architecture, mapping determines the execution time and power dissipation of the tasks running with nominal voltages. The goal of the mapping optimization step is to distribute the tasks among the processors, such that the energy dissipation is minimized and feasible designs in terms of timing constraints are achieved. Task mapping has been intensively researched over the last decade. It belongs to the class of NP-hard problems (Garey and Johnson, 1979). One effective way to address this problem is the usage of genetic algorithms, as in Dick and Jha (1998), Dick and Jha (1999), and Schmitz et al. (2004).

In a genetic algorithm-based task mapping approach solution candidates (potential mappings) are encoded into mapping strings, as shown in Figure 12.3. Each gene in these strings describes a mapping of a task to a processing element. For instance, task τ_4 in Figure 12.3 is mapped to CPU0. As typical in all genetic algorithms, ranking, selection, crossover, mutation, and offspring insertion are applied in order to evolve an initial solution pool. The key feature of this algorithm is the invocation of a genetic list scheduling algorithm for each mapping candidate in order to calculate the fitness function that guides the optimization. The genetic scheduling algorithm will try to

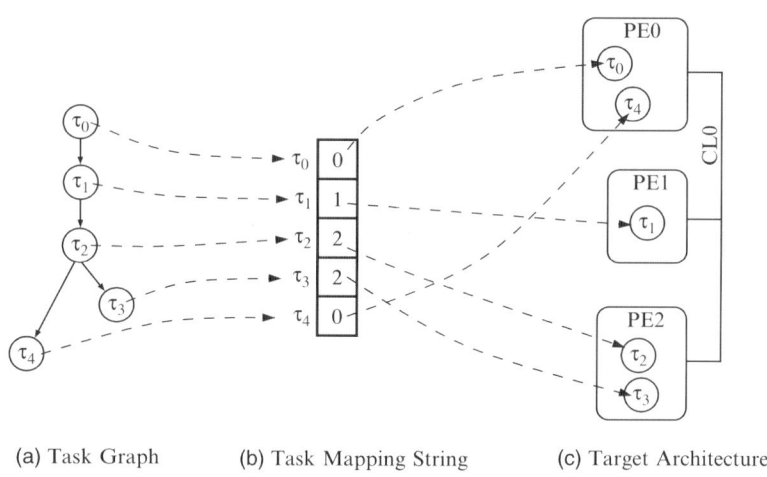

(a) Task Graph   (b) Task Mapping String   (c) Target Architecture

Fig. 12.3   Task mapping string describing the mapping of five tasks to an architecture

find for a given mapping, the schedule that respects all the task deadlines and, furthermore, has the minimum energy.

4.2   Genetic Scheduling Algorithm

The problem of statically scheduling tasks with dependencies on multiprocessor systems has been intensively studied (Adam et al., 1974; Wu and Gajski, 1990; Oh and Ha, 1996; Prakash and Parker, 1992; Sih and Lee, 1993; Bjørn-Jørgensen and Madsen, 1997; Kwok and Ahmad, 1999). Due to the computational complexity (this problem belongs to the class of NP-hard problems [Garey and Johnson 1979]), most scheduling techniques rely on various heuristic methods. In particular, one of the most widely used heuristics is list scheduling (LS). List scheduling algorithms take scheduling decisions based on task priorities. They maintain, for each processor, a ready list that contains the tasks that are ready to be scheduled. A task is considered to be ready, if all its predecessors (given by the task graph) have finished their execution. The static schedule is constructed by scheduling the ready task with the highest priority as soon as the eligible processor becomes available. Thereby, the assignment of priorities defines the task execution order.

The basic idea behind list scheduling is shown in Figure 12.4, which outlines the construction of a schedule for a single processor system. Consider the task graph with annotated priorities from Figure 12.4a. In the initial scheduling step all tasks with no ingoing edges are placed into a ready list,[1] as shown in Figure 12.4b, Step 1. For this particular example, in the first step task τ_0 is added to the ready list. Being the only task in the ready list, task τ_0 is sched-

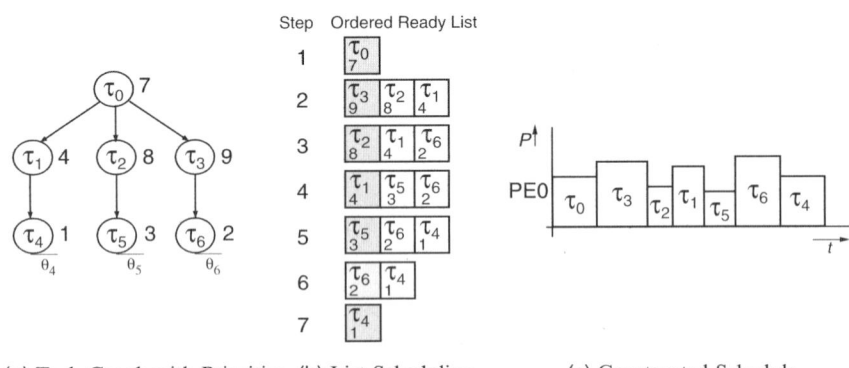

(a) Task Graph with Priorities   (b) List Scheduling   (c) Constructed Schedule

Fig. 12.4   List scheduling

---

[1] For multiprocessor systems consisting of several processors and communication links, a ready list is introduced for each component. Furthermore, priorities are additionally assigned to communications.

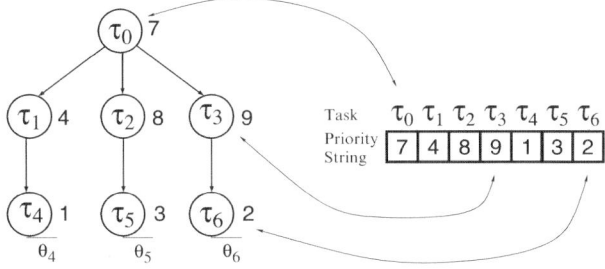

Fig. 12.5 Task priority encoding into a priority string

uled. After its execution has finished, the tasks τ_1, τ_2, and τ_3 become eligible for scheduling (due to their data dependency on τ_0); hence, they are placed into the ready list in decreasing order of their priorities (Scheduling Step 2). At this point τ_3 represents the ready task with the highest priority (9), so it is scheduled in Step 2. Having scheduled task τ_3, task τ_5 becomes ready and thus it is placed into the ready list, according to its priority. These scheduling procedure is repeated until no tasks are left in the ready list. Since each scheduling step schedules one task, seven iterations are necessary. The final schedule is shown in Figure 12.4c.

Clearly, different assignments of priorities result in different schedules. This is where the genetic algorithm comes into the play. Genetic list scheduling approaches combine fast constructive list scheduling techniques with the optimization power of genetic algorithms. By encoding the task priorities into a priority string, as shown in Figure 12.5, it becomes possible to apply a genetic algorithm-based optimization. The genetic algorithm aims to find an assignment of priorities that leads to a schedule of high quality in terms of timing behavior and exploitable slack time. The main principles behind the genetic list scheduling algorithm are illustrated in Figure 12.6. These principles are based on two strategies: crossover and mutation.

Crossover Example: Out of an initial population pool that contains six priority candidate strings, the strings 1 and 3 are selected. Offsprings are produced by replacing parts of the first parent string with parts of the second parent string. Hence, crossover results in two new offsprings (child 1 and child 2). These new priorities are used to schedule the tasks, in order to determine their quality. According to the quality, the produced strings are inserted into the solution pool. By selecting high quality strings for crossover, the chances to evolve priority strings of even higher quality are increased. The mating of two strings is carried out with respect to an arbitrarily selected crossover point.

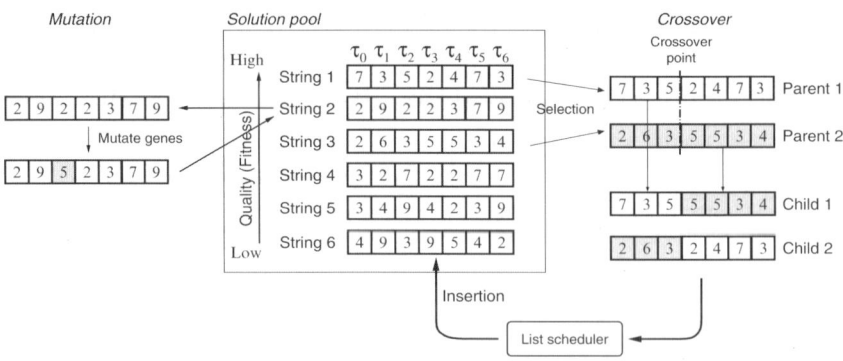

Fig. 12.6   Principle behind the genetic list scheduling algorithm

Mutation Example: In order to enter into unexplored regions of the scheduling space, the genetic algorithm mutates individuals of the solution pool occasionally (with a low probability). The mutation is carried out by randomly changing genes of a randomly selected string. For instance, in Figure 12.6 string 2 is selected and its third gene is manipulated. The modified string is then reinserted into the solution pool.

Both crossover and mutation are applied during the iterative execution of the genetic algorithm. The algorithm terminates after a stop criterion is fulfilled (e.g., a bound of the number of consecutive generations that did not improve significantly the solution).

A fitness function is used for evaluating the quality of a schedule. The fitness function captures the energy of a certain schedule. After the list scheduling has constructed a schedule for a given set of priorities, the algorithm proceeds by passing this schedule to a voltage selection algorithm that identifies the task voltages that minimize the energy dissipation. After performing the voltage selection, the fitness F_S of each schedule candidate is calculated using the following equation:

$$F_S = \underbrace{\left(\sum_{\tau_k \in \mathcal{T}} E(k) \right)}_{\text{Energy dissipation}} \tag{12.6}$$

As we have seen in this section, voltage selection is the core of the global energy optimization. In the rest of the chapter we will concentrate on several voltage selection algorithms.

5. Motivational Example

5.1 Optimizing the Dynamic and Leakage Energy

The approaches presented in this chapter achieve energy efficiency by performing the simultaneous scaling of the supply voltage V_{dd} and body-bias voltage V_{bs}. To illustrate the advantage of this simultaneous scaling over supply voltage scaling only, consider the following example. Figure 12.7 shows two optimal voltage schedules for a set of three tasks (τ_1, τ_2, and τ_3), executing in two possible voltage modes. While the first schedule relies on V_{dd} scaling only (i.e., V_{bs} is kept constant), the second schedule corresponds to the simultaneous scaling of V_{dd} and V_{bs}. Please note that the figures depict the dynamic and the leakage power dissipation as a function of time. For simplicity we neglect transition overheads in this example. Further, we consider processor parameters that correspond to CMOS technology ($<70\,nm$) which leads to a leakage power consumption close to 40% of the total power consumed (at the mode with the highest performance).

Let us consider the first schedule in which the tasks are executed either at $V_{dd1} = 1.8\,V$, or $V_{dd2} = 1.5\,V$, while V_{bs1} and V_{bs2} are kept at $0\,V$. In accordance, the system dissipates $P_{dyn1} = 100\,mW$ and $P_{leak1} = 75\,mW$ in mode 1 running at $700\,MHz$, while $P_{dyn2} = 49\,mW$ and $P_{leak2} = 45\,mW$ in mode 2 running at $525\,MHz$, as observable from the figure. We have also indicated the individual energy consumed in each of the active modes, separating between dynamic and leakage energy. The total leakage and dynamic energies of the schedule in Figure 12.7a are $13.56\,\mu J$ and $16.17\,\mu J$, respectively. This results in a total energy consumption of $29.73\,\mu J$.

Consider now the schedule given in Figure 12.7b, where tasks are executed at two different voltage settings for V_{dd} and V_{bs} ($m_1 = (1.8\,V, 0\,V)$ and $m_2 = (1.5\,V, -0.4\,V)$). Since the voltage settings for mode m_1 did not change, the system runs at $700\,MHz$ and dissipates $P_{dyn1} = 100\,mW$ and

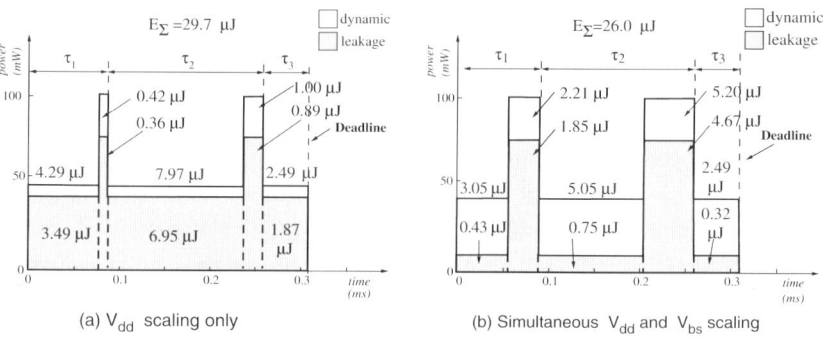

(a) V_{dd} scaling only   (b) Simultaneous V_{dd} and V_{bs} scaling

Fig. 12.7 Influence of V_{bs} scaling

$P_{leak1} = 75\,\mathrm{mW}$. In mode m_2 the system performs at $480\,\mathrm{MHz}$ and dissipates $P_{dyn2} = 49\,\mathrm{mW}$ and $P_{leak2} = 5\,\mathrm{mW}$. There are two main differences to observe compared to the schedule in Figure 12.7a. Firstly, the leakage power consumption during mode m_2 is considerably smaller than in Figure 12.7a; this is due to the fact that in mode m_2 the leakage is reduced through a body-bias voltage of $-0.4\,\mathrm{V}$ (see Equation 12.2). Secondly, the high voltage mode m_1 is active for a longer time; this can be explained by the fact that scaling V_{bs} during mode m_2 requires the reduction of the operational frequency (see Equation 12.3). Hence, in order to meet the system deadline, high performance mode m_1 has to compensate for this delay. Nevertheless, the total leakage and dynamic energies result in $8.02\,\mu\mathrm{J}$ and $18.00\,\mu\mathrm{J}$, respectively. Although here the dynamic energy was increased from $16.17\,\mu\mathrm{J}$ to $18.0\,\mu\mathrm{J}$, compared to the first schedule, the leakage was reduced from $13.56\,\mu\mathrm{J}$ to $8.02\,\mu\mathrm{J}$. The overall energy dissipation becomes then $26.02\,\mu\mathrm{J}$, a reduction by 12.5%. This example illustrates the advantage of simultaneous V_{dd} and V_{bs} scaling compared to V_{dd} scaling only.

5.2   Considering the Transition Overheads

To demonstrate the influence of the transition overheads in terms of energy and delay, consider the following example. For clarity reasons we restrict ourselves here to a single processor system that offers three voltage modes, $m_1 = (1.8\,\mathrm{V}, -0.3\,\mathrm{V})$, $m_2 = (1.5\,\mathrm{V}, -0.45\,\mathrm{V})$, and $m_3 = (1.2\,\mathrm{V}, -0.8\,\mathrm{V})$, where $m_z = (V_{dd_z}, V_{bs_z})$. The rail and substrate capacitance are given as $C_r = 10\,\mu\mathrm{F}$ and $C_s = 40\,\mu\mathrm{F}$. The processor needs to execute two consecutive tasks (τ_1 and τ_2) with a deadline of $0.225\,\mathrm{ms}$. Figure 12.8a shows a possible voltage schedule. Each of the two tasks is executed in two different modes: task τ_1 executes first in mode m_2 and then in mode m_1, while task τ_2 is initially executed in mode m_3 and then in mode m_2. The total energy consumption of this schedule is the sum of the energy dissipation in each mode $E = 9 + 15 + 4.5 + 7.5 = 36\,\mu\mathrm{J}$. However, if this voltage schedule is applied to a *real* voltage-scalable processor, the resulting schedule will be affected by transition overheads, as shown in Figure 12.8b. Here the processor requires a given time to adapt to the new execution mode. During this adaption no computations can be performed (Intel, 2000; AMD, 2000), i.e., the task execution is delayed, which, in turn, increases the schedule length such that the imposed deadline is violated. Moreover, transitions do not only require time, they also cause an additional energy dissipation. For instance, in the given schedule, the first transition overhead O_1 from mode m_2 and m_1 requires an energy of $10\,\mu\mathrm{F} \cdot (1.8\,\mathrm{V} - 1.5\,\mathrm{V})^2 + 40\,\mu\mathrm{F} \cdot (0.3\,\mathrm{V} - 0.45\,\mathrm{V})^2 = 1.8\,\mu\mathrm{J}$, based on Equation 12.4. Similarly, the energy overheads for transitions O_2 and O_3 can be calculated as $13.6\,\mu\mathrm{J}$ and $5.8\,\mu\mathrm{J}$, respectively. The overall energy dissipation of the realistic schedule shown in Figure 12.8b accumulates to $36 + 1.8 + 13.6 + 5.8 = 57.2\,\mu J$.

Let us consider the possibility of ordering the modes, as given in Figure 12.8c. Compared to the schedule in Figure 12.8a, the mode activation order in Figure 12.8c has been swapped for both tasks. As long as the transition overheads are neglected, the energy consumption of the two schedules is identical. However, applying the second activation order to a real processor would result in the schedule shown in Figure 12.8d. It can be observed that this schedule exhibits only two mode transitions (O_1 and O_3) within the tasks (intraswitches), while the switch between the two tasks (interswitch) has been eliminated. The overall energy consumption has been reduced to $E = 43.6\,\mu J$, a reduction by 23.8% compared to the schedule given in Figure 12.8b. Further, the elimination of transition O_2 reduces the overall schedule length, such that the imposed deadline is satisfied. With this example we have illustrated the effects that transition overheads can have on the energy consumption and the timing behavior and the impact of taking them into consideration when elaborating the voltage schedule.

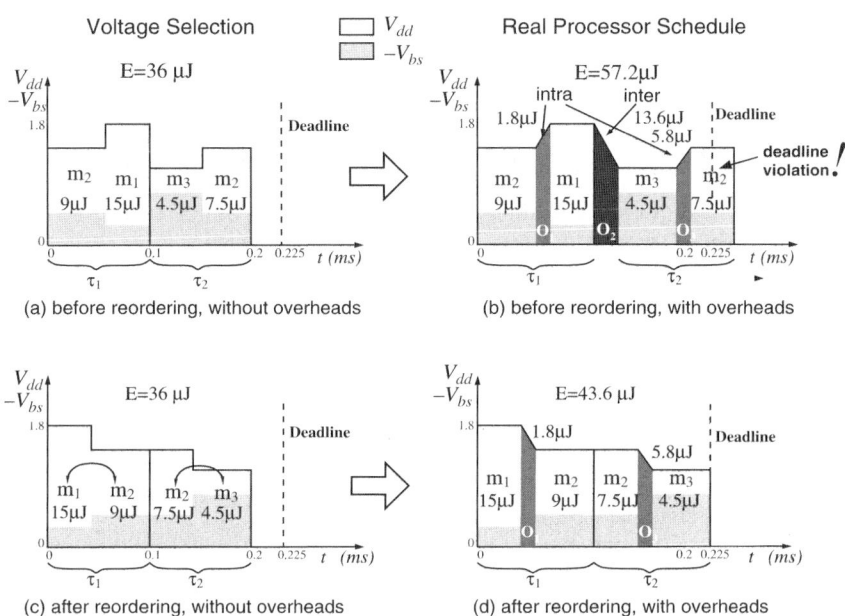

Fig. 12.8 Influence of transition overheads

6.   Problem Formulation

The voltage selection step of the global optimization from Figure 12.2 gets as input a mapped and scheduled task graph. Such a task graph is illustrated in Figure 12.1c. In this section, we define the actual problem solved during the voltage selection step.

Consider a set of tasks with precedence constraints $\mathcal{T} = \{\tau_i\}$ which have been mapped and scheduled on a set of variable voltage processors. For each task τ_i its deadline dl_i, its worst case number of clock cycles to be executed NC_i, and the switched capacitance C_{eff_i} are given. Each processor can vary its supply voltage V_{dd} and body-bias voltage V_{bs} within certain continuous ranges (for the continuous problem), or, within a set of discrete voltages pairs $m_z = \{(V_{dd_z}, V_{bs_z})\}$ (for the discrete problem). The power dissipation (leakage and dynamic) and the cycle time (processor speed) depend on the selected voltage pair (mode). Tasks are executed cycle by cycle, and each cycle can potentially execute at a different voltage pair, i.e., at a different speed. Our goal is to find voltage pair assignments for each task such that the individual task deadlines are met and the total energy consumption is minimal. Furthermore, whenever the processor has to alter the settings for V_{dd} and/or V_{bs}, a transition overhead in terms of energy and time is required (see Equations 12.4 and 12.5).

For reasons of clarity we introduce the following four distinctive problems which will be considered in this chapter: (a) continuous voltage scaling with no consideration of transition overheads (CNOH), (b) continuous voltage scaling with consideration of transition overheads (COH), (c) discrete voltage scaling with no consideration of transition overheads (DNOH), and (d) discrete voltage scaling with consideration of transition overheads (DOH).

7.   Optimal Continuous Voltage Selection

In this section we consider that supply and body-bias voltage of the processors in the system can be selected within a certain continuous range. We first formulate the problem neglecting the transition overheads (Section 7.1, CNOH) and then extend this formulation to include the energy and delay overheads (Section 7.2, COH).

7.1   Continuous Voltage Selection without Overheads (CNOH)

The continuous voltage selection problem, excluding the consideration of transition overheads (the CNOH problem), is solved using the following nonlinear problem formulation:

Minimize

$$\sum_{k=1}^{|\mathcal{T}|} \left(\underbrace{NC_k \cdot C_{eff_k} \cdot V_{dd_k}^2}_{E_{dyn_k}} + \underbrace{L_g(K_3 \cdot V_{dd_k} \cdot e^{K_4 \cdot V_{dd_k}} \cdot e^{K_5 \cdot V_{bs_k}} + I_{Ju} \cdot |V_{bs_k}|) \cdot t_k}_{E_{leak_k}} \right)$$

(12.7)

subject to

$$t_k = NC_k \cdot \frac{(K_6 \cdot L_d \cdot V_{dd_k})}{((1+K_1) \cdot V_{dd_k} + K_2 \cdot V_{bs_k} - V_{th_1})^\alpha}$$

(12.8)

$$D_k + t_k \;\leq\; D_l \quad \forall(k,l) \in \mathcal{E}$$ (12.9)

$$D_k + t_k \;\leq\; dl_k \quad \forall \, \tau_k \text{ that have a deadline}$$ (12.10)

$$D_k \;\geq\; 0$$ (12.11)

$$V_{dd_{min}} \leq V_{dd_k} \leq V_{dd_{max}} \quad \text{and} \quad V_{bs_{min}} \leq V_{bs_k} \leq V_{bs_{max}}$$ (12.12)

The variables that need to be determined are the task execution times t_k, the task start times D_k as well as the voltages V_{dd_k} and V_{bs_k}. The total energy consumption, which is the sum of dynamic and leakage energy, has to be minimized, as in Equation 12.7.[2] The minimization has to comply with the following relations and constraints. The task execution time has to be equivalent to the number of clock cycles of the task multiplied by the circuit delay for a particular V_{dd_k} and V_{bs_k} setting, as expressed by Equation 12.8. Given the execution time of the tasks, it becomes possible to express the precedence constraints between tasks (Equation 12.9), i.e., a task τ_l can only start its execution after all its predecessor tasks τ_k have finished their execution $(D_k + t_k)$. Predecessors of task τ_l are all tasks τ_k for which there exists an edge $(k,l) \in \mathcal{E}$ in the mapped and scheduled task graph. Similarly, tasks with deadlines have to be completed $(D_k + t_k)$ before their deadlines dl_k are exceeded (Equation 12.10). Task start times have to be positive (Equation 12.11) and the imposed voltage ranges should be respected (Equation 12.12). It should be noted that the objective (Equation 12.7) as well as the task execution time (Equation 12.8) are convex functions. Hence, the problem falls into the class of general convex nonlinear optimization problems. Such problems can be efficiently solved in polynomial time (given an arbitrary precision $\epsilon > 0$) (Nesterov and Nemirovskii, 1994).

---

[2]Please note that *abs* and *max* operations cannot be used directly in mathematical programming, yet there exist standard techniques to overcome this limitation by equivalent formulations (Andrei et al. 2003; Williams, 1999).

7.2 Continuous Voltage Selection with Overheads (COH)

In this section we modify the previous formulation in order to take transition overheads into account (COH problem). The following formulation highlights the modifications:
Minimize

$$\underbrace{\sum_{k=1}^{|\mathcal{T}|}(E_{dyn_k}+E_{leak_k})}_{\text{Task energy dissipation}} + \underbrace{\sum_{(k,j)\in\mathcal{E}^\bullet}\epsilon_{k,j}}_{\text{Transition energy overhead}} \qquad (12.13)$$

subject to

$$D_k + t_k + \delta_{k,j} \leq D_j \quad \forall (k,j) \in \mathcal{E}^\bullet \qquad (12.14)$$

$$\delta_{k,j} = \max(p_{Vdd}\cdot|V_{dd_k}-V_{dd_j}|, p_{Vbs}\cdot|V_{bs_k}-V_{bs_j}|) \qquad (12.15)$$

The objective function Equation 12.13 now additionally accounts for the transition overheads in terms of energy. The energy overheads can be calculated according to Equation 12.4 for all consecutive tasks τ_k and τ_j on the same processor ($\mathcal{E}^\bullet$ is defined in Section 11.2). However, scaling voltages does not only require energy but it introduces delay overheads as well. Therefore, we introduce an additional constraint similar to Equation 12.9, which states that a task τ_j can only start after the execution of its predecessor τ_k ($D_k + t_k$) on the same processor and after the new voltage mode is reached ($\delta_{k,j}$). This constraint is given in Equation 12.14. The delay penalties $\delta_{k,j}$ are introduced as a set of new variables and are constrained subject to Equation 12.15. Similar to the CNOH formulation, the COH model is a convex nonlinear problem, i.e., it can be solved in polynomial time.

8. Optimal Discrete Voltage Selection

The approaches presented in the previous section provide a theoretical upper bound on the possible energy savings. In reality, however, processors are restricted to a discrete set of V_{dd} and V_{bs} voltage pairs. In this section we investigate the discrete voltage selection problem without and with the consideration of overheads. The complexity of the discrete voltage selection problem will be also analyzed.

8.1 Problem Complexity

Theorem 1. The discrete voltage selection problem is NP-hard.

Proof. We proof by restriction. The discrete time-cost trade-off (DTCT) problem is known to be NP-hard (De et al., 1997). By restricting the discrete voltage selection problem (DNOH) to contain only tasks that require an execution of one clock cycle, it becomes identical to the DTCT problem. Hence, DTCT $\in$ DNOH which leads to the conclusion DNOH $\in$ NP. $\qquad\square$

The exact details of the proof are given in (Andrei et al., 2003). Note that the problem remains NP-hard, even if we restrict it to supply voltage selection (without adaptive body-biasing) and even if transition overheads are neglected. It should be noted that this finding renders the conclusion of Kwon and Kim (2005)[3] impossible, which states that the discrete voltage selection problem (considered in Kwon and Kim (2005) without body-biasing and overheads) can be solved optimally in polynomial time.

8.2 Discrete Voltage Selection without Overheads (DNOH)

In the following we will give a mixed integer linear programming (MILP) formulation for the discrete voltage selection problem without overheads (DNOH). We consider that processors can run in different modes $m \in \mathcal{M}$. Each mode m is characterized by a voltage pair (V_{dd_m}, V_{bs_m}), which determines the operational frequency f_m, the normalized dynamic power P_{dnom_m}, and the leakage power dissipation P_{leak_m}. The frequency and the leakage power are given by Equations 12.3 and 12.2, respectively. The normalized dynamic power is given by $P_{dnom_m} = f_m \cdot V_{dd_m}^2$. Accordingly, the dynamic power of a task τ_k operating in mode m is computed as $C_{eff_k} \cdot P_{dnom_m}$. Based on these definitions, the problem is formulated as follows:
Minimize

$$\sum_{k=1}^{|\mathcal{T}|} \sum_{m \in \mathcal{M}} \left(C_{eff_k} \cdot P_{dnom_m} \cdot t_{k,m} + P_{leak_m} \cdot t_{k,m} \right) \qquad (12.16)$$

subject to

$$D_k + \sum_{m \in \mathcal{M}} t_{k,m} \leq dl_k \qquad (12.17)$$

$$D_k + \sum_{m \in \mathcal{M}} t_{k,m} \leq D_l \quad \forall (k,l) \in \mathcal{E} \qquad (12.18)$$

$$c_{k,m} = t_{k,m} \cdot f_m \quad \text{and} \quad \sum_{m \in \mathcal{M}} c_{k,m} = NC_k \quad c_{k,m} \in \mathbb{N} \qquad (12.19)$$

$$D_k \geq 0 \quad \text{and} \quad t_{k,m} \geq 0 \qquad (12.20)$$

The total energy consumption, expressed by Equation 12.16, is given by two sums. The inner sum indicates the energy dissipated by an individual task τ_k, depending on the time $t_{k,m}$ spent in each mode m. The outer sum adds up the energy of all tasks. Unlike the continuous voltage selection case, we do not obtain the voltage V_{dd} and V_{bs} directly, but rather we find out how much

---

[3]The flaw in Kwon and Kim (2005) lies in the fact that the number of clock cycles spent in a mode is not restricted to be integer.

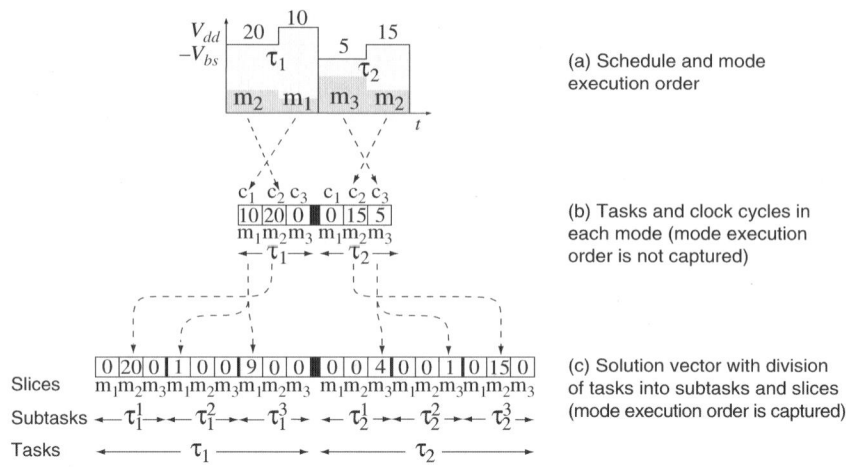

(a) Schedule and mode execution order

(b) Tasks and clock cycles in each mode (mode execution order is not captured)

(c) Solution vector with division of tasks into subtasks and slices (mode execution order is captured)

Fig. 12.9   Discrete mode model

time to spend in each of the modes. Therefore, task execution time $t_{k,m}$ and the number of clock cycles $c_{k,m}$ spent within a mode become the variables in the MILP formulation. The number of clock cycles has to be an integer and hence $c_{k,m}$ is restricted to the integer domain. We exemplify this model graphically in Figure 12.9a and b. The first figure shows the schedule of two tasks executing each at two different voltage settings (two modes out of three possible modes). Task τ_1 executes for 20 clock cycles in mode m_2 and for 10 clock cycles in m_1, while task τ_2 runs for 5 clock cycles in m_3 and 15 clock cycles in m_2. The same is captured in Figure 12.9b in what we call a mode model. The modes that are not active during a task's run-time have the corresponding time and number of clock cycles 0 (mode m_3 for τ_1 and m_1 for τ_2). The overall execution time of task τ_k is given as the sum of the times spent in each mode ($\sum_{m \in \mathcal{M}} t_{k,m}$). Equation 12.17 ensures that all the deadlines are met and Equation 12.18 maintains the correct execution order given by the precedence relations. The relation between execution time and number of clock cycles as well as the requirement to execute all clock cycles of a task are expressed in Equation 12.19. Additionally, task start times D_k and task execution times have to be equal or larger than zero, as given in Equation 12.20.

8.3   Discrete Voltage Selection with Overheads (DOH)

Let us proceed now with the incorporation of transition overheads into the MILP formulation given in Section 8.2. The order in which the modes are activated has an influence on the transition overheads, as we have illustrated in Section 5.2. Nevertheless, the formulation in Section 8.2 omits information regarding the activation order of modes. The model in Figure 12.9b does not capture the order in which modes are activated, it solely expresses how many clock cycles are spent in each mode. We introduce the following extensions needed in order to take both delay and energy overheads into account. Given m operational modes, the execution of a single task τ_k can be subdivided into m subtasks $\tau_k^s, s = 1, \ldots, m$. Each subtask is executed in one and only one of the m modes. Subtasks are further subdivided into m slices, each corresponding to a mode. This results in $m \cdot m$ slices for each task. Figure 12.9c depicts this model, showing that task τ_1 runs first in mode m_2, then in mode m_1, and that τ_2 runs first in mode m_3, then in mode m_2. This ordering is captured by the subtasks: the first subtask of τ_1 executes 20 clock cycles in mode m_2, the second subtask executes one clock cycle in m_1, and the remaining 9 cycles are executed by the last subtask in mode m_1; τ_2 executes in its first subtask 4 clock cycles in mode m_3, 1 clock cycle is executed during the second subtask in mode m_3, and the last subtask executes 15 clock cycles in the mode m_2. Note that there is no overhead between subsequent subtasks that run in the same mode. The following gives the modified MILP formulation:

Minimize

$$\sum_{k=1}^{|\mathcal{T}|} \sum_{s \in \mathcal{M}} \sum_{m \in \mathcal{M}} \left(C_{eff_k} \cdot P_{dnom_m} \cdot t_{k,s,m} + P_{leak_m} \cdot t_{k,s,m} \right)$$

$$\underbrace{\phantom{\sum_{k=1}^{|\mathcal{T}|} \sum_{s \in \mathcal{M}} \sum_{m \in \mathcal{M}}}}_{\text{Task energy dissipation}}$$

$$+ \sum_{k=1}^{|\mathcal{T}|} \sum_{s \in \mathcal{M}} \sum_{i \in \mathcal{M}} \sum_{j \in \mathcal{M}} \left(b_{k,s,i,j} \cdot EP_{i,j} \right) \qquad (12.21)$$

$$\underbrace{}_{\text{Transition energy overhead}}$$

subject to

$$\delta_k = \sum_{s \in \mathcal{M}^*} \sum_{i \in \mathcal{M}} \sum_{j \in \mathcal{M}} b_{k,s,i,j} \cdot DP_{i,j} \qquad (12.22)$$

$$\delta_{k,l} = \sum_{i \in \mathcal{M}} \sum_{j \in \mathcal{M}} b_{k,m,i,j} \cdot DP_{i,j} \quad \text{where } (k,l) \in \mathcal{E}^\bullet \qquad (12.23)$$

$$D_k + \sum_{s \in \mathcal{M}} \sum_{m \in \mathcal{M}} t_{k,s,m} + \delta_k \leq dl_k \qquad (12.24)$$

$$D_k + \sum_{s \in \mathcal{M}} \sum_{m \in \mathcal{M}} t_{k,s,m} + \delta_k + \delta_{pl,l} \leq D_l \quad \forall (k,l) \in \mathcal{E}, (pl,l) \in \mathcal{E}^\bullet \qquad (12.25)$$

$$c_{k,s,i} = t_{k,s,i} \cdot f_i \quad s \in \mathcal{M}, i \in \mathcal{M}, c_{k,s,i} \in \mathbb{N} \qquad (12.26)$$

$$\sum_{s \in \mathcal{M}} \sum_{i \in \mathcal{M}} c_{k,s,i} = NC_k \qquad (12.27)$$

In order to capture the energy overheads in the objective function (Equation 12.21), we introduce the boolean variables $b_{k,s,i,j}$. In addition, we introduce an energy penalty matrix EP, which contains the energy overheads for all possible mode transitions, i.e., $EP_{i,j}$ denotes the energy overhead necessary to change from mode i to j. These energy overheads are precomputed based on the available modes (voltage pairs) and Equation 12.4. The overall energy overhead is given by all intratask and intertask transitions. The intratask and intertask delay overheads, given in Equations 12.22 and 12.23, are calculated based on a delay penalty matrix $DP_{i,j}$, which, similar to the energy penalty matrix, can be precomputed based on the available modes and Equation 12.5. For a task τ_k and for each of its subtasks τ_k^s, except the last one, the variable $b_{k,s,i,j} = 1$ if mode i of subtask τ_k^s and mode j of τ_k^{s+1} are both active (s in $1, \ldots, |\mathcal{M}| - 1$, i, j in $1, \ldots, m$). These are used in order to capture the intratask overheads, as in Equation 12.22. For intertask overheads, we are interested in the last mode of task τ_k and the first mode of the subsequent task τ_l (running on the same processor). Therefore, $b_{k,m,i,j} = 1$ if the mode i of the last subtask τ_k^m and the mode j of first subtask τ_l^1 are both active. For the example given in Figure 12.9c, $b_{1,1,2,1}$, $b_{1,2,1,1}$, $b_{1,3,1,3}$, $b_{2,1,3,3}$, $b_{2,2,3,2}$ are all 1 and the rest are 0. The computation of the b variables is performed using the auxiliary a variables. The details regarding this computation are omitted from the model and are available in Andrei et al. (2003). Deadlines and precedence relations, taking the delay overheads into account, have to be respected according to Equations 12.24 and 12.25. Here $\sum_{s \in \mathcal{M}} \sum_{m \in \mathcal{M}} t_{k,s,m}$ represents the total execution time of a task τ_k, based on the number of cycles in each of the subtasks and modes. Equations 12.26 and 12.27 are a reformulation of Equation 12.19, which expresses the relation between the execution time and the number of clock cycles and the requirement to execute all clock cycles of a task. To ease the explanation, the above given MILP formulation has been simplified to a certain degree. In particular, we have omitted here details on the computation of the b variables as well as the constraints that make sure that one and only one mode must be used by a subtask. The complete MILP model can be found in Andrei et al. (2003).

8.4   Discrete Voltage Selection Heuristic

As shown earlier, discrete voltage selection is NP-hard. Thus, solving it using the presented MILP formulation for large instances is time consuming. In the following, a heuristic to effectively solve the discrete voltage selection problem is proposed. The main idea behind this heuristic is to perform a continuous voltage selection (as outlined in Section 7). As a result of this calculation, for each task, a continuous voltage pair $(V_{dd_{con}}, V_{bs_{con}})$ as well as the corresponding frequency f_{con} will be determined. Using the approach

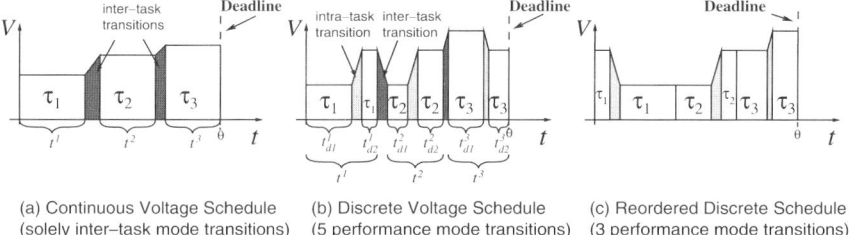

(a) Continuous Voltage Schedule (solely inter–task mode transitions) (b) Discrete Voltage Schedule (5 performance mode transitions) (c) Reordered Discrete Schedule (3 performance mode transitions)

Fig. 12.10 VS heuristic: performance mode reordering

introduced in Ishihara and Yasuura (1998), for each task the two surrounding discrete performance modes are chosen such that $f_{d1} < f_{con} < f_{d2}$. That is, the execution of a task is split into two regions with t_{d1} and t_{d2} being the execution times in the mode with f_{d1} and f_{d2}, respectively. Figure 12.10a and b indicate this transformation for an application with three tasks. In the continuous scaling case, Figure 12.10a, each of the tasks executes at a single voltage setting, i.e., the voltages are changed only between tasks. In the discrete case, the voltage setting is changed during the task execution. Of course, the required time overhead δ_i for the mode change has to be considered as well, i.e., $t_i = t_{d1}^i + t_{d2}^i + \delta_i$, where t_i is the execution time with continuous voltage setting of the task τ_i. In general, executing activities in two performance modes, determined as above, leads to close to optimal discrete voltage selection. Having determined the discrete performance mode settings, the intertask transition overheads are reduced by reordering the mode sequence of each task. We reorder the modes in a greedy manner, such that the intertask overhead between consecutive tasks is minimized. This is outlined in Figure 12.10c. While this reordering technique is optimal for processors that offer two performance modes, this is not true for components with three or more modes. Nevertheless, as demonstrated by our experiments, this heuristic is fast and efficient. Of course, the additional slack produced as a result of the reduced transition times is exploited as well.

9. Experimental Results

In order to validate the real-world applicability of the techniques presented in this chapter, a set of experiments have been perfomed, using a GSM voice codec. Details regarding this application can be found in Schmitz et al. (2004). The GSM voice codec consists of 87 tasks and 137 dependencies and is considered to run on an architecture composed of three processing elements with two voltage modes $((1.8\,\text{V}, -0.1\,\text{V})$ and $(1.0\,\text{V}, -0.6))$. At the highest voltage mode, the application reveals a deadline slack close to 10%. Switching overheads are characterized by $C_r = 1\,\mu\text{F}$, $C_s = 4\,\mu\text{F}$, $p_{Vdd} = 10\,\mu\text{s/V}$, and

Table 12.1   Optimization results for the GSM codec

Approach	E_{dyn} (mJ)	E_{leak} (mJ)	ϵ (mJ)	E_{total} (mJ)	Reduction (%)
Nominal	1.34	0.620	non	1.962	–
DVDDNOH	1.232	0.560	0.047	1.793	8.7
DVDDOH	1.193	0.560	0.0031	1.753	10.7
DNOH	1.300	0.230	0.048	1.531	22
DOH	1.257	0.230	0.0049	1.487	24.3
Heuristic	1.291	0.260	0.009	1.56	22.2

$p_{Vbs} = 10\,\mu s/V$. Table 12.1 shows the resulting energy consumptions in terms of dynamic E_{dyn}, leakage E_{leak}, overhead ϵ, and total energy E_{total} (Columns 2–5). Each line represents a different voltage selection approach. Line 2 (nominal) is used as a baseline and corresponds to an execution at the nominal voltages. The lines 3 and 4 give the results for the classical V_{dd} selection, without (DVDDNOH) and with (DVDDOH) the consideration of overheads. As can be seen in the table, the consideration of overheads achieves higher energy saving (10.7%) than the overhead neglecting optimization (8.7%). The results given in lines 5 and 6 correspond to the combined V_{dd} and V_{bs} selection schemes. Please note again the distinction between overheads neglecting (DNOH) and overhead considering (DOH) approaches. If the overheads are neglected, the energy consumption can be reduced by 22%, yet taking the overheads into account results in an reduction of 24.3%, solely achieved by decreasing the transition overheads. Compared to the classical voltage selection scheme, the combined selection achieved a further reduction of 14%. The last line shows the results of the proposed heuristic approach. Although the result does not match the optimal one given in line 6, it should be noted that such heuristic techniques are needed when dealing with problems of larger complexity (increased number of voltage modes and tasks). In the GSM application, although the number of tasks is relatively large, only two voltage modes have been considered. Therefore the optimal solutions could be obtained for the DOH problem.

10.   Related Work

There has been a considerable amount of work on dynamic voltage selection. Yao et al. (1995) proposed the first DVS approach for single processor systems which can dynamically change the supply voltage over a continuous range. Ishihara and Yasuura (1998) modeled the discrete voltage selection problem using an integer linear programming (ILP) formulation. Kwon and Kim (2005) proposed a linear programming (LP) solution for the discrete voltage selection problem with uniform and non-uniform switched capacitance. Although this work gives the impression that the problem can be solved optimally in polynomial time, it was shown in this chapter that the discrete voltage selection problem is in fact strongly NP-hard and, hence, no optimal solution can be found in polynomial time, for example using LP. Dynamic voltage selection has also been successfully applied to heterogeneous distributed systems, in which numerous processing elements interact via a communication infrastructure (Gruian and Kuchcinski, 2001; Luo and Jha, 2003; Schmitz and Al-Hashimi, 2001; Andrei et al., 2004a; Andrei et al., 2004a). Zhang et al. (2002) approached continuous supply voltage selection in distributed systems using an ILP formulation. They solved the discrete version of the problem through an approximation.

As opposed to approaches that scale only the supply voltage V_{dd} and neglect leakage power consumption, Kim and Roy (2002) proposed an adaptive body-biasing approach (in their work referred to as dynamic V_{th} scaling) for active leakage power reduction. They demonstrate that the efficiency of ABB will become, with advancing CMOS technology, comparable to DVS. Duarte et al. (2002) analyze the effectiveness of supply and threshold voltage selection, and show that simultaneously adjusting both voltages provides the highest savings. Martin et al. (2002) presented an approach for combined dynamic voltage selection and adaptive body-biasing.

The importance of the voltage transition overheads has been highlighted in Hong et al. (1998), Mochocki et al. (2002), Zhang et al. (2003), and Andrei et al. (2004a).

A similar problem for continuous voltage selection has been recently formulated in Yan et al. (2005). However, it is solved using a suboptimal heuristic.

11.   Summary

Energy reduction techniques, such as dynamic supply voltage scaling and adaptive body-biasing, can be effectively exploited for multiprocessor systems. In this chapter, we have presented the task mapping and scheduling problem, together with different notions of combined dynamic supply voltage scaling and adaptive body-biasing. These include the consideration of transition overheads as well as the discretization of the supply and threshold voltage levels. It was demonstrated that nonlinear programming and mixed integer linear programming formulations can be used to solve these problems. Furthermore, the NP-hardness of the discrete voltage selection case was shown, and a heuristic to efficiently solve the problem has been proposed.

References

Adam, T., Chandy, K., and Dickson, J. (1974). A Comparison of List Scheduling for Parallel Processing Systems. *Journal of Communications of the ACM*, 17(12):685–690.

AMD (2000). Mobile AMD Athlon™4, Processor Model 6 CPGA Data Sheet. Publication No. 24319 Rev E.

Andrei, A., Schmitz, M., Eles, P., Peng, Z., and Al-Hashimi, B. (2004a). Overhead-Conscious Voltage Selection for Dynamic and Leakage Power Reduction of Time-Constraint Systems. In *Proceedings of the Design, Automation and Test in Europe Conference (DATE04)*, pages 518–523.

Andrei, A., Schmitz, M., Eles, P., Peng, Z., and Hashimi, B. Al (2004b). Simultaneous Communication and Processor Voltage Scaling for Dynamic and Leakage Energy Reduction in Time-Constrained Systems. In *ICCAD*, pages 362–369.

Andrei, A., Schmitz, M.T., Eles, P., and Peng, Z. (2003). Overhead-Conscious Voltage Selection for Dynamic and Leakage Energy Reduction of Time-Constrained Systems. Technical report, Linkoping University, Department of Computer and Information Science, Sweden.

Bjørn-Jørgensen, Peter and Madsen, Jan (1997). Critical Path Driven Cosynthesis for Heterogeneous Target Architectures. In *Proceedings of the International Workshop on Hardware/Software Codesign*, pages 15–19.

Borkar, S. (1999). Design Challenges of Technology Scaling. *IEEE Micro*, pages 23–29.

Chandrakasan, A.P. and Brodersen, R.W. (1995). *Low Power Digital CMOS Design*. Kluwer Academic Publisher.

De, P., Dunne, E., Ghosh, J., and Wells, C. (1997). Complexity of the Discrete Time–Cost Tradeoff Problem for Project Networks. *Operations Research*, 45(2):302–306.

Dick, R. and Jha, N.K. (1999). MOCSYN: Multiobjective Core-Based Single-Chip System Synthesis. In *Proceedings of the Design, Automation and Test in Europe Conference (DATE99)*, pages 263–270.

Dick, Robert P. and Jha, Niraj K. (1998). MOGAC: A Multiobjective Genetic Algorithm for Hardware–Software Co-Synthesis of Distributed Embedded Systems. *IEEE Transactions on Computer-Aided Design*, 17(10):920–935.

Duarte, D., Vijaykrishnan, N., Irwin, M., Kim, H., and McFarland, G. (2002). Impact of Scaling on the Effectiveness of Dynamic Power Reduction. In *Proceedings of the ICCD*, pages 382–387.

Garey, M.R. and Johnson, D.S. (1979). *Computers and Intractability: A Guide to the Theory of NP-Completeness*. W.H. Freeman and Company.

Gruian, F. and Kuchcinski, K. (2001). LEneS: Task Scheduling for Low-Energy Systems Using Variable Supply Voltage Processors. In *Proceedings of the ASP-DAC'01*, pages 449–455.

Hong, Inki, Qu, Gang, Potkonjak, Miodrag, and Srivastava, Mani B. (1998). Synthesis Techniques for Low-Power Hard Real-Time Systems on Variable Voltage Processors. In *Proceedings of the Real-Time Systems Symposium*, pages 178–187.

Intel (2000). Intel® XScale™ Core, Developer's Manual.

Ishihara, Tohru and Yasuura, Hiroto (1998). Voltage Scheduling Problem for Dynamically Variable Voltage Processors. In *Proceedings of the International Symposium on Low Power Electronics and Design (ISLPED'98)*, pages 197–202.

Kim, C. and Roy, K. (2002). Dynamic Vth Scaling Scheme for Active Leakage Power Reduction. In *Proceedings of the Design, Automation and Test in Europe Conference (DATE02)*, pages 163–167.

Kwok, Yu-Kwong and Ahmad, Ishfaq (1999). Static Scheduling Algorithms for Allocating Directed Task Graphs to Multiprocessors. *ACM Computing Surveys*, 31(4):406–471.

Kwon, W. and Kim, T. (2005). Optimal Voltage Allocation Techniques for Dynamically Variable Voltage Processors. *ACM Transactions on Embedded Computing Systems*, 4(1):211–230.

Luo, J. and Jha, N. (2003). Power-profile Driven Variable Voltage Scaling for Heterogeneous Distributed Real-Time Embedded Systems. In *Proceedings of the VLSI'03*, pages 369–375.

Martin, S., Flautner, K., Mudge, T., and Blaauw, D. (2002). Combined Dynamic Voltage Scaling and Adaptive Body Biasing for Lower Power Microprocessors under Dynamic Workloads. In *Proceedings of the ICCAD-02*, pages 721–725.

Mochocki, B., Hu, X., and Quan, G. (2002). A Realistic Variable Voltage Scheduling Model for Real-Time Applications. In *Proceedings of the ICCAD-02*, pages 726–731.

Nesterov, Y. and Nemirovskii, A. (1994). *Interior-Point Polynomial Algorithms in Convex Programming*. Studies in Applied Mathematics.

Oh, Hyunok and Ha, Soonhoi (1996). A Static Scheduling Heuristic for Heterogeneous Processors. In *2nd International EuroPar Conference Vol. II*.

Prakash, S. and Parker, A. (1992). SOS: Synthesis of Application-Specific Heterogeneous Multiprocessor Systems. *Journal of Parallel & Distributed Computing*, pages 338–351.

Schmitz, M. T., Al-Hashimi, B., and Eles, P. (2004). *System-Level Design Techniques for Energy-Efficient Embedded Systems*. Kluwer Academic Publisher.

Schmitz, Marcus T. and Al-Hashimi, Bashir M. (2001). Considering Power Variations of DVS Processing Elements for Energy Minimisation in Distributed Systems. In *International Symposium System Synthesis (ISSS'01)*, pages 250–255.

Sih, Gilbert C. and Lee, Edward A. (1993). A Compile-Time Scheduling Heuristic for Interconnection-Constrained Heterogeneous Processor Architectures. *IEEE Transactions on Parallel and Distributed Systems*, 4(2):175–187.

Williams, H. P. (1999). *Model Building in Mathematical Programming*. Wiley.

Wu, M. and Gajski, D. (1990). Hypertool: A Programming Aid for Message-Passing Systems. *IEEE Transactions on Parallel and Distributed Systems*, 1(3):330–343.

Yan, L., Luo, J., and Jha, N. (2005). Joint Dynamic Voltage Scaling and Adpative Body Biasing for Heterogeneous Distributed Real-Time Embedded Systems,. *IEEE Transactions on Computer-Aided Design*, 27(7):1030–1041.

Yao, F., Demers, A., and Shenker, S. (1995). A Scheduling Model for Reduced CPU Energy. *IEEE FOCS*.

Zhang, Y., Hu, X., and Chen, D. (2002). Task Scheduling and Voltage Selection for Energy Minimization. In *Proceedings of the IEEE DAC'02*, pages 183–188.

Zhang, Y., Hu, X., and Chen, D. (2003). Energy Minimization of Real-Time Tasks on Variable Voltage Processors with Transition Energy Overhead. In *Proceedings ASP-DAC'03*, pages 65–70.

IV

Compiler Techniques

Chapter 13

Compilation Techniques for Power, Energy, and Thermal Management

Ulrich Kremer
Department of Computer Science
Rutgers University
New Brunswick, New Jersey 08903
USA

Abstract   In addition to hardware and operating system directed techniques, compiler-directed power, energy, and thermal management has gained increasing importance. This chapter discusses the potential benefits of compiler-based approaches to solve the power/energy/thermal management problem. The ability of the compiler to reshape program behavior through aggressive, whole program optimizations, and to predict future program behaviors can give it an advantage over hardware and operating systems techniques. This chapter introduces several optimization metrics, together with state-of-the-art optimizations that target these metrics.

Keywords: compiler technology, power optimization, energy optimization, thermal optimization

1.   Optimizing Compilers

Optimizing compilers may either be static, trace-based, or dynamic. A static compiler takes the program text as input and generates a low-level program representation that can be executed on a particular target machine. A trace-based scheme uses information from representative executions of a program, typically through program instrumentation, to perform program optimizations. Finally, a fully dynamic compiler is invoked during program execution to improve the quality of parts of the program under execution. In this introductory chapter, we will mainly focus on static compilation strategies, i.e., what-

J. Henkel and S. Parameswaran (eds.), Designing Embedded Processors – A Low Power Perspective,
287–303.

ever a compiler knowns about a program has to be encoded in the program text itself. Recently, researchers have started looking at fully dynamic compilation techniques (e.g., [7, 31]).

The compiler knows about the semantics of all language features and typically tries to improve the quality of the overall program by applying compiler optimizations. Such optimizations are typically data or program transformations that result in better code without changing the semantics of the program. In other words, any program transformation has to be correct, and the correctness of the transformation has to verifiable at compile time. This is not an easy task, and depending on the transformation may require different program analyses with substantial computational requirements. These analyses are only able to approximate all possible program executions, resulting in the conservative, but safe application of transformations. This means that many opportunities for optimizations may be missed if the correctness of a transformation depends on conditions that can only be determined at runtime, such as a particular value of a variable.

However, even given all the limitations discussed above, optimizing compilers have been the key to making high-level languages viable, i.e., providing a high-level abstraction together with efficient, compiler-generated code. In principle, hardware and OS based program improvement strategies face the same challenges as compiler optimizations. However, the tradeoff decisions are different based on the acceptable cost of an optimization and the availability of information about dynamic program behavior. Hardware and OS techniques are performed at runtime where more accurate knowledge about control flow and program values may be available. Here, correctness and profitability checks result in execution time overheads, and therefore need to be rather inexpensive. Profitability analyses typically use a limited window of past program behavior to predict future behavior. In contrast, in a static compiler, most of the correctness and profitability checks are done at compiler time, i.e., not at program execution time, allowing more aggressive program transformations in terms of affected scope and required analyses. Since the entire program is available to the compiler, future program behavior may be predicted more accurately in the cases where static analysis techniques are effective. As mentioned above, purely static compilers do not perform well in cases where program behavior depends on dynamic values that cannot be determined or approximated at compile time. However, in many cases, the necessary dynamic information can be derived at compile time or code optimization alternatives are limited, allowing the appropriate alternative to be selected at runtime based on compiler generated tests. The ability of the compiler to reshape program behavior through aggressive whole program analyses and transformations that is a key advantage over hardware and OS techniques, exposing optimization opportunities that were not available before. In addition, aggressive whole program analyses allow optimizations with high-runtime overheads which typically require a larger scope in order to assess their profitability.

The goal of this chapter is to illustrate the potential of compiler optimizations for power, energy, and thermal management. We are not implying that the discussed optimizations have to be always performed at the compiler level. As we mentioned above, there are tradeoffs in terms of runtime and compile time overheads, as well as precision issues due to the conservative assumptions on which compiler optimizations are based. In the following sections, we will discuss several optimization goals, i.e., optimization metrics, together with state-of-the-art optimizations that target these metrics.

2.   Optimization Metrics

Optimizing compilers need underlying performance models and metrics to be able to transform the program code to achieve a specific optimization goal. These models and metrics guide the compiler to make selections among program transformation alternatives. If one optimization goal subsumes another, there is no need to develop separate models and metrics for the subsumed models. In this section we address the question whether power, energy, performance, and temperature should be considered separate compiler optimization goals or not.

2.1   Power vs. Energy

A common metric for power dissipation is that of *activity level* at any given point during program execution. Peak power is the maximum activity level, and step power the difference in activity levels of successive program points. The *total amount of activities* within a program region is the consumed overall energy. In this metric, the energy of a program execution is the same as its average power times the program's execution time. Reducing the peak power and step power (dI/dt problem) is desirable since it can increase the reliability and lifetime of high-performance architectures, or allows a program to execute under a preset power supply limit. In the following discussion, we concentrate on CPU activities, i.e., the goal of the compiler is to reduce the CPU power dissipation and CPU energy consumption. Other system components not covered here are memory, controllers, and I/O devices (e.g., disk or wireless network cards).

Instruction Level Power/Energy Model.   An optimizing compiler may define the CPU activity level as the number of instructions executed at a given point in time. This model assumes that (1) a fixed amount of power is associated with each executed instruction, (2) the power dissipation of an instruction is largely independent of its particular operand values or other

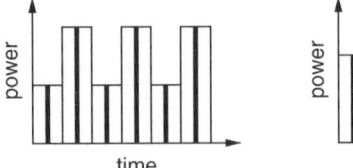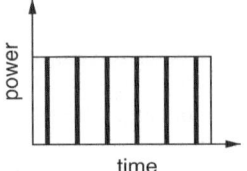

Fig. 13.1   Optimizing for power vs. energy: two possible power profiles of an example program region. The overall energy consumption remains the same, but the peak power and step power (dI/dt problem) are different

```
for (i = 1; i < N; i++) {
    a[i] = a[i-1] * 2.9;          /* S1 */
    b[i] = b[i-1] * 7.3;          /* S2 */
    c[i] = a[i-1] + b[i-1];       /* S3 */
    d{i] = d[i-1] + c[i];         /* S4 */
    }
```

Fig. 13.2   An example loop with four instructions/operations

executing instructions, and (3) not executing an instruction, i.e., keeping a functional unit idle, saves power and energy. The latter assumption holds for architectures that implement gated switching (e.g., clock gating [6]). By re-ordering or rescheduling instructions, for instance in a VLIW or superscalar architecture, the initial power profile of a program region as shown on the left of Figure 13.1 may ideally be transformed into the one shown on the right. While peak power and step power are different for both profiles, the energy usage is the same.

The example loop in Figure 13.2 illustrates a peak power and step power reduction optimization similar to the one discussed in [33]. First, let us assume that the compiler target is a VLIW architecture that allows two floating point additions and two floating point multiplications to be specified in the same VLIW instruction, i.e., to be executed at the same time. All operations are assumed to take a single cycle and the same amount of power. The goal of the instruction scheduler is to pack the four operations (S1, S2, S3, S4) of the loop body into as few VLIW instructions as possible while respecting the data dependences between the operations. There are no loop independent dependences between operations S1, S2, and S3, allowing them to be executed in any order within the loop body, or to be executed in parallel. There is a loop independent dependence between S3 and S4, requiring that S3 needs to be executed before S4. A straightforward scheduler that is not power aware may pack S1, S2, and S3 into one VLIW instruction, followed by S4 in the second instruction. The power profile of this schedule is 3 power units for the first instruction, and 1 power unit for the second, resulting in a peak power of 3, and a step power of 2. In contrast, a power aware scheduler takes the power characteristics of each instruction into consideration while grouping operations into VLIW instructions. Existing list scheduling and modulo scheduling techniques can be augmented to include a power model [26, 28, 33]. The result is a schedule that maps S3 and S4 into separate VLIW instructions, and then distribute S1 and S2 across the two instructions. For example, S1 and S3 may be packed into the first VLIW instruction, followed by S2 and S4 in the second instruction. The power dissipation of both instructions is 2 power units, resulting in a power profile with peak power of 2 and a step power of 0. The energy consumption of the two discussed power profiles is the same according to our power and energy metrics. A compiler transformation that in particular addresses the dI/dt problem is discussed in [7]. The technique inserts redundant operations to smoothen the overall power profile, thereby increasing the energy consumption.

In a superscalar architecture, the scheduling of instructions is partially performed at runtime. The hardware selects appropriate operations from an instruction window over the sequential stream of instructions to be executed next. Data dependences between the instructions in the window are considered when scheduling instructions within the window. A compiler may induce a particular grouping of instructions by making sure that two instructions that should not be executed at the same time will be considered in different instruction windows. In the example shown in Figure 13.2, a source-level transformation can move S2 to the end of the instruction sequence. Assuming that our target superscalar architecture only considers three instructions at a time, the initial instruction window will contain S1, S3, and S4, resulting in the issuing of instructions S1 and S3. S4 cannot be issued due to its dependence on S3. Then, the window will be moved and S2 will become available for scheduling, and S4 and S2 will be issued together. The reader should note that the presented description of a superscalar architecture is rather simplistic. However, it shows how a compiler can indirectly achieve the desired optimization goal by appropriately reshaping the structure of a program based on the knowledge of the behavior of the underlying target architecture.

Bit-Level Power/Energy Model. For power models based on bit-level switching activities as its activity notion, rescheduling instructions may also target overall energy usage by grouping instructions based on their particular bit patterns. Minimizing bit-level switching activities in every cycle reduces the power dissipation for the executed instruction, and therefore the average power, which corresponds to the consumed energy. In addition to instruction scheduling, a careful selection of register names in the code generation phase of a compiler can result in code sequences that have bit patterns with less switching activities, for instance due to the reuse of "similar" register names [16, 20]. The same idea of reducing switching activities, this time for fetching similar VLIW instructions, has been discussed in [24].

Given the code example shown in Figure 13.2, the power aware instruction scheduler discussed in the previous section grouped the operation S1 and S3 into the first VLIW instruction, and S2 and S4 into the second. However, the scheduler did not specify which particular functional unit should execute the operations. In the example, we assumed that the VLIW architecture has two floating point addition and two floating point multiplication units. The mapping of an operation to a particular unit can be specified by selecting the appropriate slot in the VLIW instruction. This selection can be formulated as a minimization problem of the Hamming distances between successive VLIW instructions [24].

Finally, due to the particular chemical characteristics of some batteries, highly varying discharge rates, i.e., varying power dissipations, may reduce the lifetime of a battery significantly. By "smoothing" the power dissipation profile of an application through instruction scheduling and reordering, the usable energy of a battery can be significantly increased [19].

2.2   Power/Energy vs. Performance

Typically, the performance of a program refers to its overall execution time. The faster a given program can be executed, the higher is its performance. Early work on optimizing compilers for power and energy management suggested that optimization transformations for performance subsume those for power and energy management [27]. Traditional optimizations such as common subexpression elimination, partial redundancy elimination, strength reduction, or dead code elimination increase the performance of a program by reducing the work to be done during program execution [2, 21]. Clearly, reducing the workload will result in energy savings. Applying the forementioned classical compiler optimizations may also have impact on the power dissipation, i.e., the program activity levels. However, the optimizations may decrease or increase the instantaneous power dissipation. For instance, eliminating "useless" work can result in more compact programs with higher activity levels, or in less dense regions of activity. The observed impact will depend on the characteristics of the program and the compiler's ability to rearrange or reschedule the remaining "useful" work.

Memory hierarchy optimizations such as loop fusion, loop tiling, and register allocation try to keep data closer to the processor since such data can be faster accessed. Keeping a value in an on-chip cache instead of an off-chip memory, or in a register instead of the cache also saves power/energy due to reduced switching activities and switching capacitance. For example, driving an I/O pin to access an off-chip data item takes much more power and energy than accessing the same data item on chip where it only may require to drive the on-chip memory bus. Keeping values closer to the processor allows for a higher CPU utilization, for instance by avoiding pipeline stalls due to memory accesses. In addition, modern architecture rely heavily on parallelism which allows the overlapping or parallel execution of multiple instructions. Automatic parallelization and software pipeline expose this parallelism, resulting in more activities that can be performed at the same time. This in turn increases the CPU activity level, i.e., its power dissipation. The impact of different performance oriented compiler optimizations on system power and energy have been discussed in several papers [5, 15, 29, 32].

There is fundamental difference in the models and metrics used for performance and those for power/energy optimizations. Many performance models have the notion of a critical path, i.e., a sequence of instructions or activities that are performed sequentially and whose execution time corresponds to the execution time of the entire application. The goal of a performance optimization is to reduce the execution time of the critical path. This is typically accomplished by taking advantage of underutilized system resources such as the cache or memory, or open slots in a VLIW instruction. If an optimization introduces activities on the non-critical path using the available system capacity, performance is not affected. Therefore, as long as these non-critical activities lead to an overall decrease of the length of the critical path (at least in most cases), the optimization is beneficial. The example code shown in Figure 13.3 illustrates the difference between a performance and power/energy optimization objective. As in our previous example, we assume that our compiler target is a VLIW architecture with slots for two floating point additions and two floating point multiplications.

A straightforward code generator may map S1, S2, S3, and S4 to separate VLIW instructions, where the instructions containing S2 and S3 are under the loop control, i.e., will be executed N times. In other words, S1, S2, S3, and S4 are on the critical path. By moving S1 down into the VLIW instruction that contains S2, and S4 up into the VLIW instruction that contains S3, this

```
x = y + k;                    /* S1 */
for (i = 1; i < N; i++) {
    a{i] = a[i-1] * b[i];     /* S2 */
    c{i] = c[i] * a[i];       /* S3 */
}
z = x + 10.0;                 /* S4 */
```

Fig. 13.3 An example loop with code motion opportunities

critical path can be shortened. While the critical path has been shortened, the activity level within the loop has be significantly increased, doubling the number of floating point operations executed during each loop iteration. For a performance oriented optimization, redundant operations are not an issue as long as they do not impact the critical path. This is not true for power/energy aware optimizations. Every operation that is performed dissipates power and therefore needs to be considered. In our example, a small to moderate performance improvement (depending on the actual value of N) by moving S1 and S4 into the loop may nearly double the power dissipation and energy consumption of the loop, assuming that floating point additions and multiplications dissipate similar power. As a rule of thumb, every operation counts for a power/energy optimization, while only the length of the critical path is important for a performance optimization. Potentially redundant operations occur also in the context of speculative execution.

Speculation performs activities "ahead of time" based on some assumptions about the future behavior of the program. If these assumptions turn out to be false, the speculatively performed work is lost, and additional work may be necessary to undo the impact of the speculative performed activities. Software prefetching is an example of such a transformation. The compiler may insert prefetch instructions for memory accesses across control branches. Assuming that the target machine allows multiple outstanding loads, this optimization can be very effective. Another example of a compiler-directed speculation is that of speculative parallelization of loops (e.g., [23]). Again, as long as the speculative activity can be hidden on the non-critical execution path, no negative impact on performance will occur. In the context of power/energy optimizations every additional, speculative activity has to be compensated for by the overall power/energy benefit of the optimization in order to make things not worse. In other words, the window of profitability has to be larger for power/energy optimizations than performance optimizations. This does not mean that speculation cannot be applied for power/energy optimizations, but suggests a less aggressive application of such a transformation by restricting it to the cases where the benefit is likely. Figure 13.4 shows one possible outcome

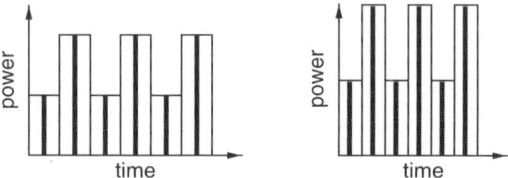

Fig. 13.4   Optimizing for power/energy vs. performance: two possible power profiles of an example program region. The performance increase leads to an increase in power, and a slight decrease in overall energy

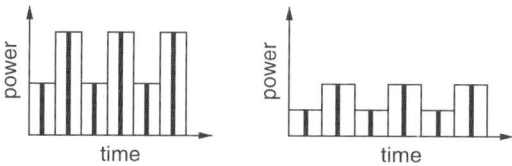

Fig. 13.5 Optimizing for power/energy vs. performance: Two possible power profiles of an example program region. The DFVS strategy increases the execution time by 25%, but substantially reduces power and energy requirements

of speculative execution. Clearly, the program performance has been improved, but at the cost of increased power demands.

An optimization that particularly targets the possible tradeoffs between power dissipation and performance is dynamic voltage and frequency scaling (DFVS). It exploits the fact that a major portion of power of CMOS circuitry scales quadratically with the supply voltage [3]. As a result, lowering the supply voltage can significantly reduce power dissipation. However, lowering the supply voltage requires a slowing down of the clock frequency in order to allow enough time for the CMOS gates to stabilize. Figure 13.5 depicts a possible tradeoff between power/energy and performance. With only a slight degradation of performance, significant power and energy savings may be achieved. Again, the actual amount of savings will depend on the particular program characteristics and characteristics of the underlying DFVS enabled architecture. Examples for relevant machine characteristics are the voltage and frequency switching overhead, and the number of supported outstanding memory references. One approach to compiler-directed frequency and voltage scaling will be discussed in more detail in this chapter. Other work on compiler support for DFVS includes an approach that aggressively fuses loops in order to achieve a more equal balance of memory operations and computation [35]. Such a balance can be beneficial since it allows larger code regions to be slowed down and take advantage of the power/energy savings induced by the corresponding voltage reduction. As a rule of thumb, is it more beneficial to slow down a large code region by a little, than several short code regions by a lot.

2.3   Power/Energy vs. Temperature

Shrinking feature sizes and increasing clock speeds has been one of the main strategies to sustain Moore's law, i.e., lead to an exponential increase in processor performance. The challenges of cooling future chips and systems have become a significant threat to this exponential growth. The cooling problem has a spatial aspect to it since the granularity of the heat source and its loca-

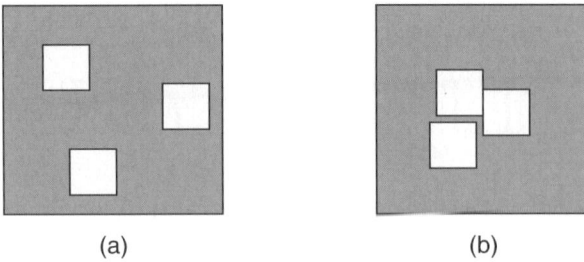

Fig. 13.6   Power density vs. temperature. The relative locations of the "hot spots" are important

tion is important for the appropriate thermal management techniques. Possible granularities include on-chip components, entire chips, boards, racks, groups of cabinets, or entire machine rooms. Depending on this granularity, different metrics may be used to describe thermal issues, for instance energy for coarser grain optimizations, and power density and physical layout for finer grain optimizations. Power density can be viewed as the *activity levels within a physical space*. Figure 13.6 depicts the power densities of three distinct regions on a chip. The main difference between the two subfigures is the location of the heat critical chip regions. While the power density metric seems to be sufficient to describe thermal behavior for heat critical regions that are sufficiently apart (left subfigure), it may be inappropriate for closer heat regions (right subfigure). Skadron et al. showed that thermo-dynamic models are needed to describe the heat build-up in such situations [14, 25].

Possible compiler techniques for thermal management include resource hibernation and spatial scheduling. In order to avoid hot spots, activities are spatially distributed, thereby avoiding excessive heat build-up. In addition, a compiler could help to alert the cooling subsystem of potential future cooling requirements, which may allow the system to concentrate its cooling resources on such developing hot spots. An example thermal optimization for VLIW architectures has been presented in [22]. If not all functional units are fully utilized in every instructions, a load balancing scheme distributes and rotates the activities among multiple functional units, avoiding thermal hot spots at individual units.

2.4   Summary

All the discussed metrics are related, but neither subsumes the other. It will depend on the particular compiler optimization what aspects of the overall system characteristics need to be considered, i.e., need to be modeled, and at which level of granularity. As a rule of thumb, *what* overall activities are performed relates to energy, *when* activities are performed relates to power, and *where* activities are performed relates to temperature. Performance optimizations typically impact the *what* and *when* of activities. However, parallelization as a performance optimization may also impact the *where*, for instance in a VLIW architecture or a multi-core system.

An efficient compilation strategy needs to consider the interactions of different system components in order to achieve overall system power and energy improvements. For example, a 10% performance penalty in return for 50% CPU power/energy savings may sound like a good idea. However, in a system where the CPU accounts for 20% of the overall power/energy budget, and another component, for instance the display, accounts for 50%, the 10% CPU performance penalty will require the display to be on 10% longer. This will translate into an overall 2% CPU power/energy savings and 5% display power/energy penalty, resulting in an overall 3% power/energy increase, which is not a good idea.

3. Future Compiler Research Directions

Compiler research for power and energy management is evolving. Such research requires platforms that expose power, energy, and thermal management features to higher software levels such as the compiler through standardized interfaces (APIs). While efforts have been made in some areas (e.g., ACPI [4]), more work needs to be done.

In addition, the lack of a reliable and effective evaluation infrastructures for power, energy, and thermal optimizations has significantly hampered compiler research. The compiler community relies mostly on physical measurements on existing target systems for a set of representative benchmarks to evaluate the benefits of a given optimization or set of optimizations. Simulation results are accepted as an indication of a potential benefit of an optimization, but are typically not considered sufficient proof that the optimization is worthwhile in practice. What is needed is an evaluation infrastructure for power, energy, and thermal optimizations that consists of a combination of physical measurements and performance modeling. Physical measurements need to include current and voltage measurements [12, 13], as well as temperature measurements. Performance models are needed for the CPU, memory subsystems, controllers, communication modules, and I/O devices such as the disk and screen. This technology is crucial to be able to understand and assess the benefits of a proposed optimization for the entire target system, subsets of system components, or single system components.

Power, energy, and thermal management may also benefit from new programming language designs that include resource requirement specifications and possible quality-of-result tradeoffs. These specifications will allow the underlying run-time system and operating system to make better tradeoff decisions during resource allocation and scheduling.

4. Techniques Covered in Subsequent Chapters

In the following, three compiler supported optimizations are introduced in more detail. These optimizations are just examples, and are presented to illustrate the potential benefits of compile time power/energy management. This list is by no means complete. In particular, we do not discuss thermal management in detail here.

After this introduction, there are three in-depth discussions of the presented power/energy optimization strategies. Compiler-directed DVFS is covered by Chung-Hsing Hsu (Los Alamos National Laboratory) and Ulrich Kremer (Rutgers University). An example of a transformation to enable resource hibernation is presented by Feihui Li, Guangyu Chen, Mahmut Kandemir (Pennsylvania Sate University), and Mustafa Karakoy (Imperial College, UK). Finally, compile-time and run-time support for computation offloading is discussed in a contribution by Chen Wang (Purdue University) and Zhiyuan Li (Purdue University).

4.1 Dynamic Voltage and Frequency Scaling

Dynamic voltage and frequency scaling (DVFS) is recognized as one of the most effective power reduction techniques. As mentioned before, it exploits the fact that a major portion of power of CMOS circuitry scales quadratically with the supply voltage [3]. Therefore, lowering the supply voltage can significantly reduce power dissipation caused by the switching activities in the CMOS gates, which is referred to as *dynamic power*. *Static power* is a technology dependent power dissipation due to leakage currents within the CMOS transistors. DVFS can only address dynamic power. The optimization strategy of hibernation can target static power and will be discussed in Section 4.2. The basic idea of hibernation is to "shut-down" an unused component rather than to "slow-down" an active component as is done by DVFS.

For non-interactive applications such as movie playing, decompression, and encryption, fast processors reduce device idle times, which in turn reduce the opportunities for power savings through hibernation strategies. In contrast, DFVS techniques are still beneficial in such cases, i.e., DFVS reduces power even when these devices are active. However, DFVS comes at the cost of performance degradation. An effective DFVS algorithm is one that intelligently determines *when* to adjust the current frequency–voltage setting (*scaling points*) and to *which* frequency–voltage setting (*scaling factors*), so that considerable savings in energy can be achieved while the required performance is still delivered.

One possible compiler-directed algorithm identifies program regions where the CPU can be slowed down with negligible performance loss [11]. It is implemented as a source-to-source level transformation using the SUIF2 [1] compiler infrastructure. The chapter by Hsu and Kremer discuss this work in more detail.

4.2   Resource Hibernation

A common approach to increase energy efficiency puts idle resources or entire devices in low-power (hibernation) states until they have to be accessed again. The transition to a lower power state usually occurs after a period of inactivity (an *inactivity threshold*), and the transition back to active state usually occurs on demand. Unfortunately, the transitions to and from the low-power state can consume significant time and energy. Nevertheless, this strategy works well when there is enough idle time to justify incurring such costs. Resource hibernation is an effective compilation technique for I/O devices (e.g., disk, network card [10]) as well as cache lines [34] and memory banks [15]. The latter optimizations target the leakage (static power) problem.

Source-level transformations can be used to reshape the program behavior such that inactivity thresholds of a device or component are extended, allow hibernation to be more effective. By allowing the compiler to give hints to the operating system about expected idle times of these components and devices, the operating system is able to issue deactivation directives earlier and activation directives just in time before the device or component is used again. In addition, the operating system can use these hints to implement the most efficient policy for the set of active processes. The results reported in [8, 9] show that on a set of streamed and non-streamed application, the reshaped programs can achieve disk energy reductions ranging from 55% to 89% (70% on average) under a sophisticated energy management policy with only a small performance degradation.

The chapter by Feihui Li et al. discusses compiler-directed hibernation of communication links for array-based applications. The discussed strategy determines the last use of a communication link at each loop nest, and generates explicit instructions that turn off these links. Links are reactivated on demand, i.e., the next time they are accessed.

4.3   Remote Task Mapping

Mobile devices come in many flavors, including laptop computers, Web-phones, pocket computers, Personal Digital Assistance (PDA), and intelligent sensors. Many such devices already have wireless communication capabilities, and we expect most future systems to have such capabilities. There are two main differences between mobile and desk-top computing systems, namely the source of the power supply and the amount of available resources. Mobile systems operate entirely on battery power most or all the time. The resources available on a mobile system can be expected to be at least one order of magnitude less than those of a "wall-powered" desk-top system with similar technology. This fact is mostly due to space, weight, and power limitations placed on mobile platforms. Such resources include the amount and speed of the processor, memory, secondary storage, and I/O. With the development of new and even more power-hungry technology, we expect this gap to widen even more. Remote task mapping is a technique that tries to off-load computation to a remote server, thereby saving power and energy on the mobile devices [17, 18, 30].

A possible compilation strategy that generates two versions of the initial application, one to be executed on the mobile device (client), and the other on a machine connected to the mobile device via a wireless network (server) [17]. The client and server codes have to be able to deal with disconnection events. The proposed compilation strategy uses checkpointing techniques to allow the client to monitor program progress on the server, and to request checkpoint data in order to reduce the performance penalty in case of a possible server and/or network failure. The reported results have been obtained by actual power measurements of an image processing application (face detection and face recognition) on three client systems, (1) the StrongARM based low-power SKIFF system developed at Compaq's Cambridge Research Laboratory, (2) Compaq's commercially available StrongARM based iPAQ H3600, and (3) a PentiumII based laptop. Initial experiments show that energy consumption can be reduced significantly, in some cases up to one order of magnitude, depending on the selected characteristics of the mobile device, remote host, and wireless network.

The following chapter by Zhiyuan Li et al. introduces a similar compilation scheme for remote mapping based on a client/server model. Their framework represents a program as a task graph and maps tasks dynamically to the server if profitable. Their technique is able to adapt to changes in server workloads and to changes of the quality of the communication link.

Acknowledgment

This work has been partially supported by NSF CAREER award #9985050. Any opinions and conclusions expressed in this chapter are those of the author, and do not necessarily reflect the view of the National Science Foundation.

References

[1] National Compiler Infrastructure (NCI) project. Overview available online at http://www-suif.stanford.edu/suif/nci/index.html., Co-funded by NSF/DARPA, 1998.

[2] A.V. Aho, R. Sethi, and J. Ullman. *Compilers: Principles, Techniques, and Tools*. Reading, MA, second edition, 1986.

[3] T. Burd and R. Brodersen. Energy efficient CMOS microprocessor design. In *The 28th Hawaii International Conference on System Sciences (HICSS-95)*, January 1995.

[4] Intel Corp., Microsoft Corp., and Toshiba Corp. ACPI implementers' guide. Draft, February 1998.

[5] V. Delaluz, M. Kandemir, N. Vijaykrishnan, and M.J. Irwin. Energy-oriented compiler optimizations for partitioned memory architectures. In *International Conference on Compilers, Architecture, and Synthesis of Embedded Systems (CASES)*, pages 138–147, San Jose, CA, November 2000.

[6] M. Donno, A. Ivaldi, L. Benini, and E. Macii. Clock-tree power optimization based on RTL clock-gating. In *Design Automation Conference (DAC)*, Anaheim, CA, June 2003.

[7] K. Hazelwood and D. Brooks. Eliminating voltage emergencies via microarchitectural voltage control feedback and dynamic optimization. In *International Symposium on Low Power Electronics and Design (ISLPED'04)*, Newport Beach, CA, August 2004.

[8] T. Heath, E. Pinheiro, J. Hom, U. Kremer, and R. Bianchini. Application transformations for energy and performance-aware device management. In *International Conference on Parallel Architectures and Compilation Techniques (PACT'02)*, Charlottesville, VA, September 2002.

[9] T. Heath, E. Pinheiro, J. Hom, U. Kremer, and R. Bianchini. Code transformations for energy-efficient device management. *IEEE Transactions on Computers*, 53, August 2004.

[10] J. Hom and U. Kremer. Energy management of virtual memory on diskless devices. In L. Benini, M. Kandemir, and J. Ramanujam, editors, *Compilers and Operating Systems for Low Power*, pages 95–113. Kluwer Academic Publishers, Norwell, MA, 2003.

[11] C-H. Hsu and U. Kremer. The design, implementation, and evaluation of a compiler algorithm for cpu energy reduction. In *ACM SIGPLAN Conference on Programming Languages, Design, and Implementation (PLDI'03)*, San Diego, CA, June 2003.

[12] C. Hu, A. Jimenez, and U. Kremer. Toward an evaluation infrastructure for power and energy optimizations. In *Workshop on High-Performance, Power-Aware Computing Systems (HP-PAC'05)*, Denver, CO, August 2005.

[13] C. Hu, J. McCabe, A. Jimenez, and U. Kremer. Infrequent basic block based program phase classification and power behavior classification. In *Workshop on Interaction between Compilers and Computer Architecture (INTERACT-10)*, Austin, TX, February 2006.

[14] W. Huang, S. Ghosh, K. Sankaranarayanan, K. Skadron, and M.R. Stan. HotSpot: Thermal modeling for cmos vlsi systems. *IEEE Transactions on Component Packaging and Manufacturing Technology*, 2005.

[15] M. Kandemir, N. Vijaykrishnan, M.J. Irwin, and W. Ye. Influence of compiler optimizations on system power. In *Design Automation Conference (DAC)*, Los Angeles, CA, June 2000.

[16] M. Kandemir, N. Vijaykrishnan, M.J. Irwin, W. Ye, and I. Demirkiran. Register relabeling: A post compilation technique for energy reduction. In *Workshop on Compilers and Operating Systems for Low Power (COLP'00)*, Philadelphia, PA, October 2000.

[17] U. Kremer, J. Hicks, and J. Rehg. A compilation framework for power and energy management on mobile computers. In *International Workshop on Languages and Compilers for Parallel Computing (LCPC'01)*, Cumberland, KT, August 2001.

[18] Z. Li, C. Wang, and R. Xu. Computation offloading to save energy on handheld devices: A partition scheme. In *International Conference on Compilers, Architectures and Synthesis for Embedded Systems (CASES 2001)*, Atlanta, GA, November 2001.

[19] T. Martin and D. Siewiorek. The impact of battery capacity and memory bandwidth on CPU speed-setting: A case study. In *International Symposium on Low Power Electronics and Design (ISLPED)*, pages 200–205, San Diego, CA, August 1999.

[20] H. Mehta, R.M. Owens, M.J. Irwin, R. Chen, and D. Gosh. Techniques for low energy software. In *International Symposium on Low Power Electronics and Design (ISLPED'97)*, pages 72–75, Monterey, CA, August 1997.

[21] S.S. Muchnick. *Advanced Compiler Design Implementation*. Morgan Kaufmann Publishers, San Franscisco, CA, 1997.

[22] M. Mutyam, F. Li, V. Narayanan, M. Kandemir, and M.J. Irwin. Compiler-directed thermal management for VLIW functional units. In *ACM SIGPLAN/SIGBED Conference on Languages, Compilers, and Tools for Embedded Systems (LCTES'06)*, Ottava, Canada, June 2006.

[23] L. Rauchwerger and D. Padua. The LRPD test: Speculative run-time parallelization for loops with privatization and reduction parallelization. In *ACM SIGPLAN 1995 Conference on Programming Language Design and Implementation (PLDI'95)*, La Jolla, CA, June 1995.

[24] D. Shin and J. Kim. An operation rearrangement technique for low-power VLIW instruction fetch. In *Workshop on Complexity-Effective Design*, Vancouver, B.C., June 2000.

[25] K. Skadron, K. Sankaranarayanan, S. Velusamy, D. Tarjan, M.R. Stan, and W. Huang. Temperature-aware microarchitecture: Modeling and implementation. *ACM Transactions on Architecture and Code Optimization*, 1(1):94–125, 2004.

[26] C.-L. Su, C.-Y. Tsui, and A.M. Despain. Low power architecture and compilation techniques for high-performance processors. In *IEEE COMPCON*, pages 489–498, San Francisco, CA, February 1994.

[27] V. Tiwari, S. Malik, A. Wolfe, and M. Lee. Instruction level power analysis and optimization of software. *Journal of VLSI Signal Processing*, 13(2/3):1–18, 1996.

[28] M.C. Toburen, T. Conte, and M. Reilly. Instruction scheduling for low power dissipation in high performance microprocessors. In *Power Driven Microarchitecture Workshop*, Barcelona, Spain, June 1998.

[29] M. Valluri and L. John. Is compiling for performance == compiling for power? In *The 5th Annual Workshop on Interaction between Compilers and Computer Architectures (INTERACT-5)*, January 2001.

[30] C. Wang and Z. Li. Parametric analysis for adaptive computation offloading. In *ACM SIGPLAN 2004 Conference on Programming Language Design and Implementation (PLDI'04)*, Washington, DC, June 2004.

[31] Q. Wu, V.J. Reddi, Y. Wu, J. Lee, D. Connors, D. Brooks, M. Martonosi, and D.W. Clark. A dynamic compilation framework for controlling microprocessor energy and performance. In *International Symposium on Microarchitecture (MICRO-38)*, Barcelona, Spain, November 2005.

[32] H. Yang, G.R. Gao, A. Marquez, G. Cai, and Z. Hu. Power and energy impact of loop transformations. In *Workshop on Compilers and Operating Systems for Low Power (COLP'01)*, Barcelona, Spain, September 2001.

[33] H.-S. Yun and J. Kim. Power-aware modulo scheduling for high-performance VLIW. In *International Symposium on Low Power Electronics and Design (ISLPED'01)*, Huntington Beach, CA, August 2001.

[34] W. Zhang, J.S. Hu, V. Degalahal, M. Kandemir, N. Vijaykrishnan, and M.J. Irwin. Compiler-directed instruction cache leakage optimization. In *International Symposium on Microarchitecture (MICRO-35)*, Istanbul, Turkey, November 2002.

[35] YK. Zhu, G. Magklis, M.L. Scott, C. Ding, and D. Albonesi. The energy impact of aggressive loop fusion. In *International Conference on Parallel Architecture and Compilation Techniques (PACT'04)*, Antibes Juan-les-Pins, France, September 2004.

Chapter 14

Compiler-Directed Dynamic CPU Frequency and Voltage Scaling

Chung-Hsing Hsu[1] and Ulrich Kremer[2]

[1] *Los Alamos National Laboratory*
Los Alamos, New Mexico 87545, USA

[2] *Department of Computer Science*
Rutgers University
New Brunswick, New Jersey 08903, USA

Abstract   This paper presents the design, implementation, and evaluation of a compiler algorithm that effectively optimizes programs for energy usage using dynamic voltage and frequency scaling (DVFS). The algorithm identifies program regions where the CPU can be slowed down with negligible performance loss, and has been implemented as a source-to-source level compiler transformation using the SUIF2 compiler infrastructure. Physical measurements on a notebook computer show that total *system* energy savings of up to 28% can be achieved with performance degradation of less than 5% for the SPEC CPU95 benchmarks. On average, the system energy and energy-delay product are reduced by 11% and 9%, respectively, with a performance slowdown of 2%.

Keywords: dynamic voltage scaling, power reduction, energy efficiency, compiler

1.   DVFS

Dynamic voltage and frequency scaling (DVFS) is one of the most effective energy reduction mechanisms for CMOS circuitry. The mechanism allows system software to increase or decrease the supply voltage and the operating frequency of a CMOS circuit at run time. Since a major portion of energy

J. Henkel and S. Parameswaran (eds.), Designing Embedded Processors – A Low Power Perspective,
305–323.

consumed by a CMOS circuit scales quadratically with the supply voltage [2] and decreasing the frequency allows us to decrease the supply voltage, we effectively reduce energy consumption.

Current generations of processor are made of CMOS circuitry. As a result, DVFS has been enabled on many mobile and desktop processors. For example, AMD has enabled the DVFS support on their Athlon 64 processor, referring to as PowerNow. Similarly, Intel has enabled the DVFS support on their Pentium M processor, referring to as SpeedStep. Each such DVFS-enabled processor provides a discrete set of voltage and frequency that system software can select from. From system software perspective, the selection is performed after writing the desired voltage and frequency values into specific registers.

DVFS is different from the traditional resource-hibernation mechanism in that DVFS can reduce energy when the system is actively running tasks. In contrast, resource hibernation reduces energy only when the system is idle. More often than not, energy management based on resource hibernation, such as running a task at the fastest processor speed followed by putting the processor into hibernation, ends up consuming more energy than energy management based on DVFS that runs the task at a slower speed. Therefore, the use of DVFS seems to be a more appropriate choice to achieve low power than resource hibernation, especially for non-interactive applications.

2.   DVFS Scheduling is Challenging

However, the use of DVFS needs to be judicious because DVFS reduces energy at the cost of performance degradation. Two types of source cause performance degradation. First, each change of voltage and frequency incurs time and energy overheads. Currently, the time overhead is on the order of hundreds of microseconds. Given that memory latency is on the order of hundreds of nanoseconds, applying DVFS at every cache miss will severely degrade system performance. Second, DVFS reudces energy by lowering frequency and thus increases the total execution time. In fact, increasing the execution time results in some devices staying active longer and consuming more energy that may mask out the savings from the DVFS-enabled devices.

In this paper, we study how we can judiciously use DVFS on a DVFS-enabled processor. We design a scheduling algorithm that can automatically and intelligently determine *when* to apply DVFS (*scaling points*) and to *which* frequency–voltage setting (*scaling factors*) so as to achieve considerable energy savings while delivering the required performance.

A judicious use of DVFS on a DVFS-enabled processor is when processor performance is *not* critial for system performance. In these occasions, the time overhead and increased execution time induced by DVFS will not perturb system performance. For exmaple, if a laptop user is reading a document for an extended period of time while running on battery power, the laptop can automatically scale down the frequency and voltage of the processor to reduce energy consumption without disturbing the user's reading pace. This is because human's reading pace is generally much slower than CPU's processing pace, and thus the processor is not a performance bottleneck in this kind of scenario.

To identify when to apply DVFS in the above scenario, CPU utilization is often used to assist the detection. CPU utilization refers to the fraction of time that the CPU spends non-idle. One can periodically poll current CPU utilization, and scale down CPU frequency and voltage when CPU utilization is low; if CPU utilization becomes high, one can scale up CPU frequency and voltage. This type of DVFS scheduling strategy is called *demand-based switching* (DBS) by Intel, which advocates the use of DBS-DVFS in data centers as CPU utilization on nights and weekends in a data center is usually low.

Unfortunately, the DBS strategy has two potential drawbacks. First, the strategy has a minimal effect on the execution of non-interactive applications. A non-interactive application tends to have a high CPU utilization throughout its entire execution. Second, the strategy sometimes leads to severe performance degradation for interactive applications. Grunwald et al. [4] recently performed a measurement study on several DBS implementations and found out noticeable performance degradation in these implementations.

3. Our DVFS Algorithm in a Nutshell

In this paper, we present a compiler-directed DVFS scheduling algorithm that addresses the two deficiencies in the DBS strategy. First, we observe that the execution of a non-interactive application oftentimes enters a phase that has a high cache miss ratio. In these phases, performance is limited by memory latency and bandwidth rather than CPU frequency. Since memory performance is not affected by changes in CPU frequency, we can reduce CPU frequency and voltage in these phases to save energy without affecting overall performance.

Second, our scheduling algorithm constructs a model that accurately captures the performance impact due to CPU frequency changes. The primary reason that DBS cannot regulate DVFS-induced performance degradation well is due to the fact that CPU utilization by itself does not provide enough information about the performance impact induced by CPU frequency changes [5]. Therefore, any DVFS scheduling algorithm based solely on CPU utilization can only provide loose control over performance degradation. Having an accurate performance-prediction model will solve the problem.

The algorithm we present in this paper identifies scaling points and determines scaling factors off-line, with the help of profile data. The scaling points are the entry and exit points of program regions that generate execution phases

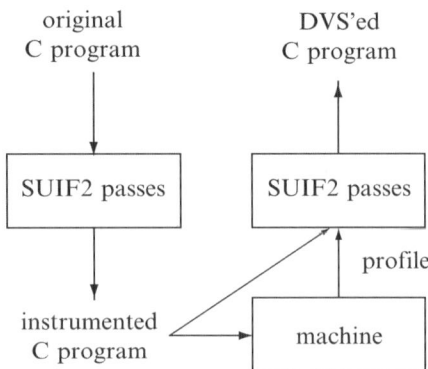

Fig. 14.1   The flow diagram of the compiler implementation

with a high cache miss ratio. Here we differentiate program regions from execution phases. A region refers to a program structure in the source code whereas a phase refers to a period of time during which a region is executed. Examples of a region include a loop nest, a call site, a called procedure, a sequence of statements, or even the entire program. Regions can be nested.

Each region in our algorithm has an associated performance model that predicts the total execution time of the region at any given CPU frequency. The region-based performance models are constructed via the use of profile data. Once the performance model for each region is constrcuted, our scheduling algorithm enumerates all possible regions and tries to determine the scaling factor for each region by answering the following question: if I execute this region at this particular CPU frequency, can I acquire considerable energy savings while still ensuring that the required performance is met. In other words, our scheduling algorithm solves a minimization problem.

Figure 14.1 shows the flow of our DVFS scheduling algorithm. The algorithm starts by instrumenting the input program at selected program regions. The instrumented code is then executed to collect performance data of these program regions. Once the profiling is done, the algorithm begins to enumerate all profiled regions and their potential combinations at each available CPU frequency and voltage. After identifying program regions and their associated scaling factors, our scheduling algorithm inserts appropriate DVFS system calls at the boundaries of the regions.

In sum, our scheduling algorithm is a profile-driven source-to-source compiler transformation.

4.   An Illustrating Example

To better understand the algorithm, let us consider how it works on an example program shown on the left in Figure 14.2. The example code is presented in

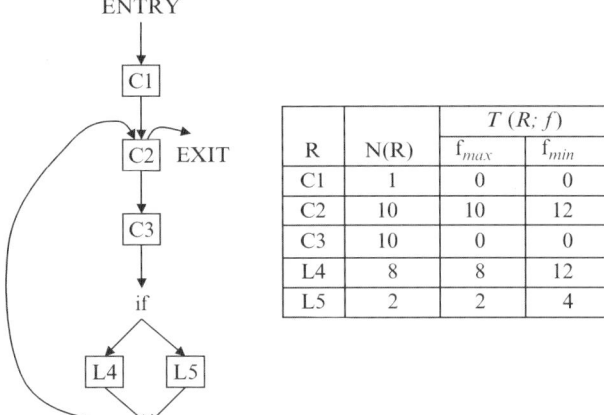

R	N(R)	$T(R; f)$	
		f_{max}	f_{min}
C1	1	0	0
C2	10	10	12
C3	10	0	0
L4	8	8	12
L5	2	2	4

Fig. 14.2   The example program is shown on the left in terms of control flow between regions. The table on the right represents the profiled data for the instrumented regions

terms of control flow between what we call *basic* regions, where each L (or C) stands for a loop nest (or a call site). Basic regions are the regions that our algorithm needs to instrument and profile. So, the first step of our algorithm is to instrument and profile basic regions of an input program.

There are two types of information that need to be collected for a basic region R: the number of times the region is executed (denoted as $N(R)$ in Figure 14.2) and the total execution time of the region running at each available CPU frequency (denoted as $T(R, f)$ for CPU frequency f). So for basic region C2, it is executed 10 times and the total execution times at two different frequencies f_{max} and f_{min} are 10 and 12, respectively.

After the algorithm completes the profiling of basic regions, it starts to consider other non-basic regions. For example, if(L4, L5) is a non-basic region that encloses basic regions L4 and L5 using an if–then–else operator. To derive the informaton of $N(R)$ and $T(R, f)$ for this non-basic region, our algorithm performs a simple calculus as follows:

$$N(\texttt{if(L4,L5)}) = N(\texttt{L4}) + N(\texttt{L5}) = 8 + 2 = 10$$

$$T(\texttt{if(L4,L5)}, f_{max}) = T(\texttt{L4}, f_{max}) + T(\texttt{L5}, f_{max}) = 8 + 2 = 10$$

$$T(\texttt{if(L4,L5)}, f_{min}) = T(\texttt{L4}, f_{min}) + T(\texttt{L5}, f_{min}) = 12 + 4 = 16$$

For completeness, we list the entire calculus in Figure 14.3.

Once the algorithm has the information of $N(R)$ and $T(R, f)$ for all regions, it begins to enumerate each every region, compute its DVFS-induced energy usage and execution time, and compare the two values against the currently best solution. For example, our algorithm will compute the energy usage E

if statement:

R: **if** () **then** R_1 **else** R_2

$$T(R, f) = T(R_1, f) + T(R_2, f)$$
$$N(R) = N(R_1) + N(R_2)$$

explicit loop structure:

R: **loop** () R_1

$$T(R, f) = T(R_1, f)$$
$N(R)$ is profiled

call site:

R: **call** F()

$$T(R, f) = T(F, f) \cdot N(R)/N(F)$$
$N(R)$ is profiled

sequence of regions:

R: **sequence**($R_1, \ldots, R_n$)

$$T(R, f) = \Sigma\{T(R_i, f) : 1 \le i \le n\}$$
$$N(R) = N(R_1) = \ldots = N(R_n)$$

procedure:

F: **procedure** F() R

$$T(F, f) = T(R, f)$$
$$N(F) = \Sigma\{N(R_i) : R_i \text{ is a call site to } F()\}$$

Fig. 14.3 The rules of deriving the table entries $T(R, f)$ in our DVFS algorithm

and the execution time T of running region L4 at f_{min} and the rest of the program at f_{max} as follows:

$$E_{f_{max}} = P_{f_{max}} \cdot T_{f_{max}}$$
$$T_{f_{max}} = 0 + 10 + 0 + 8 + 2 = 20$$
$$E = E_{f_{max}} - (P_{f_{max}} - P_{f_{min}}) \cdot T(\text{L4}, f_{min}) + E_{dvfs} \cdot 2 \cdot N(\text{L4})$$
$$T = T_{f_{max}} + [T(\text{L4}, f_{min}) - T(\text{L4}, f_{max})] + T_{dvfs} \cdot 2 \cdot N(\text{L4})$$

where $E_{f_{max}}$ and $T_{f_{max}}$ represent the energy usage and the execution time of running the entire program at f_{max}, respectively; E_{dvfs} and T_{dvfs} represent the energy and time overheads of performing DVFS, respectively; finally, $P_{f_{max}}$ and $P_{f_{min}}$ represent power consumption at f_{max} and f_{min}, respectively.

In our algorithm, a solution is considered better if it is valid and consumes less energy (a smaller E value). A solution is considered valid if $T \le (1 + r) \cdot T_{f_{max}}$ for a user-tunable parameter r. The default r value in our implementation is 5%. That is, a user is willing to sacrifice at most 5% performance for lower energy consumption. Apparently, the larger the r value, the more the energy savings.

After all the regions are enumerated, the best solution is identified. The final step of the algorithm is to insert corresponding DVFS system calls at the boundaries of the selected region(s).

5.   Design and Implementation Issues

In the previous sections we intentionally omitted the details of the design and implementation of our DVFS scheduling algorithm. There are in fact several decisions that we need to make in order to ensure our algorithm implementable and cost-effective. Since each decision has some effect on the effectiveness of the algorithm, we record here our decision process for the design and implementation of the algorithm.

5.1   What is a Region

A program region in our implementation is chosen to be a *single-entry-single-exit* (SESE) program structure. While this definition may seem too restrictive, it enables us to have a simple computation of the DVFS overheads. Specifically, the algorithm is able to count the number of DVFS system calls to a region R as $2 \cdot N(R)$ if the DVFS calls are only placed at the entry and the exit of the region. Experiments have shown that this definition works well in practice.

Our implementation categorizes a region as either basic or non-basic. A basic region is the minimal program structure that our implementation needs to identify. Our implementation identifies two kinds of basic regions: call sites and explicit loop structures. Explicit loop structures include `for` and `while` loops. In contrast, loops based on `goto`'s are considered as implicit loop structures and are not identified in our current implementation because to identify this type of loop structure, an additional expensive analysis on the control flow graph is required.

Our current implementation only accepts several types of non-basic region. For example, `if(L4,L5)` in Figure 14.2 is considered as a region, so are `seq(C2,C3)` and `seq(C2,C3,if(L4,L5))`. On the other hand, `seq(C1,C2)` is not considered as a region since it has two entry points, one at the entry of `C1` and the other at the entry of `C2`. Similarly, `seq(C3,L4)` is not a region because of the two exit points.

Our current implementation does not consider `seq(if(L4,L5),L2)` as a region either. While it satisfies the SESE property, it does not satisfy our forward sequencing restriction, i.e., for program construct

$$\textbf{loop}() \ \textbf{sequence}(R_1, \ldots, R_n)$$

the regions composed of $R_i \rightarrow R_{i+1} \rightarrow \cdots \rightarrow R_j$ for $j \geq i$ are considered, but not the "wrap-around" regions, $R_i \rightarrow \cdots \rightarrow R_n \rightarrow R_1 \rightarrow \cdots \rightarrow R_j$ for $j < i$. The reason for us to impose this restriction is to reduce the number of regions that need to be examined in the algorithm.

5.2 How Many Regions to Slow Down

Our implementation chooses only *one* region to slow down. This restriction can certainly be relaxed to allow multiple regions to be slowed down (at different factors). However, in doing so the algorithm is required to solve a zero–one integer linear programming problem (ZILP) where each zero–one variable represents whether the region R is executed at frequency f or not. Moreover, the estimation of the DVFS overheads becomes more complicated. A transition graph between regions needs to be constructed to estimate the DVFS overheads, and the construction is not trivial.

Furthermore, solving ZILP problems is time-consuming. ZILP problems are in general considered hard problems due to the combinatorial aspect of integer programming. Experiences tell us that when the number of edges in a transition graph exceeds over 50, the solver has a hard time to solve it within a few hours. This happens to 8 out of the 10 SPEC CFP95 benchmarks. There, the medium on the number of edges is over 100. More information on the multi-region DVFS scheduling algorithm and its comparison with the single-region version can be found in [7].

5.3 What Region to Pick

Our implementation imposes an additional restriction on the region that can be slowed down. We enforce the region size to be sufficiently large as determined by the following equation:

$$T(R, f_{max})/T(P, f_{max}) \geq \rho$$

where parameter ρ is a design parameter for the compiler specifying how large the region should be to be considered "sufficiently large". In our default setting, ρ is set to 20%.

We impose this restriction for two reasons. First, the resitrction makes sure that the region takes longer time to execute than a single DVFS system call. Second, our experience has suggested that executing a larger region (in time) at a higher frequency often has less performance impact than executing a smaller region at a lower frequency. For example, without the size restriction, our implementation will select a small region in the SPEC CFP95 benchmark turb3d to slow down, resulting in 5% more performance degradation than the user's perference. With the size restriction, such dissatisfaction does not happen.

5.4   Why Compiler-Directed

The determination of scaling points and scaling factors does not need to be done off-line. They can be determined on-line as well. We chose to determine scaling points and scaling factors off-line through compiler analysis for two reasons. First, in many cases, a compiler is able to not only predict but also *shape* the execution behavior of a program, giving compilers an advantage over runtime-based techniques. In other words, a compiler can reshape the execution behavior of a program in such a way that the opportunities for DVFS are created or magnified.

Second, determine scaling points and scaling factors manually is challenging. The medium on the number of candidate regions for SPEC CFP95 benchmarks is over 10,000. It will be difficult to manually identify which region to slow down for the maximum benefit. In contrast, our implementation can identify the best scheduling policy in less than 1 min. Within this 1 min, our implementation has performed, among other passes, an interprocedural analysis pass that extracts the strongly connected components in the call graph, visits them in a bottom-up fashion, and computes the values of $T(R, f)$ and $N(R)$.

5.5   Is Profile-Driven Necessary

The proposed DVFS scheduling algorithm is profile-driven. Profile-driven compiler optimization has its advantages and disadvantages. On the negative side, a program optimized with respect to one data input or machine configuration may not work well on another data input or machine configuration. Profile-driven optimization also increases the compilation time caused by profiling. On the positive side, profiling captures more system effects which a compiler model may have difficulty to model and is more generally applicable. Without a good (i.e., simple and accurate) compiler model, the quality of the optimized code will be compromised. Yet, the complex interaction between all components in a computer system makes it difficult to derive a good compiler model.

We chose profile-driven approach primarily due to the difficulty in deriving a good compiler model to estimate $T(R, f)$. Our early work [9] proposed a semi-compiler model that allows us to profile $T(R, f_{max})$ and estimate the rest of $T(R, f)$ analytically. This compile model involves the computation of the memory stall time and requires the help from performance counters. Unfortunately, for the target system we experimented on (described later), we had a hard time relating the counted events to the actual performance. This is in particular due to the lack of documentation and desired event types. Hence, our current implementation measures the wall-clock times directly for $T(R, f)$.

6.   Evaluation Strategy

To evaluate the effectiveness of our profiled-based, compiler-directed DVFS scheduling algorithm, we implemented the algorithm using the SUIF2 compiler infrastructure [13] and performed physical measurements on a notebook computer with a DVFS-enabled high-performance processor. The following is the detailed description about the hardware and software platforms we used to perform evaluation, as well as the benchmark choice and the measurement setup.

6.1   Hardware Platform

The hardware platform is a Compaq Presario 715US notebook computer. We chose a notebook computer as our hardware platform because at the time of evaluation, only notebook computers are equipped with DVFS-enabled processors. We chose this particular notebook computer due to the fact that the computer is equipped with a high-performance processor, AMD's Athlon 4. AMD's Athlon 4 is a high-performance processor since it provides on-chip floating-point units and memory level parallelsim (i.e., allow multiple outstanding cache misses at the same time), two salient features found in many high-performance processors. In contrast, Intel's Xscale processors do not have floating-point units. Transmeta's Crusoe processors have floating-point units but do not provide memory level parallelism. Recently, Intel provides another line of processor, Pentium M, that has the two features.

AMD's Athlon 4 processor is a 3-way superscalar out-of-order decode and execution decoupled computing engine with dynamic branch prediction and speculative execution. It contains a 64-KB instruction cache, a 64-KB data cache, a full-speed on-die 256-KB level-two exclusive cache, and a hardware data prefetching unit. In terms of DVFS support, the processor exports seven different frequency–voltage combinations, all under the control of software. Table 14.1 lists the main configuration of the Presario 715US notebook computer.

6.2   Software Platform

The Linux 2.4.18 kernel was installed on the Presario 715US notebook computer. All the benchmarks were compiled by the GNU compilers using optimization level -O2. DVFS support is done through user-level system calls. The input of a DVFS call is the desired frequency. The call will find the corresponding voltage from Table 14.1, and write both frequency and voltage values into machine-specific registers. The processor adjusts the CPU clock frequency and supply voltage accordingly. The time required for each such transition is less than 100 µs [1].

Table 14.1   The configuration of the Compaq Presario 715US notebook computer

Spec	Compaq Presario 715US
CPU	AMD mobile Athlon 4
f	600–1200 MHz
V	1.15–1.45 V
Front side bus	DDR 100 MHz
memory	256 MB PC-133
graphics	VIA 16 MB
LCD display	14.1-in 1024x768
disk	20 GB

f (MHz)	V (V)
600	1.15
700	1.20
800	1.25
900	1.30
1000	1.35
1100	1.40
1200	1.45

To profile the values of $T(R, f)$ and $N(R)$ as required by our DVFS algorithm, we also implemented another user-level system call. The input of such a call is the region number. For $T(R, f)$, a high-resolution timer is needed to measure the elapsed time between two such system calls. We did this by reading out the current cycle counter value on a per-process basis. For $N(R)$, the system call implementation maintains a table indexed by the region number and increments the appropriate table entry. The cost of each system call is approximately 50 ns.

6.3   Benchmark Choices

The SPEC CFP95 benchmark suite was used for experiments because of its variety of CPU boundedness. We found that a less CPU-bound application has potentially more energy reduction than a more CPU-bound application for the same performance requirement in the form of relative slowdown. In order to investigate the impact of CPU boundedness to the effectiveness of our DVFS algorithm, we chose SPEC CFP95 benchmarks as our benchmark choices.

To justify our decision in the choice of benchmarks, we computed the CPU boundedness of the entire SPEC CPU95 benchmark suite using a metric proposed in [8]. The metric β_{cpu} is defined as a ratio between zero and one, with one being extremely CPU-bound. Table 14.2 shows the CPU boundedness for each benchmark in the CPU95 suite. We can see from the table that the CFP95 benchmark suite has a *wider* range of CPU boundedness than the CINT95 benchmark suite. Recent studies [11, 12] have shown that multimedia benchmarks are typically more CPU-bound than SPEC CINT95 benchmarks. This is partially due to the fact that many multimedia benchmarks have excellent cache locality, despite of their generally larger memory footprints.

Table 14.2   The potential DVFS energy savings and performance slopes of the `SPEC95` benchmarks

SPEC CFP95		SPECint95	
Benchmark	β_{cpu}	**Benchmark**	β_{cpu}
swim	0.04	compress	0.47
tomcatv	0.06	vortex	0.70
hydro2d	0.13	gcc	0.83
su2cor	0.17	ijpeg	0.95
applu	0.30	li	1.00
apsi	0.37	perl	0.98
mgrid	0.45	go	1.00
wave5	0.57	m88ksim	1.00
turb3d	0.75		
fpppp	1.00		

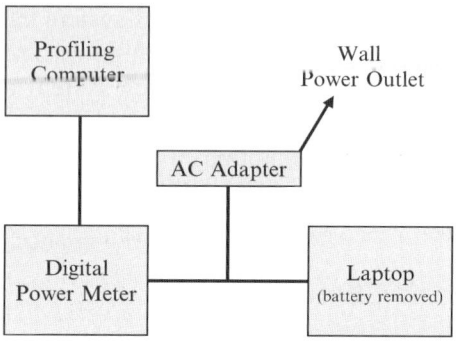

Fig. 14.4   The experimental setup

Measurement Setup.   We performed several experiments with our DVFS algorithm and measured the actual energy consumption of the system through a digital power meter. The power meter, a Yokogawa WT110 [10], sent power measurement data every 250 ms to another computer, which stores them in a log for later use. The power meter samples current and voltage at a rate of 26 μs. During the measurements, the battery was removed to eliminate any noise from battery recharging process. The power dissipation of the AC adapter was excluded. Figure 14.4 shows the measurement setup.

In terms of experiments, all the benchmarks were run to completion. The comparisons were done by executing the benchmark with the reference data set. The training data set was used for the profiling phase of the DVFS algorithm. In addition, we set the time (T_{dvfs}) and the energy (E_{dvfs}) overheads of performing DVFS to be 20 μs and 0 joules. Note that this setting does not

Table 14.3 The input parameters for our algorithm

Parameter	Value
$T(R, f)$	Profiled
$N(R)$	Profiled
P_f	$V_f^2 \cdot f$
T_{trans}	20 μs
P_{trans}	0 W
r	5%
ρ	20%

faithfully reflect the real machine characteristics. However, we did this on purpose to demonstrate that as long as T_{dvfs} is sufficiently large to prevent the DVFS-induced time overhead from becoming a major factor in determining system performance, the accuracy of T_{dvfs} and E_{dvfs} is not critical. Table 14.3 lists the parameter settings of our DVFS algorithm used in the experiments.

7. Experimental Results

In this section we present the experimental results in three aspects: the compilation time, the effectiveness of our DVFS algorithm in the form of DVFS-induced energy savings and performance slowdown, and finally the impact of different training inputs. For SPEC CFP95 benchmarks, our algorithm takes minutes to run and results in the total *system* energy savings of up to 28% with less than a 5%n performance slowdown. On average, the system energy is reduced by 11% with a 2% performance slowdown.

7.1 The Compilation Time

The compilation time of our algorithm is in the order of minutes. Table 14.4 lists the timing spent in each phase. The instrumentation phase takes 7–157 s which includes the times of converting to and from the SUIF2 intermediate representation plus the time of selecting program locations to instrument. The sub-phase of selecting program locations to instrument contributes 6–13% of the total compilation time for the instrumentation phase, with 9% on average. In other words, the conversion between the input C program and the SUIF2 representation is very expensive. On the other hand, in the selection phase, the dominating sub-phase is the process of evaluating all candidate regions for the best region. It accounts for 74–98% of the total compilation time for the phase, with the average 83%.

Benchmarks apsi, turb3d and fpppp took the longest compilation times among all benchmarks. The long compilation times can be attributed to

Table 14.4   The compilation time (in seconds) of our algorithm in various phases

	Total compilation time	Instrumentation phase	Profiling phase	Selection phase
swim	34	7	8	19
tomcatv	173	4	158	11
hydro2d	340	44	173	123
su2cor	403	37	257	109
applu	284	83	13	188
apsi	1264	157	40	1067
mgrid	190	10	152	28
wave5	544	151	48	345
turb3d	1839	39	268	1532
fpppp	1628	82	11	1535

the large number of candidate regions and the cost of finding these candidate regions. The current implementation enumerates all possible sequences of a statement list for finding candidate regions. As a result, the cost can be characterized by the number of statement sequences tried. For benchmarks apsi and fpppp, the selection phase evaluated 77,335 and 51,025 candidate regions, respectively. In contrast, only 2,044–27,535 candidate regions were evaluated for other benchmarks. The selection phase looked 290,299–340,448 statement sequences for the three benchmarks. It only looked 3,955–118,775 combinations for other benchmarks. Clearly, there is room for improvement in our compiler algorithm.

7.2   Effectiveness

The effectiveness of our DVFS algorithm in the form of relative execution time (T_r) and system energy consumption (E_r) to the same program running on a DVFS-disabled system (i.e., running at the peak frequency and voltage) is shown in Table 14.5. From the table we can see that the system energy savings of up to 28% can be achieved. On average, the system energy savings is about 11% with a 2.15% performance slowdown. This 11% system energy savings can be translated into the 22% processor energy savings. Moreover, our DVFS algorithm tightly regulates DVFS-induced performance slowdown for every benchmark in the CFP95 suite to be within 5% as directed by the user. (We set the maximum performance slowdown r to be 5% in the experiments.) This tight regulation of performance slowdown distinguishes our DVFS algorithm from many other DVFS algorithms such as DBS that cannot tightly control DVFS-induced performance penalties.

Table 14.5   The relative execution time and system energy usage for the SPEC CFP95 benchmarks using training input `train.in`.

Benchmark	β_{cpu}	T_r (%)	E_r (%)
swim	R/600	102.93	76.88
tomcatv	R/800	101.18	72.05
hydro2d	R/900	102.21	78.70
su2cor	R/700	100.43	86.37
applu	R/900	104.72	87.52
apsi	R/1100	100.94	97.67
mgrid	R/1100	101.13	98.67
wave5	R/1100	104.32	94.83
turb3d	R/1100	103.65	97.19
fpppp	P/1200	100.0	100.0
Average		102.15	88.99

As we justified earlier that the reason why we chose to use the SPEC CFP95 benchmark suite is to investigate the impact of CPU boundedness to the effectiveness of our DVFS algorithm. Here we present our investigation results. We found that the DVFS-induced energy savings of a benchmark using our compiler algorithm correlates negatively with its CPU boundedness. For example, our algorithm was able to identify that benchmark `fpppp` is extremely CPU-bound (i.e., its $\beta_{cpu} = 1$ is equal to one) and therefore cannot be slowed down without significant performance penalties.

More interestingly. there is a big gap in terms of energy consumption at $\beta_{cpu} = 0.3$. What this gap indicates is that if an application is CPU-bound, our algorithm becomes less effective due to its characteristics in findings a program region to slow down. As a result, other types of DVFS algorithm are required to better exploit DVFS opportunities in CPU-bound a[[;ocatopms/ Pme camdodate os DVFS scheduling algorithm base on phase detection at run time, e.g.,[5]. (Recall that a region refers to a period of time during which a region is executed.) DVFS algorithms based on phase detection can create different schedules for different schedules for differnt phases even thought these are the results of executing the same program region [14].

7.3   Different Training Inputs

A common question for profile-based algorithms is how much the quality of the results is affected by the different training inputs. In this section, we evaluate such an impact using another training data set `std.in` developed by Burger [3]. For SPEC CFP95 benchmarks, since they all have a common structure of an initialization phase followed by repetitions of a computation phase, in most

Table 14.6   The relative execution time and system energy usage for SPEC CFP95 benchmarks using training input `std.in`

Benchmark	std.in	Selection	T_r
swim	ref,900→45	P/700	101.10
tomcatv	ref,750→62	Same	101.18
hydro2d	ref,200→6	Same	102.21
su2cor	ref,40→5	R/800	107.31
applu	ref,300→5	R/900	103.61
apsi	ref,960→6	R/900	105.00
mgrid	test,40→4	R/1000	102.91
wave5	ref,40→10	R/800	102.33
turb3d	ref,111→2	R/1100	106.16
fpppp	Train	Same	100.00

cases data set `std.in` uses the same reference data set but reduces the number of repetitions. Table 14.6 column "`std.in`" gives the definition of data set `std.in` for SPEC CFP95 benchmarks. For example, data set `std.in` for benchmark `swim` uses the same reference data set but reduces the repetitions from 900 down to 45. Note that benchmarks `mgrid` and `fpppp` do not use the reference data set as input.

The column "selection" in Table 14.6 lists the slow-down strategy our compiler finds if data set `std.in` is used as the training input. Except those benchmark rows marked "same", our compiler algorithm found different regions to slow down at different speed if using `std.in`. However, as shown in Figure 14.5, the quality of slow-down strategies using different training inputs is quite similar in terms of energy usage.

For benchmarks `swim` and `su2cor`, using data set `std.in` as training input seems to produce better results. This is due to the different CPU boundness values provided by the two different data sets. For the reference data set of `swim`, the CPU boundness β_{cpu} is 0.04. In contrast, data set `std.in` provides $\beta_{cpu} = 0.07$ and data set `train.in` provides $\beta_{cpu} = 0.19$. Since data set `std.in` more closely models the CPU boundness of the reference data set, our algorithm was able to find a better slow-down strategy. Similarly, the reference data set of `su2cor` provides $\beta_{cpu} = 0.17$, while the data sets `std.in` and `train.in` provides $\beta_{cpu} = 0.25$ and 0.47, respectively. As a result, our algorithm was able to exploit more memory boundness of the benchmark and produced a more energy-efficient DVFS schedule.

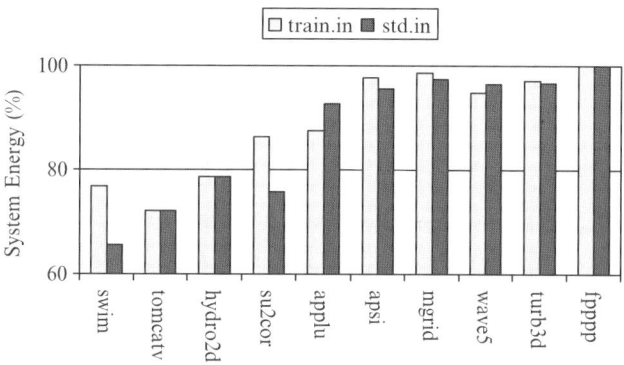

Fig. 14.5   The relative energy usage of our compiler approach using training inputs `train.in` and `std.in` and the potential energy savings

8.   Conclusions and Future Work

In this paper we have discussed a novel compiler algorithm that effectively utilizes dynamic voltage scaling to save energy. The algorithm picks a single region to be executed at a lower performance level without introducing serious performance degradation. A prototype implementation based on the SUIF2 compiler infrastructure was used to evaluate the algorithm on the SPEC CFP95 benchmarks. Physical measurements showed that significant system energy savings in the range of 0% to 28% can be achieved with performance penalties less than 5% for the SPEC CFP95 benchmarks. On average, the energy consumption is reduced by 11% with 2.15% performance slowdown.

As we said earlier that a compiler-directed DVFS scheduling algorithm has the advantage over runtime-based techniques by being able to shape the execution behavior of a program in such a way that the opportunities for DVFS are created or magnified. This part of research is still in its infancy stage. For example, we are not yet sure about the impact of performance-oriented locality optimizations on the DVFS opportunities. On one hand, locality optimizations try to reduce the memory stalls to improve performance whereas our DVFS algorithm exploits memory stalls for energy reduction. On the other hand, improving performance may improve energy efficiency as energy consumption is proportional to the execution time. An early work [6] has shown that there are still plenty of opportunities to apply our DVFS algorithm to the highly optimized codes. It is also observed that in some cases the less successful optimization lead to higher energy savings.

Acknowledgments

This research was performed while one of the authors was at Rutgers University and was partially supported by NSF CAREER award CCR-9985050.

References

[1] Inc. Advanced Micro Devices. Mobile AMD athlon 4 processor model 6 CPGA data sheet. Publication 24319, November 2001.

[2] T. Burd and R. Brodersen. Energy efficient CMOS microprocessor design. In *the 28th Hawaii International Conference on System Sciences (HICSS-95)*, January 1995.

[3] D. Burger. *Hardware Techniques to Improve the Performance of the Processor/Memory Interface*. PhD thesis, Computer Sciences Department, University of Wisconsin-Madison, 1998.

[4] D. Grunwald, P. Levis, K. Farkas, C. Morrey III, and M. Neufeld. Policies for dynamic clock scheduling. In *Proceedings of the 4th Symposium on Operating System Design and Implementation (OSDI)*, October 2000.

[5] C. Hsu and W. Feng. Effective dynamic voltage scaling through cpu-boundedness detection. In *Workshop on Power-Aware Computer Systems (PACS)*, December 2004.

[6] C. Hsu and U. Kremer. Dynamic voltage and frequency scaling for scientific applications. In *Proceedings of the 14th Annual Workshop on Languages and Compilers for Parallel Computing (LCPC 2001)*, August 2001.

[7] C. Hsu and U. Kremer. Single region vs. multiple regions: A comparison of different compiler-directed dynamic voltage scheduling approaches. In *Workshop on Power-Aware Computer Systems (PACS'02)*, 2002.

[8] C. Hsu and U. Kremer. The design, implementation, and evaluation of a compiler algorithm for CPU energy reduction. In *Proceedings of the ACM SIGPLAN Conference on Programming Languages Design and Implementation (PLDI)*, June 2003.

[9] C. Hsu, U. Kremer, and M. Hsiao. Compiler-directed dynamic frequency and voltage scheduling. In *Workshop on Power-Aware Computer Systems (PACS)*, November 2000.

[10] K. Masahiro, K. Kazuo, H. Kazuo, and O. Eiichi. WT110/WT130 digital power meters. Yokogawa Technical Report 22, 1996.

[11] N. Slingerland and A. Smith. Cache performance for multimedia applications. In *Proceedings of the 15th IEEE International Conference on Supercomputing*, June 2001.

[12] S. Sohoni, Z. Xu, R. Min, and Y. Hu. A study of memory system performance of multimedia applications. In *ACM Joint International Conference on Measurement & Modeling of Computer Systems (SIGMETRICS)*, June 2001.

[13] SUIF2 Stanford University Intermediate Format. http://suif.stanford.edu.

[14] F. Xie, M. Martonosi, and S. Malik. Compile time dynamic voltage scaling settings: Opportunities and limits. In *Proceedings of the ACM SIGPLAN Conference on Programming Languages Design and Implementation (PLDI)*, June 2003.

Chapter 15

Link Idle Period Exploitation for Network Power Management

Feihui Li[1], Guangyu Chen[1], Mahmut Kandemir, and Mustafa Karakoy[2]

[1] *CSE Department*
Pennsylvania State University
University Park, PA 16802
USA

[2] *Department of Computing*
Imperial College
London, SW7 2AZ
United Kingdom

Abstract   Network power optimization is becoming increasingly important as the sizes of the data manipulated by parallel applications and the complexity of interprocessor data communications are continuously increasing. Several hardware-based schemes have been proposed in the past for reducing network power consumption, either by turning off unused communication links or by lowering voltage/frequency in links with low usage. While the prior research shows that these schemes can be effective in certain cases, they share the common drawback of not being able to predict the link active and idle times very accurately. This paper, instead, proposes a compiler-based scheme that determines the last use of communication links at each loop nest and inserts explicit link turn-off calls in the application source. Specifically, for each loop nest, the compiler inserts a turn-off call per communication link. Each turned-off link is reactivated upon the next access to it. We automated this approach within a parallelizing compiler and applied it to eight array-intensive embedded applications. Our experimental analysis reveals that the proposed approach is very promising from both performance and power perspectives. In particular, it saves more energy than a pure hardware-based scheme while incurring much less performance penalty than the latter.

Keywords: link; power; idle; loop.

J. Henkel and S. Parameswaran (eds.), Designing Embedded Processors – A Low Power Perspective,
325–345.
© 2007 *Springer.*

1.   Introduction

Increasing energy consumption of parallel architectures makes network power consumption an important target for optimization. While network performance has received a lot of attention in the past from both hardware and software communities, network power optimization is a relatively new topic (Agarwal, 1991; Patel, 1997; Benini and Micheli, 2001; Duato et al., 2002; Kim et al., 2003; Eisley and Peh, 2004). To date, several hardware-based schemes have been proposed for reducing network power consumption, by either turning off unused communication links or lowering voltage/frequency in links with low usage (Kim et al., 2003; Soteriou and Peh, 2004; Gupta et al., 2003; Kim and Horowitz, 2002). While the prior research shows that the hardware-based network power optimization schemes can be very effective in certain cases (Raghunathan et al., 2003), they share a common drawback that they cannot extract the high level network usage patterns (e.g., the order and timing of link usages). As a result, they may be inefficient in certain cases by reacting too late to the variations in network usage patterns.

The main goal of this paper is to explore whether a software-directed approach to network power management is possible. For this purpose, we first collected data on network energy consumption of several array-intensive embedded applications. These data show that an overwhelming portion of the link energy consumption in a mesh architecture is spent during idle periods, as opposed to the active periods where the link is used for communication. In other words, the link utilization is low, due to effective parallelization of these applications, which reduces both the number and volume of interprocessor data messages. Our experimental analysis also shows that a significant portion of this idle time energy occurs for each loop nest after the last use of the links, i.e., the period between the last use of the link and the end of the loop nest execution. Based on this observation, we propose a *compiler-directed* communication link shut-down (turn-off) strategy for reducing network energy. The proposed strategy takes as input a parallelized application code (using explicit message-passing directives such as those supported by MPI [Gropp et al., 1994]). It first analyzes the interprocessor communication pattern of the input code and identifies the link access pattern, i.e., the order and timing of communication link usage. It then inserts explicit link turn-off calls in the application source. Specifically, for each loop nest in the application code, the compiler inserts a call that shuts down the links after their last uses. Each turned-off link is reactivated upon the next access to it. Our approach targets at small networks that are used by a single application at a time.

We automated this approach within a parallelizing compiler (Banerjee et al., 1995) and applied it to eight array-intensive embedded applications. Our experimental analysis reveals that the proposed approach (i) saves significant amount of link energy when averaged over all the eight codes tested (the average link energy savings with our default simulation parameters is around 44.6%), (ii) performs competitively as compared to a hypothetical (ideal) scheme that can eliminate all last idle time energies of the communication links, and (iii) performs better than a pure hardware-based power optimization scheme, based on link shut-down, from both energy and performance angles. Note that, since our work is compiler-directed, it is different from all published work on network power optimization.

The next section describes our experimental setup and simulation methodology. Section 3 quantifies the energy consumption in last idle periods and show that this energy constitutes a significant fraction of the total link energy spent in idle periods. Section 4 explains the network abstraction used by our approach. It also discusses the architectural support needed by our compiler-based scheme. Section 5 presents our compiler-based approach for exploiting this information by turning off the communication links that completed their last uses (communications) in a given loop nest. Section 6 gives experimental data showing that the compiler approach is very effective in practice. Finally, Section 7 summarizes our major observations and gives pointers for future research directions.

2. Experimental Setup

While we focus on a mesh architecture in this paper (see Figure 15.4), the proposed approach is very general and is applicable to other types of networks as well. The only requirement we have for using our compiler-based approach is that the routing algorithm used should be static, as we need to extract the link usage information at compile-time.

We implemented our approach within the Paradigm compiler (Banerjee et al., 1995), using a customized front-end. This compiler takes a sequential code, and produces an optimized message-passing parallel program with calls to the selected communication library and the Paradigm run-time system. It applies several optimizations for reducing the number and volume of interprocessor communication. Consequently, the resulting code is highly optimized. We applied our approach (detailed in Section 5), after well-known communication optimizations such as message vectorization, message coalescing, and message aggregation (Hiranandani et al., 1992). The additional increase in compilation time as a result of our optimizations was about 65% when averaged over all the applications tested.

To compute the network energy consumption, we use a modified version of the model proposed in Eisley and Peh (2004). The default values of the important experimental parameters used are listed in Table 15.1. Most of the values in this table are based on Eisley and Peh (2004), Shang et al. (2003) and Kim et al.

Table 15.1   Default values of our simulation parameters

Parameter	Value
Mesh size	4×4
Active link energy consumption	10.2 pJ/bit
Idle link energy consumption	8.5 pJ/bit
Link power-up delay	100 µsec
Hardware link shut-down threshold	150 µsec
Processor frequency	1 GHz
Packet header size	3 flits
Flit size	39 bits

Table 15.2   Benchmarks used in the experiments

Benchmark name	Benchmark description	Input size (KB)	Execution cycles (M)	Energy consmp. (mJ)
Newton	Newton iteration	266.4	367.5	128.8
Smooth	3D image smoothening	427.3	752.4	406.8
Courant	2D shallow water equations	387.0	487.5	287.2
Mgrid	Multigrid solver	96.6	163.0	90.5
27point	3D isotropic stencil algorithm	521.1	905.8	456.1
SOR	Successive overrelaxation	187.4	238.8	88.9
ID	Image dithering	411.4	597.2	310.1
Bounce	Game program	495.1	704.7	338.7

(2003). We embedded this model within a simulation environment. This environment takes as input a network description and the application executable and generates as output the energy and performance numbers. Since the prior research Kim et al. (2003) shows that communication link power constitutes a significant fraction of the overall network power, our focus in this paper is on link power optimization.

The important characteristic of the benchmarks used in this paper are presented in Table 15.2. The second column gives a description of each benchmark code and the next one shows the amount of input data used for executing the benchmark. The fourth column gives the number of execution cycles for each benchmark when no link power optimization is applied. Finally, the last column of the table gives the total link energy consumption, again when no optimization is applied.

3.   Quantification of Last Idle Times

Figure 15.1 shows the communication link energy distribution between active periods and idle periods. An active period corresponds to the case where the link is active in communicating a message, whereas the idle period represents the case where the link is not used for communication. One can observe from this graph that the most of link energy consumption is spent during idle periods (78.4% on an average). This interesting result can be explained as follows. In message-passing architectures, the application code is generally parallelized in such a fashion that the inter-processor communication is minimized as much as possible. To achieve this, communication optimizations such as message vectorization, message aggregation, and message coalescing (Hiranandani et al., 1992) are used. As a consequence of this, most of the communication links are idle at a given time, and are responsible from a significant portion of the link energy consumption.

Figure 15.2 depicts the access pattern for a given communication link in a sample program fragment with three separate loop nests. In this figure, the time is assumed to progress from left to right, and each active period of the link is specified using a pulse (rectangle) whose width represents the length of the active period. The remaining periods are idle periods. One active and one idle periods are marked for the first loop nest. Our focus is on idle periods T_1, T_2, and T_3. A common characteristic of these three idle periods is that they are the *last idle periods* (*times*) in each nest, and represent the time frame between the last use of the link and the end of the loop nest. We collected statistics on these last idle periods and present in Figure 15.3 the energy contribution of these last idle periods to the total energy consumption of all idle periods. In Figure 15.2,

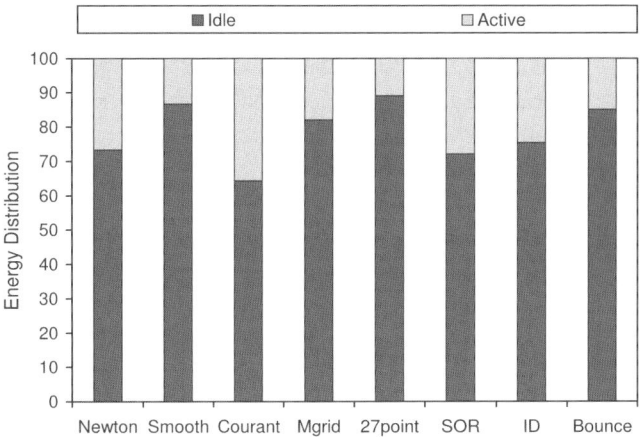

Fig. 15.1   Distribution of link active and idle energies

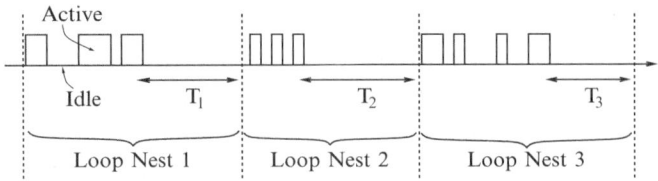

Fig. 15.2   An example link access pattern for a program with three loop nests

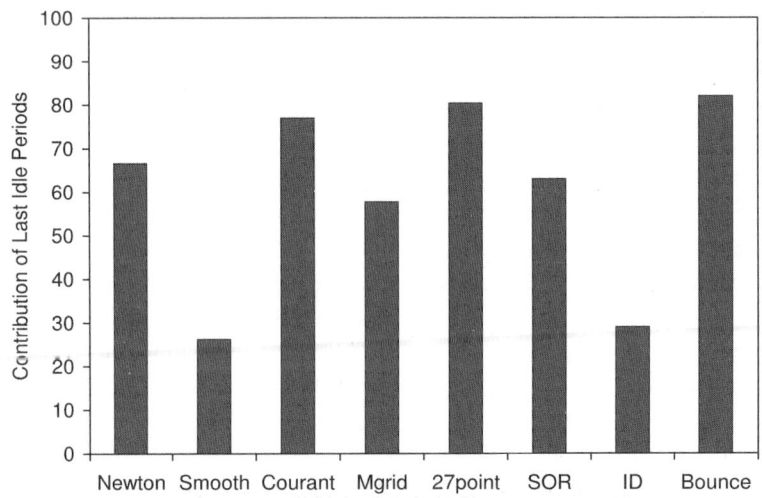

Fig. 15.3   Energy contribution of the last idle periods (times), when accumulated over all the loop nests in the application

for example, this contribution amounts to $\{(E_1 + E_2 + E_3)/E_{\text{total-idle}}\} \times 100$, where E_1, E_2, and E_3 correspond to the energy consumptions in idle periods T_1, T_2, and T_3, respectively. $E_{\text{total-idle}}$, on the other hand, captures the total idle time energy when considering all three nests in the program (including the energies consumed in the last idle times as well). An interesting observation from Figure 15.3 is that a very significant portion of the energy spent in idle times are spent in the last idle times, specifically, 60.2% when averaged over the eight benchmarks. This indicates that a scheme that can exploit these last idle periods can be very useful in practice. The next section presents such a scheme embedded within a parallelizing compiler.

4.   Network Abstraction and Hardware Support

In this section, we discuss the network abstraction our compiler-directed approach uses and the architectural support needed. We focus on an $M \times N$

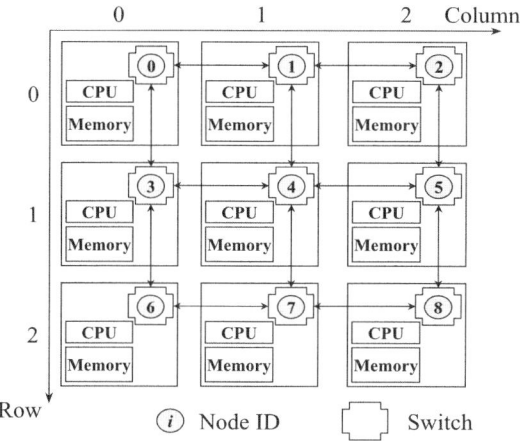

Fig. 15.4 A 3×3 mesh network architecture

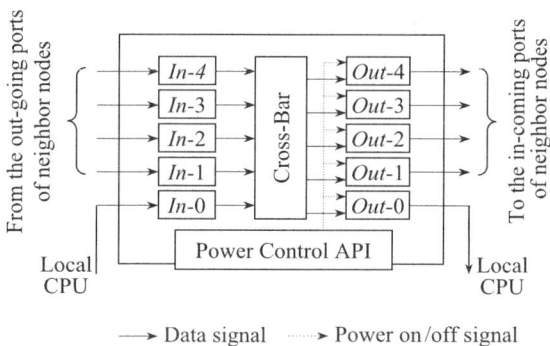

Fig. 15.5 The structure of a switch

(M rows, N columns) mesh architecture[1] as depicted in Figure 15.4. Each node in the mesh consists of a processor, a memory module, and a switch.[2] The node at the $i^{\text{th}} (i = 0, 1, \ldots, M - 1)$ row and $j^{\text{th}} (j = 0, 1, \ldots, N - 1)$ column is labeled with an integer ID: $i \times N + j$.

Figure 15.5 gives the structure of a switch. Each switch has five incoming ports (In-0 through In-4) and five out-going ports (Out-0 through Out-4). The ports In-0 and Out-0 are connected to the local processor (the processor in the same node as the switch). The remaining four incoming ports and four out-going ports are connected to the switches in the neighboring nodes by a set

---

[1]Our approach can also be used with other types of architectures. We will elaborate on this issue later in the paper.
[2]Unless a confusion occurs, we use the terms "node" and "processor" interchangeably in our discussion.

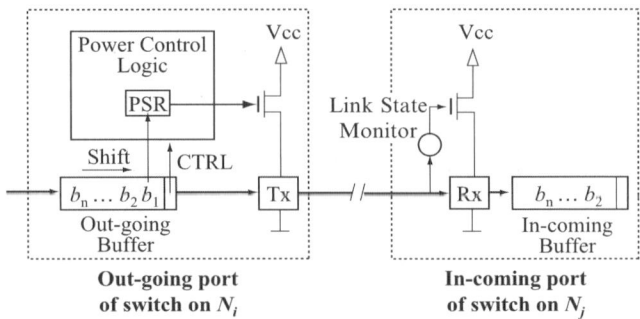

Fig. 15.6 The structure of a link. PSR: power state register. The out-going buffer contains a power control message with power control bit vector $(b_1 b_2 \dots b_n)$. The first bit, b_1, is shifted out of the buffer and stored in PSR. The rest of this vector, $(b_2 \dots b_n)$, is transferred to the in-coming port of this link, and is subsequently forwarded to the next link

of wires. A switch also provides a power control API that allows the local processor to turn on/off each out-going port of this switch. As will be discussed later, when an out-port is turned on/off, its corresponding in-port (in the switch of a neighbor node) is also turned on/off.

In this paper, we define $link(i, j)$ as the directed physical connection from a node (N_i) to one of its neighbors (N_j). We refer to nodes N_i and N_j as the sender and receiver of $link(i, j)$, respectively. In a mesh, each pair of adjacent nodes, N_i and N_j, are connected by a pair of links, namely, $link(i, j)$ and $link(j, i)$. Each link consists of a pair of ports (an out-going port of the sender switch and an in-coming port of the receiver switch) and the wires that connect these two ports.

Figure 15.6 shows the structure of a communication link from node N_i and N_j. A message to be transferred by a link is first stored in the buffer of the out-going port of this link. The power control logic in the out-going port checks the one-bit "CTRL" in the header of this message. CTRL = 0 indicates that the message in the buffer is a *data message* that carries data that will be used by an application process. This message is forwarded without any modification (i.e., as it is) from one link to another until it arrives at its destination, and the contents of this message does not affect the power state of any communication link on its path. CTRL = 1, on the other hand, indicates that the message in the buffer is a *power control message*. The body of this message contains a *vector*, each bit of which controls the power state of a link in the path from the source node to the destination node. This message is discarded by the switch on the destination node, and thus it is never received by any application process. When the power control logic detects that the message in the buffer is a power control message, it signals the buffer to shift the entire message body (not including the header) by one bit. The first bit of the message body, which is shifted out of the buffer, is stored in the one-bit *power state register* of the power control logic. The shifted message is then forwarded to the in-coming port, which is on the other end of this link. After that, the power control logic sets the power state of the Tx unit based on the value of the power state register. Specifically, if the value of the power state register is one, the power control logic turns off the Tx unit immediately; otherwise, the Tx unit remains in the active state. The link state monitor of the in-coming port monitors the state of the Tx unit. It turns on/off the Rx unit when it detects that the corresponding Tx unit has been turned on/off. In addition to controlling the power state of a link using power control messages, a switch also provides a programming interface for the local processor to turn on/off each out-going port of this switch without sending any message.

A parallel program consists a set of parallel processes running on different nodes of the mesh. A process sends messages to another process through a logical connection (or connection for short). A logical connection consists of multiple links if the sender and receiver processes are running on two nodes that are not adjacent to each other. We use $C(i, j)$ to denote the set of links in the connection from the source node N_i to the destination node N_j. Since a connection can be unambiguously identified by the set of links used in this connection, we also use $C(i, j)$ to denote the connection from N_i to N_j. Note that, using the power control messages, we can control the power state of each link in any connection.

5. Compiler Support

In this section, we discuss the details of our compiler algorithm that exploits the last idle periods of communication links. We focus on the array-based, loop-intensive embedded programs parallelized over a mesh architecture. Such a program consists of a set of parallel processes running on different nodes in a mesh. Each process consists a set of loop nests. These processes communicate with each other through inter-node connections. We assume that all the communication links in the mesh are initially turned off. A link used by a process is turned on automatically upon its first use. Our compiler inserts explicit link turn-off instructions (calls) in each loop nest to turn off the communication links whose last usage has taken place.

Before inserting the link turn-off instructions, we first break large loop nests in the given program into a set of smaller loop nests (referred to as "sub-nests") such that the sets of links used by each pair of consecutive sub-nests are different. For a loop nest that uses different communication links during different sets of iterations, splitting it into sub-nests such that each pair of successive sub-nests use different sets of links increases the opportunities for link

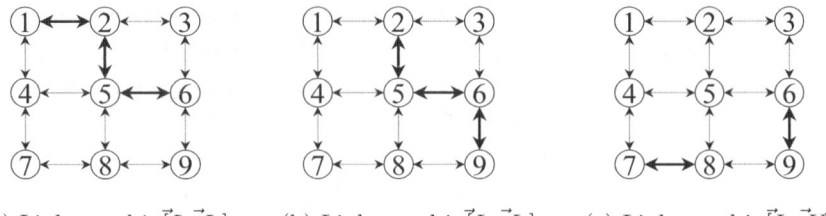

(a) Links used in $[\vec{L}, \vec{I_1}]$.   (b) Links used in $[\vec{I_1}, \vec{I_2}]$.   (c) Links used in $[\vec{I_2}, \vec{U}]$.

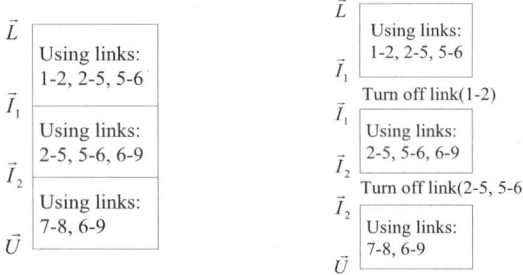

(d) Original loop nest.   (e) Sub-nests with link turnoff calls.

Fig. 15.7   Splitting a loop nest with iteration space $[\vec{L}, \vec{U}]$ into three sub-nests with iteration spaces: $[\vec{U}, \vec{I_1}]$, $(\vec{I_1}, \vec{I_2}]$, and $(\vec{I_2}, \vec{U}]$, respectively

turn-off. Further, splitting large loop nests into smaller sub-nests allows us to turn off links earlier since we can now turn off links at the end of each sub-nests, instead of waiting for the entire loop nest to terminate.

Therefore, our compiler-based link energy optimization is performed in two steps. In the first step, we split loop nests into smaller loop nests (sub-nests). In the second step, we insert code at the end of each sub-nest to turn off the communication links that are not used in the sub-nests that follow the current sub-nest. Figure 15.7 shows an example scenario. In this scenario, we show how three different parts (sub-nests) of a loop iteration space ($\vec{L}$: lower bound, and $\vec{U}$: upper bound) exercise the different set of communication links. We also show the link turn-off instructions that are inserted by our compiler. It is important to note that after each sub-nest, we turn off the links that will not be needed by the following sub-nests.

5.1   Splitting Loop Nests

We now present our compiler algorithm for splitting a loop nest L into a set of sub-nests $\{L_1, L_2, \ldots, L_n\}$. These sub-nests are executed in the order of $L_1, L_2, \ldots, L_n$ so that the order in which the iterations of L are executed (which are now distributed into sub-nests) is not changed.[3] Assuming that the sets of links used in the loop nests $\mathcal{L}_i$ and $\mathcal{L}_{i+1}$ are $LK(\mathcal{L}_i)$ and $LK(\mathcal{L}_{i+1})$, respectively, we can turn off the links in the set $LK(\mathcal{L}_i) - LK(\mathcal{L}_{i+1})$ to conserve energy without significantly degrading the performance of the application since the links we turn off are not used in the following loop nests. Our goal is to split a given loop nest $\mathcal{L}$ such that we can keep more links in the power-off mode for longer period. In addition, since increasing the number of loop nests increases the size of the application (which may have an adverse impact on code memory management), we split a loop nest only at certain points such that each pair of successive sub-nests use different sets of links. Note that, if two successive sub-nests use the same set of links, we cannot turn off any link at the end of the first sub-nest. Our approach handles each loop nest in the application code one by one, and in the following discussion, we explain our approach for nest $\mathcal{L}$.

We focus on a loop nest $\mathcal{L}$, which can be expressed in an abstract form as follows:

$$\mathcal{L}: \text{ for } \vec{I} \in [\vec{L}, \vec{U}] \ \{\text{Body}\}.$$

In this loop nest, $\vec{L}$ and $\vec{U}$ are the lower and upper bound vectors, and $\vec{I}$ is the iteration vector.[4] In this paper, we use $[\vec{L}, \vec{U}], (\vec{L}, \vec{U}], [\vec{L}, \vec{U})$, and $(\vec{L}, \vec{U})$ to denote the sets $\{\vec{I} | \vec{L} \preceq \vec{I} \preceq \vec{U}\}, \{\vec{I} | \vec{L} \prec \vec{I} \preceq \vec{U}\}, \{\vec{I} | \vec{L} \preceq \vec{I} \prec \vec{U}\}$, and $\{\vec{I} | \vec{L} \prec \vec{I} \prec \vec{U}\}$, respectively, where $\preceq$ and $\prec$ denote lexicographic ordering on vectors.

Let us assume that $S(\mathcal{L}) = \{s_1, s_2, \ldots, s_n\}$ is the set of message-sending instructions in the body of loop nest $\mathcal{L}$. Each message-sending instruction, s_i, has the form "send $(d_i(p), m)$", where m is the message to be sent, p is the id of the node that executes this loop nest, and function $d_i(p)$ gives the id of the destination node for message m. s_i uses connection $C(p, d_i(p))$. Note that $d_i(p)$ is an invariant since p cannot be changed after we assign the loop nest to a node. It is possible that, in the application code we optimize, there might exist loop nests that the destination node for a send instruction is not an invariant. In this case, we do not optimize the loop nest under consideration.

The set of nodes to which node p sends message during the execution of loop nest $\mathcal{L}$ can be expressed as:

$$D(p, \mathcal{L}) = \{d_i(p) | s_i \in S(\mathcal{L})\}.$$

We use X_i to denote the set of iterations at which s_i is executed. If the execution of s_i does not depend on any conditional branch instruction, we have X_i

---

[3]That is, loop splitting does not affect any data dependence in the loop nest, i.e., it is always legal.
[4]Vector $\vec{I}$ keeps the loop indices from the outermost position to the innermost position. $\vec{L}$ and $\vec{U}$ are also defined as vectors and each contains an entry for each loop index, again from the outer most position to the inner most position.

equal to the set containing all iterations of loop nest $\mathcal{L}$. In this paper, we only consider X_i sets that can be expressed using Presburger expressions (Pugh, 1994).[5] The set of iterations at which connection $C(p,d)$ (where $d \in D(p,\mathcal{L})$) is used can be computed as:

$$U(p,d) = \bigcup_{d_i(p)=d} X_i.$$

Since X_i can be expressed using Presburger expressions, $U(p,d)$ can also be expressed in terms of Presburger expressions. We define $\vec{I}_{min}(p,d)$ and $\vec{I}_{max}(p,d)$ as the first and last iteration vectors for connection $C(p,d)$; i.e., they can be computed as:

$$\vec{I}_{min}(p,d) = \max_{\vec{I} \in U(p,d)} \vec{I},$$

$$\vec{I}_{max}(p,d) = \max_{\vec{I} \in U(p,d)} \vec{I}.$$

Now, we can define the splitting set for loop $\mathcal{L}$ as:

$$SP(p,\mathcal{L}) = \{\vec{I}_{min}(p,d)|d \in D(p,\mathcal{L})\}$$
$$\cup \{\vec{I}_{max}(p,d)|d \in D(p,\mathcal{L})\} \cup \{\vec{L},\vec{U}\}.$$

We further assume that:

$$SP(p,\mathcal{L}) = \{\vec{I}_1,\vec{I}_2,\ldots,\vec{I}_n\},$$

where $n = |SP(p,\mathcal{L})|$ and $\vec{I}_1 \prec \vec{I}_2 \prec \ldots \prec \vec{I}_n$. Since $\vec{L},\vec{U} \in SP(p,\mathcal{L})$, we have $\vec{I}_1 = \vec{L}$ and $\vec{I}_n = \vec{U}$. Using the vectors in splitting set $SP(p,\mathcal{L})$, we can now split loop nest $\mathcal{L}$ into $n-1$ smaller loop nests (sub-nests) as follows:

$$\mathcal{L}_1: \qquad \text{for } \vec{I} \in [\vec{I}_1,\vec{I}_2] \ \{\text{Body}\}$$
$$\mathcal{L}_2: \qquad \text{for } \vec{I} \in (\vec{I}_2,\vec{I}_3] \ \{\text{Body}\}$$
$$\ldots\ldots$$
$$\mathcal{L}_n: \qquad \text{for } \vec{I} \in (\vec{I}_{n-1},\vec{I}_n] \ \{\text{Body}\}.$$

We handle the conditional statements that may occur within $\mathcal{L}$ conservatively. Specifically, if a message statement can access a communication link depending on the run-time value of some condition, we conservatively assume at compile-time that the said statement will access that link. This assumption helps us reduce the potential performance penalty associated with link turn-offs.

---

[5]Presburger formula is a class of logical formulas which can be built from affine constraints over integer variables, the logical connectives ($\vee, \wedge$, and $\neg$), and the existential and universal quantifiers ($\exists$ and $\forall$). The Omega Library is an example tool that manipulates integer tuple relations and sets, which are described using Presburger formulas.

5.2 Inserting Link Turn-off Instructions

Our compiler inserts calls (instructions) between each pair of successive loop nests, $\mathcal{L}_i$ and $\mathcal{L}_{i+1}$, to turn off the communication links that have been used in $\mathcal{L}_i$ and will not be used in $\mathcal{L}_{i+1}$. The set of links that are used in a given loop nest $\mathcal{L}_i$ running on node p can be determined as follows:

$$LK(p, \mathcal{L}_i) = \bigcup_{d \in D(p, \mathcal{L}_i)} C(p, d),$$

where $D(p, \mathcal{L}_i)$ is the set of nodes to which node p sends message within loop nest $\mathcal{L}_i$. For a pair of consecutive loop nests, $\mathcal{L}_i$ and $\mathcal{L}_{i+1}$, the set of communication links that can be turned off after $\mathcal{L}_i$ is: $LK(p, \mathcal{L}_i) - LK(p, \mathcal{L}_{i+1})$. If $LK(p, \mathcal{L}_i) - LK(p, \mathcal{L}_{i+1}) = \phi$, no link can be turned off after $\mathcal{L}_i$. In this case, we merge loop nests $\mathcal{L}_i$ and $\mathcal{L}_{i+1}$ if they were extracted from the same loop nest in the previous step.[6]

Figure 15.8 gives the pseudo code for our algorithm. This algorithm has two phases. In the first phase, we insert code between each pair of successive loop nests if there are links that can be turned off at that point. In the second phase, we merge successive loop nests if no link turn-off instruction is inserted between them.

```
Input: A program consisting of loop nests L₁, L₂, ..., Lₘ;
Output: A program augmented with explicit link turnoff calls;
// Phase-1
for each pair of consecutive loop nests Lᵢ and Lᵢ₊₁ {
    if (LK(p, Lᵢ) − LK(p, Lᵢ₊₁) ≠ φ) {
        assume D(p, Lᵢ) − D(p, Lᵢ₊₁) = {d₁, d₂, ..., dₙ};
        Sₙ = LK(p, Lᵢ₊₁);
        for i = n − 1 to 1 step − 1 { Sᵢ = Sᵢ₊₁ ∪ C(p, dᵢ₊₁); }
        for i = 1 to n
            if(C(p, dᵢ) − Sᵢ ≠ φ) {
                m.type = "CTRL";
                m.body = power_control_vector(p, dᵢ, Sᵢ);
                insert "send(dᵢ, m)" to the end of Lᵢ
            }
    }
}
// Phase-2
for each pair of consecutive loop nests Lᵢ and Lᵢ₊₁ {
    if(LK(p, Lᵢ) − LK(p, Lᵢ₊₁) ≠ φ) { merge Lᵢ and Lᵢ₊₁ }
}

power_control_vector(p, dᵢ, Sᵢ) {
    assume that a message from node p to dᵢ is transfered along links l₁, l₂, ..., lₖ
    for i = 1 to k { if(lᵢ ∈ Sᵢ) bᵢ = 0; else bᵢ = 1; }
    return vector (b₁, b₂, ..., bₖ);
}
```

Fig. 15.8   Algorithm for inserting link turn-off instructions

---

[6]This is because in this case loop splitting would only increase code size without any energy benefits.

In our algorithm, we generate link turn-off calls between loop nest L_i and L_{i+1} if $LK(p, L_i) - LK(p, L_{i+1}) \neq \phi$. The link turn-off instructions turn off the communication links by sending power control messages. Let us assume:

$$D(p, L_i) = \{d_1, d_2, \ldots, d_n\}.$$

We send power control messages $m_1, m_2, \ldots, m_n$ to nodes $d_1, d_2, \ldots, d_n$, respectively; to turn off the links in the set $LK(p, L_i) - LK(p, L_{i+1})$. Each message m_i carries a link control vector that specifies the power state for each link in the connection from node p to d_i. To each node $d_i \in D(p, L_i)$, the power control message m_i turns off the links in the following set:

$$F(p, d_i) = C(p, d_i) - (LK(p, L_{i+1}) \cup \bigcup_{j=i+1}^{n} C(p, d_j)).$$

Since we have:

$$LK(p, L_i) - LK(p, L_{i+1}) = \bigcup_{i=1}^{n} F(p, d_i),$$

control messages $m_1, m_2, \ldots, m_n$ together turn off all the links in $LK(p, L_i) - LK(p, L_{i+1})$. Message m_i does not turn off links in $C(p, d_i) \cap LK(p, L_{i+1})$ because these links will be used in L_{i+1}. Message m_i does not turn off links in $C(p, d_i) \cap C(p, d_j)$ where $i < j \leq n$ either because these links are used by another power control message m_j that will be sent after m_i. Therefore, in the link control vector for message m_i, only those bits corresponding to the links in $F(p, d_i)$ are set to 1.

5.3   Example

We now give an example to show how our approach works. Figure 15.9a shows the code of a loop nest L scheduled to be run on node p of an $M \times M$ mesh network. In this loop nest, node p sends messages to nodes $p - M - 1, p - M$, and $p - M + 1$. Figure 15.9b shows node p and its neighbors. The links marked using bold lines are the links used by p to send messages. The first column of the table in Figure 15.9c lists the connections that will be used by node p during the execution of L; the second column of this table shows the links in each connection and the last two columns give the first and last use vectors for each connection. Using the technique presented in Section 5.1, we split loop nest L into three sub-nests, L_1, L_2, and L_3, as shown in Figure 15.9d. Since $link(p, p - M - 1)$ in the connection $C(p, p - M - 1)$ is used in sub-nest L_1 but not used in sub-nest L_2, we insert instructions to send a power control message to turn it off after L_1. Loop nest L_2 uses only one link, $link(p, p - M)$, and this link is also used in the loop nest L_3 that immediately follows L_2. As a result,

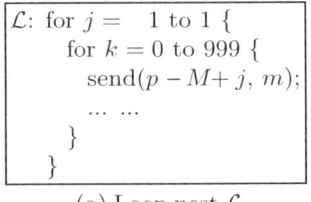

(a) Loop nest $\mathcal{L}$.

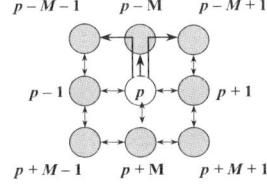

(b) Node p and its neighbors.

connection	links	$\vec{I}_{min}$	$\vec{I}_{max}$
$C(p, p-M-1)$	$link(p, p-M)$ $link(p-M, p-M-1)$	$\begin{pmatrix} -1 \\ 0 \end{pmatrix}$	$\begin{pmatrix} -1 \\ 999 \end{pmatrix}$
$C(p, p-M)$	$link(p, p-M)$	$\begin{pmatrix} 0 \\ 0 \end{pmatrix}$	$\begin{pmatrix} 0 \\ 999 \end{pmatrix}$
$C(p, p-M+1)$	$link(p, p-M)$ $link(p-M, p-M+1)$	$\begin{pmatrix} 1 \\ 0 \end{pmatrix}$	$\begin{pmatrix} 1 \\ 999 \end{pmatrix}$

(c) Connections used by node p during $\mathcal{L}$.

```
L₁: j = −1;
      for k = 1 to 1000 {
        send(p − M + j, m);
        ... ...
      }
L₂: j = 0;
      for k = 1 to 1000 {
        send(p − M + j, m);
        ... ...
      }
L₃: j = 1;
      for k = 1 to 1000 {
        send(p − M + j, m);
        ... ...
      }
```

(d) Code after splitting $\mathcal{L}$ into three sub-nests.

```
L'₁: j = −1;
      for k = 1 to 1000 {
        send(p + 1, m);
        ... ...
      }
      m.type = "CTRL";
      m.body = (0, 1);
      send(p − M − 1, m);
L'₂: for j = 0 to 1 {
      for k = 1 to 1000 {
        send(p − M + j, m);
        ... ...
      }
    }
      m.type = "CTRL";
      m.body = (1, 1);
      send(p − M + 1, m);
```

(e) The optimized code with explicit link turnoff instructions inserted.

Fig. 15.9   Example

we cannot turn off any communication links after sub-nest $\mathcal{L}_2$. Therefore, we merge loop nests $\mathcal{L}_2$ and $\mathcal{L}_3$ into one loop $\mathcal{L}'_2$. Finally, by using the approach discussed in Section 5.2 to the loop nests shown in Figure 15.9d, we obtain the optimized code shown in Figure 15.9e.

5.4   Discussion

It should be noted that our optimization approach is also applicable to network topologies other than meshes without significant modification. However, our approach requires the following two conditions to be satisfied. First, the message routing in the network must be static. That is, the set of links used to transfer a message from one node to another must be determined at compile-time. In practice, static routing algorithms are widely used in commercial systems due to their ease of implementation. Also, many NoC-based systems use static routing due to its energy benefits over dynamic (adaptive) routing. Second, the message-passing behavior of the application must be predictable at compilation time. Fortunately, most parallel embedded applications with regular array access patterns satisfy this requirement, as they are array/loop-intensive codes with rare conditional flow of execution.

Another issue is that link energy optimizer performs link turn-off analysis on the code of each process of a given program individually. Since we do not use any interprocess information during our analysis, a process may turn off some links that are still needed by other processes. Since a link that has been turned off can be automatically turned on when it is next used, turning off links mistakenly does not cause any error in the program. The performance and energy penalties, however, are captured in our experimental results.

6.   Experimental Results

In this section, we present the results from our experiments showing how our approach performs in practice. We also compare our approach to a pure hardware-based link shut-down scheme. This hardware schemes turns off a communication link after 150 µsec from the point at which the link has become idle (see Table 15.1). It predicts the length of a link idleness using a history-based mechanism. Specifically, for each link, it assumes that the length of the current idle period would be the same as that of the previous idle period.

Figure 15.10 gives the normalized energy consumption with different schemes. All the values are normalized with respect to the last column of Table 15.2. The first bar represents the normalized link energy when all last idle time energies are eliminated. This in a sense represents the best case scenario (i.e., the ideal case) for any optimization that targets energy consumption in last idle periods.[7] The second bar for each benchmark, on the other hand, shows the normalized energy consumption achieved by our compiler-driven approach, explained in Section 5. We see from these results that the

---

[7]Note, however, that one can potentially do even better in theory by considering and exploiting other idle periods as well. In practice, however, this is difficult as the remaining idle periods are really very short to take any advantage of.

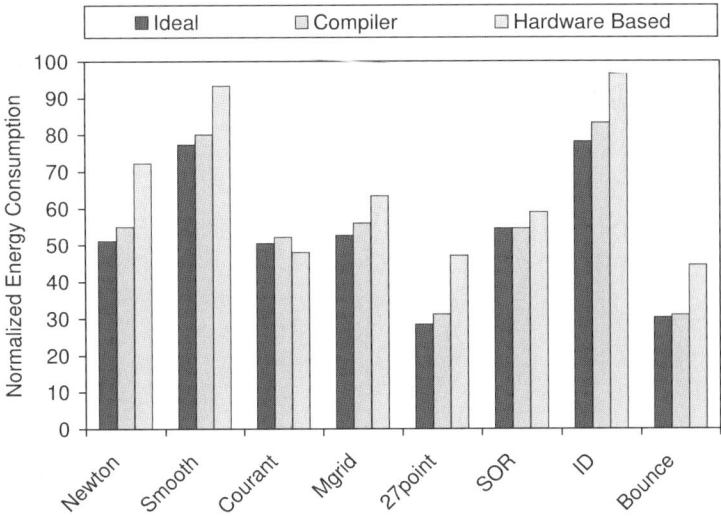

Fig. 15.10 Normalized link energy consumptions with different schemes

average energy savings with the ideal scheme and our approach are 47.1% and 44.6%, respectively. These results clearly indicate that the compiler approach is very effective in practice and achieves most of what it tries to optimize. In particular, we see from Figure 15.10 that our approach generates very similar results with the ideal scheme in benchmarks such as SOR and Bounce. The difference between our approach and the ideal scheme in the remaining benchmarks is due to the conservative nature of the compiler algorithm. Recall from our discussion in Section 5 that when there is an IF-statement within the loop body, the compiler inserts the turn-off calls conservatively based on the longest branch (predicted at compile-time). When, during execution, the shorter branch is taken, this compiler inserted turn-off instruction is invoked a bit later than the optimum time, and this leads to a slight loss in potential power savings.

Let us now compare our compiler-based approach to the hardware-based approach described above. The third bar, for each benchmark, in Figure 15.10 gives the normalized energy consumption when the hardware-based approach is employed. We see that it reduces the link energy consumption of our benchmarks by 34.4% on the average. We also see that the compiler approach generates better energy savings than the hardware scheme in all the benchmark codes except one (Courant). The main reason is that the hardware scheme cannot accurately predict the time when a link becomes idle, and this in most cases causes a loss in potential energy savings. The reason that the hardware-based scheme outperforms our approach in Courant is the fact that it exploits numerous small idle periods that are not targeted by our approach.

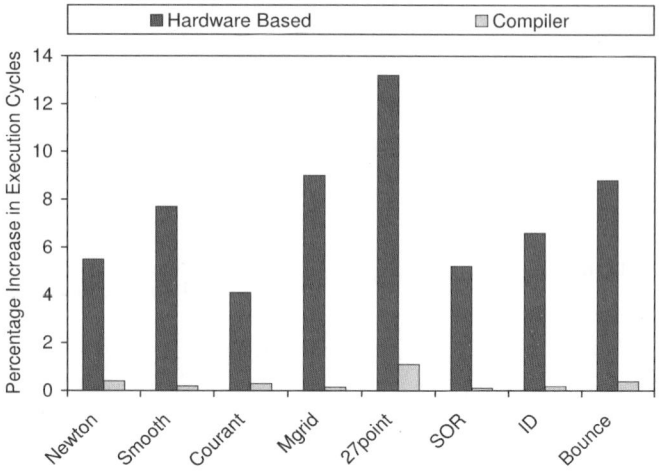

Fig. 15.11 Percentage performance degradations with different schemes

We next evaluate the performance behavior of our approach and compare it against the hardware-based scheme from the performance angle. The results are presented in Figure 15.11. Each bar in this figure gives the percentage increase in original execution cycles over the case when no power management is employed (see fourth column of Table 15.2). We see that the average degradation in performance is 0.36% and 7.51% with our approach and the hardware-based scheme, respectively. The main reason for this behavior is the fact that the hardware-based scheme cannot accurately predict the next time a given turned-off communication link will be used again, and this in turn translates to degradation in performance.

It is also important to evaluate the influence of network size. Figure 15.12 gives the results with different mesh sizes. Recall that the default mesh size used so far in our experimental evaluation was 4×4. These results show that our approach saves more energy as we increase the network size. This is because increasing network size generally increases the number of idle links and the duration of idleness for a given link in the network. Since our approach is oriented towards exploiting idle times, this results in larger energy savings. To sum up, we can conclude that our approach generates better results than the hardware-based scheme in terms of both energy and performance. In addition, our power savings increase with the increased number of processors (i.e., our approach scales very well).

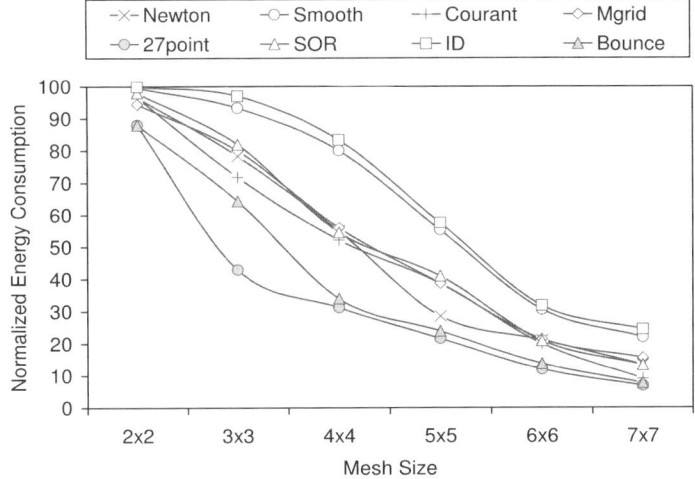

Fig. 15.12   Impact of network size on energy consumption

7.   Concluding Remarks

Reducing power consumption of networks is an important optimization goal in many application domains, ranging from large-scale simulation codes to embedded multi-media applications. Most of the prior efforts on network power optimization are hardware-based schemes. These schemes are predictive by definition as they control communication link status based on observations made in the past. Since prediction may not be very accurate most of the time, these hardware approaches can result in significant overheads in terms of both performance and power. This paper proposes a compiler-driven approach to link voltage management. In this approach, an optimizing compiler analyzes the application code and identifies the last use of a link at each loop nest. It exploits this information by inserting explicit link turn-off instructions after the last use of the link. We implemented this approach within an optimizing compiler and conducted experiments with eight array-intensive applications. Our experiments reveal that the proposed approach (i) saves significant amount of link energy when averaged over all the eight codes tested (the average savings with our default simulation parameters is around 44.6%), (ii) performs competitively as compared to a hypothetical scheme that can eliminate all last idle time energies, and (iii) performs better than a pure hardware-based power optimization scheme from both energy and performance angles. Our next goal is to extend this scheme so that it can focus on multiple loop nests at a time. Work is also underway in developing new code restructuring techniques to lengthen the duration of link idle periods.

Acknowledgments

This work is supported in part by a grant from GSRC. A preliminary version of this paper appears in EMSOFT 2005 Proceedings (Li et al., 2005). This paper enhances the EMSOFT paper by presenting details of our approach.

References

A. Agarwal. Limits on interconnection network performance. *IEEE Transaction on Parallel and Distribution Systems,* 2(4), October 1991.

P. Banerjee, J.A. Chandy, M. Gupta, E.W. Hodges IV, J.G. Holm, A. Lain, D.J. Palermo, S. Ramaswamy, and E. Su. The PARADIGM compiler for distributed-memory Multicomputers. *IEEE Computer,* 28(10):37–47, October 1995.

L. Benini and G.D. Micheli. Powering networks on chips: energy-efficient and reliable interconnect design for SoCs. In *Proceedings of the 14th International Symposium on Systems Synthesis,* 2001.

J.B. Duato, S. Yalamanchili, and L. Ni. *Interconnection Networks.* Morgan Kaufmann Publishers, 2002.

N. Eisley and L.-S. Peh. High-level power analysis of on-chip networks. In *Proceedings of the 7th International Conference on Compilers, Architectures and Synthesis for Embedded Systems,* September 2004.

W. Gropp, E. Lusk, and A. Skjellum. *Using MPI: Portable Parallel Programming with the Message-Passing Interface.* MIT Press, 1994.

P. Gupta, L. Zhong, and N.K. Jha. A high-level interconnect power model for design space exploration. In *Proceedings of the IEEE/ACM International Conference on Computer-Aided Design,* 2003.

S. Hiranandani, K. Kennedy, and C.-W. Tseng. Compiling Fortran D for MIMD distributed-memory machines. *Communications of the ACM,* 35(8):66–80, August 1992.

E.J. Kim, K.H. Yum, G. Link, N. Vijaykrishnan, M. Kandemir, M.J. Irwin, M. Yousif, and C.R. Das. Energy optimization-techniques in cluster interconnects. In *Proceedings of the International Symposium on Low Power Electronics and Design,* August 2003.

J.S. Kim, M.B. Taylor, J. Miller, and D. Wentzlaff. Energy characterization of a tiled architecture processor with on-chip networks. In *Proceedings of the International Symposium on Low Power Electronics and Design,* August 2003.

J. Kim and M. Horowitz. Adaptive supply serial links with sub-1V operation and per-pin clock recovery. In *Proceedings of International Solid-State Circuits Conference,* February 2002.

F. Li, G. Chen, M. Kandemir, and M. Karakoy. Exploiting last idle periods of links for network power management. In *Proceedings of ACM Conference on Embedded Software*, September 2005.

C.S. Patel. Power constrained design of multiprocessor interconnection networks. In *Proceedings of the International Conference on Computer Design*, Washington, DC, USA, 1997.

W. Pugh. Counting solutions to Presburger formulas: how and why. In *Proceedings of the ACM SIGPLAN Conference on Programming Language Design and Implementation*, Orlando, Florida, 1994.

V. Raghunathan, M.B. Srivastava, and R.K. Gupta. A survey of techniques for energy efficient on-chip communication. In *Proceedings of the 40th Conference on Design Automation*, 2003.

D.A. Reed and D.C. Grunwald. The performance of multicomputer interconnection networks. *IEEE Transaction on Computers*, 20(6), June 1987.

L. Shang, L.-S. Peh, and N.K. Jha. Dynamic voltage scaling with links for power optimization of interconnection networks. In *Proceedings of High Performance Computer Architecture*, February 2003.

V. Soteriou and L.-S. Peh. Design space exploration of power-aware on/off interconnection networks. In *Proceedings of the 22nd International Conference on Computer Design*, October 2004.

Chapter 16

Remote Task Mapping

Zhiyuan Li[1] and Cheng Wang[2]

[1]*Department of Computer Science*
Purdue University
West Lafayette
Indiana

[2]*Programming Systems Laboratory*
Intel Corporation
Santa Clara
California

Abstract   Widespread deployment of wireless LANs offer opportunities for users of hand-held devices to access not only public information over the internet but also resources on their own desktop computers or trusted servers. Such resources include data, storage, and CPU, among others. The discussion in this chapter focuses on the issue of remote task mapping for the purpose of offloading computational tasks from the resource-constrained handheld devices to the resource-rich desktop computers and servers. The main objective is to reduce both the time and energy required to accomplish the tasks. Compiler techniques used to analyze an ordinary program before transforming it into an efficient client-server distributed program are presented, along with a set of experimental results.

Keywords: Computation offloading, client-server environments, task partitioning

1.   Computation Offloading on Handheld Devices

The idea of computation offloading is to move computational tasks from one machine to another in order to reduce either the resource consumption or the execution time on the former [5, 10, 12]. This can be a practical approach to saving both execution time and energy on networked embedded systems. For

J. Henkel and S. Parameswaran (eds.), Designing Embedded Processors – A Low Power Perspective,
347–370.

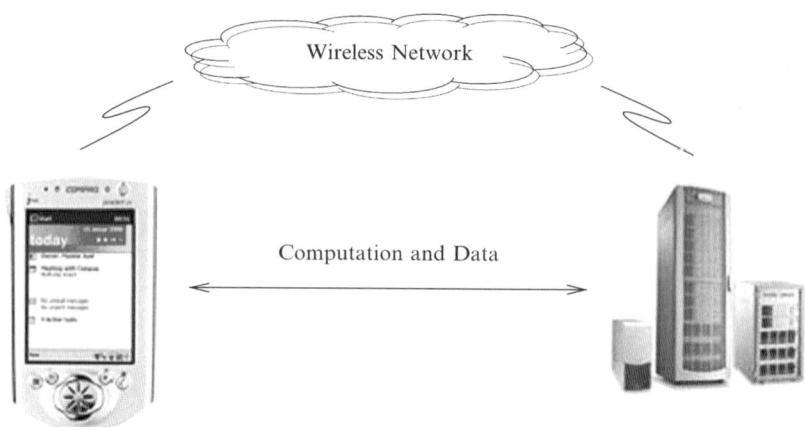

Fig. 16.1   Offloading through a wireless network

example (Figure 16.1), computation tasks can be offloaded through a wireless network from a handheld PDA to a desktop machine or a server to save energy on the PDA [5, 12].

The input data needed by an offloaded task must be transferred to the machine which computes it. In return, the output produced by the task must be transferred back to the device where the application was originally launched. These data transfer activities consume energy. To make offloading worthwhile, one must make a careful tradeoff between the saving in computation and the overhead in communication. The problem of *task assignment* is to determine which tasks (if any) should be offloaded such that the total energy cost for the given application is minimized.

In order to effectively take advantage of computation offloading, several issues must be addressed.

The dynamic nature of the environment.   The communication environment of the PDAs and the program workload may change from time to time. The network bandwidth may change when the handheld device moves around or when network contention increases. The program workload and communication requirements may both vary when running under different inputs and options. Often such variations are difficult to predict prior to the time when the application is launched on the device. Therefore, it is more practical to make decisions on task assignment at run time when sufficient information about the workload and communication requirements becomes available.

Correctness of message passing.   If the given application was originally written to run on a single device, then the program must be transformed into a form of distributed program in order to offload the tasks. The most essential part of the transformation is to insert message passing between the client device and the server. Considering the complex data dependence and control flows in a large application, it is highly desirable to have a software tool that can automatically partition the program into tasks and insert the necessary message passing according to the optimal task assignment.

The insertion of message passing may be complicated by the lack of precise information about the data dependences among different tasks. Under such circumstances, a run-time supporting mechanism is needed to make sure the necessary data transfer take place correctly.

In the following discussions, we first give an overview of the work involved when partitioning an ordinary program into a client–server execution form. We then present a run-time data registeration mechanism to ensure correct data transfer. The cost of such conservative data transfer is taken into account when formulating the task assignment problem. We present an optimal assignment scheme which adapts to the workload and communication environment.

1.1   An Overview

Figure 16.2 shows a possible framework of the automatic transformation tool mentioned above for computation offloading. The framework contains two parts, namely static analysis and run-time scheduling. The static analysis performs offline program analysis and transformation. The transformed distributed program performs self-scheduling at run time. As a result, the computation is partitioned between the handheld device and the server.

The static analysis consists of three steps: the cost modeling, task-assignment analysis and program transformation. The cost-modeling step analyzes the given program (written in an ordinary imperative language such as C). It first generates a task control flow graph ($TCFG$), where the entire program is divided into tasks which are represented as nodes in the $TCFG$. The edges in $TCFG$ represent how the program execution transits from one task to another. In this step, we also extract cost parameters which are associated with the nodes and edges. We then build a mathematical system of constraints over the $TCFG$. In the step of task-assignment analysis, we analyze the cost of different task schedules expressed as functions of input parameters. We then solve a *parametric task scheduling problem* to derive a task-assignment solution. This solution prescribes the optimal task scheduling decision according to different run-time parameters. The step of the program transformation finally generates the distribute program which self-schedules its execution according to program run-time parameters.

At run-time, the distributed program checks the current run-time parameters in order to determine the optimal task scheduling. The distributed program

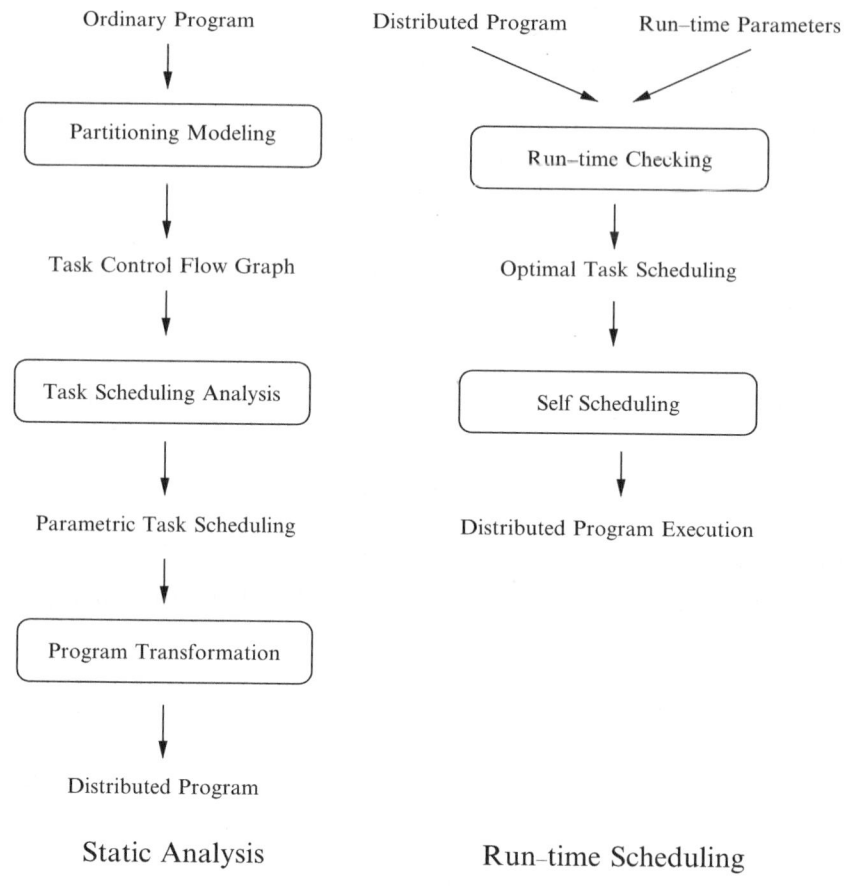

Ordinary Program   Distributed Program   Run–time Parameters

Partitioning Modeling   Run–time Checking

Task Control Flow Graph   Optimal Task Scheduling

Task Scheduling Analysis   Self Scheduling

Parametric Task Scheduling   Distributed Program Execution

Program Transformation

Distributed Program

Static Analysis   Run–time Scheduling

Fig. 16.2 Workflow of the proposed framework

then self-schedules its distributed program execution between the client and the server according to the optimal task scheduling decision.

To transform an ordinary program into a distributed version for computation offloading, we must guarantee the correct control and data flow relations in the original program for all possible program execution contexts. Difficulties arise when programs contain run-time control and data flow information which is undecidable at compile time.

We take an approach which makes use of both the static information and the run-time information. Statically, we produce an abstraction of the given program in which all memory references are mapped to references to *abstract memory locations*. The program abstraction contains only statically available information. We then perform partition analysis on the program abstraction to statically determine the task allocations and data transfers of abstract memory locations subject to the control and data flow defined over the program abstraction. Our program transformation will insert efficient run-time bookkeeping codes for the correct mapping between abstract memory locations and run-time physical memory. The resulting distributed program guarantees the correct control and data flow at run time.

1.2 The Execution Model

The given program is decomposed into tasks. The tasks can be defined at various levels. A task can be a basic block, a loop, a function or a group of closely related functions. Tasks assigned to the server are called *server tasks* and the remaining tasks are called *client tasks* which run on the handheld device. The program transformation must guarantee correct execution in the distributed environment under all possible execution contexts.

Conceptually, a client task and a server task are allowed to run simultaneously on different machines as long as all data dependence relations between these two tasks are maintained to satisfy the original program semantics. In practice, due to the significant computation speed gap between the PDA and the server, the application does not lose any tangible speed advantage if we require that only one task from the same application runs at any given time. This requirement makes both data transfers and task assignment much easier to handle. In the absence of simultaneous tasks, program execution simply follows the original sequential control flow.

During the program execution, the control and data transfer are implemented by message passing. As explained earlier, at any moment, only one host (active host) performs the computation. The other host (passive host) waits in a message processing function which acts in accordance to the incoming messages. The active host sends a message to start a task on the opposite host. Upon receiving the message, the receiver becomes active by starting the execution of corresponding task. Meanwhile, the sender becomes passive by blocking its current task execution and entering the message processing function. The active host becomes passive by sending a message to the opposite host indicating the termination of its tasks. Upon receiving such a message, the passive host becomes active by exiting the message processing function and resuming the blocked task execution.

Two mechanisms (shown in Figure 16.3b) are used to perform run-time bookkeeping to ensure the correct mapping between the abstract memory locations and their physical memory. The *registration mechanism* keeps track of the local mapping between abstract memory locations and their corresponding physical memory with registration table. Entries in the registration table are indexed by the abstract memory location ID for lookup. Each entry in the table contains a list of memory addresses for that abstract memory location. The

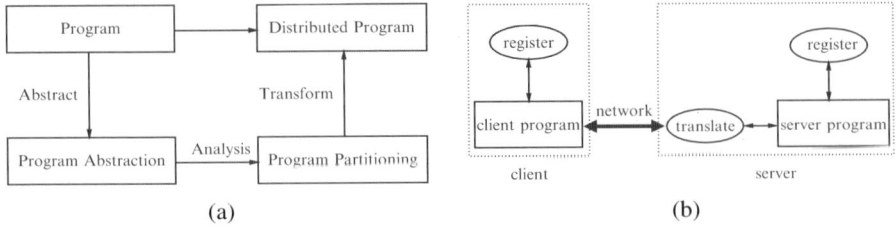

Fig. 16.3   A computation offloading scheme

translation mechanism keeps track of the mapping of the same data between different hosts. Only the server has the translation mechanism which translates the data representation back and forth for the server and the handheld device. For translation of data addresses, we maintain a mapping table on the server. Entries in the mapping table contains the mapping of memory addresses for the same data on the server and the handheld device. To reduce the run-time overhead, the registration and translation mechanism only apply to memory locations that are accessed on both hosts. Both the registration table and the translation table are updated at run time only when new (stack or heap) memory is allocated or released.

With the registration and translation mechanisms discussed above, we can now transfer data safely between the server and handheld device. We use two methods for data transfer, namely the *push* method and the *pull* method. The push method sends abstract memory locations to the opposite host. The pull method lets the intended receiver make a request for modified abstract memory locations from the opposite host.

1.3   Message Passing

The program execution control and data transfer between the server and the client are implemented by message passing. At any moment, only one host (active host) performs the computation. Meanwhile, the other host (passive host) blocks its task execution and enters the *wait_for_message* function which waits and acts in accordance to the incoming messages. The *wait_for_message* function takes the form shown in Figure 16.4, where $MESSAGEn$ is a message identifier and $PROCESS\_MESSAGEn$ is the message processing primitive for $MESSAGEn$. The active host sends messages to the passive host indicating the termination of tasks, which will cause the passive host to exit the *wait_for_message* function and resume the blocked task execution.

```
wait_for_message() {

    while(1) {
        wait for message;
        switch(message) {
            case MESSAGE1:
                PROCESS_MESSAGE1;

            case MESSAGE2:
                PROCESS_MESSAGE2;

            ...

            case MESSAGEn:
                PROCESS_MESSAGEn;
        }
    }
}
```

Fig. 16.4 The *wait_for_message* function

1.4   A Number of Primitives

All the messages passed between the server and the client can be divided into three classes: *information messages*, *data messages* and *scheduling messages*. Information messages are used to exchange data allocation information between the server and the client. Data messages are used to transfer data for data consistency. Scheduling messages are used to schedule task execution.

Next, we present the primitives used in the transformed program. In all the definitions of primitives, we use the following conventions:

- d is an abstract memory location ID.

- dv and ds are data value and data size, respectively.

- p and p' are data (or function) addresses on this and opposite hosts respectively.

- ps and pc are data (or function) addresses on server and client respectively.

- in and out are function parameter and return value.

The Information Messages.   Two mechanisms dynamically maintain the correct mapping between the abstract memory locations and the real memory locations. The registration mechanism uses the registration table to keep track

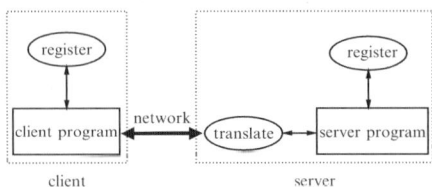

Fig. 16.5   Distributed program execution

of the local mapping between the abstract memory locations and their corresponding real memory locations. The translation mechanism uses the mapping table to keep track of the mapping of the real memory locations between different hosts.

Under our scheme, the execution of the distributed program can be illustrated by the diagram in Figure 16.5. Both hosts perform the registration. However, only the server performs the translation. The data on the server are first translated and then sent to the client, and the data on the client are first sent to the server and then translated on the server. In this way, we can greatly reduce the run-time cost on the handheld device for data representation translation because the server has much more computation power than the client. We also restrict registration and translation mechanism to memory locations that *may be accessed on both hosts* to reduce run-time cost. Other memory locations do not need registration and transformation.

For functions and global variables, the local mapping between their abstract memory locations and their real memory locations is trivial. Data registration is needed only for the memory locations which are *allocated at run time* on the stack or the heap. Entries in the registration table are indexed by the abstract memory location ID for fast lookup. Each entry in this table contains a list of data addresses and data sizes for the specific abstract memory location. Figure 16.6 shows an example of registration tables, where p_m_n is the address and s_m_n is the size of the nth real memory location corresponding to the abstract memory location m.

The translation mechanism translates the internal representations of data which are transfered between two machines. This includes the conversion between data endians, as well as the memory addresses contained in pointer variables. For memory address translation, we maintain a mapping table on the server. Figure 16.7 illustrates such a mapping table. Each entry in the table is a tuple (d_n, p_s_n, p_c_n) where p_s_n is a real memory location on the server and p_c_n is its corresponding real memory location on the client. The element d_n is the abstract memory location ID which serves as a link between the mapping table and the registration table. Such a link is needed when the data is released.

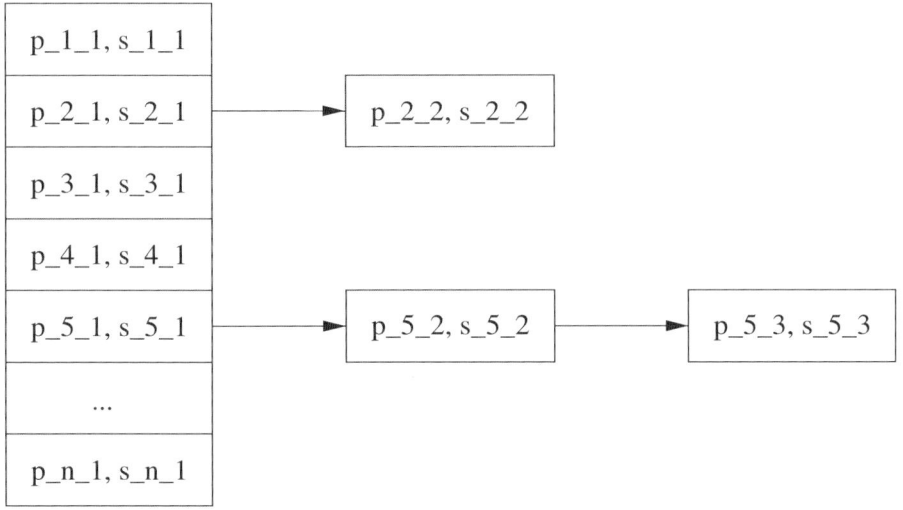

Fig. 16.6 A registration table

d_1	p_s_1	p_c_1
d_2	p_s_2	p_c_2
d_3	p_s_3	p_c_3
...	...	...
d_n	p_s_n	p_c_n

Fig. 16.7 A mapping table

At program startup, the registration table is empty and the mapping table contains only functions and global variables which are referenced on both the client and the server. Both the registration table and the translation table are updated at run time when new memory locations are allocated or released. Two types of information messages are used to maintain the registration table and the mapping table. The *ALLOC_DATA* message is sent when a new real memory location is allocated, and the *RELEASE_DATA* message is sent when a real memory location is released.

Figure 16.8 shows the primitives for data allocation. For each shared data item, the server need to know the memory address on the client in order to insert the entry in the mapping table. Hence, after the *ALLOC_DATA* message is sent by the server, the client must send a reply. Figure 16.9 shows the

```
#define ALLOC(d, p, ds)
   insert registration table with (d, p, ds);
   send a ALLOC_DATA message with (d, p, ds);
   #ifdef SERVER /* server version */
      receive a reply with p';
      insert mapping table with (d, p, p');
   #endif

#define PROCESS_ALLOC_DATA(d, p', ds)
   allocate memory of size ds and get address p;
   insert registration table with (d, p, ds);
   #ifdef SERVER /* server version */
      insert mapping table with (d, p, p');
   #else /* client version */
      send a reply with p;
   #endif
```

Fig. 16.8   Primitives for data allocation

```
#define RELEASE(p)
   #ifdef SERVER /* server */
      lookup mapping table to get (d, p, p');
      send a RELEASE_DATA message with (d, p');
      delete registration table with (d, p, *);
      delete mapping table with (d, p, p');
   #else
      send a RELEASE_DATA message with (*, p);
      receive a reply with d;
      delete registration table with (d, p, *);
   #endif

#define PROCESS_RELEASE_DATA(d, pc)
   #ifdef SERVER /* server */
      lookup mapping table to get (d, ps, pc);
      send a reply with d;
      free memory at address ps;
      delete registration table with (d, ps, *);
      delete mapping table with (d, ps, pc);
   #else
      free memory at address pc;
      delete registration table with (d, pc, *);
   #endif
```

Fig. 16.9   Primitives for data release

primitives for data release. The *RELEASE_DATA* message sent by the client requires a reply from the server in order for the client to determine which entry in the registration table should be deleted.

The Data Messages.   We use two methods for data transfer, namely the *push* method and the *pull* method. The push method sends a piece of modified data to the opposite host. The pull method lets the intended receiver make a request for the modified data from the opposite host. Which method to use depends on the task mapping and the data transfer point. Recall that at any moment, only one host computes. At a data transfer point, if the data sender is computing, then the push method is used. If the data receiver is computing, then the pull method must be used.

Two kinds of data messages are used for data transfer. The *PUSH_DATA* messages are used for sending data with the push method and the *PULL_DATA* messages are used for receiving data with the pull method. Figure 16.10 shows the primitives to send data with the push method and Figure 16.11 shows the primitives to receive data with the pull method. The server version and the client version are different because data representation is translated on the server only.

For a piece of data which is represented by d, there are two ways to get the data address p and the data size s. If d is a function or a global variable, the (p, s) can be determined at compile time. Otherwise, we find (p, s) in the registration table at run time using d as the index. We should note that we may get a list of (p, s) corresponding to d. All items in the list must be transfered.

The Scheduling Messages.   The scheduling messages allow computational tasks to be dynamically scheduled at run time. Two kinds of scheduling messages are used for task scheduling. The *START_TASK* message starts a task on the opposite host. Upon receiving this message, the receiver starts the execution of the named task. Meanwhile, the sender blocks its current task and invokes the $wait\_for\_message$ function. The *END_TASK* message terminates a task. Upon its reception, the receiver exits the $wait\_for\_message$ function and resumes the previously blocked execution.

Figure 16.12 shows the primitives for task scheduling. Here we use the function address p instead of the abstract memory location ID d because we want to handle the function calls by pointers.

1.5   Code Generation

Using the program transformation primitives defined above, the program transformation takes the following steps.

```
#define PUSH(d)
   get (p, ds) of d;
   #ifdef SERVER
      lookup mapping table to get (d, p, p');
      translate data at address p to the client version dv;
      send a PUSH_DATA message with (p', ds, dv);
   #else
      get dv at address p;
      send a PUSH_DATA message with (p, ds, dv);
   #endif

#define PROCESS_PUSH_DATA(pc, ds, dv)
   #ifdef SERVER
      lookup mapping table to get (d, ps, pc);
      translate dv to the server version and
            write the result to address ps;
   #else
      write dv to address pc;
   #endif
```

Fig. 16.10   Primitives for the push method

```
#define PULL(d)
   get (p, ds) of d;
   #ifdef SERVER
      lookup mapping table to get (d, p, p');
      send PULL_DATA with (p', ds);
      receive a reply with dv;
      translate dv to the server version and and
            write the result to address p;
   #else
      send PULL_DATA with (p, ds);
      receive a reply with dv;
      write dv to address p;
   #endif

#define PROCESS_PULL_DATA(pc, ds)
   #ifdef SERVER
      lookup mapping table to get (d, ps, pc);
      translate data at address ps to the client version dv;
   #else
      get dv at address pc;
   #endif
   send a reply with dv;
```

Fig. 16.11   Primitives for the pull method

```
#define SCHEDULE(p, in)
    #ifdef SERVER
       lookup mapping table to get (d, p, p');
       translate in;
       send START_TASK with (d, *, in);
    #else
       send START_TASK with (*, p, in);
    #endif
    return wait_for_message ();

#define PROCESS_START_TASK(d, pc, in)
    #ifdef SERVER
       lookup mapping table to get (d, ps, pc);
       translate in;
       call function d with in and get result out;
       translate out;
    #else
       call function d with in and get result out;
    #endif
    send END_TASK with (d, out);

#define PROCESS_END_TASK(d, out)
    #ifdef SERVER
       translate out;
       return with out;
    else
       return with out;
    #endif
```

Fig. 16.12   Primitives for task scheduling

1. We let the start-up function of the server program issue a call to the function *wait_for_message*.

2. We let the start-up function of the client program make the run-time task scheduling decision and select a server to offload tasks based on current system status. The client then send a remote program execution call to launch the server program with an schedule ID. This ID will be used by both the server and the client to follow that specific task schedule. Before the client program terminates, it sends a *END_TASK* message to terminate the server program.

3. In both the server program and the client program, at each call site of a procedure mapped to the opposite host, we replace the function call by applying the *SCHEDULE* primitive to that function.

4. In both the server program and the client program, after each run-time (heap or stack) allocation (or release) of data on the heap or the stack that are accessed by both hosts, we insert an *ALLOC* (or a *RELEASE*) primitive.

5. In both the server program and the client program, at each data transfer point, we insert a *PUSH* or a *PULL* primitive according to the data transfer direction.

We assume that the client program and the server program are compiled separately and pre-installed on the client and the server. If we wish to be able to use a server which does not have the pre-installed server program, we can let the client carry server versions of the pre-compiled server program. At run time, we may transmit the program to a dynamically acquired server after proper authentication. Standard mobile codes, such as Java bytecode, make such run-time binding much easier. The techniques presented in this chapter can be extended for Java bytecode execution.

We use the program shown in Figure 16.13 as an example. For simplicity, we do not show the code that makes run-time scheduling decisions. We assume that the task schedule is already fixed such that the function *main* and *print* are mapped to the client, and the function *foo* is mapped to the server. Furthermore, the abstract memory location *A6* is shared by both hosts and it is transfered from the server to the client before function *print* is called. The resulting distributed program is as shown in Figure 16.13.

Figure 16.14 illustrates the event trace of the distributed program execution. The dashed lines between the client and the server represent message passing. The dotted lines indicate certain repeated patterns which are omitted.

1.6   Task-Assignment Analysis

We now discuss the mathematical framework for making the task-assignment decisions. In such a framework, we construct a constraint system in which mapping decisions of individual tasks, data accesses and data validity states are represented simultaneously in the system formulation. Besides the computation cost and data communication cost, the cost analysis also considers the cost for the run-time bookkeeping which is necessary for correct distributed execution.

Under such a constraint system, a polynomial time algorithm is available to find the optimal program partition for the given program input data [12].

Program Abstraction.   We build a directed control-flow graph $CFG = (V, E)$ for the program such that each vertex $v \in V$ is a basic block and each edge $e = (v_i, v_j) \in E$ represents the fact that v_j may be executed immediately after v_i.

```
 1:  int main() {
 2:      #ifdef SERVER
 3:          wait_for_message();
 4:      #else
 5:          launch server program;
 6:          SCHEDULE(foo, NULL);
 7:          send END_TASK message;
 8:      #endif
 9:  }
10:  void print(struct node *p) {
11:      struct node next*;
12:
13:      while(p) {
14:          next = p->next;
15:          printf("%d\n", p->data);
16:          free(p);
17:          RELEASE(p);
18:          p = next;
19:      }
20:  }
```

Fig. 16.13 An example program

```
21:  void foo() {
22:      struct node *head, *pnode;
23:      int i;
24:
25:      head = NULL;
26:      for( i = 0; i < 3; i++ ) {
27:          pnode = ( struct node * )
28:              malloc( sizeof
29:              ( struct node ) );
30:          ALLOC(A6, pnode, 8);
31:          pnode->next = head;
32:          pnode->data = i;
33:          head = pnode;
34:      }
35:
36:      PUSH(A6);
37:      SCHEDULE(print, head);
38:  }
```

Fig. 16.14 The framework of the generated distributed program

We abstract all the memory accessed (including code and data) by program at run time by a finite set of *typed abstract memory locations*. The abstraction of run-time memory is a common approach used by pointer analysis techniques [1, 6] to get conservative but safe point-to relations. The type information is needed to maintain the correct data endians and data addresses during data transfer between the server and the handheld device. Each memory address is represented by a unique abstract memory location, although an abstract memory location may represent multiple memory addresses because a single reference in the program may cause multiple memory references at run time. We use D to denote the set of all abstract memory locations d.

With point-to analysis, we conservatively identify the references to abstract memory locations at each program point. For example, to get the value of $*x$, the program reads x as well as the abstract memory locations which x points to. To write into $*x$, the program reads x and writes all the abstract memory locations which x points to.

The Constraint System.   Dynamically, each basic block v and each flow edge e can have many execution instances. We define $f(v_i, v_j)$ as the execution count for the flow edge $e = (v_i, v_j)$ and $g(v)$ as the execution count for basic block v. The values of f and g may vary in different runs of the same program when using different input data. However, the following constraint always holds.

Constraint 1: For any basic block $v_i \in V$,

$$g(v_i) = \sum_{e=(v_i,v_j)\in E} f(v_i, v_j) = \sum_{e=(v_j,v_i)\in E} f(v_j, v_i)$$

Since each task will be mapped either to the server or to the client, but not both, the task mapping can be represented by a boolean function M such that $M(v)$ indicates whether basic block v is mapped to server. We define the M function for basic blocks because the computation of a basic block has little variance. All the instructions in a basic block always execute together.

By program semantics, the mapping of certain basic blocks may be fixed. For example, in interactive applications, basic blocks containing certain I/O functions are required to execute on the handheld device. Moreover, to partition the computation at a level higher than a basic block, certain basic blocks are required to be mapped together. For example, for function-level partitioning, all the basic blocks in one function are required to be mapped together. The following constraint reflects these requirements.

Constraint 2: If basic block v is required to execute on the handheld, then $M(v) = 0$. If basic block v is required to execute on the server, then $M(v) = 1$.

If basic block v_1 and basic block v_2 are required to be mapped together, then $M(v_1) \iff M(v_2)$.

We use two boolean variables to represent data access information for abstract memory locations d such that $N_s(d)$ indicate whether d is accessed on server and $N_c(d)$ indicate whether d is *not* accessed on handheld. By definition, we have:

Constraint 3: If d is accessed within basic block v then $M(v) \Rightarrow N_s(d)$ and $N_c(d) \Rightarrow M(v)$.

Distributed share memory (DSM) systems [11] keep track of the run-time data validity states to determine the data transfers at run-time. We analyze data validity states statically for the abstract memory locations to determine the data transfers statically. Due to constraint 1, it is obvious that inserting data transfers on control flow edges is always no worse, sometimes better, than inserting them in basic blocks. We consider the validity states of abstract memory location d before the entry and after the exit of each basic block v such that: $V_{si}(v, d)$ indicates whether the copy of d on server is valid before the entry of v. $V_{so}(v, d)$ indicates whether the copy of d on server is valid after the exit of v. $V_{ci}(v, d)$ indicates whether the copy of d on client is *not* valid before the entry of v. $V_{co}(v, d)$ indicates whether the copy of d on client is *not* valid after the exit of v.

For data consistency, the local copy of d must be valid before any read operation on d. After a write operation on d, the copy of d on the current host becomes valid and the copy of d on the opposite host becomes invalid. If there is no write operation on d within a basic block v, then the local copy of d is valid after the exit of basic block v only if it is valid before the entry of v. In cases where d is possibly or partially written in a basic block, we conservatively require d to be valid before the write. Otherwise, d may be inconsistent after the write. The following constraint is introduced for data consistency.

Constraint 4: If basic block v has a (possibly or definitely) upward exposed read of d, then $M(v) \Rightarrow V_{si}(v, d)$ and $V_{ci}(v, d) \Rightarrow M(v)$. If d is (possibly or definitely) written in basic block v, then $V_{so}(v, d) \iff M(v)$ and $M(v) \iff V_{co}(v, d)$. If d is definitely not written in basic block v, then $V_{so}(v, d) \Rightarrow V_{si}(v, d)$ and $V_{ci}(v, d) \Rightarrow V_{co}(v, d)$. If d is possibly or partially written in basic block v, then $M(v) \Rightarrow V_{si}(v, d)$ and $V_{ci}(v, d) \Rightarrow M(v)$.

Cost Analysis.   There are four kinds of costs in our computation offloading scheme: computation cost for task execution, scheduling cost for task scheduling, bookkeeping cost for run-time bookkeeping, and communication cost for data transfer.

If we associate a computation cost $c_c(v)$ with each execution instance of basic block v running on the client, and a computation cost $c_s(v)$ with each execution instance of basic block v running on server, we get the total computation cost:

$$\sum_{v \in V} M(v) c_s(v) g(v) + \neg M(v) c_c(v) g(v) \qquad (16.1)$$

If we associate scheduling cost c_r with each instance of task scheduling from the client to the server, and a scheduling cost c_l with each instance of task scheduling from the server to the client, we get the total scheduling cost:

$$\sum_{(v_i, v_j) \in E} \neg M(v_i) M(v_j) c_r f(v_i, v_j) + \neg M(v_j) M(v_i) c_l f(v_i, v_j) \qquad (16.2)$$

Our program transformation discussed later performs run-time bookkeeping for allocations and releases of data *accessed by both hosts*. We assume each data release corresponds to a previous data allocation, so we include the release cost in the allocation cost. If we associate a bookkeeping cost c_a with each instance of data allocation, and let $A(v)$ denote the set of abstract memory locations d that are allocated in basic block v, we get the total bookkeeping cost:

$$\sum_{d \in D, d \in A(v)} \neg N_c(d) N_s(d) c_a g(v) \qquad (16.3)$$

We can derive the data transfer information from data validity states. On each edge (v_i, v_j), if $V_{so}(v_i, d) = 0$ and $V_{si}(v_j, d) = 1$, then the copy of d on the server is invalid after the exit of v_i but becomes valid before the entry of v_j. So there is a data transfer of d from the client to the server on edge (v_i, v_j). Similarly, if $V_{co}(v_i, d) = 1$ and $V_{ci}(v_j, d) = 0$, then there is a data transfer of d from the server to the client on edge (v_i, v_j). If we associate a communication cost $c_d(d)$ with each instance of data transfer of d from the server to the client, and a communication cost $c_u(d)$ with each instance of data transfer of d from the client to the server, then the total communication cost is:

$$\sum_{(v_i, v_j) \in E} \neg V_{so}(v_i, d) V_{si}(v_j, d) c_u(d) f(v_i, v_j)$$

$$+ \quad \neg V_{ci}(v_j, d) V_{co}(v_i, d) c_d(d) f(v_i, v_j) \qquad (16.4)$$

Partition Algorithm. For the given program input data, edge profiling techniques [2] can easily get f and g in formulas (1) – (4). The optimal program partitioning problem can then be expressed as:

Problem 1: Finding boolean values for M, N_s, N_c, V_{si}, V_{so}, V_{ci} and V_{co} subject to constraints 2 – 4 and minimize the sum of total cost (1) – (4).

Problem 1 can be reduced to a single-source single-sink **min-cut network flow problem** [3] which can be solved in polynomial time. It is possible that the optimal program partitions vary with different program inputs. Problem 1 can be treated as a parametric problem with parameters f and g that satisfy constraint 1. However, the parametric problem 1 can be reduced from a **2-path problem** [3] and is hence NP-hard. We omit the proofs of these claims due to the space limit.

We use an option-clustering heuristic for the parametric problem 1. Our heuristic groups a training set of options into a relatively small number of clusters and prepares one partition for each cluster such that, for any option in the training set, the cost difference between the prepared partition and its optimal partition is within a given error-tolerance ratio. For a program with r independent options, all the possible program options form an r-dimensional option space. We divide the whole option space into subspaces according to the clustering of the training set. At run time, we check to see which subspace the option belongs to, and the program will run based on the corresponding partition. We omit the details in the paper due to the space limit.

1.7 Experiments

In order to show the benefit of the computation offloading scheme presented above, experimental data is given below. The handheld device used to collect the experimental data is an HP IPAQ 3970 Pocket PC which has a 400 MHZ Intel XScale processor. The Server is a P4 2GHz Dell Precision 340 machine. We run Linux on both machines. The wireless connection is through a Lucent Orinoco (WaveLan) Golden 11Mbps PCMCIA card inserted into a PCMCIA expansion pack for the IPAQ. Besides the program execution time, we also measure the program energy consumption. We connect an HP 3459A high precision digital multimeter to measure the current drawn by the handheld device during program execution. In order to get a reliable and accurate reading, we disconnect the batteries from both IPAQ and the extension pack and we use an external 5V DC power supply instead.

We implement our computation offloading scheme in GCC. A pointer analysis similar to [1] is used to get the point-to information. Such information is then used to identify references to the abstract memory locations. We partition the program at function level and restrict all the I/O functions to execute on the handheld during the computation offloading.

The Cost Parameters.   In our experiments, we model the cost as the program execution time during task partitioning. We estimate the execution time of each instruction by averaging over repeated execution of that instruction. We then get the cost c_s and c_c for each basic block by adding the execution time of the instructions in the same basic block. For small data size, c_d and c_u are simply the measured network latency time. For large data transfer, we also consider the network transfer time which can be calculated with the data size and network bandwidth. We obtain other cost c_a, c_d and c_u by physical measurement. Each cost item is averaged over a large number of synthesized workloads. In our experiments, with our cost model, the difference of the whole program execution time between the measured result and estimated cost is in range of 10%–30%. We will consider more accurate cost models [4, 7] in our future work.

Computation Offloading Result.   Figure 16.15 shows the performance for several programs. We compare the results between two versions for each program. One version is the original program running completely on the handheld device. The other version is partitioned between the handheld device and the server which is obtained by applying our partition algorithm for that particular execution. To generate the machine code (for both the server and the handheld), all the programs, including the transformed ones, are compiled using the GCC compiler with the $-O2$ optimization level. Figure 16.16 shows the measured results of energy consumption which we can see, are proportional to the performance.

We should note that not all the options for these programs can get benefits from computation offloading. Here we only show the results for a subset of options that can benefit from computation offloading. In each figure, we append the program names by the execution option and by the number of repeated execution. The program SUSAN performs photo processing. We run this program using the option $-s$ (smoothing) $-e$ (recognizing edges). The program RASTA performs speech recognition. It can generate results in two formats: binary data (default) and ASCII data (option $-A$). The program EPIC compresses graphics. Program gnuplot runs interactively with a command interface, and the options are specified for various commands. We generate three-dimensional figures with the *splot* command, and *x*x−y*y* is the plotted function. The option *eps* means generating an eps file, and *X11* means generating X11 events for display. For figures with the same options, we use different input files.

It is difficult to measure the time spent on bookkeeping. We count the operations instead. Figure 16.17 shows the estimated bookkeeping overhead ratio for the test programs. This figure only shows the results for execution time. The results for energy consumption are similar. The average run-time bookkeeping overhead is about 13%.

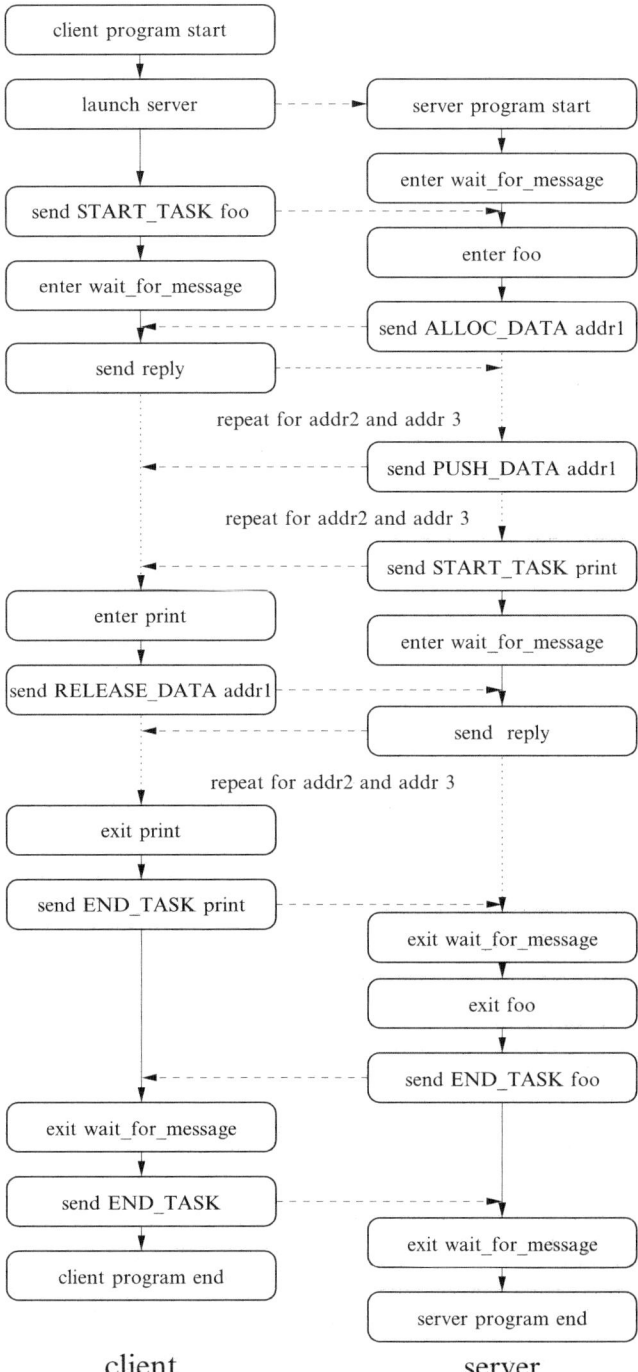

Fig. 16.15 An event trace of distributed program execution

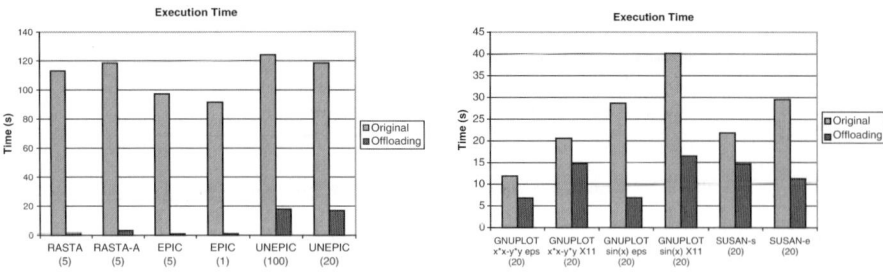

Fig. 16.16   Performance

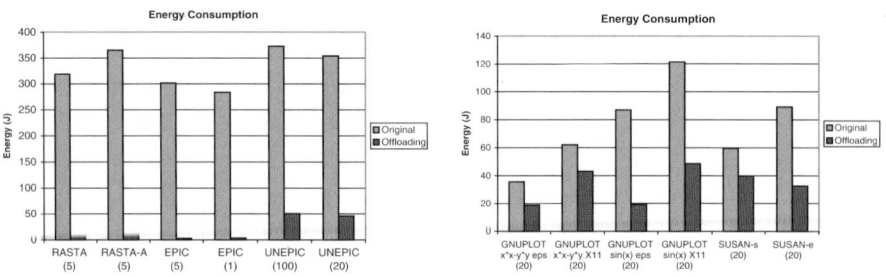

Fig. 16.17   Energy consumption

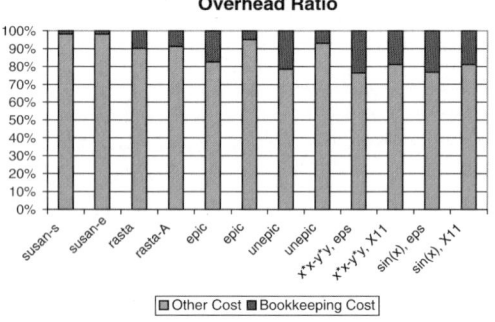

Fig. 16.18   Bookkeeping overhead

We should note that optimal program partitions vary with different program inputs. The optimal partition for one option may slow down the program execution for other inputs. Figure 16.18a shows the program speedup for different inputs of SUSAN using the program partition got by the option *-s*. With our option-clustering heuristic, we can group the program options into two clusters and prepare one partition for each cluster. The resulting program speedup is shown in Figure 16.18b.

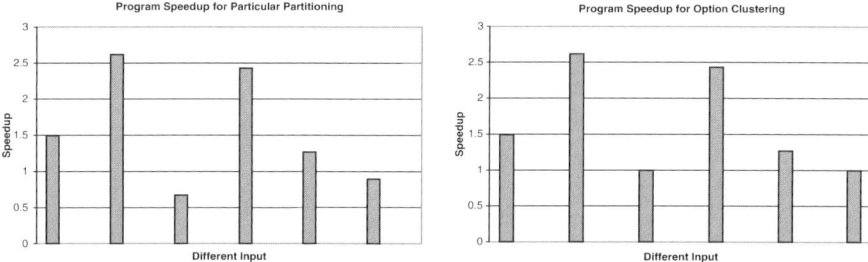

Fig. 16.19   Experimental result for option clustering

1.8   Related Work

Li, Wang and Xu [8, 9] propose a partition scheme for computation offloading which results in significant energy saving for half of the multimeter benchmark programs. The scheme is based on profiling information about computation time of procedures and inter-procedural data communications and it minimizes the cost and guarantees correct communication within a specific execution context only. Their work does not study the important issue of how to guarantee the correct distributed execution under all possible execution contexts. Neither does it consider the different partitions for different program inputs.

Kremer, Hicks and Rehg [5] introduce a compilation framework for power management on handheld computing devices through remote task execution. Their paper does not address details about the data consistency issue. They only consider the procedure calls in the main routine as the candidates for remote execution and they evaluate the profitability of remote execution for each individual task separately. For tasks that share common data, evaluating the profitability of remote execution of each individual task separately may result in the overcount of data communication cost. Using their approach, due to the overcount of data communication cost, only 7 out of the 12 testing programs in the previous section can get benefits from computation offloading.

References

[1] L. Andersen. Program analysis and specialization for the C programming language. *PhD thesis, DIKU, University of Copenhagen*, 1994.

[2] T. Ball and J. R. Larus. Optimally profiling and tracing programs. *ACM Transactions on Programming Languages and Systems*, 1992.

[3] J. Bang-Jensen and G. Gutin. Digraphs: Theory, algorithms, and applications. *Springer-Verlag, London*, 2001.

[4] J. Engblom and A. Ermedahl. Modeling complex flows for worst-case execution time analysis. *Proc. of RTSS'00, 21st IEEE Real-Time Systems Symposium*, 2000.

[5] U. Kremer, J. Hicks, and J. M. Rehg. A compilation framework for power and energy management on mobile computers. *14th International Workshop on Parallel Computing (LCPC'01)*, August 2001.

[6] R. P. Wilson and M. S. Lam. Efficient context-sensitive pointer analysis for C programs. *In Proceedings of the ACM SIGPLAN '95 Conference on Programming Language Design and Implementation*, June 1995.

[7] Y.-T. S. Li, S. Malik, and A. Wolfe. Efficient microarchitecture modeling and path analysis for real-time software. *IEEE Real-Time Systems Symposium*, 1996.

[8] Z. Li, C. Wang, and R. Xu. Computation offloading to save energy on handheld devices: A partition scheme. *Proc. of International Conference on Compilers, Architectures and Synthesis for Embedded Systems (CASES)*, pages 238–246, 2001.

[9] Z. Li, C. Wang, and R. Xu. Task allocation for distributed multimedia processing on wirelessly networked handheld devices. *Proc. of 16th International Parallel and Distributed Processing Symposium (IPDPS)*, 2002.

[10] A. Rudenko, P. Reiher, G. J. Popek, and G. H. Kuenning. Saving portable computer battery power through remote process execution. *Mobile Computing and Communications Review*, 2(1):19–26, January 1998.

[11] C. Amza, A. Cox, S. Dwarkadas, P. Keleher, H. Lu, R. Rajamony, W. Yu, and W. Zwaenepoel. Treadmarks: Shared memory computing on networks of workstations. *IEEE Computer*, 29(2):18–28, 1996.

[12] C. Wang and Z. Li. Parametric analysis for adaptive computation offloading. *Proc. of ACM SIGPLAN Symposium on Programming Languages Design and Implementation (PLDI)*, June 2004.

V

Multi-Processors

Chapter 17

A Power and Energy Perspective on MultiProcessors

Grant Martin
Tensilica, Inc.
Santa Clara
California
USA

Abstract   In the past few years, we have seen the rise of multiprocessor and multicore approaches to system-on-chip and processor design, driven by performance, power dissipation, and energy consumption motivations. In fact, there is considerable confusion over the various architectural choices for multiprocessor systems, and the primary design methods that are appropriate to each choice. The techniques for processor design and power and energy reduction that have been discussed in previous chapters of this book are often orthogonal to the basic architectural choices in multiprocessor systems, so that these techniques can be combined together in designing multiprocessor or multicore SoCs. In this chapter, we will review the main approaches to multiprocessor architecture and their impact on power and energy. We will examine the leading approaches for low-power design and their relationship to multiprocessor based design. In particular, we will review the approach to heterogeneous, asymmetric multiprocessor systems based on networks of configurable and extensible processors and the profound impact on power and energy reduction offered by this approach. Finally, we will point to two key methods discussed in more detail in subsequent chapters: design space exploration based on abstract power-performance modeling and network-on-chip communications architectures.

Keywords: multiprocessor systems; multicore; symmetric multiprocessing; asymmetric multiprocessing; homogenous multiprocessors; heterogeneous multiprocessors; interconnect architecture; buses, memory hierarchy, network-on-chip.

1.   Introduction

Recently there has been a tremendous flurry of interest in the embedded systems and electronic product design worlds in what has been variously called "multicore" and "multiprocessor" design (Gibbs, 2004). One can rarely go a

J. Henkel and S. Parameswaran (eds.), Designing Embedded Processors – A Low Power Perspective,
373–389.
© 2007 *Springer.*

week without reading in the electronics press a report on multicore design or multiprocessor system-on-chip (MPSoC) (Merritt, 2005, Maxfield, 2006). The active talk in the industry is all about programming models for multicore systems, energy and power reductions with multicore and multiple processors (Tremblay, 2005), and effective design and integration methods given the complexity of such products (Jerraya and Wolf, 2005). However, there has not yet emerged a clear consensus on the differences between different types of multicore and multiprocessor design, one that would make it easier for designers to choose the most appropriate multiprocessing approach for a particular design problem. Acronyms abound: SMP (symmetric multiprocessing), SMT (simultaneous multithreading), AMP (asymmetric multiprocessing), CMP (chip multiprocessing), CMT (chip multithreading), MPSoC, and multi-clustering, for just a few. The distinctions between many of these remain unclear. First, let us outline the variety of multiprocessing approaches and set them in the context of their impact on various design goals, including, but not limited to, reducing peak power and overall energy consumption.

1.1   Multicore and Multiprocessor Definitions

The following definitions are based on those in Wikipedia (en.wikipedia.org) as of 27 December, 2005. They have been modified and extended by the author to reflect the current practices in embedded SoCs.

Multicore Processor.   Processor combining two or more independent processors into a single package, usually a single integrated circuit. Hence "dual core" contains two independent processors, "quad-core" four processors, and so on. Multicore processors often support thread-level parallelism (TLP) without including multiple microprocessors in separate physical packages. Wikipedia calls this form of TLP, "chip-level multiprocessing" or CMP, although the focus on TLP makes this definition unnecessarily restrictive.

Examples of dual core processors include offerings from IBM based on the Power processor, AMD and Intel. SUN's "Niagara" multicore processor may include up to eight cores and may be packaged up in two, four, and eight core offerings. Each core may have up to four threads for a total of 32 threads in a maximal Niagara.

Symmetric Multiprocessing or SMP.   Multiprocessor computer architecture where two or more identical processors are connected to a single shared main memory. Many common multiprocessor systems today use an SMP architecture. SMP systems also often share second and third level cache, and interconnect. Having a large coherent memory space is important to allow flexible task or thread assignment and re-assignment to processors.

SMP systems allow any processor to work on any task no matter where the data for that task is located in memory; with proper operating system support, SMP systems can easily move tasks between processors to balance the work load efficiently. On the downside, if memory is much slower than the processors accessing them, then processors tend to spend a considerable amount of time waiting for data to arrive from memory. Thus a unified coherent memory hierarchy including caches is essential to performance to avoid multiple processors being starved. In addition, fast hardware-based context switching may be extremely important to allow threads to be run if their data is available as soon as another thread stalls.

The multicore systems discussed previously often are SMP systems. In addition, large scientific computing with hundreds, thousands, or tens of thousands of processors, such as the IBM "BlueXXX" machines, which are intended for large scale weather, earth, or nuclear explosion simulations, are usually large SMP machines with extremely fast IO subsystems and large coherent memory subsystems. On these systems, the applications change faster than the machines, and the applications often exhibit a large amount of data-wise or thread-wise concurrency which can be exposed in a general way to take advantage of an SMP architecture.

One example of an embedded SMP system is the ARM-NEC 4xARM MPCore system announced in 2005.

Simultaneous Multithreading or SMT. A technique for improving the overall efficiency of the CPU. SMT permits multiple independent threads of execution to better utilize the resources provided by modern processor architectures. SMT is often best mapped into an SMP system which allows threads to be mapped to any processor and rely on the large coherent memory space to be flexibly mapped and run. HW support for thread context switching often allows threads stalled on memory access to be replaced in one or at most a few cycles by a thread ready to run, and thus improve overall system efficiency. The key issue in SMT design is finding an appropriate and human-tractable programming model. Twenty to forty years of research into parallelism in programming have failed to turn up any good general purpose automatic methods to find concurrency in arbitrary applications. The best general methods are to provide Application Programming Interfaces (APIs) to allow programmers to expose thread-level concurrency manually. APIs such as OpenMP and Message Passing Interface (MPI) are examples.

Asymmetric Multiprocessing or AMP. In Asymmetric multiprocessing, multiple processors with different characteristics are combined into a chip multiprocessing system. There are several types of AMP architecture possible.

A good reference for one type of AMP is Kumar et al. (2005). In this model, the cores have a single instruction set architecture (ISA), but are still heterogeneous in terms of resources, performance, and power consumption. This allows the right application to be mapped into the right core to meet realtime latency and throughput requirements while minimizing power consumption. This is contrasted with multi-ISA multicore architectures, such as the IBM/Toshiba/Sony Cell processor which combines control processing using a POWER core with data-intensive processing for video and other streaming data applications via multiple processing elements. Here the different ISA processors are used for different parts of the application.

Perhaps the most interesting type of AMP architecture is one composed of multiple processors with different ISAs in which the ISAs have been extended for particular applications or portions of applications via instruction set extension, and the cores configured for the coarse-grained characteristics of the application set (such as including DSP features, SIMD instructions, MAC and/or multiplier units, and a variety of local and system memories). The Tensilica Xtensa LX processor (Rowen and Leibson, 2004) is a good example of a configurable and extensible processor core that can be tailored to applications and portions of applications using this strategy.

AMP is an especially good architectural strategy for many embedded applications that have intensive dataflow-style processing, and whose characteristics and indeed source code are known or well-characterized in advance. This includes many signal and image processing applications, for example in the audio and video domain.

1.2   Power/Energy Drivers for Multiprocessor and Multicore Architectures

So what are the key driving factors behind the move to multiprocessor or multicore architectures in SoC or integrated processors? How do they relate to power consumption, heat dissipation, and overall energy consumption? We can identify three key trends in applications:

1. Performance drives the first group of applications. While performance is the driver, avoiding unnecessary, or indeed, infeasible heat (localized power dissipation) on the device has become a critical design issue. This is especially important for high end servers and very high end applications. Both getting rid of excessive heat (or avoiding it) and the overall energy consumption impose a severe cost constraint – the first in the form of expensive packaging and cooling costs; the second in the power requirements of large server farms or very large multiprocessing configurations.

2. The second group is driven primarily by energy cost – power consumption integrated over time. This is especially important for portable battery-operated devices with increasing applications load, significant media processing, and the very slow rate of improvement in battery energy storage over the last decade or more. At the same time, it is necessary to ensure adequate performance for the applications, which is helped by media applications requiring distinctive plateaus of performance for specific resolutions and compression factors. This is a secondary criterion.

3. The third group is driven by a desire to increase performance of desktop or consumer applications by recognizing some level of inherent application concurrency, while keeping maximum performance requirements bounded and while being powered through the mains. Lower energy consumption is useful but a secondary criterion, and may come as a byproduct of optimizing application performance.

How do multicore or multiprocessor architectures reduce peak power and overall energy consumption? First and foremost, they exploit the nonlinearities involved in power and energy consumption vs. frequency and voltage. Since active power has a linear relationship with frequency, at first glance, the idea of replacing a single processor running at frequency F with two processors running at $F/2$ seems to get us nowhere, especially if the two cores require access to the same data and complex cache coherency and memory hierarchy structures, consuming additional power, are required to let both processors run efficiently. However, running at a higher frequency in general requires higher operating voltages (or lowering the frequency may allow the voltage to be lowered), and since active power has a nonlinear relationship with voltage (voltage is squared), then two processors running at frequency $F/2$, operating at a voltage less than is required to power the single processor, may represent substantial reductions in peak power and overall energy consumption indeed. Voltage may be reduced statically at design time, or through runtime based dynamic techniques, as described later.

Merely reducing frequency and voltage is only one dimension of reducing power and energy through multicores. The second dimension is that of making more efficient use of the computing resources in a world where memory access times have been reduced at a rate far less than the increase in processor speeds. This has resulted in a huge imbalance between the number of cycles required to access main memory in a device and the amount of data and instructions required to keep the processor busy. Complex memory hierarchies starting with caches have helped in this regard. Multithreading, especially with hardware support that allows fast context switching between stalled threads and those ready to run, also help. But given relative application or thread independence, or data-wise parallelism, it is often much easier to keep two, four, or more cores busy on their local applications or portions thereof, and thus maximize the use of the processing resources and the amount of work done per each portion of energy consumed, than to do so with a single processor architecture. This is especially so in applications where AMP is the natural computational model.

A final dimension lies in further exploiting AMP for those applications best suited to it. Experience with configurable, extensible processors (Wei and Rowen, 2005) have shown that instruction extension, which maximizes application work done per instruction executed, is highly correlated with reducing the amount of energy consumed per useful application computation. By tailoring instruction function, data size, and concurrent operations to the application, the number of instructions executed per application computation is greatly reduced – sometimes for the tight loop kernels in the application, by one to two orders of magnitude. Since fewer instructions are executed, the amount of instruction fetching is greatly reduced. Since each instruction is highly tuned to the application, the amount of wasted computation is greatly reduced (e.g. if the data items have 17 bits of precision, multiplying two 32-bit numbers when only a 17×17 bit multiplication is needed consumes a lot of extra energy). By executing multiple operations in parallel, more is done per unit of energy. This has a profound impact on energy and power. Indeed, the potential energy savings are sufficient justification to exploit a very high degree of AMP in many wireless, portable, battery-powered embedded consumer applications.

1.3   Classifying Multiprocessor Architectures

Table 17.1 illustrates a number of multiprocessor architectures and relates them to the different applications discussed above, and the driving power and energy factors.

Table 17.1   Multiprocessor approaches and impact on power and energy

Applications	Approach	Architecture	Performance	Heat	Energy
Server, general	SMP	Multicore	Maximize	First	Second
Embedded control	SMP–SMT	Multicore	Improve	Second	Important
Embedded dataflow	AMP	Multiprocessor	Maximize	Second	Highest
Desktop and graphics	AMP	Multiprocessor	Maximize	Second	Second

As discussed, examples of the first class in the table include IBM Power PC dual core, AMD dual core, Intel dual core, and SUN Niagara. Examples of the second class include the ARM-NEC 4X ARM SMT MPCore. Examples of the third class include the classical wireless baseband processor – RISC plus DSP; now complemented by media processors. Examples of the fourth class include the Cell processor for graphics.

As well as the focus on multiprocessors themselves, the heterogeneous AMP approaches also include what are called SoC "platforms" dedicated to specific application spaces. These include devices such as TI's OMAP family of chips for wireless and mobile applications, including baseband processors, media processors, and application chips; ST's Nomadik chipsets for the same domains; and Philips Nexperia platforms for both wireless and digital multimedia applications (e.g. set-top boxes, high end TVs).

2. A Survey of Multiprocessor Approaches for Low-Power, Low-Energy Design

2.1 Basic Techniques

As discussed by Irwin et al. (2005), there are four basic categories of techniques for designing low-power, low-energy MPSoCs. The following is a precis of the techniques as discussed in Irwin et al. (2005), along with some additional techniques. Mudge (2000) gives a good overview as well.

Energy-Aware Processor Design. The main techniques used here include

1. Reducing active energy: This is first achieved by the reduction of supply voltage to meet the requirements of the application space (Mudge, 2000). In addition, different voltages can be used for different processors, depending on their role in the design (this makes more sense in an AMP approach, although the single-ISA, multicore approach of Kumar et al. (2005) uses the same underlying ISA but with different implementations that could run at different operating frequencies and use multiple voltages). At runtime, the use of dynamic voltage and frequency scaling (DVFS) (Choi et al., 2004) gives more flexibility and a chance to tune the dynamic operations of the various processors to the immediate application needs for computation. Some have proposed continuous DVFS schemes, others a fine-grained or stepped DVFS approach. Indeed, it has been suggested that only two choices of voltage/frequency for each processor are sufficient to allow most of the gain from this technique to be realized: a "fast" and "slow" mode. Of course, coupled with all of this is the possibility of putting the processors to sleep – a powered-down or powered-off mode – for periods when the processing is not needed at all – thus reducing standby energy.

2. Reducing standby energy: Or leakage power, which gets worse as we continue IC process scaling to 90 nm, 65 nm, and beyond. Here there are several ideas extrapolated from general and single-processor design (Blaauw et al., 2002). This includes increasing threshold voltage, which reduces threshold leakage current by a large amount. This also increases gate delay. The idea of doing this dynamically at runtime has also been mooted, as well as having several multiple threshold voltages, similar to the multiple supply voltage idea. Finally, sending processors to sleep will reduce standby energy consumption tremendously – or they can be powered off as need be.

3. Configuring and extending instruction set: This will be discussed in further detail later.

Energy-Aware Memory System Design.   Memory subsystems are a large consumer of energy, especially in processor subsystems, and MPSoC both increases the problem and gives a number of possible architectural solutions to it.

1. Reducing active energy: Here there are many ideas from single processor design that can be applied to multiprocessors, centered around reducing cache energy consumption: partitioning, hierarchy, prediction, filtering, and compression (Kim et al., 2003). Of course software optimizations which reduce memory accesses, as discussed earlier, are also important.

2. Reducing standby energy: Leakage is of course a big problem in memory design (Kim et al., 2003). Shutting down all or portions of caches and memories when not needed is possible, although since this loses state, re-awakening will have energy and performance costs in resetting state. Some have suggested retaining state while in a reduced energy mode by using a small state-preserving supply voltage (Flautner et al., 2002). Optimizing cache and memory usage both statically at design time and dynamically at run time by monitoring usage patterns, is an important area for optimization.

3. Cache and memory architecture: Architectural choices – from single multi-ported, high energy consuming shared cache for all processors to a variety of private to semi-private (locally shared) caches and memories – can have a big impact on memory consumption. Here use of AMP techniques as opposed to SMP techniques can allow specific processor local memories and caches to be optimized to the very specific needs of the applications they are meant for. An important adjunct to deciding on optimal MP architectures in this context is the ability to do design space exploration and measure the cost, power, and area impact of a variety of such architectural choices (Rowen and Leibson, 2004).

4. Reducing snoop energy: In SMP architectures, snooping by cache controllers is used to ensure cache data coherence. There are a number of detailed microarchitectural techniques used to optimize snoop energy consumption (Ekman et al., 2002), including using dedicated tag arrays, and serialization of tag and data array accesses, as well as special structures and different schemes to order the snooping activities.

Energy-Aware Communication System Design.   The classical on-chip communications architecture is an on-chip bus, such as ARM AMBA AHB/APB, IBM CoreConnect, or SONICS OCP-based bus, or a hierarchy of buses bridged by special logic blocks. Buses are used to access global resources, such as shared memories, to access peripheral blocks and HW accelerators. As large shared resources, buses consume more energy than dedicated communications resources (point to point) for a given communication need, but offer cost, design time, and flexibility advantages. Advanced communications structures such as network-on-chip, exploiting packet-switching concepts, hold some promise to either replace or supplement classical buses. Asynchronous design techniques may help reduce energy consumption in general, including that ascribed to on-chip buses. There are also other ideas:

1. Bus encoding: Rather than a full set of bus wiring, special encoding schemes have been devised that minimize switching activity and thus dynamic power consumption. The amount of power saving depends heavily on the characteristics of the data stream (its entropy rate) or address stream. Schemes for both more random data, and highly correlated data have been devised. (Certain communication patterns, such as sampling external sensor data, or the address stream of a processor, may be highly correlated from one data item to the next.) (Cheng and Pedram, 2002).

2. Low Swing Signaling: This scheme, that has been used off-chip as well, has been investigated on chip for potential savings in the face of noise and reliability concerns (Ferretti and Beerel (2001)).

3. Advanced interconnect architectures: Here network-on-chip (NoC) packet-switch architectures are being explored in order to determine an efficient, scalable architecture that will not run out of communications bandwidth as the complexity and number of IP blocks rises. Bus-based architectures, if flat, become seriously congested as complexity rises; avoiding this leads to complex hierarchical bus schemes. The NoC theory is that packet-switching approaches will scale much more gradually with complexity and utilize global wiring resources more efficiently. Different packetizing architectures and schemes will have an effect on overall energy consumption. In addition, reliability concerns for overall communication will also affect energy consumption if there are

many detection/correction/recovery/retransmission problems. Probably the most important thing to provide designers exploring such approaches are models and tools to allow the performance, power, and cost tradeoffs of different schemes and architectures to be explored (Vitkovski et al., 2004).

The other kind of communications architecture to explore is a greater use of private point to point communications structures, such as dedicated HW FIFOs between processors, in place of always using shared resources whether buses or networks. For high bandwidth data passing between well-known tasks on dedicated processors, using private HW queues may save considerable energy as compared to using shared buses, for example, making the increase in design area worthwhile. For this to be effective, the processor microarchitecture must provide advanced non-blocking access between the queues and the processor datapath, as for example is discussed in (Rowen and Leibson, 2004; Leibson and Kim, 2005).

Energy-Aware Software.   Energy-awareness has an important role in designing software tasks and the networks of tasks that make up applications. To the greatest extent possible, the software subsystem should be able to actively control the hardware resources so that resources not required can be put to sleep, powered down, or if needed at a lower performance level, controlled in execution speed and rate. An adaptive software-based energy control scheme, supported by HW controls over resources, can significantly reduce both standby and active energy consumption.

The other aspect is more subtle: determining the right computational model for the software and the right interplay with architecture. Again, the more that is known about the applications *a priori*, the more appropriate an AMP architecture is, as opposed to the general purpose SMP approach.

2.2   Formal Control of DVFS for CMP

In the research reported by Juang et al. (2005), ad-hoc control over multiprocessor SoC using dynamic voltage and frequency scaling (DVFS) is replaced by a more formalized energy management scheme that takes a full-chip perspective on the energy management problem. A first analysis of independent DVFS in CMP, using local information, reveals that local-only control leads to significant oscillations in processor activity and speed. The limitations of local DVFS control lead to the development of a coordinated DVFS control algorithm that is based on knowledge of the activity and workload over the whole set of CMP "tiles". This "gestalt" approach results in important improvements in the energy-delay products of parallelized applications.

2.3   Use of Transactional Memory in Multiprocessor Systems

In the research reported by Moreshet et al. (2005), the use of access to shared memory in multiprocessor systems is examined in light of energy consumption. Traditional mechanisms for such access are via locks, which provide mutual exclusion and atomic access. As the paper shows, locking mechanisms may result in many memory accesses to gain access to a shared variable, wasting considerable energy and performance in the process. Transactional memory systems, introduced by previous researchers, are speculative mechanisms that assume successful access to shared variables without conflict, proceed on that basis, and abort if conflict is detected before the transaction finishes. Such an approach can be implemented in hardware, software, or a mixture of both. The research develops a transactional memory solution that can be targeted for general multiprocessors, and using an appropriate simulation model, studies the energy consumption tradeoffs of locking vs. transaction memories, where the savings can be considerable (on the order of 80% for the specific mechanisms studied). This is a useful example of the kind of techniques that are appropriate for consideration in architecting MP systems for power.

3.   Asymmetric Multiprocessing

We next consider the energy and power advantages of AMP – asymmetric multiprocessing – in more detail, motivated by the key idea of using configurable and extensible processors as the basis for an AMP system.

3.1   Multiprocessor Systems of Configurable, Extensible Processors

Traditionally, designers have thought of hardware as being a lower power solution for many design functions than software running on a processor. This is due to thinking about processors as general purpose execution engines for general purpose software, implementing a fixed and static instruction set architecture (ISA) that is targeted by compilers typically used for non-real-time software. Focused, dedicated hardware is clearly more efficient than compiled general purpose software, and consumes less energy as well as offering higher performance, in this thinking.

However, this kind of thinking has been motivated by the thought that ISAs in general are fixed and immutable. Of course, in the past, it has not been easy to tailor processors for specific tasks, by either configuring them in coarse-grained ways (structurally), or in fine-grained ways (by adding specific new instructions to the datapath). This has led to the approach that to accelerate computation on a processor, the best way is to crank up the frequency of operation. As we have seen with uni-processors, increasing frequency as process technologies have scaled has led to ever-increasing problems with power density and energy consumption, and is not a reasonable approach for many embedded applications, especially those that are battery-powered. Furthermore, many embedded applications, such as media processing (audio and video), tend to require fixed rates of processing and move in steps from one plateau of processor requirements to the next as codec standards change and new codecs are introduced. This implies that if a processor could be customized to a specific codec's computational requirements, it may not be necessary to increase processing rate.

Other kinds of "traditional" thinking about processors include the fact that it is often extremely difficult to partition applications onto multiple processors, particularly in a general way. As discussed previously, decades of research that attempts to find ways to expose parallelism or concurrency in applications have in general failed to deliver reliable ways of doing this automatically or programmatically. Most successful parallelized applications have relied on manual methods – software or algorithm designers have had to expose the concurrency explicitly in the code through use of pragmas or API calls.

Interestingly, however, many media processing applications have a natural "pipeline" parallelism in which frames or packets can be processed in stages in a more streaming style. This kind of parallelism lends itself to dedicated processors to handle portions of the media processing task, and a dataflow style of application and algorithm development in which frames are processed in discrete pipeline stages. If these can be mapped to specific optimized processors, then we have an algorithm that can take maximum use of asymmetric multiprocessing and of configurable, extensible processing. The final comment on this is that such concurrency requires the ability to acquire and use sufficient multi-processors for the task. If restrictions are placed on the natural concurrency that can be exploited, because you do not have enough processors to handle all the tasks, or because they are forced to be more general-purpose than required, then it becomes much harder to multi-task the processors that you do have for real-time embedded applications.

We have briefly discussed the advantages of using configurable, extensible processors such as those discussed in Rowen and Leibson (2004) as the basis for building AMP subsystems oriented to specific tasks. When you have special instructions available for specific application code, the necessary computations can be carried out using fewer instructions, reducing instruction fetches and allowing you to finish earlier, and go to sleep, or run slower to meet a fixed rate, and thus lower frequency, or voltage, or both. With multi-operation instructions based on the application code, using data sized appropriately, less wasteful work is carried out to meet the application computational needs. Of course, processors dedicated to specific applications can go to sleep when not required.

And by tailoring structural or coarse-grained parameters to the tasks, such as amounts and kind of instruction and data memory, communications resources, and the presence or absence of special processing units, further improvements in peak power consumption and energy are possible.

In an AMP system, when the workload is changing or unpredictable, one can utilize strategies such as DVFS to match the processing capabilities to the required workload while minimizing energy consumption. When tasks runs with known throughput requirements, one is better off by picking the lowest frequency, and lowest concomitant voltage that is suitable, to get the job done just in time.

Experiments with the Artisan Metro 0.13 μm library and a configurable, extensible processor cores (Wei and Rowen, 2005) indicate that while frequency scaling from 240 MHz down to 40 MHz (suitable, due to processor instruction extension, for many applications) resulted in a 6X reduction in energy consumption, by also scaling the voltage down to a level adequate to support the frequency reduction, further huge gains are possible. In fact, by scaling voltage from the nominal 1.08 V at 240 MHz to an adequate 0.6 V at 40 MHz, a total 19X energy reduction, to accomplish the needed computations, is possible – an energy savings of 95%.

So this kind of AMP strategy is definitely not a multicore strategy as with the SMP approaches. It is definitely an "MP-SoC" approach, with multiple processors, tasks mapped to processors that have been optimized, exploiting natural application parallelism. These processors are run as slowly as possibly, as fast as necessary, without wasting work on general instructions where a highly optimized one is possible. Communications can utilize a variety of schemes including direct FIFO communications where that results in overall energy savings. And higher level controls over processors, including sleep and power-off modes and dynamic voltage and frequency scaling are all possible, to further reduce energy consumption to a very low level.

4.   Techniques Covered in Subsequent Chapters

The following chapters cover some specific MP and multicore related techniques and the impact on architectures and design methodologies by bringing in considerations of power and energy to complement the more traditional focus on performance and cost.

4.1   Power-Performance Modeling and Design for Heterogeneous Multiprocessors

In this follow-on chapter, JoAnn Paul and Brett Meyer discuss the issues of modeling heterogeneous multiprocessor systems to facilitate design space exploration in order to optimize the performance and power consumption of such systems for a variety of applications. However, performance itself is a multiply measured concept – single numbers for latency, throughput, or average execution time are insuffient to evaluate the suitability of a particular solution in the design space for a given set of applications.

Paul and Meyer then describe their modeling environment known as MESH that enables performance simulation of single-chip heterogeneous multiprocessor systems at a more abstract thread level, rather than at the instruction level that is more typical of microarchitectural simulators. MESH provides a set of modeling elements: processing elements, application threads, and scheduling and arbitration capabilities to define the co-ordination and co-operation between processing elements. Using this basic framework, extensions for modeling energy and power have been added to MESH, including both energy based on execution of thread fragments and the memory subsystem.

This then leads to an ability to carry out power-performance design evaluation in the MESH framework, thus permitting designers to undertake systematic design space exploration of architectural and application alternatives. The chapter describes experiments with a number of application kernels exhibiting characteristics across a wide range of data patterns (streaming, sporadic, timed, and periodic), on microarchitectures containing a variety of processing cores. Three different core types – ARM7, ARM11, and Infineon TriCore1 – were used to build families of heterogeneous platforms, with many processing elements – ranging from 18 to 36 processors. Both performance and energy consumption were analyzed using this framework. The value of design space exploration is demonstrated in that the optimal system for the applications was a non-obvious combination of processor types and numbers.

Readers are encouraged to read the chapter in detail for further information.

4.2   System-Level Design of Network-on-Chip Architectures

In this follow-on chapter, Karam Chatha and Krishnan Srinivasan start with the assumption of an advanced interconnect structure for MPSoC design: a network-on-chip (NOC). The key challenge is then to design the right NOC architecture to meet the communications needs of potentially hundreds of cores, satisfying performance constraints and physical design challenges, while at the same time minimizing energy consumption.

Both regular and custom NOC topologies are possible, and the choice of which topology is best rests on many assumptions. For example, general purpose processor platforms may best utilize a regular topology since the application needs are not known *a priori*. Highly application-specific approaches may favor a custom interconnect approach, since deeper knowledge of the applications and use domains is possible, and processing elements, storage elements (memories), and the communications infrastructure can all be highly tuned to deliver the maximum application performance for a given amount of energy consumption. NoC architectures which have been customized may be a better communications infrastructure than traditional bus-based ones for custom topologies.

Designing an efficient NoC, whether regular or custom, requires efficient NoC routers, and the chapter describes in some detail a router architecture that has been designed in detail and characterized for use. An NOC design methodology is presented, involving systematic study of power and energy used in a variety of NoC microarchitectures, the study of application traffic traces to determine latency and bandwidth requirements, and tools for generating either regular or custom floorplans and mapping of cores to routers in the network to best meet application requirements, whether in general or highly specific. The analysis and optimization methods used in these tools are discussed in detail. These are then illustrated on a number of benchmarks, drawn from media encoding and decoding, and both custom and mesh topologies are studied to determine optimal configurations from a power basis (since the approach meets the performance requirements inherently). Customized topologies show clear superiority when compared to mesh topologies in terms of power consumption and router area requirements, reinforcing the desirability of designing highly application-specific MPSoC, and the interconnect architecture to support it.

Readers are encouraged to read the chapter in detail for further information.

5.   Conclusion

In this chapter we have discussed the variety of architectural approaches for multiprocessor and multicore design, and classified them according to their ability to satisfy various segments of the overall design space. We have discussed a number of concepts and design techniques for reducing peak power and overall energy consumption, many derived from single-processor and single-core based design, and reviewed how they might be applied to MPSoC.

In particular, we have focused on the advantages for power and energy that asymmetric multiprocessing (AMP) approaches offer for designing an MPSoC system highly tuned to particular applications, while offering maximum performance for each unit of energy expended. The advantages of AMP approaches couple with the concept of configurable and extensible processor design, leading to large networks of highly tuned heterogeneous processors on which the various applications and their tasks have been mapped.

The following two chapters show in more detail two significant areas important to the design of energy and power-effective MPSoC: design space exploration of the performance and power tradeoffs of the various multiprocessor architectural choices; and analysis and exploration of interconnect structures using in this case an advanced network-on-chip approach.

These chapters underline the importance of approaching MPSoC design for low power in a systematic fashion, using appropriate models, simulations, and analysis techniques.

References

Blaauw, David, Martin, Steve, Flautner, Krisztian and Mudge, Trevor, (2002) Leakage Current Reduction in VLSI Systems, *Journal of Circuits, Systems and Computers*, Volume 11, Number 6, pp. 621–636.

Cheng, Wei-Chung and Pedram, Massoud (2002) Power-optimal Encoding for a DRAM Address Bus, *IEEE Transactions on VLSI Systems*, Volume 10, Number 2, April 2002, pp. 109–118.

Choi, Kihwan, Lee, Wonbok, Soma, Ramakrishna, and Pedram, Massoud (2004) Dynamic Voltage and Frequency Scaling Under a Precise Energy Model Considering Variable and Fixed Components of the System Power Distribution, *ICCAD*, pp. 29–34.

Ekman, Magnus, Dahlgren, Fredrik, and Stenstrom, Per. (2002) TLB and Snoop Energy-Reduction Using Virtual Caches in Low-Power Chip-Multiprocessors, *ISLPED*, pp. 243–246.

Ferretti, Marcos and Beerel, Peter A. (2001) Low Swing Signaling Using an Dynamic Diode-Connected Driver, *ESSCIRC*, September, 2001.

Flautner, Krisztian, Kim, Nam Sung, Martin, Steve, Blaauw, David and Mudge, Trevor. (2002) Drowsy Caches: Simple Techniques for Reducing Leakage Power, *ISCA*, pp. 148–157.

Gibbs, W. Wayt. (2004) A Split at the Core, *Scientific American*, November, 2004, pp. 96–101.

Irwin, Mary Jane, Benini, Luca, Vijaykrishnan, N., and Kandemir, Mahmut. (2005) Techniques for Designing Energy-Aware MPSoCs, Chapter 2, pp. 21–47, in Jerraya, Ahmed and Wolf, Wayne (2005), Editors, *Multiprocessor Systems-on-Chip*. San Francisco, California: Elsevier Morgan Kaufmann.

Jerraya, Ahmed and Wolf, Wayne (2005). Editors. *Multiprocessor Systems-on-Chip*. San Francisco, California: Elsevier Morgan Kaufmann.

Juang, Philo, Wu, Qiang, Peh, Li-Shiuan, Martonosi, Margaret, and Clark, Douglas W. (2005) Coordinated, Distributed, Formal Energy Management of Chip Multiprocessors, *ISLPED*, pp. 127–130.

Kim, Nam Sung, Blaauw, David, and Mudge, Trevor (2003) Leakage Power Optimization Techniques for Ultra Deep Sub-Micro Multi-Level Caches, *ICCAD*, pp. 627–632.

Kim, Soontae, Vijaykrishnan, N., Kandemir, Mahmut, Sivasubramaniam, Anand, and Irwin, Mary Jane (2003) Partitioned Instruction Cache Architecture for Energy Efficiency, *ACM Transactions on Embedded Computing Systems*, Volume 2, Number 2, May, 2003, pp. 163–185.

Kumar, Rakesh, Tullsen, Dean M., Jouppi, Norman P., and Ranganathan, Parthasarathy (2005) Heterogeneous Chip Multiprocessors, *IEEE Computer*, November, 2005, Volume 38, Number 11, pp. 32–38.

Leibson, Steve and Kim, James (2005) Configurable processors: a new era in chip design, *IEEE Computer*, July, 2005, pp. 51–59.

Maxfield, Clive (2006) Reconfigurable cores boost processor power, *EETimes*, February 20, 2006, pp. 43–44, 52, 54.

Merritt, Rick (2005) Multicore ties programmers in knots, *EETimes*, October 24, 2005, pp. 1, 20, 22.

Moreshet, Tali, Bahar, R. Iris, and Herlihy, Maurice (2005) Energy Reduction in Multiprocessor Systems Using Transactional Memory, *ISLPED*, pp. 331–334.

Mudge, Trevor (2000) Power: A First Class Design Constraint for Future Architectures, *High Performance Computer Conference*, pp. 215–224.

Rowen, Chris, and Leibson, Steve (2004) *Engineering the Complex SOC.* Upper Saddle River, New Jersey: Prentice-Hall PTR.

Tremblay, Marc (2005)In search of cool computing, *EDN*, September 15, 2005, pp. 77–78.

Vitkovski, Arseni, Haukilahti, Raimo, Jantsch, Axel, and Nilsson, Erland (2004) Low-power and Error Coding for Network-on-Chip Traffic, *IEEE Norchip*.

Wei, John and Rowen, Chris (2005) Implementing low-power configurable processors: practical options and tradeoffs, *Design Automation Conference*, Anaheim, California, pp. 706–711.

Chapter 18

System-Level Design
of Network-on-Chip Architectures*

Karam S. Chatha and Krishnan Srinivasan
Department of Computer Science and Engineering
Arizona State University
Tempe, AZ, 85287-5406

Abstract   Multi-processor System-on-Chip (MPSoC) architectures in future will be implemented in less than 50 nm technology and include tens to hundreds of processing element blocks operating in the multi-GHz range. The on-chip interconnection network will be a key factor in determining the performance and power
consumption of these multi-core devices. Packet switched interconnection networks or Network-on-Chip (NoC) has emerged as an attractive alternative to traditional bus-based architectures for satisfying the communication requirements
of these MPSoC architectures. The key challenge in NoC design is to produce
a complex, high performance and low energy architecture under tight time to
market requirements. The NoC architectures would support the communication
demands of hundreds of cores under stringent performance constraints. In addition to the complexity, the NoC designers would also have to contend with the
physical challenges of design in nanoscale technologies. The NoC design problem would entail a joint optimization of the system-level floorplan and power
consumption of the network. All these factors coupled with the requirement for
short turn around times raises the need for an intellectual property (IP) re-use
methodology that is well supported with design and optimization techniques, and
performance evaluation models. This chapter introduces the concept of NoC and
presents the various elements of the IP-based system-level methodology required
for its design.

Keywords: network-on-chip, multi-processor system-on-chip, interconnection architecture,
system-level design, performance evaluation.

*Partial funding provided by grants from National Science Foundation (CAREER Award CCF-0546462,
IIS-0308268) and Consortium for Embedded Systems.

J. Henkel and S. Parameswaran (eds.), Designing Embedded Processors – A Low Power Perspective,
391–422.
© 2007 *Springer.*

1.   Introduction

1.1   Multi-Processor System-on-Chip (MPSoC) Architectures

Multi-processor System-on-Chip (MPSoC) architectures are high performance VLSI circuits that utilize multiple processors, memory units, custom cores, and buses on the same chip. As discussed in the introductory chapter on "Multi-processor Embedded Systems", MPSoC architectures incorporate both symmetric multiprocessing (SMP) and asymmetric multiprocessing (AMP) approaches. MPSoC devices have evolved in response to two primary needs. The short design turn around times of MPSoC architectures encourage intellectual property (IP) re-use rather than design from scratch. The increased complexity and performance requirements of target application domains necessitate the usage of multiple cores.

The International Technology Roadmap for Semiconductors (ITRS) [1] predicts that in the future MPSoC will be implemented in less than 50 nm technology. It will include tens to hundreds of heterogeneous processing element blocks operating in the multi-GHz range. The processing and storage cores will communicate with each other in Gbits/s. The MPSoC designer will have to contend with the high power consumption challenge of nanoscale technology. Technology scaling leads to an exponential growth in the leakage or standby power consumption. This factor, coupled with the increase in the transistor counts, results in higher power densities and die temperatures. The increased die temperature adversely effects the device performance, reduces its reliability, and further increases the leakage power consumption.

The MPSoC designers will address the high power dissipation challenge by dividing the cores into multiple voltage and clock islands. The power consumption of each such island would be controlled dynamically. Therefore, an MPSoC architecture will be a maximally parallel implementation in the globally asynchronous locally synchronous (GALS) model. The various cores will operate independently under local clocks, and communicate with each other in an asynchronous manner. The global communication pattern will be fully distributed with little or no coordination. In the GALS based MPSoC architectures, the interconnection network will be a very important factor in determining the system performance and power consumption.

1.2   Interconnection Woes and Network-on-Chip (NoC)

Technology scaling has also led to an increase in the interconnect delay with respect to the gate delay. In the future, global signal delays will span multiple clock cycles [2, 3]. Signal integrity will also be compromised due to increased RLC effects [4–6]. The increased RLC effects will lead to an increase in power consumption. The large signal propagation delays will make the bus based synchronous and shared communication architectures impractical. Further, the bus based architectures are inherently unscalable and are not suitable as interconnection infrastructure for future MPSoC architectures.

A radical shift in the design of on-chip interconnection architectures is required for realizing GALS based multi-core devices. Packet switched interconnection networks or Networks-on-Chip (NoC) (as shown in Figure 18.1) are an attractive alternative to traditional bus-based architectures [7–10]. The routers in the NoC communicate with the various cores through a network interface unit. The data transfer between the routers occurs in an asynchronous manner. NoC supports high communication bandwidth by distributing the propagation delay across multiple routers, thus pipelining the signal transmission. Packet switching networks also enable error correction schemes that can be applied towards improving the signal integrity. Quality of service can also be ensured by distinguishing between different classes of traffic.

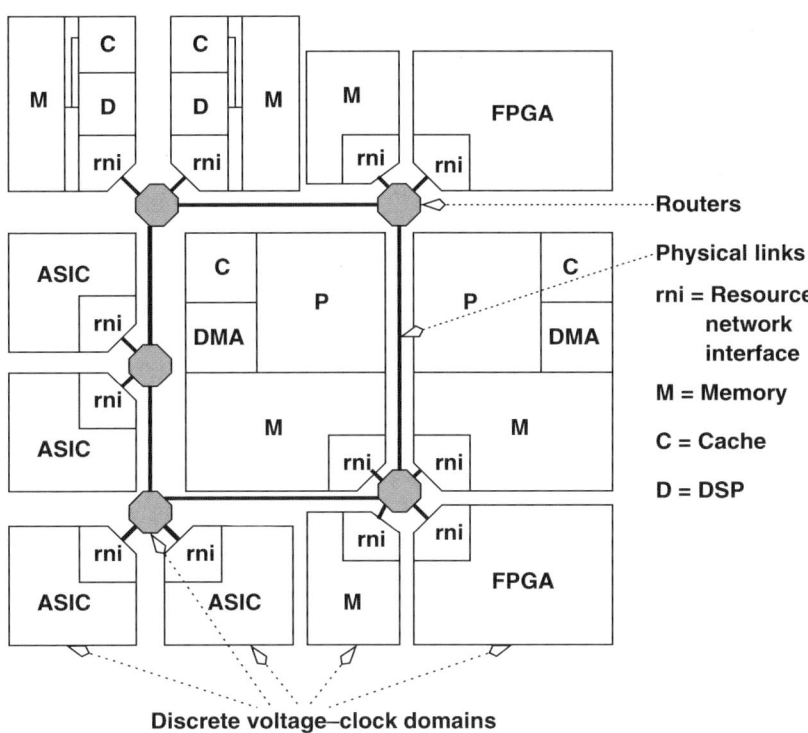

Fig. 18.1 An MPSoC architecture with NoC

1.3   IP-based Methodology for NoC Design

The key challenge in MPSoC design is to produce a complex, high performance and low energy architecture under tight time to market requirements. The NoC architectures in future MPSoCs would need to support the communication demands of hundreds of cores under stringent performance constraints. The topology that defines the structure of the network and the routing strategy for transferring data across the network would need to be optimized for the target application.

In addition to the complexity, the NoC designers would also have to contend with the physical challenges of design in nanoscale technologies. The designers would need to minimize the power consumption in the network. Our previous work [11] on power evaluation of NoC architectures demonstrates that physical links will consume greater than 30% of the communication power. The power consumption in the physical links is dependent on their length and bandwidth of traffic flowing through them. The length of the physical links in the communication network is a function of the system-level floorplan.

It is clear that design and optimization of the NoC architecture must be integrated with the system-level floorplanning. Thus, NoC design problem entails joint optimization of the system-level floorplan and power consumption of the network. All these factors coupled with the requirement for short turn around times raises the need for an IP re-use methodology that is well supported with performance evaluation models and design tools.

The system-level design of future MPSoC architectures would consist of two stages: computation architecture design and NoC architecture design (see Figure 18.2). The computation architecture design stage would determine the processing (instruction processors, custom ASIC, configurable fabric) and memory (cache, SRAM, DRAM) cores that would be part of the system-level architecture. Computation architecture design has been extensively addressed by the system-level design community. The chapter focuses on the system-level design and optimization of the NoC architecture. In particular, we address the following two topics: NoC router architecture and characterization, and design and optimization techniques for NoC.[1] System-level designers will have a choice of two categories of NoC architectures based on their topology:

- *Regular Topologies:* MPSoC architectures that are aimed at third-party software based customization will be developed with a large number of programmable processors (not necessarily homogeneous). Such architectures are particularly amenable for implementation with regular topologies. Pre-designed regular topologies such as mesh, torus, hypercube, and so on also offer the advantage of shorter design times.

---

[1]The work discussed in this chapter is derived from our publications in DATE'04 [11], IEEE Transactions on VLSI [12], and ISLPED'05 [28].

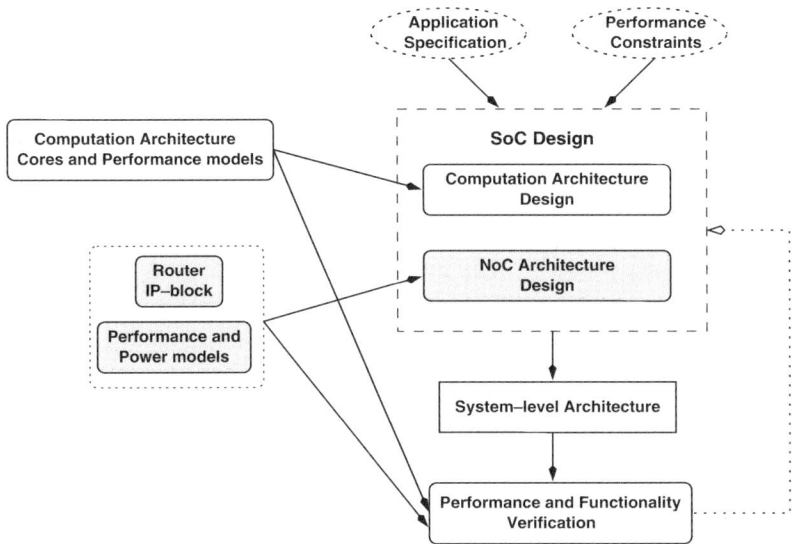

Fig. 18.2   System-level design flow

- *Custom topologies:* Application specific MPSoC architectures that are aimed at high performance application domains would contain numerous heterogeneous computing cores (such as programable processors, application specific integrated circuits, reconfigurable fabric) and memory units. Each core would implement a limited set of application domain functionality. Therefore, the inter-core communication traffic would depict well defined patterns as determined by the functionality of the communicating cores in the specific domain. For such designs, the application specific custom NoC architecture has been demonstrated to be superior to regular topologies in terms of energy and area [12, 13].

In the chapter we will address design solutions for both custom and regular topologies.

The chapter is organized as follows: Section 2 presents the NoC architecture, and its power and performance characterization, Section 3 discusses design techniques for NoC architectures with regular and custom topologies, Section 4 discusses other related research, and Section 5 concludes the chapter.

2.   NoC Router Architecture and Characterization

In this section we begin by presenting the architecture of a router that can be utilized in a two-dimensional mesh based NoC topology. Next, we discuss the development of a cycle accurate model by addressing the power and performance characterization of the router, and finally, present some results that give insights into the system-level NoC design problem.

2.1 NoC Router Architecture

A router that can be utilized in a two-dimensional mesh based topology consists of five input/output ports: four ports communicate with neighboring routers, and a fifth port is connected to a processing or storage unit through the network interface block. Data that needs to be transmitted between source and destination is partitioned into fixed length packets which are in turn broken down into flits or words. A flit is the smallest unit of data that is transferred in parallel between two routers. A packet consists of three kinds of flits – the header flit, the data flit and the tail flit, that are differentiated by two bits of control information. The header flit contains information of the destination router (X,Y) for each packet.

The router architecture is shown in Figure 18.3. The router consists of five unit routers to communicate in X-minus, X-plus, Y-minus, Y-plus directions, and with the processor. Unit routers inside a single router are connected through a 5×5 crossbar. Data is transferred across routers or between the processor and the corresponding router by a four phase (two clock cycle) asynchronous handshaking protocol. A single unit router is highlighted in the right half of Figure 18.3. It consists of input and output link controllers, input and output virtual channels, a header decoder, and an arbiter.

Data arrives at an input virtual channel of an unit router from either the previous router or the processor connected to the same router. The header decoder decodes the header flit of the packet after receiving data from the input virtual channel, decides the packet's destination direction (X−, X+, Y−, Y+, processor), and sends a request to the arbiter of the unit router in the corresponding direction for access to the crossbar. Once the grant is received, the header decoder starts sending data from the input to the output virtual channel through the crossbar. There is a round robin priority based mechanism for each of the following tasks:

- Selection of an input virtual channel by the header decoder.

- Selection of an output virtual channel by the arbiter.

- Grant of the crossbar to the header decoder by arbiter.

- Selection of the output virtual channel by the link controller.

The header decoder implements the dimension order (X–Y) routing strategy [14]. This is a deadlock free routing strategy where the destination node and the current node is compared for each packet for calculating the X-offset and the Y-offset. The packet is routed in the X-direction first till the X-offset is

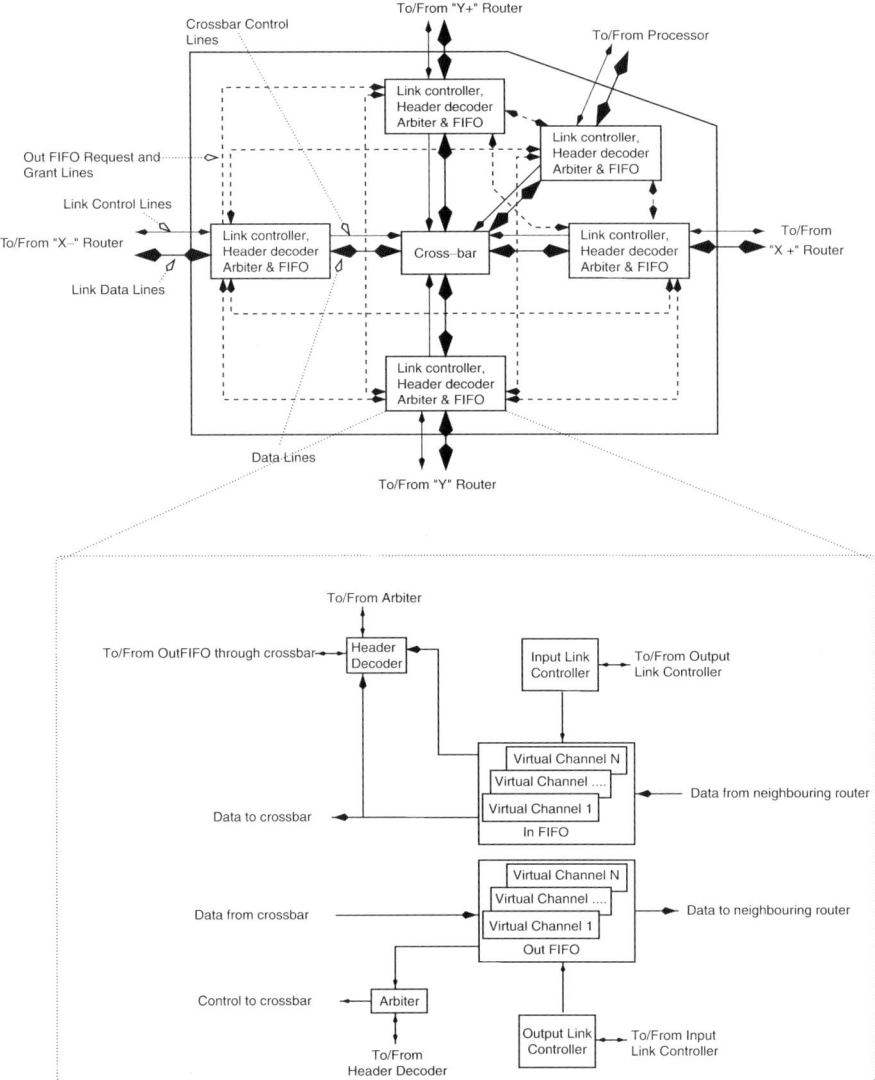

Fig. 18.3 Router architecture for mesh topology

zero and then in the Y-direction. The header decoder communicates with only one arbiter in any one direction at a particular instance of time. It decodes the header flit of a packet from the non-empty input virtual channel and sends a request to the corresponding arbiter to obtain the crossbar connection. Once the connection is established, the header decoder starts sending data from the input virtual channel to the output virtual channel through the crossbar. If the communication gets blocked due to (a) the input virtual channel being empty or (b) the output virtual channel being full, the header decoder gives up the crossbar connection. It then selects the next non-empty input virtual channel for data transfer.

2.2   Power and Performance Characterization

Characterization of Physical Links.   The energy consumption and performance of a physical link is determined by its width (number of bits of data and control signals), length, and capacitive load. In nanoscale technologies, individual wires are modelled by distributed RLC expressions for accurate description of their physical characteristics [4, 15, 16]. Sotiriadis et al. [15] and Taylor et al. [16] have proposed three wire models for evaluating the power consumption in n-bit wide physical links. The links are characterized in sets of three, two, and single wires, respectively. The three and two wire sets include the distributed RLC effects and cross-coupling capacitances, while the single wire model only includes the distributed RLC effects. We modelled the physical links for 180 nm, 130 nm and 100 nm technologies. We obtained energy values for 64 (8×8), 16 (4×4), and 4 (2×2) different switching combinations for the three, two, and single wire sets, respectively. The wire sets were characterized for varying lengths of 1000 μm to 5000 μm in increments of 500 μm.

We included the link characterization values as a table in our performance model. The energy consumed by a n-bit wide link can be calculated from the energy consumed by the three, two, and single wire sets of similar length. For example, consider the 9-bit (odd) wide link shown in the left hand side of Figure 18.4. The total switching energy consumed by the links can be calculated by adding the switching energy consumed by the three wire sets S0, S1, S2, and S3, and subtracting the energy consumed by single wire links A, B, and C, respectively. In the case of an 8-bit (even) wide link shown in the right hand side of Figure 18.4, the energy consumed by two wire set S3 is included

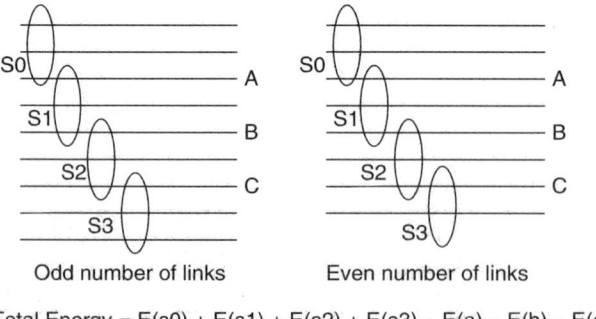

Odd number of links   Even number of links

Total Energy = E(s0) + E(s1) + E(s2) + E(s3)− E(a)− E(b)− E(c)

Fig. 18.4   Performance evaluation of links

in the calculation. The length of the physical link which is a major factor in determining its power consumption and performance is specified by the designer.

Characterization of Router Architecture.   The router architecture was first implemented in VHDL as a register transfer level (RTL) design. The basic building blocks for the router were extracted and synthesized with Synopsys Design Compiler (DC). Standard cell libraries files were given as inputs to Synopsys DC. The Verilog gate netlist obtained from this step was imported as an input to Silicon Ensemble, a tool by Cadence, that performs automatic place and route. Library files (*lef files*) were also given as inputs to Silicon Ensemble. The intermediate format (*gds2*) obtained from this step was then imported into Cadence Virtuoso and Cadence Analog Artist (see Figure 18.5). Hspice netlist obtained from this step was then simulated to obtain the timing and energy values for 180 nm, 130 nm, and 100 nm technologies. The energy values obtained for all the individual components were then back-annotated as look up tables in the RTL specification to construct the cycle accurate model. The characterization tables for the physical links were also included into the model. The clock width for simulating the models in the various technologies were obtained from the Hspice simulation results.

The following section utilizes the cycle accurate simulator to obtain insights into the NoC design problem.

2.3   Elements of NoC Design Process

Experimental Set-up.   We evaluated the performance and power consumption of a 4×4 mesh based NoC architecture. The performance of NoC can be defined by the average latency (in clock cycles) of packet transmission from source to sink, and the acceptance rate or supported bandwidth (in bits/cycle/node or packets/cycle/node). The power consumption of the NoC is given by the sum of the dynamic and leakage power consumption (in watts). All these values are a function of the injection rate or offered bandwidth (in bits/cycle/node or packets/cycle/node) that is injected from various source nodes. In the experiment, the width of the physical links (consequently the width of input and output FIFO, and crossbar) is 32 bits, number of virtual channels is 2, depth of virtual channels is 4, and the number of flits in the packet is 8 (packet size 256 bits). As the neighboring link controllers utilize a 2-clock cycle hand-shaking protocol to transfer a flit, the maximum bandwidth that can be supported over the physical links is 16 bits/cycle or 0.0625 packets/cycle. The clock period was conservatively assumed to be 3 ns. The traffic model that we assumed for the experimentation was uniformly distributed mean delay to random destinations that were also uniformly distributed. The simulator was first allowed to stabilize for 1,000 clock cycles, and the data was collected over the next 10,000 cycles.

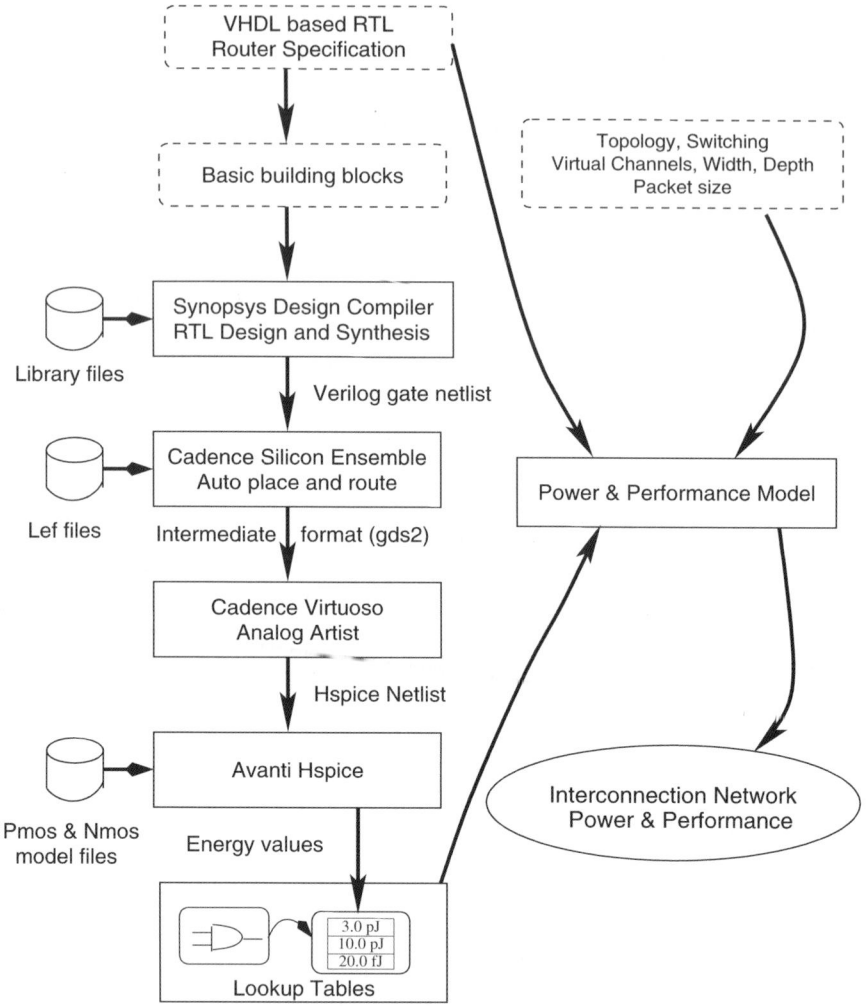

Fig. 18.5   Design flow for characterization

Power Consumption in Physical Links.   Figure 18.6 plots the percentage of power consumption of the various components of a mesh based router in two technologies, 180 nm and 100 nm at 0.055 packets/cycle/node. The length of the physical link in 180 nm and 100 nm was assumed to be 4.5 mm (similar to the MIT RAW architecture [17]) and 2.5 mm (by linear scaling from 4.5 mm), respectively. Technology scaling has the greatest affect on the dynamic energy consumption of the physical links. The power consumption in the links rises from 25% of the power consumption in 180 nm to 34% of the power consumption in 100 nm. Therefore, the design of the interconnection network

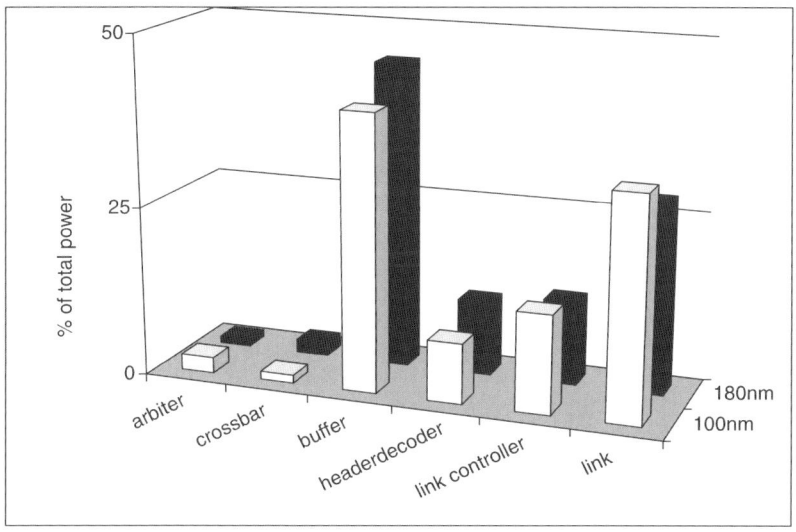

Fig. 18.6 Energy per component

in nanoscale technologies must account for the power consumption due to the physical links. The power consumption in the physical links varies linearly with both link length and bandwidth of traffic. During the design of the interconnection network topology, the bandwidth of traffic can be determined by the communication traces that would be routed through a particular physical link. However, determining the length of the physical link requires the generation of the system-level floorplan. In other words, design of the interconnection network must be integrated with the system-level floorplan.

Power Consumption in Router Ports.   The power consumption of the router can be analyzed by considering a particular port of the router in isolation. We split the router port into two sets of components: namely the input port (link controller, header decoder, input virtual channels and crossbar), and the output port (arbiter, output virtual channels, and link controller). We analyzed the variations in power consumption at the input and output ports with injected bandwidth in 180 nm by utilizing the cycle accurate simulator. Figure 18.7 shows the variation in the input port power with injected bandwidth. The power consumption in the input port shows a linear variation with respect to injected bandwidth. The output power consumption also shows a similar trend. Thus, the overall power consumption in the NoC can be optimized by minimizing the cumulative bandwidth of traffic flowing through the network.

Network Latency.   Figure 18.8 plots the average latency for transmitting the packets from source to destination on the x-axis versus the injection rate

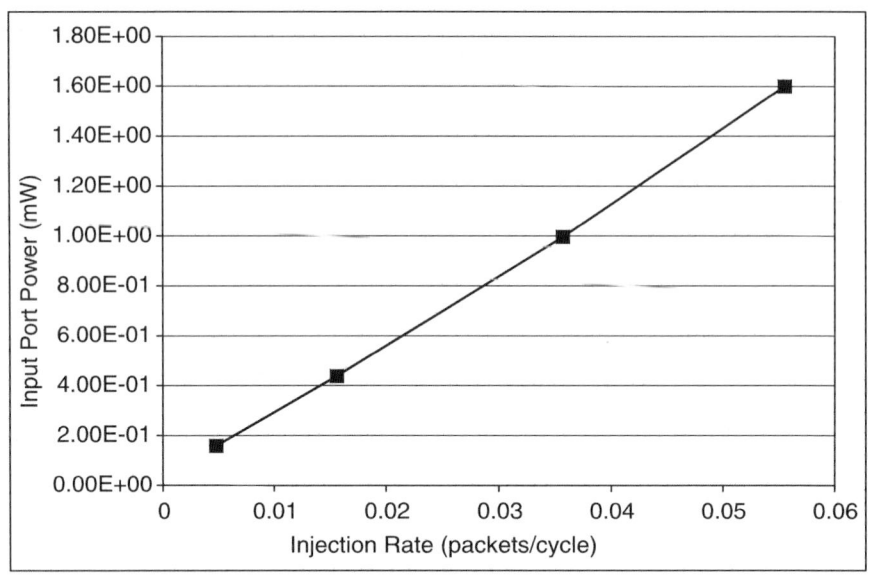

Fig. 18.7   Input port power consumption

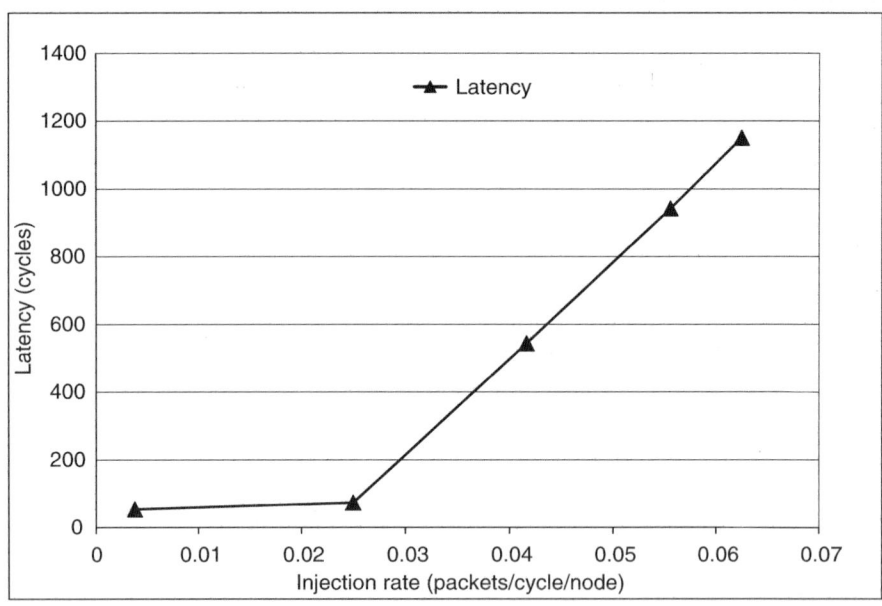

Fig. 18.8   Average latency

on the y-axis. The average latency remains fairly constant as long as the network is not congested. Once the network is congested the average latency rises steeply. A similar trend is observed when we consider the average latency of packets that traverse two hops. Thus if the network is designed such that it is largely operated in the uncongested mode, the network latency constraint can be expressed in terms of router hops (such as 1 or 2) instead of an absolute number (such as 100 cycles).

Regular and Custom NoC Design Problems. The system-level regular and custom interconnection architecture design problems are shown in Figure 18.9. The inputs to the problems are computation architecture specification with communication constraints, characterized library of NoC IP blocks, and floorplanning constraints. Additionally, the regular NoC design problem also accepts a designer specified topology specification as input. The computation architecture is defined by a communication trace graph (CTG) $G(V,E)$ where the various nodes represent processing and memory elements shown by rectangular blocks labelled as "P/M" in the top of the figure. Each "P/M" block is uniquely identified by a node number "$n\ i$" as denoted within each rectangle. The directed edges between any two blocks represent the communication traces. The communication traces are annotated as "$Cm(B,L)$" where

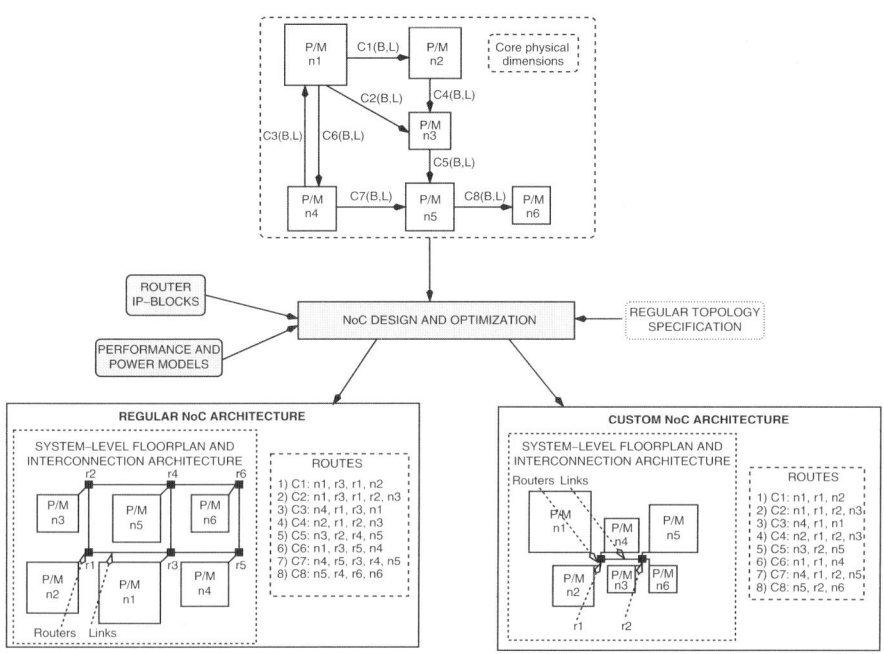

Fig. 18.9   System-level NoC design problems

"*m*" represents the trace number, "B" represents the bandwidth requirement, and "L" is the latency constraint in number of hops. The bandwidth and latency requirements of a communication trace can be obtained by profiling the system-level specification in the context of overall application performance requirements. The floorplanning constraints consist of the physical dimensions of the cores in the computation architecture, and die dimension constraints.

The latency constraints imposed by the traffic traces specify the maximum number of hops allowed for the trace. Bandwidth requirements and latency constraints of communication traces can be viewed as mutually independent. A trace such as a signalling event or a cache miss is not expected to have a high bandwidth requirement, but is bound by tight latency constraints. On the other hand, many non-critical multimedia streams have high bandwidth requirement, and their latency is bound only by the period constraint of the application. NoC design has to perform a trade-off between placing high bandwidth traffic traces close to each other such that power consumption is minimized, and placing tight latency traces close to satisfy the performance constraints.

The regular NoC design problem outputs a system-level floorplan, a mapping of the cores to the various routers and a static route of the communication traces on the network. The output of the custom NoC design problem is a system-level floorplan of the final design, application specific topology of the NoC, a mapping of the cores on the routers, and a static route of the communication traces on the network such that the performance and floorplanning constraints are satisfied. The optimization objective of both the design problems is to minimize the power consumption due to the communication network subject to the bandwidth, and latency constraints.

3.   Design and Optimization Techniques for NoC

We address the complexity of NoC design by dividing it into the following three stages and developing optimization techniques for each stage:

- *Floorplanning of the computation architecture cores:* In the case of custom topologies this stage generates the system-level floorplan of the computation architecture. Several researchers have proposed wire efficient VLSI layout schemes for regular interconnection networks [18–22]. These schemes are topology specific and exploit the regularity in the graph structure. Thus, in the case of regular topologies existing schemes can be utilized to generate an efficient layout of solely the network architecture. The actual computation architecture cores would then be placed on the open locations on the layout. The placement for regular NoC architectures also specifies the mapping of the cores to the routers in the regular NoC topology. The optimization objective function for both types of topologies captures the trade-off between minimizing

the NoC power consumption and satisfaction of the performance constraints.

- *Allocation of routers and mapping of cores to routers:* This stage is performed only for a custom interconnection network with an objective of
  minimizing the power consumption of the NoC.

- *Topology design and route generation:* In the case of custom topologies
  this stage generates the NoC topology, and also the custom routes of the
  communication traces on the interconnection architecture. The objective
  is to minimize the communication power consumption subject to the
  bandwidth constraints on the architecture. For the regular interconnection network, this stage only generates a unique route if the target router
  architecture supports a custom routing strategy.

In the following sections we describe optimal mixed integer linear programming (MILP) and heuristic based optimization strategies for the three stages
that can be utilized for design of both regular and custom NoC architectures.

3.1   MILP based Approach

System-Level Floorplanning.   At the floorplanning stage, the power consumption due to the interconnection architecture can be abstracted as the power
required to perform communication via point to point physical links between
communicating cores. Although, such a cost function does not include the
router power consumption, it is a true representation of the power consumption
due to the physical links. However, inclusion of only the power consumption in
the cost function ignores the performance requirements on the communication traces. Bandwidth constraints on the communication traces can be satisfied by finding alternative routes or adding more interconnection architecture
resources. However, satisfying latency constraints is more difficult if the cores
are placed wide apart. In addition to minimizing power and latency, it is also
important that the layouts consume minimum area. Therefore, we specify our
minimization goal as a linear combination of the power-latency function, and
the area of the layout. Mathematically, we minimize

$$\alpha \cdot \left[\sum_{\forall e(u,v) \in E} dist(u,v) \cdot \Psi_l \cdot \frac{\omega(e)}{\sigma^2(e)} \right] + \beta \cdot [X_{max} + Y_{max}]$$

where $dist(u,v)$ is the distance between the cores u and v, α and β are constants, $\omega(e)$ is the bandwidth of traffic between two cores, $\sigma(e)$ is the latency
constraint between two cores, and X_{max} and Y_{max} represent the die boundaries
in positive X- and Y- directions, respectively. Note that minimizing X_{max} and
Y_{max} minimizes the die area that is given by $X_{max} \times Y_{max}$. A number of other
researchers have proposed MILP formulations for floorplanning [23–25] that

can be combined with our NoC specific optimization function. The existing techniques can also be modified to generate floorplans for mesh based topologies that conform to a regular grid layout.

NoC Topology and Route Generation.   The power consumption of the NoC is dependent upon the length of the physical links in the architecture. We utilize the floorplan from the previous stage to select router locations, and thus determine inter router, and node to router distances. By intelligently determining router locations, we can reduce the size of the MILP formulation and thus, reduce its runtime.

Initially, we create a bounding box for each node. A bounding box is a rectangular enclosure of the node such that the bounding boxes of two adjacent nodes abut each other. For example, in Figure 18.10a the bounding box of node 4 extends to the top boundary of node 3, and that of node 10 extends to the top boundary of node 12, and to the left boundary of node 9. In the figure, all other bounding boxes are the co-ordinates of the respective nodes.

The second step is the placement of routers. We exploit the fact that the dimensions of routers are much smaller than that of the processing cores [26]. Therefore, once the bounding boxes are generated, we place routers at the nodes of the channel intersection graph [23] formed by the bounding boxes. A channel intersection graph is a graph in which the bounding boxes form the edges, and the intersection of two perpendicular boundaries forms a node. In Figure 18.10b, the black circles depict the placement of routers at the channel intersections.

Finally, we remove all redundant routers. We remove all routers that are:

- placed along the perimeter of the layout and

- placed less than a specified distance apart.

The motivation for removing routers can be explained as follows. The routers in the perimeter are not likely to be utilized, and are redundant. Similarly, if two

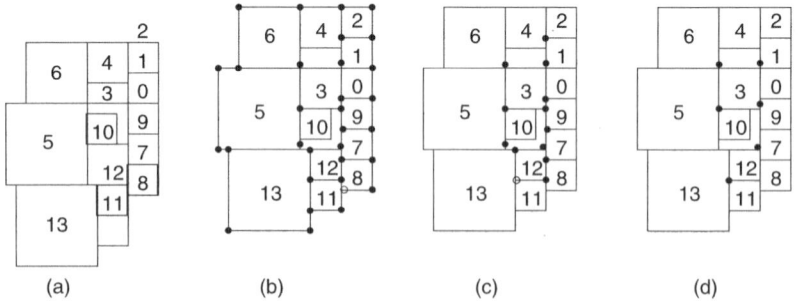

(a)   (b)   (c)   (d)

Fig. 18.10   Example of router allocation for custom topology

routers are placed very close to each other, one of them becomes redundant as it is unlikely to be utilized in the final topology. Figures 18.10c and 18.10d show the example layout after the removal of perimeter routers and redundant routers, respectively.

We can further minimize the size of the formulation by limiting the number of routers that a node can be mapped to. Since the layout places communicating nodes close to each other, it is very unlikely that an optimal solution will have a node that is mapped to a router located at a large distance away from it. Further, the length of the physical link is also dependent on the signal propagation time which is constrained by the clock period of the source router. Therefore, for each node, we consider only those routers that are within a certain maximum distance from it. The distance is specified by the designer.

Let $\mathcal{R}_i$ denote the set of routers available to node v_i. Since we know the location of the routers, we can determine the shortest distance from a node v_i to the routers in $\mathcal{R}_i$. By the same argument, we can also determine all inter router distances.

The objective function of the formulation is minimization of the communication power. The power consumption in the NoC is the sum of the power consumed by the routers and the physical links. The power consumed by the routers is given by the product of the bandwidth of data flowing through the ports and the characterization function that specifies the power consumption per unit bandwidth. Similarly, the power consumption in a physical link is a product of the bandwidth of data flowing through the link, length of the link, and the characterization function that specifies the power consumption per unit bandwidth per unit length.

In the following paragraphs, we describe the necessary constraints for the MILP formulation. We refer the reader to our paper [12] for a detailed description of the various equations.

- *Port capacity constraint*: The bandwidth usage of an input or output port should not exceed its capacity.

- *Port-to-port mapping constraint*: A port can be mapped to one core, or to any one port that belongs to a different router.

- *Core-to-port mapping constraint*: A core should be mapped exactly to one port.

- *Traffic routing constraints*: The traffic routing constraints discussed below ensure that for every traffic trace in the computation architecture there exists a path in the NoC.

    1. If a core is mapped to a port of a router, all traffic emanating from that core has to enter that port. Similarly, all traffic terminating at that core should leave from that port.

2. If a core is mapped to a port of a router, no traffic from any other core can either enter or leave that port.

3. If a traffic enters a port of the router, it should not enter from any other port of that router. Similarly, if a traffic leaves a port of a router, it should not leave from any other port of that router. This constraint ensures that the traffic does not get split across multiple ports.

4. If a traffic enters a port of a router, it has to leave from exactly one of the other ports of that router. In the same way, if a traffic leaves a port of a router, it must have entered from exactly one of the other ports of that router. This constraint ensures the conservation of flow of traffic.

5. If two ports of different routers are connected, traffic leaving from one port should enter the other, and vice versa.

6. If two ports of different routers are connected, a traffic can leave exactly one of the two ports. Similarly, a traffic can enter only one of the two ports.

7. If a traffic enters a port of a router, that port must be mapped to a core or to a port of a different router. Similarly, if a traffic leaves a port of a router, that port must be mapped to a core, or to a port of a different router.

- *Latency constraint*: The latency constraint refers to the maximum number of hops that is allowed to route the traffic from a source node to a sink node. For example, a latency of 2 means that the traffic can pass through at most two routers.

The MILP formulation for NoC topology and route generation is constrained by exponentially increasing solution times for large communication trace graphs. A clustering based heuristic technique can be utilized [12] for reducing the solution times. The technique as the name implies first divides the layout into clusters, then generates the topology for each individual cluster by utilizing the MILP formulation, and then merges the various clusters to generate the final solution.

Results.   We generated custom and mesh based NoC architectures for several multimedia benchmarks by utilizing our MILP based formulation. The benchmarks included mp3 audio encoder, mp3 audio decoder, H.263 video encoder, and H.263 video decoder algorithms. In addition, we obtained six other designs by combinations of two applications from the above mentioned benchmarks. The benchmarks are shown in Table 18.1. The communication

Table 18.1 Benchmark characteristics

Graph	Graph ID	Nodes	Edges
mp3 decoder	G1	5	3
263 encoder	G2	7	8
mp3 encoder	G3	8	9
263 decoder	G4	9	8
263 enc mp3 dec	G5	12	12
mp3 enc mp3 dec	G6	13	12
263 dec mp3 dec	G7	14	16
263 enc mp3 enc	G8	15	17
263 enc 263 dec	G9	16	16
263 dec mp3 enc	G10	17	17

Table 18.2 Comparison of custom and mesh topologies

No.	Graph	Power (μW)			Routers		
		Cluster	Mesh	Ratio	Cluster	Mesh	Ratio
1	G1	2.622	7.363	2.80	1	5	5
2	G2	108.3	291.4	2.69	2	7	3.5
3	G3	5.7	10.51	1.84	2	8	4
4	G4	5.722	12.51	2.18	3	9	3
5	G5	110.4	273.7	2.47	4	12	3
6	G6	8.157	18.02	2.21	5	13	2.6
7	G7	8.535	22.27	2.60	5	14	2.8
8	G8	155.2	277.0	1.78	5	15	3
9	G9	115.6	296.7	2.56	5	16	3.2
10	G10	11.54	28.63	2.15	6	17	2.8

trace graphs for the benchmarks were obtained from the work presented by Hu et al. [27].

We compared the customized NoC designs for application specific MPSoC architectures against solutions with mesh based NoC topologies. Table 18.2 shows the results of the comparative study for each benchmark application. The regular mesh based topology on an average consumes over 2.3 times more power and requires 3.5 times as many routers as a customized topology. The pre-designed physical connections in the mesh based topologies force the communication traces to pass through more routers, thus, leading to the increased power consumption.

3.2   Heuristic Approach

In this section, we discuss our technique for design of low power mesh based on-chip interconnection architectures, henceforth called MOCA [28]. MOCA operates in two phases. In the first phase, it invokes a min-cut based slicing tree generation technique to map cores on to the different routers of the mesh. In the second phase, MOCA invokes a hierarchical router that generates routes for all the communication traces.

3.3   MOCA Phase I: Core to Router Mapping

The MOCA core mapper (CM) takes a CTG and a mesh topology as inputs, and maps the cores to different routers of the given mesh. The mesh is assumed to be placed in the first quadrant of the X–Y plane with the routers placed at unit distance apart. The mapping of the cores to routers assigns a unique (x, y) co-ordinate to each core that corresponds to the router at that particular location. We determine the coordinates of the cores by recursively invoking the partitioning technique proposed by Fiduccia and Mattheyses [29] (FM) to solve the graph equicut problem. The input to the equicut problem is a graph with weights on its edges. The solution is a partition of the graph such that the cumulative weight of the edges crossing the partition is minimized. Each partitioning step divides the mesh and the CTG into two halves to generate a slicing tree. On algorithm completion, the intermediate nodes of the tree are the directions of each cut (horizontal or vertical) and the leaf nodes are the cores.

During the invocation of the FM based partitioner, we utilize dummy nodes [30] to bias the partitions. In the FM algorithm, the edges across partitions are ignored in subsequent equicut sub-problems. Consequently, neighboring nodes of a CTG that are in different partitions can drift further apart. Introduction of a dummy node for each cut-edge, and its placement at the cut boundary can prevent this phenomenon. CM also adds m additional nodes to the existing n nodes in the CTG such $n + m$ is a power of 4. This modification ensures that every recursive call to FM divides the nodes into two equal halves. CM also modifies the edge weight identical to the MILP based approach to reflect the trade-off between the power minimization objective and satisfaction of latency constraints.

Figures 18.11 to 18.13 give examples of the input CTG, slicing tree, and various stages of the algorithm execution, respectively. In Figure 18.12, the first cut is a vertical cut, the two children cuts are horizontal, and so on. The position of the node in the tree indicates its coordinates after successive partitioning steps. The leaves of the tree are occupied by the nodes of CTG. The empty circles in the Figure 18.13 denote the m additional nodes. The black circles denote the dummy nodes. In Figure 18.13b, traces A–D and A–E are captured

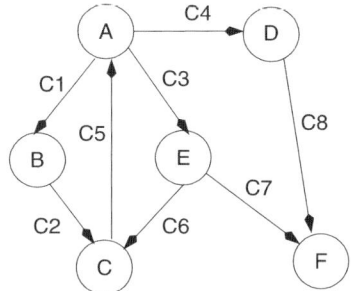

Fig. 18.11   Example CTG

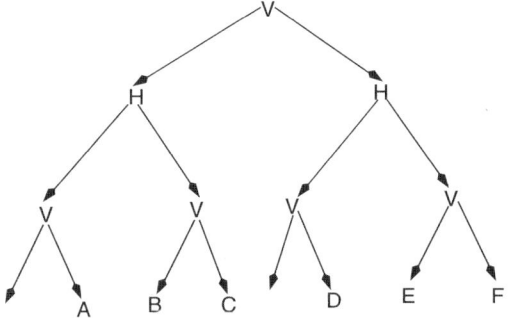

Fig. 18.12   Partitioning based slicing tree

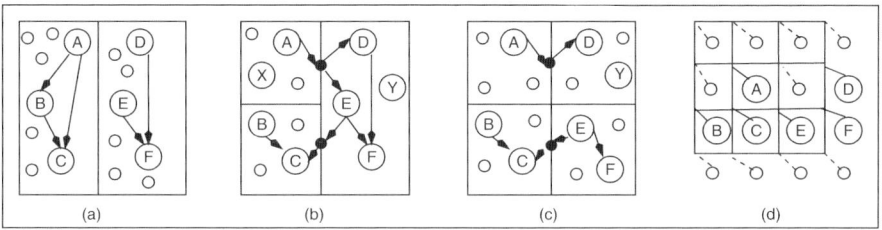

Fig. 18.13   Output of MOCA Phase I: core to router mapping

by the dummy node on the top half plane, and trace C–E is captured by the dummy node on the bottom half plane.

3.4   MOCA Phase II: Route Generation

The MOCA route generator (RG) is a novel technique that operates on the slicing tree to formulate a unique route for each communication trace. RG also operates in two stages namely, RG hierarchical router (RG_{hier}) that generates a route for every trace by traversing the slicing tree, and RG shortest path router (RG_{sp}) that searches for a minimal distance route for a communication trace that was not routed by RG_{hier}.

RG_{hier} attempts to find a minimal path from the source to the destination for each traffic trace. RG_{hier} recursively traverses the slicing tree generated by CM and routes traces that cross the cut at each level of the tree. The input to RG_{hier} is the layout, and a direction of cut (horizontal or vertical). RG_{hier} returns a unique route for every traffic trace in the CTG. RG_{hier} also returns a list of traffic traces that could not be routed due to violation in either the bandwidth or latency constraints.

RG_{hier} considers two partitions of the given mesh as defined by CM. A partial route for the any trace that crosses the cut is generated by mapping it on the routers adjacent to the cut. This step is equivalent to assigning the traces to physical links that are across the cut subject to the bandwidth constraints. Clearly, this is a knapsack problem and is known to be NP-complete. RG_{hier} routes traces on the respective routers by considering the traces in a decreasing order of their bandwidth requirement. The pair of routers that are connected to the physical links affected by the cut are considered for routing the trace. The trace is routed through the pair of routers that is closest to the source, and can support the traffic without bandwidth violation. Once the trace is routed, the partial route of the trace in the set is updated.

Figure 18.14 gives an example of the algorithm execution on the mapping shown in Figure 18.13d. In the figure, the black squares represent the routers, the solid lines represent the links, and the labelled circles represent the cores from Figure 18.11. The dotted lines in Figure 18.14 refer to the successive cut lines that are generated during the algorithm execution. The NoC architecture at the end of the routing stages is shown in Figure 18.14g. As an example of the traces that are chosen to be routed across a cut, Figure 18.14b maps traces (A,B), (A,C), and (A,E) across the horizontal cut in the left hand side.

We state the following theorem (see [28] for proof) regarding the optimality of RG_{hier}.

Theorem 10.1. *RG_{hier} finds a minimal path if bandwidth constraints are not violated.*

It is possible that RG_{hier} cannot find minimal paths for some traces without violating the performance constraints. If a trace cannot be mapped due to such violations, the partial route for the trace is removed, and the trace is added to a list of unmapped traces. RG_{sp} is invoked for each traffic trace that is left unmapped at the end of RG_{hier} stage. RG_{sp} attempts to find alternate routes for these traces. For each trace e that is yet to be routed, RG_{sp} sweeps the links L of the mesh, and assigns an edge weight of ∞ to all links that would see a bandwidth violation on the ports constituting the links, if the trace was routed

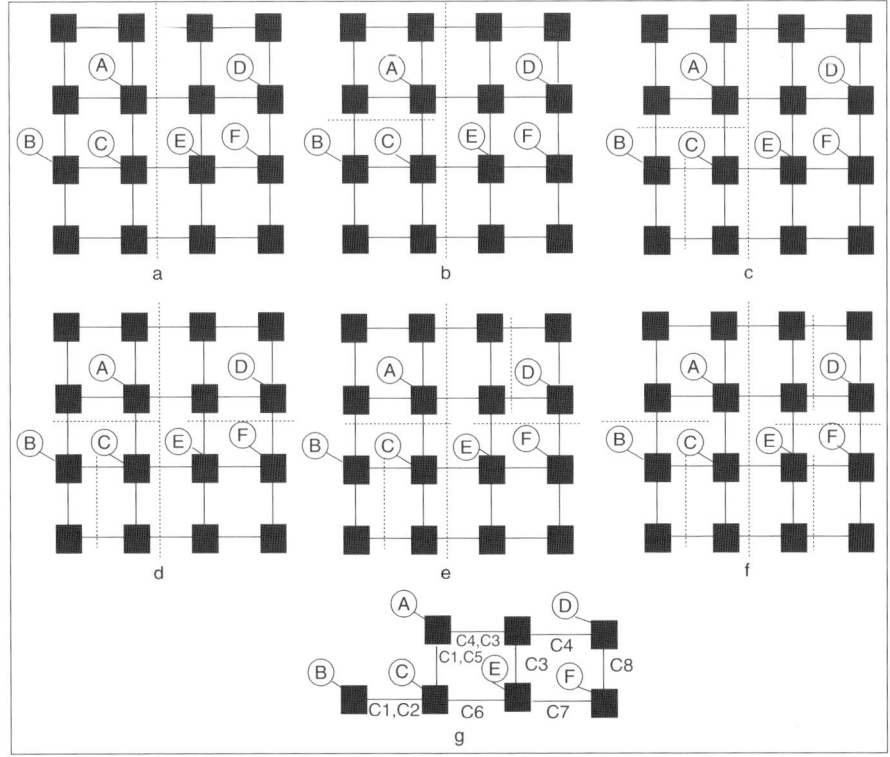

Fig. 18.14 MOCA Phase II: Route generation

through that link. These links are not utilized to generate the route for the trace. This step is followed by invoking the Dijkstra's shortest path algorithm to find a route for the trace on the mesh.

Let n be the number of nodes in the CTG, u be the number of nodes in the mesh, e be the number of edges in the CTG, and f be the number of links in the mesh. MOCA adds m nodes such that $n + m = u$. In a square mesh based network with $u = 2^{2p}$ for $p \geq 1$, $f = 2(u - u^{\frac{1}{2}})$. The overall complexity of CM is $O(ue)$, RG_{hier} is $eu^{\frac{1}{2}} log_2(u)$, and RG_{sp} is $O(e(3u - 2u^{\frac{1}{2}}))$. The overall complexity of MOCA is given by $O(max(eu, eu^{\frac{1}{2}} log_2(u), e(3u - 2u^{\frac{1}{2}}))) = O(eu)$.

Results.   Figure 18.15 presents the trace graph for MPEG4 decoder [13]. The labels of the edges denote bandwidth requirement in Mbps and latency constraint in router hops, respectively. Figures 18.16 and 18.17 present the mesh based NoC architectures for MPEG4 generated by MOCA for the latency constrained and latency unconstrained case, respectively. Since MOCA gives

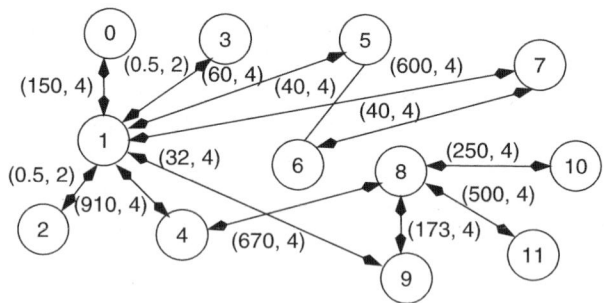

Fig. 18.15   CTG for MPEG4 decoder

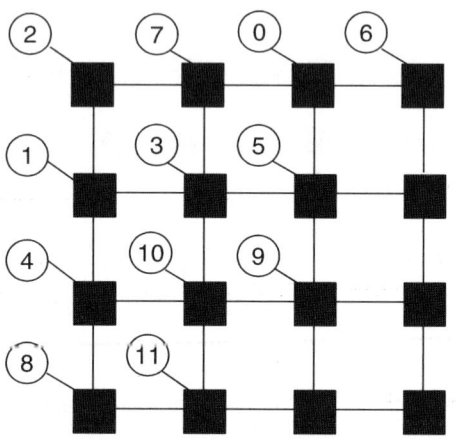

Fig. 18.16   MPEG4 with latency constraints

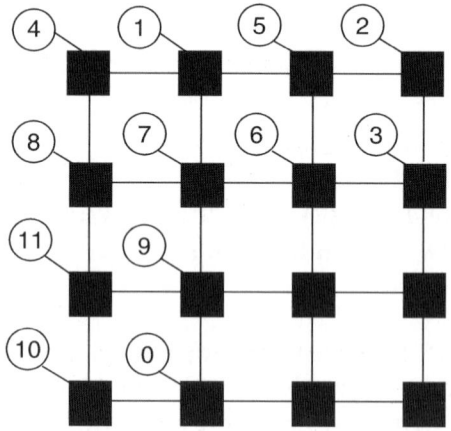

Fig. 18.17   MPEG4 without latency constraints

higher priority to latency, traces with tight latency are routed through minimum hops. For example, in the case of MPEG4 decoder, trace (1,2) is routed in only two hops due to its tight latency (two hops). Note that when latency is not a constraint, the same trace is routed in three hops due to its low bandwidth requirement.

4. Related Research

Packet switching networks have been successfully applied toward computer networks [31] and shared memory multi-processor networks [14]. The designers of shared memory multi-processor system have developed a large body of work for micro-interconnection networks [14, 32–36]. NoC are distinguished from micro-interconnection networks by three significant characteristics.

- Shared memory multi-processor system interconnects are generic solutions that are incorporated as commercial off-the-shelf (COTS) components. In contrast, NoC can be optimized (in terms of topology, routing strategy, and so on) at design time for the target application.

- The existing research on micro-interconnection networks has primarily focused on performance optimization. NoC, on the other hand are optimized for both power consumption and performance.

- Shared memory interconnects are constrained by the pin count on chip packages. NoC can potentially exploit much larger routing resources that are available on the layout.

As discussed in the previous section, the design of NoC requires the development of i) router IP-blocks, ii) performance and power evaluation models, and iii) design and optimization techniques. In the following sections we discuss the previous work in each of these categories.

4.1   NoC Router Architectures

SPIN [37, 38] and Proteo [39] were early router architectures that supported best effort service class. The MIT RAW processor interconnection architecture [17], Nostrum [40, 41], AEthereal [42, 43], and Vellanki et al. [44, 45] proposed router architectures that supported a guaranteed throughput service class in addition to best effort traffic. Vellanki et al. [44, 45], and Zimmer et al. [46] evaluated error correction schemes and high reliability service classes for NoC architectures, respectively. Worm et al. [47], Chen et al. [48], and Nilsson et al. [49] proposed architectural extensions for dynamic power optimizations aimed at physical links, buffers, and clocking strategies, respectively.

4.2   NoC Performance and Power Consumption Models

The research on performance models for micro-interconnection networks [14, 32, 34, 35] did not evaluate the power consumption characteristics of the architecture. The existing tools on system-level power and performance evaluation [50, 51] are targeted towards uniprocessor environments with shared bus architectures.

Pamunuwa et al. [52] and Bolotin et al. [53] proposed analytical models for performance and cost estimation of NoC architectures. Ye et al. [54] analyzed the power consumption in the switch fabrics of network routers. Wang et al. [55] proposed a word (or in interconnection network terminology "flit") level power-performance simulator for interconnection network. Both the models [54, 55] estimate only the dynamic power consumption due to switching capacitance of various data-path units in the network. Hence, they do not calculate power consumption due to leakage (or standby) power, control units, and RLC effects in physical links, all of whose contributions are substantial in nanoscale technologies. In the chapter we discussed a cycle accurate power and performance evaluation model [11] that accurately accounts for the contributions due to these factors. Recently, Chan et al. [56] also proposed a cycle accurate model for NoC power and performance evaluation that was characterized by linear regression based techniques.

4.3   NoC Design and Optimization Techniques

Existing research have primarily concentrated on regular mesh based topologies [27, 28, 57–59]. The design problem for custom NoC architectures was first proposed by Srinivasan et al. [60]. Since then several alternative techniques have been proposed for custom NoC design [61–65]. In [66], the authors presented a heuristic technique for constraint driven communication architecture design that has at most two routers between each source and sink. Hence, their problem formulation did not address routing.

5.   Conclusion

The chapter introduced the concept of NoC and discussed an IP-based system-level methodology for its design. As part of the IP-based methodology the chapter discussed architecture of a router IP-block, power and performance models, and design optimization techniques for NoC. There are several open challenges for NoC design. The existing research on NoC router architectures has primarily addressed the problem for specific routers, and not in the context of system-level design. Consequently, almost all existing research (with the exception of SPIN which considered fat tree topologies) has focused on routers that can be applied to generic mesh based topologies. The router architectures are mostly fixed architectures with limited customization capabilities. Flexible router IP-blocks that can be optimized at design time are an open research problem. In terms of the NoC design optimization techniques, existing research has primarily addressed the problem in isolation. Integrated optimization techniques for co-design of the computation architecture along with interconnection architecture is an open research topic. As the NoC design problem requires floorplanning, thermal effects must also be considered along with performance and power consumption. Existing techniques have not addressed this problem.

References

[1] SIA. International Technical Roadmap for Semiconductors. Technical report, http://public.itrs.net/, 2004.

[2] D. Sylvester and K. Keutzer. A Global Wiring Paradigm for Deep Submicron Design. *IEEE Transactions on Computer Aided Design of Integrated Circuits and Systems*, pages 242–252, February 2000.

[3] R. Ho, K. Mai, and M. Horowitz. The Future of Wires. In *Proceedings of IEEE*, pages 490–504, April 2001.

[4] J. Davis and D. Meindl. Compact Distributed RLC Interconnect Models – Part I: Single Line Transient, Time Delay and Overshoot Expressions. *IEEE Transactions on Electron Devices*, 47(11):2068–2077, November 2000.

[5] J. Davis and D. Meindl. Compact Distributed RLC Interconnect Models – Part II: Coupled Line Transient Expressions and Peak Crosstalk in Multilevel Networks. *IEEE Transactions on Electron Devices*, 47(11):2078–2087, November 2000.

[6] D. Sylvester and K. Keutzer. Impact of Small Process Geometries on Microarchitectures in Systems on a Chip. In *Proceedings of the IEEE*, pages 467–484, April 2001.

[7] A. Hemani, A. Jantsch, S. Kumar, A. Postula, J. Oberg, and M. Millberg. Network on Chip: An Architecture for Billion Transistor Era. In *Proceedings of IEEE NorChip Conference*, November 2000.

[8] M. Sgroi, M. Sheets, A. Mihal, K. Keutzer, S. Malik, J. Rabaey, and A. Sangiovanni-Vincentelli. Addressing the System-on-a-Chip Interconnect Woes Through Communication-Based Design. In *Proceedings of Design Automation Conference*, pages 667–672, June 2001.

[9] L. Benini and G. De-Micheli. Networks on Chips: A New SoC Paradigm. *IEEE Computer*, pages 70–78, January 2002.

[10] W.J. Dally and B. Towles. Route Packet, Not Wires: On-Chip Interconnection Networks. In *Proceedings of DAC*, June 2002.

[11] N. Banerjee, P. Vellanki, and K.S. Chatha. A Power and Performance Model for Network-on-Chip Architectures . In *Proceedings of DATE*, Paris, France, February 2004.

[12] Krishnan Srinivasan, Karam S. Chatha, and Goran Konjevod. "Linear Programming based Techniques for Synthesis of Network-on-Chip Architectures". Accepted for *IEEE Transactions on VLSI*, 2006.

[13] A. Jalabert, S. Murali, L. Benini, and G. De-Micheli. xpipesCompiler: A Tool for Instantiating Application Specific Networks on Chip. In *DATE*, 2004.

[14] J. Duato, S. Yalamanchili, and L. Ni. *Interconnection Networks, an Engineering Approach*. IEEE Computer Society, 1997.

[15] P. Sotiriadis and A. Chandrakasan. A Bus Energy Model for Deep Submicron Technology. *IEEE Transactions on Very Large Scale Integration (VLSI) Systems*, 10(3):341–350, June 2002.

[16] Clark N. Taylor, Sujit Dey and Yi Zhao. Modeling and Minimization of Interconnect Energy Dissipation in Nanometer Technologies. In *Proceedings of DAC*, June 2001.

[17] M.B. Taylor, J. Kim, J. Miller, D. Wentzlaff, F. Ghodrat, B. Greenwald, H. Hoffman, P. Johnson, J.-W. Lee, W. Lee, A. Ma, A. Saraf, M. Seneski, N. Shnidman, V. Strumpen, M. Frank, S. Amarasinghe, and A. Agrawal. The RAW Microprocessor: A Computational Fabric for Software Circuits and General-Purpose Programs. *IEEE Micro*, pages 25–35, March–April 2002.

[18] Chi-Hsiang Yeh, Emmanouel A. Varvarigos, and Behrooz Parhami. Multilayer VLSI Layout for Interconnection Networks. In *International Conference on Parallel Processing*, pages 33–40, 2000.

[19] Ronald I. Greenberg and Lee Guan. On the Area of Hypercube Layouts. *Information Processing Letters*, 84(1):41–46, 2002.

[20] Shimon Even and Roni Kupershtok. Layout Area of the Hypercube: (Extended Abstract). In *SODA '02: Proceedings of the Thirteenth Annual ACM-SIAM Symposium on Discrete Algorithms*, pages 366–371, Philadelphia, PA, USA, 2002. Society for Industrial and Applied Mathematics.

[21] Chi-Hsiang Yeh, Emmanouel A. Varvarigos, and Behrooz Parhami. Efficient VLSI Layouts of Hypercubic Networks. In *Proceedings of Symposium on Frontiers of Massively Parallel Computation*, pages 98–105, February 1999.

[22] Andre DeHon. Multilayer Layout for Butterfly Fat-Tree. In *Twelfth Annual ACM Symposium on Parallel Algorithms and Architectures*, July 2000.

[23] S.M. Sait and H. Youssef. *VLSI Physical Design Automation: Theory and Practice*. McGraw-Hill Inc, 1994.

[24] P. Chen, and E.S. Kuh. Floorplan Sizing by Linear Programming Approximation. In *Proceedings of DAC*, Los Angeles, California, June 2000.

[25] J.G. Kim, and Y.D. Kim. A Linear Programming Based Algorithm for Floorplanning in VLSI Design. *IEEE Transactions on CAD*, 22(5), 2003.

[26] W.J. Dally and C.L. Seitz. Deadlock-free Message Routing in Multi-Processor Interconnection Networks. *IEEE Transactions on Computers*, C-36(5):547–553, 1987.

[27] J. Hu and R. Marculescu. Energy-Aware Mapping for Tile-based NoC Architectures Under Performance Constraints. In *ASP-DAC*, 2003.

[28] Krishnan Srinivasan and Karam S. Chatha. A Technique for Low Energy Mapping and Routing in Network-on-Chip Architectures. In *Proceedings of International Symposium in Low Power Electronic Design*, San Diego, CA, August 2005.

[29] C.M. Fiduccia and R.M. Mattheyses. A Linear-Time Heuristic for Improving Network Partitions. In *Proceedings of DAC*, 1982.

[30] Naveed Sherwani. *Algorithms for VLSI Physical Design Automation*. Kluwer Academic Publishers, 1995.

[31] A. Leon-Garcia and I. Widjaja. *Communication Networks, Fundamental Concepts and Key Architectures*. McGraw Hill, 2000.

[32] H.J. Seigel. A Model of SIMD Machines and a Comparison of Various Interconnection Networks. *IEEE Transactions on Computers*, 28(12):907–917, December 1979.

[33] L.N. Bhuyan (editor). Special Issue: Interconnection Networks. *IEEE Computer*, June 1987.

[34] W.J. Dally. Performance Analysis of k-ary n-cube Interconnection Network. *IEEE Transactions on Computers*, 39(6):775–785, June 1990.

[35] J.F. Draper and J. Ghosh. A Comprehensive Analytical Model for Wormhole Routing in Multicomputer Systems. *Journal of Parallel and Distributed Computing*, 23:202–214, 1994.

[36] W.B. Ligon III and U. Ramachandran. Towards a More Realistic Performance Evaluation of Interconnection Networks. *IEEE Transactions on Parallel and Distributed Systems*, 8(7):681–694, July 1997.

[37] A. Andriahantenaina and A. Greiner. Micro-Network for SoC: Implementation of a 32-Port SPIN Network. In *DATE*, Munich, Germany, March 2003.

[38] A. Andriahantenaina, H. Charlery, A. Greiner, L. Mortiez, and C. A. Zeferino. SPIN: A Scalable, Packet Switched, on-Chip Micro-Network. In *DATE*, Munich, Germany, March 2003.

[39] D. Siguenza-Tortosa and J. Nurmi. Proteo: A New Approach to Network-on-Chip. In *Proceedings of IASTED International Conference on Communication Systems and Network*, Malaga, Spain, 2002.

[40] Mikael Millberg, Erland Nilsson, Rikard Thid, Shashi Kumar, and Axel Jantsch. The Nostrum Backbone – A Communication Protocol Stack for Networks on Chip. In *VLSI Design Conference*, Mumbai, India, January 2004.

[41] Mikael Millberg, Erland Nilsson, Rikard Thid, and Axel Jantsch. Guaranteed Bandwidth Using Looped Containers in Temporally Disjoint Networks within the Nostrum Network on Chip.. In *DATE*, pages 890–895, February 2004.

[42] John Dielissen, Andrei Rădulescu, Kees Goossens, and Edwin Rijpkema. Concepts and Implementation of the Philips Network-on-Chip. In *IP-Based SOC Design*, November 2003.

[43] E. Rijpkema, K.G.W. Goossens, and A. Radulescu. Trade Offs in the Design of a Router with Both Guaranteed Best-Effort Services for Networks on Chip. In *DATE*, 2004.

[44] Praveen Vellanki, Nilanjan Banerjee, and Karam S. Chatha. Quality of Service and Error Control Techniques for Network-on-Chip Architectures. In *Proceedings of IEEE Great Lakes Symposium on VLSI (GLSVLSI)*, Boston, MA, April 2004.

[45] Praveen Vellanki, Nilanjan Banerjee, and Karam S. Chatha. "Quality-of-Service and Error Control Techniques for Mesh based Network-on-Chip Architectures". *Integration, The VLSI Journal*, 38:353–382, January 2005.

[46] H. Zimmer and A. Jantsch. A Fault Model Notation and Error-Control Scheme for Switch-to-Switch Buses in a Network-on-Chip. In *ISSS/CODES*, 2003.

[47] F. Worm, P. Ienne, P. Thiran, and G. De Micheli. An Adaptive Low-Power Transmission Scheme for On-Chip Networks. In *Proceedings of ISSS*, Kyoto, Japan, 2002.

[48] X. Chen and L-S Peh. Leakage Power Modeling and Optimization in Interconnection Networks. In *Proceedings of ISLPED*, Seoul, Korea, 2003.

[49] E. Nilsson and J. Oberg. Reducing Power and Latency in 2-D Mesh NoC Using Globally Pseudochronous and Locally Synchronous Clocking. In *Proceedings of ISSS-CODES*, 2004.

[50] David Brooks, Vivek Tiwari, and Margaret Martonosi. Wattch: a Framework for Architectural-Level Power Analysis and Optimizations. In *International Symposium on Computer Architecture*, pages 83–94, 2000.

[51] W. Ye, N. Vijaykrishna, M. Kandemir, and M.J. Irwin. The Design and Use of SimplePower: A Cycle-Accurate Energy Estimation Tool. In *Proceedings of Design Automation Conference*, June 2000.

[52] D. Pamunuwa, J. Oberg, L.R. Zheng, M. Millberg, A. Jantsch, and H. Tenhunen. Layout, Performance and Power Trade-Offs in Mesh-based Network-on-Chip Architectures. In *IFIP International Conference on Very Large Scale Integration(VLSI-SOC)*, Darmstadt, Germany, pages 362–367, December 2003.

[53] Evgeny Bolotin, Israel Cidon, Ran Ginosar, and Avinoam Kolodny. Cost Considerations in Network-on-Chip. In *Integration – the VLSI journal*, November 2003.

[54] T.T. Ye, L. Benini, and G. De Micheli. Analysis of Power Consumption on Switch Fabrics in Network Routers. In *Proceedings of DAC*, 2002.

[55] H.-S. Wang, L.-S. Peh, and S. Malik. Orion: A Power-Performance Simulator for Interconnection Network. In *International Symposium on Microarchitecture*, Istanbul, Turkey, November 2002.

[56] J. Chan and S. Parameswaran. "NoCEE: Energy Macro-Model Extraction Methodology for Network-on-Chip Routers". In *Proceedings of International Conference on Computer-Aided Design*, San Jose, CA, November 2005.

[57] J. Hu and Radu Marculescu. Exploiting the Routing Flexibility for Energy/Performance Aware Mapping of Regular NoC Architectures. In *DATE*, 2003.

[58] S. Murali and G. De Micheli. Bandwidth-Constrained Mapping of Cores onto NoC Architectures. In *DATE*, 2004.

[59] G. Ascia, V. Catania, and M. Palesi. Multi-Objective Mapping for Mesh-based NoC Architectures. In *Proceedings of ISSS-CODES*, 2004.

[60] K. Srinivasan, K.S. Chatha, and Goran Konjevod. Linear Programming based Techniques for Synthesis of Network-on-Chip Architectures. In *Proceedings of IEEE International Conference on Computer Design (ICCD)*, San Jose, USA, October 2004.

[61] Krishnan Srinivasan and Karam S. Chatha. SAGA: Synthesis Technique for Guaranteed Throughput NoC Architectures. In *Proceedings of ACM/IEEE Asia-South Pacific Design Automation Conference (ASP-DAC)*, Shanghai, China, January 2005.

[62] S. Murali, L. Benini, and G. De Micheli. Mapping and Physical Planning of Networks-on-Chip Architectures with Quality-of-Service Guarantees. In *Proceedings of ASPDAC*, 2005.

[63] U.Y. Ogras and R. Marculescu. Energy- and Performance-Driven NoC Communication Architecture Synthesis Using a Decomposition Approach. In *Proceedings of DATE*, Munich, Germany, 2005.

[64] Krishnan Srinivasan, Karam S. Chatha, and Goran Konjevod. An Automated Technique for Topology and Route Generation of Application Specific on-Chip Interconnection Networks. In *Proceedings of International Conference on Computer-Aided Design*, San Jose, CA, November 2005.

[65] K. Srinivasan and K.S. Chatha. A Low Complexity Heuristic for Design of Custom Network-on-Chip Architectures. In *Proceedings of Design Automation and Test in Europe*, Munich, Germany, March 2006.

[66] A. Pinto, L. P. Carloni, and A.L. Sangiovanni-Vincentelli. Efficient Synthesis of Networks on Chip. In *ICCD*, 2003.

Chapter 19

Power-Performance Modeling and Design for Heterogeneous Multiprocessors

JoAnn M. Paul[1] and Brett H. Meyer[2]
[1]*Virginia Tech*
[2]*Carnegie Mellon University*

Abstract   As single-chip systems are increasingly composed of heterogeneous multi-processors an opportunity exists to explore new levels of low-power design. At the chip/system-level any processor is capable of executing any program (or task) with only differences in performance. When the system executes a variety of different task sets (loading), the problem becomes one of establishing the cost and benefit of matching task types to processor types under anticipated task loads on the system. This includes not only static task mapping, but dynamic scheduling decisions as well as the selection of the most appropriate set of processors for the system. In this chapter, we consider what models are appropriate to establish system-level power-performance trade-offs and propose some early design strategies in this new level of design.

Keywords: heterogeneous multiprocessor; low-power design; system-level design.

1.   Introduction

Power and performance are interrelated. For example, a collection of simpler processors executing at lower frequencies might maintain the performance of a single, more powerful processor by leveraging parallel execution. Presumably the system comprised of the collection of simpler processors could also execute at higher frequencies and so the parallel architecture has more performance potential – processors are simplified, or voltages are turned down, in order to trade off performance for power.

Thus, in order to do meaningful power reduction, performance must, first, be modeled. Therein lies one difficulty of next-generation system-level design.

423

J. Henkel and S. Parameswaran (eds.), Designing Embedded Processors – A Low Power Perspective,
423–448.

For many of these systems, performance will not distill to a single number for latency, throughput, average execution time or some constant performance constraint as in a real-time system or hardware design. Performance-based design evaluation becomes richer as a result of the integration of diverse application types onto the same chip in the creation of new kinds of systems.

Another limitation of power reduction at the system level has been the time to develop and simulate models of many processors at the instruction set simulator level. And yet, intuitively, the ability to efficiently model, manipulate, and simulate system-level design abstractions can ideally lead to new design strategies and new insights into how power-performance trade-offs may be manipulated at the system level.

This chapter describes prior work in high-level power-performance modeling and simulation, to include insight into the necessary level of power annotation for an existing performance simulator. It then takes that prior work one step further to explore how system-level performance modeling of future designs is interrelated with evaluation of overall system power consumption.

2.   The design space

Driving the need to explore system-level power-performance design are, simultaneously, the underlying complexity – and heterogeneity – of both the architecture and the application space. It will soon be possible to place a hundred ARM-equivalent processors on single chips. The diversity of available processing elements (PEs), communication strategies, and memories leads the designer to consider the heterogeneity in both performance and power optimization. Beyond that, however, it is clear that the applications, themselves, will have unprecedented forms of heterogeneous parallelism.

Applications that previously used to be considered in isolation are beginning to be integrated for simultaneous execution on the same computing device, creating entirely new application sets. A common example of this kind of system is the cell phone which can also serve as a web browser and, with more sophisticated HCI, a personal digital assistant (PDA) executing functionality previously associated with laptops or even desktop computers. For these designs, performance can no longer be distilled to a single number for worse-case or average behavior, but will instead result from trade-offs between application sets that simultaneously execute in the system in a variety of situations [1]. The need to also reduce power consumption only increases the complexity of strategically approaching the design space, as well as evaluating the meaning of a "good" design.

We refer to single-chip systems that contain diverse PEs in order to carry out a common objective behavior as "single-chip heterogeneous multiprocessors (SCHMs)." In doing so, we do not wish to add to the alphabet soup of system-level design, but rather to try to be specific about our target systems. They have a finite-sized design space (e.g., that of a chip), where the dominant physical design elements are processors rather than registers [2], and where there is a strong likelihood that those PEs will be heterogeneous. SCHMs are similar to Asymmetric Multiprocessors, as described in the introductory chapter by Grant Martin. The reader is referred to that chapter for a more complete taxonomy. SCHMs are unique in their chip-level programmer's view that relates operating modes to resources. By contrast, other approaches presume an operating system or universal scheduling strategy.

Because the primary design elements of SCHMs are processors, the possibility exists that tasks suited for one processor type may, under certain circumstances, execute on another processor type. The only assumptions required are that the processors are Turing complete (which is quite reasonable) and it may be necessary (or desirable) to have multiple, compiled copies of some tasks on hand. Because the chip is finite, system-level designers can assess the overall cost of taking advantage of executing tasks on processor resources not ideally performance suited to them, as compared to the cost of having portions of the chip be idle in some circumstances. The decision of which thread executes on which PE can be made at the chip level because the chip is presumed to have a finite number of operating modes. Thus, power and performance can be optimized over a finite collection of PEs and a finite set of operating modes.

A primary motivation for designing SCHMs is to exploit new design strategies at the system level to save power. There are two well-known examples for this: substituting multiple PEs at lower clock rates for one at a higher clock rate, and turning off PEs when they are not needed. Both of these approaches may become even more important as leakage current begins to dominate power consumption and instead of reducing transitions, the goal is to reduce the number of unnecessary, powered-on transistors. Here, the use of collections of smaller, simpler cooperative PEs executing at lower frequencies is attractive for three reasons. First, transistors that support performance for general-purpose computation are eliminated as processor types can be more suited to application types. Second, the cost of multiplexing many unrelated or loosely related threads on a high-speed processor can be eliminated by instead executing the threads in parallel. Third, the intelligent management of collections of processors that can satisfy peak loading, but be turned off otherwise, can satisfy different levels of demand on the system while reducing overall power consumption.

None of these strategies can be exploited or even understood, unless system-level design features can be easily manipulated.

3.   MESH

In this section, we summarize how high-level modeling of the SCHM design space has been achieved in the Modeling Environment for Software and Hardware (MESH). In so doing, we will be introducing new modeling elements to some readers at the same time we describe how power modeling is achieved in MESH.

3.1   MESH as a Performance Simulator

MESH can be thought of as a thread-level simulator for an SCHM instead of an instruction-level simulator for a microarchitecture. Using MESH, we have had success exploring SCHM design well above the level of the ISS, where designers manipulate threads, processors, scheduling and communications strategies instead of instructions, functional units, and registers. [3–6] We have compared our performance results with ISS-level models and found reasonable accuracy while execution is typically two orders of magnitude faster than ISS-level models. [6]

In MESH, system performance is simulated by resolving software execution into physical timing using high-level models of processor capabilities, with thread and message sequence determined by schedulers. MESH captures the different sequencing of software and hardware, resolved though scheduling and communication with new design elements. These elements are based upon thread types which capture the contributions to overall system performance of both logical and physical sequencing [7] as well as the layered resolution of the two fundamental kinds of sequencing, also captured as thread-based design elements.

Figure 19.1 illustrates the primitive modeling elements of MESH at a high level. This view provides the essential modeling elements of design of SCHMs

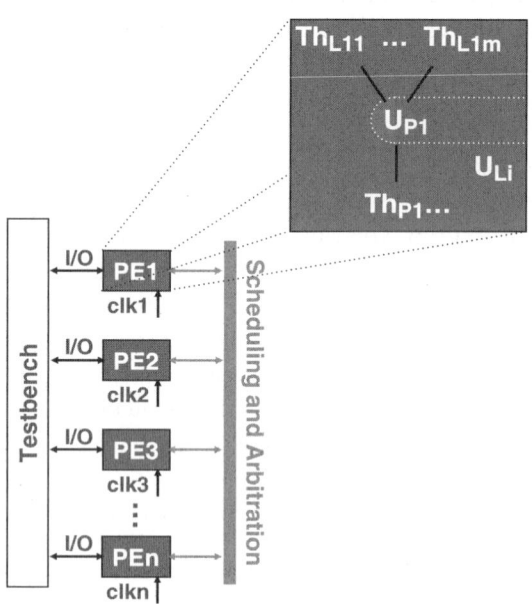

Fig. 19.1   Layered logical and physical design elements in MESH

at the thread level as opposed to the instruction or cycle level. At the lower left of the figure an SCHM is illustrated interacting with some external testbench. The SCHM includes n PEs, with n separate clocks. The cooperation of the PEs in the SCHM is illustrated as a layer of "scheduling and arbitration," shown as a solid, thick line. Many overlapping layers of scheduling and arbitration may exist in SCHMs; the single domain is shown to simplify the illustration of the main design elements of MESH. Also, the label "scheduling and arbitration" may represent networked communications as well as shared busses; the label in the figure simply illustrates processors grouped into a logical, cooperative domain.

In Figure 19.1, $PE1$ is expanded into a MESH "thread stack." The thread stack illustrates all of the design elements contributed by $PE1$ (each PE contributes a similar thread stack to the SCHM). At the bottom is a model of the physical capabilities of the processor resource, modeled as physical thread Th_{P1}. Each unique PE represents a different physical capability within the SCHM, which can be programmed to carry out concurrent behavior. This programming is captured as a collection of logical threads, $Th_{L11}, \ldots, Th_{L1m}$, shown at the top of the expanded view. The logical threads represent the untimed, logical sequencing of software. This captures the effects of software executing on hardware, where physical timing is not determined until the software executes on a model of (or a real) physical machine. Each PE may execute an unbounded number of software (logical) threads, with scheduling decisions potentially made dynamically in response to different datasets. This per-processor, thread-level decision making is modeled by a physical scheduler, U_{P1}.

Cooperation amongst the various PEs, including the exchange of state across private memory spaces as well as communications and task arbitration, is modeled by the cooperative, or logical scheduling domain, U_{Li}. Because U_{Li} represents a cooperative scheduling domain, the threads that execute on any given resource may also be eligible to execute on any other resource that belongs to the same logical scheduling domain. Significantly, this is true even if the processors are heterogeneous.

Because U_{Li} captures the penalty of resource sharing and cooperation [6] we can model what is perhaps the key design challenge of the design of SCHMs at a greatly reduced level of modeling and simulation detail – the trade-off between few powerful, complex, PEs that execute more threads locally, or more less-powerful, simpler, PEs with the cost of global cooperation.

The basic types of threads in MESH are summarized in Table 19.1.

Table 19.1   Basic MESH thread types

Th_{Lij}	One of j logical threads (software) that will execute on processor i
Th_{Pi}	A model of the ith physical resource in the system, such as a processor
U_{Pi}	A scheduler that selects logical threads intended to execute on resource Th_{Pi}
U_{Li}	A logical scheduler that can schedule M threads to N heterogeneous resources

3.2   Energy Modeling in MESH

In extending MESH to model power as well as performance, our hypothesis was the intuitive assumption that different program fragments, which are finer grained than threads, would have different instruction mixes, and so require the annotation of different power consumption values in the MESH threads. This is not what we found.

Our experimentation confirmed the observation that a low proportion of system power is dissipated in the actual execution of instructions; the majority of the power is dissipated in actions that occur for each instruction, such as fetch and decode. [8] Additionally, compilation plays a key role in smoothing out the types of instructions that execute on a processor. In retrospect, this is not surprising, since compilers use patterns to generate code. It is unlikely, therefore, that an ARM executing gcc-compiled code would require more power signatures when executing most real applications. We conclude that when modeling the power consumption of compiler-generated threads executing on processor cores on SCHMs, a single power signature may be used to represent all compiled programs for a given processor with reasonable accuracy. [9]

Since energy is power $\times$ time, this simplification of power modeling leads to system run-time (i.e. performance) being the dominant factor to be dynamically calculated at this level of simulation. Interestingly from both our experimental results and as leakage current comes to dominate, the key to both power modeling and power management at this level of design may be the ability to characterize how long a processor is on or off, rather than what it is processing. The only exception may be when a processor is in some sort of sleep or idle state, in which case it is processing a pathological set of instructions or reduced to even less activity, or when custom hardware is included well above the traditional instruction level. In both cases, power annotation may be a by-product of the design customization instead of a number automatically determined from source code.

Based upon these results, we have added extensions to MESH that estimate energy consumption, resulting in a first-generation high-level power-performance simulator. Energy is calculated in MESH as the power signature associated with a given fragment, multiplied by the time that fragment is executed over the duration of simulation, i.e., the performance of the fragment. All processors are modeled with at least one power signature. While the average power dissipation (signature) of a given fragment is assumed to be constant over the fragment, the execution time of all fragments can be data-dependent, affecting the execution time and thus the power dissipation calculated by MESH. Total system power is then summed over all PEs.

In MESH we currently include the ability to model time penalties associated with memory contention. [6] Since our power calibration currently includes off-chip memory access penalty, our memory access overhead is worse than for PE cores on a single chip. We have observed that reference models (or reference designs) for the calibration and verification of high-level power modeling of on-chip interprocessor coordination and interactions may be more important than instruction-accurate processor power models.

3.3 Power-Performance Design Evaluation in MESH

Spatial voltage scaling was previously introduced as one way the SCHM design space might be narrowed for one example system. [9] Here, we only summarize those results for the sake of motivating, through a relatively simple example, how system-level models can result in entirely new ways of thinking about designs as well as entirely new ways of simultaneously optimizing power and performance.

Spatial voltage scaling is inspired by a blend of dynamic voltage scaling and *processor-rich design*. We describe processor-rich design as the well-known hypothesis that n processors executing at a frequency f/n (and thus a lower voltage) can be substituted for a single processor executing at frequency f for nearly equivalent overall performance and greatly reduced overall power. Of course, this is the case only if the application lends itself to parallel execution and if the cost of the parallelization does not exceed the benefits. Dynamic voltage scaling is a technique that matches the execution frequency (and so voltage level) of a given processor to the demands of the application at runtime. [10] When performance is less critical, processor voltage is scaled down so that it takes longer for a portion of the application to execute. The result is a finite set of voltage levels and a dynamic decision-making strategy for when the different voltage levels available in the processor are applied. Thus, instead of substituting n processors executing at f/n in place of a single processor executing at f, spatial voltage scaling seeks to find not only n, but a set of f_i, one for each processor in the system. Thus, n processors may be substituted for a single processor where the set of processors has a set of frequencies, $\{f_1, \ldots, f_n\}$. The frequency of each processor in the system is fixed at design time, potentially resulting in a heterogeneous set of frequencies on homogeneous processor types. In reality, if the frequency reduction is drastic, a processor would likely be replaced by a simpler processor consisting of a less complex architecture.

Our experimental system is a very simple model of a "document management system." Image and text compression are each potentially required in a given document, but with highly variable numbers and content. The baseline architecture consists of one ARM and one DSP, each executing at 233 MHz with a 1.65 V supply voltage, consuming 4.49 W and 4.62 W for the ARM and DSP respectively. The DSP is modeled based on a Renesas M32R that contains various DSP class instructions such as a MAC; the power consumption of the DSP is modeled after that of the ARM, but adjusted to account for the presence of DSP class instructions.

The baseline software design consists of three thread types that can execute on either processor type. The text portions are processed by a thread that runs *gzip* compression. Each image compression requires two threads: *wavelet* transform and *zerotree* quantization. The execution time of a given *gzip* job is approximately twice as fast on an ARM as it is on a DSP. Conversely, the execution time of the *wavelet* is approximately twice as fast on a DSP as on an ARM. An ARM and a DSP are equally adept at *quantization*.

MESH, coupled with the SVS design strategy, enabled us to converge on a hybrid statically/spatially scaled system consisting of four ARM/DSP pairs capable of dynamic shutdown: two pairs operating at 120 MHz, one pair at 80 MHz, and one pair at 40 MHz. The final design was discovered using SVS and dynamic shutdown. It reduces both power and latency over a baseline design that used dynamic shutdown but did not use SVS. It achieves an average of 66% energy improvement over a baseline case and an average of 15% latency improvement. The latency-energy product of our final design is a 49% improvement over the next best performer. The optimal design is a hybrid statically/spatially scaled system. Its discovery required a combination of a strategic way of exploring the design space, enabled by high-level simulation.

Motivated by this result, we extended our experimentation to include a more diverse set of behavior applications and processor types.

4.   Performance Evaluation of SCHMs

A design is only as good as the techniques used to evaluate its goodness. For traditional embedded designs, this means meeting a performance specification. The design meets the performance criteria or it does not. For general purpose, programmable designs, this means evaluation against a set of representative programs that model the behavior the design, once programmed, is intended to carry out. The set of representative programs is referred to as benchmark suite. In the uniprocessor world, SPEC CPU2000 [11] is a popular set of benchmark applications; SPLASH-2 [12] is the analog in the multi-processor world. Design evaluation is the overall, average performance of the programmed design, where each program is evaluated on the architecture one at a time and each program is presumed to not persist in the system.

By contrast, an SCHM cannot be performance evaluated using either one of these evaluation techniques. This observation is critical if designers are to decide if any given solution is good or not. [3]

Our evaluation scheme for SCHMs is illustrated in Figure 19.2. SCHM design, like computer architecture, starts with a set of application-programs, $S = \{A_1, \ldots, A_n\}$. For a given SCHM there are k sets of up to m applications, S_1 through S_k. Each application in a set S_i executes concurrently with other applications in the set; there are k sets of applications to represent the variety of different operating modes of the system. Each set may have a different amount of applications; $n_i \leq m$, but each set will also (likely) have overlapping applications, leading to the persistence of some applications across different system operating modes. Thus, *time*, a variable missing from the computer architecture approach to system evaluation, is an important aspect of system evaluation, since overlapping sets of applications execute at different points in time and compete for a finite set of resources.

When executed, each application in an application set S_i processes associated input data from the set $F = \{D_1(t), \ldots, D_m(t)\}$. This set contains representative input data for all application sets; input data D_i is only relevant when application A_i is in the application set currently executing. Unlike the input data sets considered by computer architects, these data sets change with time. This also contributes to making time an important element of system evaluation, because applications may behave in different ways at different points in

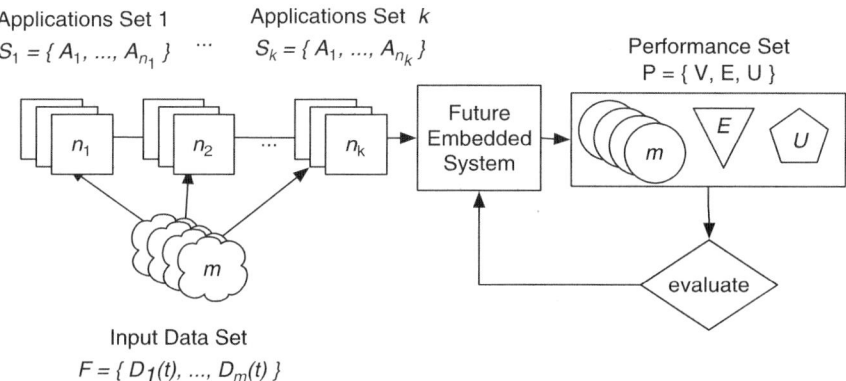

Applications Set 1   Applications Set k
$S_1 = \{A_1, \ldots, A_{n_1}\}$ $\cdots$ $S_k = \{A_1, \ldots, A_{n_k}\}$

Performance Set
$P = \{V, E, U\}$

Input Data Set
$F = \{D_1(t), \ldots, D_m(t)\}$

Fig. 19.2   SCHM design evaluation

time based upon differences in their input sets. Interestingly, this modeling of system inputs over time has an analogy to the testbench of a hardware description language.

The result of executing S with input data F is a set of performance values, $P = \{\{V_1, \ldots, V_m\}, E, U\}$, composed of a set of application performance values (one for each application), V (latency, throughput, slack), system energy consumed, E, and system utilization, U, while executing S with F, over t. More performance variables are possible.

To add to the complexity – and richness – of the design space, the concurrently executing applications within an SCHM arise from application types with different classifications of performance, discussed next.

4.1   Application Performance Classification

Historically, embedded systems have been designed by dividing applications into only two classes: periodic tasks that have both a predictable arrival behavior and a predictable deadline – and everything else. This simple classification is inadequate in evaluating a design space where applications can be classified into a far richer set of performance classes, and those classes, in turn, can be used to identify new forms of optimal designs.

We classify applications by *arrival characteristics and timing, input data content dependence*, and *input data size variation*. In so doing, our purpose is not to create yet another way of classifying the performance of computer systems; nor do we wish to generate any controversy as to whether this is the one, true classification scheme for computer systems. Our purpose is strictly utilitarian. We will use our classification scheme to show how simultaneous consideration of performance-heterogeneous application types can lead to optimal power-performance designs that designers would not otherwise find.

4.2   Arrival Characteristics and Timing

We divide applications into four arrival and timing classes, *periodic, sporadic, timed*, and *streaming*.

Periodic tasks have real-time deadlines such as control updates or sensor monitoring. Examples of applications that do not need to be executed periodically and also do not have real-time deadlines include web browsers and email clients. These programs may be requested by a user at any time, and their execution is not at all constrained by a deadline or deadlines. These sorts of tasks we will denote *sporadic*.

None of the remaining tasks arrive periodically, but all of them have real-time deadlines. These deadlines can take two forms: there may be a single, end-to-end deadline for the whole application, or there may be a series of deadlines that an application must satisfy each time that it is executed.

Timed applications are applications that must be completed as a whole within a certain time frame for correctness or user satisfaction. Examples of this include human–computer interface processing such as handwritten character recognition or voice-activated command recognition. In both of these cases, performance is simply measured as a timely response, start to finish; there are no intermediate conditions or time-based checkpoints that must be satisfied for correctness.

Streaming applications must satisfy recurring, intermediate deadlines. Examples include a cell phone call and MPEG encoding and decoding. In each case, the application must sequentially process a number of frames (voice or video), and each frame must be individually, completely processed before a deadline.

Input Data Content Dependence.   An application is *content independent* (CI) if the application executes for the same length of time regardless of the values of the data being operated on. Examples of this are text compression and filtering, applications that all do some fixed amount of work per fixed amount of data. If an application is content independent, it may be possible to correlate some other property of the input data, for instance, its size, to the execution time of the application and thereby improve the performance of the system by enabling the scheduler to assign the task to a less powerful, and hence less power hungry processor.

An application is *content dependent* (CD) if the values of an input data set have an impact (independent of the size of the input data set) on the execution time of an application. An example of a CD application is MPEG4 decoding. Depending on how much motion there is in a series of frames, decoding the motion estimation frames takes a greater or lesser amount of time. If an application is content dependent, designers are forced to use WCET figures to safely ensure that deadlines are met; however, this does not rule out the possibility of also using other factors, like input data size variation, to improve the quality of the WCET calculation.

Input Data Size Variation.   Application input data size variation, like input data content dependence, is also classified in one of two ways. An application is *size-constant* (SC) if it always processes input data sets of the same size. An example SC application is face recognition, since the input data is a fixed-resolution image (assuming some other application performs pre-processing to locate faces, etc.). If an application is known to be size-constant, then file size cannot be correlated with execution time except in trivial cases, but this relieves the scheduler or designer from considering this parameter when making scheduling decisions or choosing PEs, respectively.

Alternatively, an application is *size-variable* (SV) if the input data processed by it changes from task instance to task instance. An example of this is file encryption; the size of the file alone determines how long the algorithm needs to execute. If an application is size-variable, it is possible that the size of the input data can be correlated to the total execution time of the application, which can assist in calculating WCET.

4.3   Observations

The most significant observation about our classification scheme is that, depending upon the granularity with which a designer is modeling behavior, behaviors can move from category to category. This shift in behavior classification based upon granularity is a phenomenon that actually results in the need for the classification scheme in the first place. At the one level any chip with a global clock might well be seen as an embedded system, where inputs of data and instruction are clocked. However, when considering applications as groups that overlap and persist in the system, it is no longer useful to view systems in this way for no insight will be gained into how to manipulate coarser-grained design decisions.

The second observation, which follows from the first, is that the way any given set of applications may be classified is not unique. This permits designers to understand how granularity groups and transforms in programmable designs while hierarchy only groups in models of physical systems.

Our classification system yields a total of 16 possible classes. Example applications for each of these 16 classes are listed in Table 19.2, which lists applications that are content independent, and in Table 19.3, which lists applications that are content dependent.

Table 19.2   Content independent example applications

	Size-constant	Size-variable
Periodic	Data sampling	N/A
Sporadic	Quantization	Compression
Timed	Repainting screen	Taking a picture
Streaming	Filtering	Decompression

Table 19.3   Content dependent example applications

	Size-constant	Size-variable
Periodic	Sensor monitoring	N/A
Sporadic	Face recognition	Quicksort
Timed	Handwritten character recognition	Speech recognition
Streaming	MPEG encoding	MPEG decoding

We will use this classification scheme as a foundation for system-level performance evaluation, developed by example in the next section.

5. Heterogeneous Performance Balance

Heterogeneous performance balance (HPB) is introduced in this chapter as a means of pointing to not only new designs, but new ways of thinking about the optimality of designs. HPB is illustrated via an example, which we loosely describe as a photographic management system in that the set of applications roughly approximates the requirements of a future system that will permit users increasing control over their photographic images.

Our target implementation is an SCHM with a fixed area budget that must not only execute a set of applications known at design time, but also must have adequate resources to execute a set of *anticipated applications* determined after design time. This captures design for programmability which will be important to virtually all future designs.

The application set includes both DSP and non-DSP tasks as well as applications with all arrival characteristics described in Section 4.2. The set also includes all possible combinations of input data characteristics described in Section 4.2. Given these applications, we set about determining the best mix of processor types assuming fixed system area.

5.1 Processing Elements

Three different processors were chosen for our experiments. The Advanced RISC Machines (ARM) ARM7TDMI (hereafter ARM7) was chosen because it is an extremely low power, but versatile processor. The ARM1136J-S (hereafter ARM11) was selected because it is an extremely high-performance microcontroller, but also because it includes media extensions that accelerate DSP applications. The Infineon TriCorel (hereafter TC1) was chosen because it was specifically designed to execute DSP tasks, but still executes non-DSP tasks relatively well. The main architectural characteristics of the three processors are summarized in Table 19.4.

The ARM7, a no-frills three-stage RISC pipeline microcontroller, is the simplest, smallest, and least power hungry of the three cores, consuming $0.26\,mm^2$

Table 19.4 PE summary

	Clock (MHz)	Power (W)	Area (mm^2)	Type	Pipeline-stages	Issue	Commit
ARM7	133	0.008	0.26	RISC	3	1	1, in order
ARM11	550	0.33	2.85	RISC-DSP	8	1	1, out-of-order
TC1	300	0.195	2.2	DSP	4	2	2, out-of-order

of die per core (1.51 mm^2 including an 8 kB L1 cache and simple five-port router with 4 kB of buffer space [13]) and 0.008 W (0.011 W including cache) when active. It has been designed to execute all applications equally well (or poorly) as it has no special architectural structures designed to improve the execution of specific types of applications. Of the three types of PEs, the ARM7 has clear advantages in terms of power consumption and core area, which are each at least an order of magnitude smaller than either of the other core types; in comparison with the energy hungry ARM11, this processor basically offers free processing power.

A RISC-DSP with a split eight-stage pipeline, the ARM11 is the largest and most power hungry of the three cores, consuming 2.85 mm^2 (5.42 mm^2 including a 32 kB L1 cache and router) of die, and 0.33 W (0.49 W including cache) when active. Though the processor is still considered *scalar* in that it only issues one instruction per cycle, it can issue instructions to two pipelines: a data processing pipeline and a load/store pipeline. The split pipeline allows data processing operations to complete out-of-order with respect to memory operations as long as there are not any interinstruction dependencies. The ARM11 ISA includes advanced arithmetic instructions, like those implemented in DSP architectures, designed to improve the performance of signal processing applications.

In contrast with the ARM11, the TC1 is designed specifically for DSP applications, which it executes better than the ARM11, but because of its lower clock speed, cannot match the ARM11 in terms of general-purpose performance. The TC1 falls between the ARM7 and ARM11 in terms of both area and power consumption, requiring 2.2 mm^2 (4.77 mm^2 including a 32 kB L1 cache and router) of die space and consuming 0.195 W (0.275 W including cache) when active. The TC1 is a *superscalar* processor: it is capable of issuing multiple (up to three) instructions per cycle. These instructions are issued to three different pipelines: a data processing pipeline, a load/store pipeline, and a loop pipeline. Though instructions are not issued out-of-order, they may complete out-of-order as in the ARM11. The advantage that the TC1 has over the ARM11 is that it executes complex arithmetic instructions in half the number of cycles. The TC1 has been designed with power and performance trade-offs in mind, and this is clear from the area and power consumption advantage it has over the ARM11, at the expense of clock frequency.

Each processor is also presumed to be equipped with sleep transistors. [14] Sleep transistors gate the voltage supply to each core, reducing the power consumption of each core to 20% of its active power consumption while it is idle with minimal performance impact. The area, active power, and idle power requirements of each core type (including cache area, cache power, and router area) are summarized in Table 19.5.

Table 19.5 Core area requirements and power consumption including L1 caches

	Area (mm^2)	Active power (W)	Idle power (W)
ARM7	1.51	0.011	0.002
ARM11	5.42	0.49	0.098
TC1	4.77	0.275	0.055

The power consumption and area requirements were derived from information available from ARM [15,16] and Infineon. [17] L1 cache power and area requirements were calculated using eCacti. [18]

Because of the architectural differences between the three cores, there is no easy way to quantify the relative performance of each of the processors. Maximum clock frequency is an inadequate performance metric since the TC1 and ARM11 include instructions far more complex than any instruction the ARM7 can execute. Further, even though both the TC1 and ARM11 may execute essentially the same complex DSP instructions, they have different pipeline depths and instruction issue rates.

Because of these difficulties, and because cycle-accurate simulators are not publicly available for these processors, the performance of each processor was characterized by considering its cycle efficiency executing a variety of application kernels. We selected four DSP kernels, FIR filtering (FIR), vector quantization (VQ), FFT butterfly (FFT), and Levinson–Durbin recursion (LDR). These particular kernels were chosen because they represent a wide variety of signal processing applications and because Infineon has made optimized TC1 assembly language code available for each algorithm. [19] The Dhrystone 2.1 benchmark was chosen to represent general integer and control flow code because, despite its weaknesses, it is a commonly used benchmark and the results are readily available for each of the three processors we chose for our experiments. [16, 17, 20]

Taking the optimized TC1 assembly code, we performed analysis by hand to determine how many cycles the kernel code (not including prolog or epilog code) would take to execute on the ARM7 and ARM11, using instruction execution and pipeline architecture information in publicly available technical reference material. [21–24] These relative cycle counts (per kernel loop iteration) we then used to determine, based on the cycle frequency of each processor, the relative performance of each processor executing each application. These results are summarized in Table 19.6, along with the relative Dhrystone 2.1 benchmark results.

Table 19.6 Relative execution rate of each task executing on each processor

	FIR	VQ	FFT	LDR	Dhrystone
ARM7	1	1	1	1	1
ARM11	12.42	12.42	9.83	9.83	5.08
TC1	13.5	13.5	10.69	10.69	3.46

Table 19.7 Known application set and classification

App.	Kernel	Arrival	Data content	Data size	Processor
1	FIR	Streaming	Independent	Constant	ARM7
2	GI	Sporadic	Dependent	Constant	ARM7
3	GI	Timed	Dependent	Variable	TC1/ARM11
4	LDR	Periodic	Independent	Variable	ARM7/TC1
5	VQ	Sporadic	Independent	Constant	TC11/ARM11

5.2 Applications and System Scheduling

We designed our experimental system to execute a heterogeneous application set, consisting of five applications with widely varying computational requirements and characteristics. The type of computation, and the classifications of each application, along with the processors that can execute each application, are summarized in Table 19.7 (general integer (GI) refers to code represented by the Dhrystone benchmark).

Leveraging information made available through our classification scheme, we implemented a modified version of a list scheduler. [25, 26] This scheduler first attempts to schedule applications 3, 4, and 5 on a lower power processor (the first processor listed in the Processor column of Table 19.7). If there are no processors of the first type available, or if that processor type would not be able to meet the deadline of the specific task instance, then a processor of the second type is used.

List scheduling saves power or improves performance, by taking advantage of lower power processors when they are available, or freeing higher-powered processors to execute other tasks. This can be done when tasks have predictable runtimes, like application 3, or when they do not, like application 4. Since application 3 is CI (content independent) and SV (size-variable), its runtime is predictable. Though the runtime is not always short enough that it can always be scheduled on an ARM7 and still meet its deadline, there are times when this is the case; scheduling this task on an ARM7 saves power if the TC1 is not used for anything else, or improves performance if there is a task that would execute on it instead. Application 4 is handled in a similar manner, except that since it is a CD (content dependent) application, it does not have a predictable runtime; a WCET is must be used.

5.3   Tiles and the Tiled System

A tile-based implementation was selected for the photo management system since the space can be parameterized based on what combinations of elements fit in a single tile. The system has a total of four tiles, or an area constraint of about 93 mm^2. Under the assumption of a 130 nm manufacturing process, each tile in the system may have up to 16 ARM7, 4 ARM11, or 4 TC1 cores, or some smaller number of cores and a 128 kB L2 cache block (eCacti was used to determine the area requirement of the L2 cache). An example system containing 16 ARM7, 4 ARM11, 4 TC1, and 256 kB of L2 cache is illustrated in Figure 19.3.

The advantages of parameterization and uniform interconnect (at the inter-tile level) come at the cost of area overhead. In Figure 19.3, this overhead is manifested as the white space present in each tile. Obviously this is an undesirable overhead, the negative impact of which has been previously explored, [27] but system-level design that uses fixed-size tiles are assumed to trade off to area for the ease of combining blocks. In systems considered in our experiments, this overhead ranges from 4% to 17%, with systems that make heavy use of TC1 cores suffering the most. This is due to the fact that even though TC1s occupy a middle ground between ARM7s and ARM11s in terms of area requirement, the area difference is not significant enough to allow more cores of this type to be included in a tile.

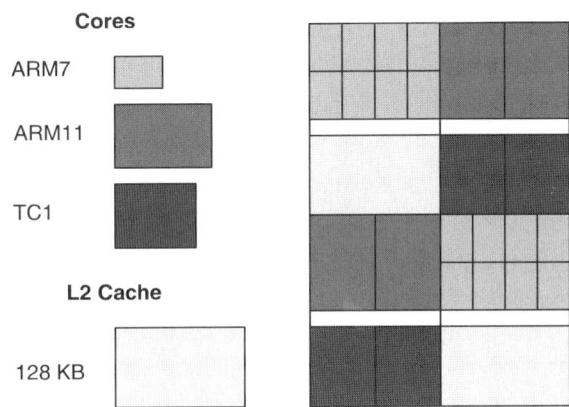

Fig. 19.3   Example system with 16 ARM7, 4 ARM11, 4 TC1, and 256 kB of L2 cache

5.4   Modeling Assumptions

For the purposes of performance, we assumed a perfect memory system, i.e., caches always have the requested data. This can be justified in part by the fact that the digital signal processing algorithms used in this work all have regular memory access patterns that could very easily be taken advantage of by a pre-fetch unit. Despite this assumption, we modeled the power dissipated in the caches. Without a pre-fetch unit, 20% of cache accesses would miss (a conservative estimate), meaning that the L2 cache must be used to service 20% of load/store instructions. Under the assumption that 30% of instructions executed are memory access instructions, [28] this means that L2 caches are accessed by 6% of all instructions.

Since each processor executes at a different rate, the L1 caches of each processor type consume a different amount of power. Though accessed by the same percentage of instructions, each processor executes at a different rate, and hence the L1 caches of different processors are accessed at different rates. If the ARM 11 accesses its L1 cache 30% of system cycles (defined as 550 MHz, the clock frequency of the ARM11), then the TC1 and ARM7 access their L1 caches 16% and 7% of system cycles respectively.

Network communication delay is not included in the system model. This is partially justified by the small impact that communication delay would have on any of the experimental architectures if it were considered under the assumption of perfect caches; in any assumption of perfect caches in a shared memory machine is the additional assumption that network latency is sufficiently hidden by pre-fetching.

Network power consumption does not represent a significant portion of the power consumed in the system. In a 130 nm process, interconnect power consumption is dominated by buffering, [29] and the power cost of buffering in our systems is 15% of the total system power or less for the overwhelming majority of architectures. Under worst case operating conditions, in the overwhelming majority of architectures interconnect power consumption constitutes 15% of total system power consumption or less. This is based on the assumption that only 6% of instructions executed by each processor lead to an L2 access. In this case, ARM7s, TC1s, and ARM11s generate one request every 17, 31, and 67 network cycles respectively. This equates to approximately 1, 0.5, and 0.25 active requests per ARM11, TC1, and ARM7 respectively at any given time. If half of these requests encounter congestion at each hop and must be buffered, then the power consumed by buffering is at most 0.3 W, less power than a single active ARM11. This is only significant for a system with 40 or more ARM7s (and hence two or fewer TC1s or ARM11s), since each ARM7 only consumes 0.011 W.

Scheduling overhead is not included in the model. A conservative estimate of the cost of resource-task pairing is 2000 cycles, and an ultraconservative assumption is that each time this occurs the scheduling information that must be gathered takes 1000 cycles to complete. Then, 55,000 pairings (typical for the 30 s simulation we conducted) translates to 0.3 s, or 1% of the total simulation time on one processor, and 1.4 uW, two orders of magnitude less than the typical average system power.

5.5   Experiments and Results

Experiments and results are described as a processor vector, where $R_{x,y,z}$ denotes an architecture with x ARM7s, y TC1s, and z ARM11s. Hence, $R_{16,2,6}$ is an architecture with 16 ARM7s, 2 TC1s, and 6 ARM11s.

A set of applications intended to model the known, design time, set as shown in Table 19.7 was executed for 30 s on 10 different architectures. The range of possible architectures is bounded by the constraints of the known applications: only those architectures capable of meeting all application deadlines were considered. These initial results [30] (not shown) were as expected. The latency of all applications changes in proportion to the number of processors in the system that can execute it. Applications 3 and 5 each benefit from systems with large numbers of high-powered processors. Application 2 on the other hand, benefits the most from systems with very few high-powered processors.

A second-order analysis (also not shown) reveals that systems $R_{16,2,6}$, $R_{16,4,4}$, $R_{16,6,2}$ strike the best balance; Applications 2, 3, and 5 achieve reasonable performance, at 75%, 32%, and 26% worse than optimal respectively (compared with 270%, 193%, and 195% worse than optimal in the worst case).

Figure 19.4 shows the energy consumed by each architecture, and the energy that would be consumed in the system were all resources utilized the whole time. The energy actually consumed during simulation varies from 27 to 45 J, or the power dissipated varied from 0.9 to 1.5 W. The clear upward trend in both the maximum energy and actual energy consumed is due to the replacement of ARM7s (0.011 W) with more power hungry TC1s (0.275 W) and ARM11s (0.49 W). The spike in actual and maximum energy at $R_{16,6,2}$ is due the number of ARM11s in that system; architectures to the right and left of it in the plot have both fewer ARM11s and more TC1s, a trade that results in lower total system energy consumption.

Application execution latency must be balanced with system power consumption. This leads to the normalized system value plot of Figure 19.5. Architecture $R_{16,2,6}$ is the architecture that best balances system performance and system power.

However, the system best equipped to handle both the known application set *and* applications not known at design time is another matter. To gain insight

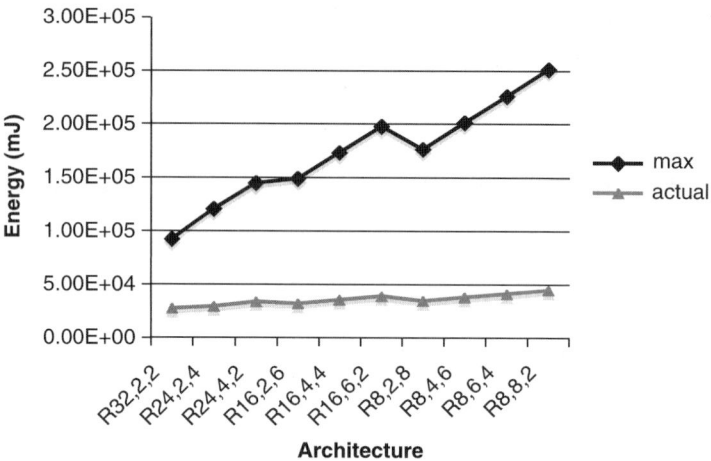

Fig. 19.4   The actual energy consumed by each system, and the energy that would be consumed at maximum utilization

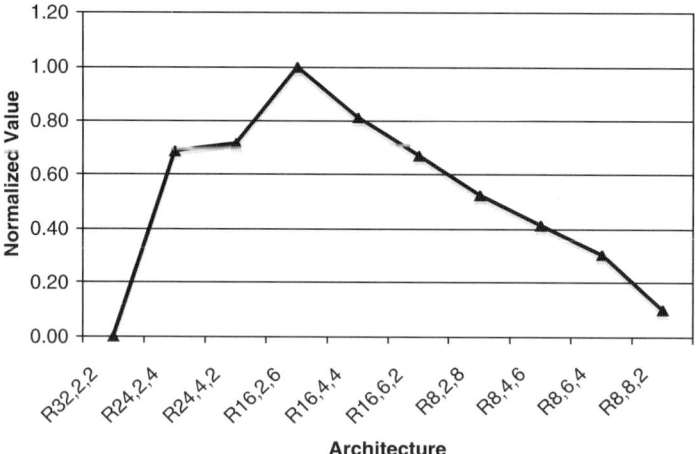

Fig. 19.5   Normalized linear system value as a function of application latency and system energy consumption

into this, systems that had the most leftover resources were considered, i.e., which systems had the most capacity leftover after applications known at design time were mapped.

Figure 19.6 illustrates the available capacity, or the unutilized resources, of each system. This is measured in equivalent unutilized ARM7s, since each system has a different set of processors. ARM11s and TC1s are converted into ARM7s based on how many ARM7s it would take to perform the same amount of work (see Table 19.6); the systems consistently have more DSP resources

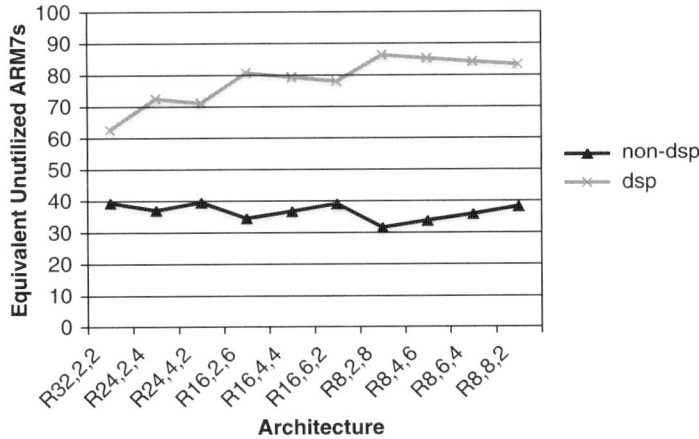

Fig. 19.6   Available resources measured in equivalent unutilized ARM7s

available because each ARM11 and TC1 is equivalent to 10 or more ARM7s, compared with 3–5 ARM7s when executing non-DSP applications. The two curves appear to be nearly mirror images because TC1s contribute more DSP resources than ARM11s, but ARM11s contribute more non-DSP resources than TC1s. Hence, the systems with the most DSP resources are those systems with the most TC1s, and therefore the systems with the most non-DSP resources are those with the most ARM11s.

What is more important than the amount of unreserved resources, however, is the cost of taking advantage of them. Figure 19.7 illustrates the cost of the unreserved resources in Figure 19.6; this figure demonstrates the dilemma in designing future systems: more resources clearly means higher power consumption, and this relationship in our experiment is clear. The second-order peaks in the DSP plot of Figure 19.7 are, again, due to the lower power consumption of the TC1 cores relative to ARM11 cores; TC1 cores offer more DSP processing power per Watt than the ARM11 cores, which make $R_{16,2,6}$ and $R_{8,2,8}$ local maximums, though neither of these architectures offer the resources to Watt efficiency of $R_{32,2,2}$, in either the non-DSP or DSP case.

Figure 19.8 combines application latency, system energy consumption, unreserved resources, and the cost of accessing them, in a single value function. Interestingly, considering unreserved resources and the cost of accessing them brings to the fore an architecture that did no better than fourth best in any of the previous system value calculations: $R_{24,2,4}$. This architecture is the best choice for systems where the unknown applications are either non-DSP applications or a combination of non-DSP and DSP applications. $R_{16,2,6}$ is the best choice if all unknown applications are DSP applications, but is a distant third

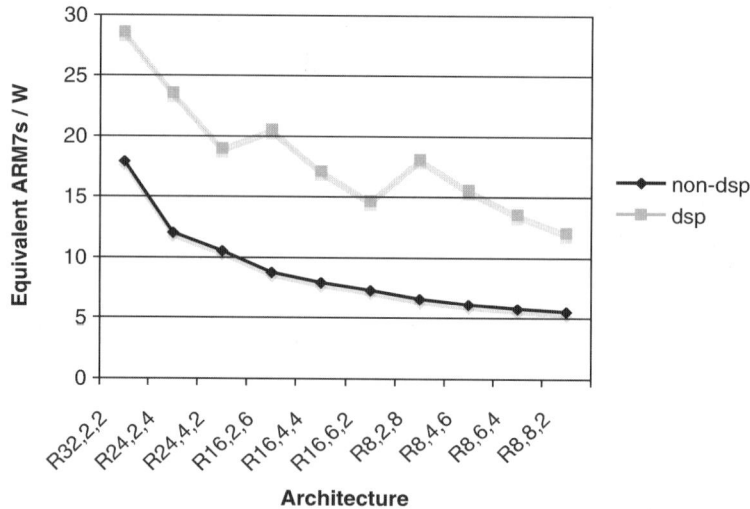

Fig. 19.7   The energy cost of utilizing the unutilized resources in each architecture

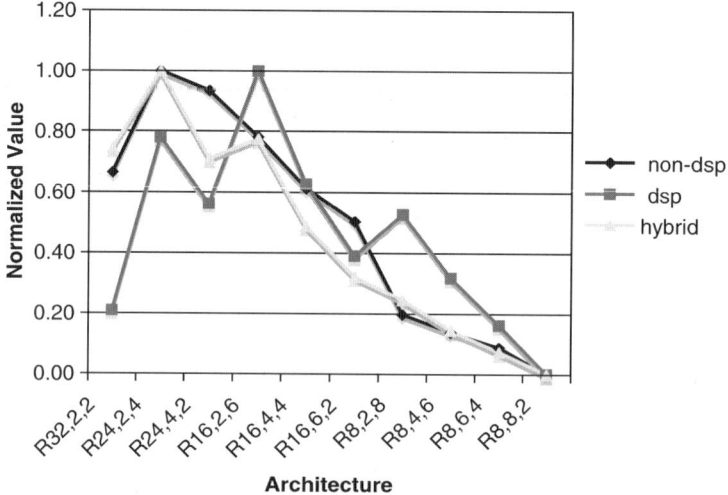

Fig. 19.8   Normalized relative value of each architecture as a function of application latency,
system energy, available resources, and cost of utilizing those resources

in each of the other categories, offering a solution more than 20% worse than
$R_{24,2,4}$ in both cases. Though $R_{16,2,6}$ is among the top performers, or at least a
reasonable compromise, its unreserved resources are clearly too expensive to
make use of (see Figure 19.7) for non-DSP applications.

5.6   Discussion

The system that was finally identified as the optimal system was not one that would have been chosen at any previous step in the analysis. $R_{24,2,4}$ balanced the known application execution, system energy consumption, system utilization, and system cost of executing anticipated applications better than the next best systems (considering a hybrid and non-DSP set of anticipated applications). In these cases, it also performed more than 20% better than the system that up until the point that system utilization was considered was the best system: $R_{16,2,6}$.

If the experimental systems were evaluated based on any one criterion, $R_{16,2,4}$ would have been a reasonable candidate. If the systems were evaluated based on everything except for the cost of system utilization by anticipated applications, $R_{16,2,4}$ would have been chosen. Instead, when all criteria are considered, the result is quite unexpected. $R_{24,2,4}$ does not initially appear to be a system that balances performance, power and utilization, but it emerged as optimal in two of the three cases we considered.

6.   Conclusions

When performance no longer distills to a single number, power can no longer be considered a trade-off against a single performance value. The single most important point of system-level design is that a wider, more diverse design space requires a more holistic definition of design optimality.

Leveraging heterogeneity at the system level, including not only architecture heterogeneity, but application heterogeneity, can lead to non-intuitive forms of optimality. The only way this heterogeneity can be leveraged in design is to develop more comprehensive performance models for future single-chip designs. For programmable designs, this means modeling a semi-persistent set of applications with variable computation times that compete for a finite set of heterogeneous processor resources.

Software and hardware are already heterogeneous models, but even more heterogeneity lies in the different performance categories the software takes on, inputs that continuously exercise the system, and the different characteristics of candidate processing elements and other system infrastructure upon which the concurrent software executes. These multifaceted forms of heterogeneity simultaneously contribute to overall system performance and are all part of the design space, and so models for each must simultaneously contribute to system simulation and design evaluation. Reduction of modeling detail while preserving and even embracing heterogeneity remain the goals of MESH. This chapter only begins to scratch the surface of what this new and exciting level of design integration will bring.

Acknowledgements

This work was supported in part by ST Microelectronics and the National Science Foundation under Grants 0607934 and 0606675. Any opinions, findings, and conclusions or recommendations expressed in this material are those of the authors and do not necessarily reflect the views of the NSF. The authors would like to thank Donald E. Thomas for his careful reading of this chapter.

References

[1] J.M. Paul. Programmer's Views of SoCs, *International Conference on Hardware/Software Codesign and System Synthesis (CODES-ISSS)*, pp. 159–161, October 2003.

[2] F. Karim, A. Mellan, A. Nguyen, U. Aydonat, T. Abdelrahman, A multi-level computing architecture for embedded multimedia applications, *IEEE Micro*, Vol. 24, pp. 56–66, 2004.

[3] J.M. Paul, D.E. Thomas, A. Bobrek, Benchmark-based design strategies for single chip heterogeneous multiprocessors, *International Conference on Hardware/Software Codesign and System Synthesis (CODES-ISSS)*, pp. 54–59, 2004.

[4] A.S. Cassidy, J.M. Paul, D.E. Thomas, Layered, multi-threaded, high level performance design, 6th *Design, Automation and Test in Europe (DATE)*, pp. 954–959, 2003.

[5] J.M. Paul, A. Bobrek, J.E. Nelson, J.J. Pieper, D.E. Thomas, Schedulers as model-based design elements in programmable heterogeneous multiprocessors, 40th *Design Automation Conference (DAC)*, pp. 408–411, 2003.

[6] A. Bobrek, J.J. Pieper, J.E. Nelson, J.M. Paul, D.E. Thomas, Modeling shared resource contention using a hybrid simulation/analytical approach, *Design, Automation and Test in Europe (DATE)*, Vol. 2, pp. 1144–1149, 2004.

[7] C.L. Seitz, System timing, *Introduction to VLSI Systems*, C. Mead, L. Conway, Eds., Reading, MA: Addison-Wesley, 1980.

[8] T. Weiyu, R. Gupta, A. Nicolau, Power savings in embedded processors through decode filter cache, 5th *Design Automation and Test in Europe (DATE)*, pp. 443–448, 2002.

[9] B.H. Meyer, J.J. Pieper, J.M. Paul, J.E. Nelson, S.M. Pieper, A.G. Rowe, Power-performance simulation and design strategies for single-chip heterogeneous multiprocessors, *IEEE Transactions on Computers*, vol. 54, Iss. 6, June 2005.

[10] T.D. Burd, T.A. Pering, A.J. Stratakos, R.W. Brodersen, A dynamic voltage scaled micro-processor system, *IEEE Journal of Solid-State Circuits*, Vol. 35, pp. 1571–1580, 2000.

[11] J.L. Henning, SPEC CPU2000: measuring CPU performance in the New Millennium, *Computer*, Vol. 33, Iss. 7, July 2000.

[12] S. Woo, M. Ohara, E. Torrie, J. Sing, A. Gupta, The SPLASH-2 programs: characterization and methodological considerations, *International Symposium on Computer Architecture 1995*, June 1995.

[13] W.J. Dally, B. Towles, Route packets, not wires: on-chip interconnection networks, *Design Automation Conference 2001*, 2001.

[14] P. Babighian, L. Benini, E. Macii, Sizing and characterization of leakage-control cells for layout-aware distributed power-gating, *Proceedings of the Design, Automation and Test in Europe (DATE)*. 2004.

[15] ARM7TDMI, http://www.arm.com/products/CPUs/ARM7TDMI.html, 2005.

[16] ARM1136J(F)-S, http://www.arm.com/products/CPUs/ARM1136JFS.html, 2005.

[17] TriCore™ 1 – 32-bit MCU-DSP Architecture, http://www.infineon.com/ cgi/ecrm.dll/ecrm/scripts/prod_ov.jsp?oid=30926&cat_oid= −83 62& stlnocount=true, 2005.

[18] M. Mamidipaka, N. Dutt, eCacti: an enhanced power estimation model for on-chip caches, CECS Technical Report #04-28, University of California Irvine, 2004.

[19] TriCore 32-bit Unified Processor DSP Kernel Benchmarks, http://www.infineon.com/cgi/ecrm.dll/ecrm/scripts/public_download.jsp?oid=45812&parent_oid=30926, 2002.

[20] Chipdir, http://www.xs4all.nl/~ganswijk/chipdir/fam/arm/, 2005.

[21] ARM7TDMI Product Overview, http://www.arm.com/pdfs/DVI0027B_7_R3.pdf, 2001

[22] ARM7TDMI (Rev 4) Technical Reference Manual, http://www.arm.com/pdfs/DDI0210B_7TDMI_R4.pdf, 2003.

[23] The ARM11 Microarchitecture, http://www.arm.com/pdfs/ARM11 MicroarchitectureWhite Paper.pdf, 2002.

[24] ARM1026EJ-S r0p2 TRM, http://www.arm.com/pdfs/DDI0211E_arm1136_r0p2_trm.pdf, 2003.

[25] T.L. Adam, K.M. Chandy, J.R. Dickson, A comparison of list schedules for parallel processing systems, *Communications of the ACM*, Vol. 17, pp. 685–690, Dec. 1974.

[26] B.A. Shirazi, A.R. Hurson, K.M. Kavi, *Scheduling and Load Balancing in Parallel and Distributed Systems*, IEEE Computer Society Press, Los Alamitos, CA, 1995.

[27] A. Jalabert, S. Murali, L. Benini, G.D. Micheli, xpipesCompiler: a tool for instantiating application specific networks on Chip, 7th *Design, Automation and Test in Europe (DATE)*, 2004.

[28] J.L. Hennessy, D.A. Patterson, *Computer Architecture*, Third Edition, Morgan Kaufmann, pp. 112, 138–9, 142, 2003.

[29] T.T. Ye, L. Benini, G.D. Micheli, Analysis of power consumption on switch fabrics in network routers, 39th *Design Automation Conference (DAC)*, 2002.

[30] B.H. Meyer, Toward a new definition of optimality for programmable embedded systems, CMU-CSSI Tech Report No. CSSI 05-04.

VI

Reconfigurable Computing

Chapter 20

Basics of Reconfigurable Computing

Reiner Hartenstein and TU Kaiserslautern
Kaiserslautern University of Technology,
Kaiserslautern, Germany

Abstract   This chapter introduces the basic concepts of Reconfigurable Computing and its disruptive impact on the classical instruction-streambased mind set of computing sciences. It illustrates the essentials of the paradigm shift by Reconfigurable Computing and explains the mechanisms behind the massive speed-ups obtained by software to configware migration.

Keywords: accelerator, ASIC, ASM, auto-sequencing memory, Bill Gates, coarsegrained, communication technology, computational density, configware, cryptology, CS curricula, CS education, datastream, design crisis, disruptive technology, DMA, DSP, dynamically reconfigurable, educational deficits, face detection fast Fourier transform, FFT, flowware, FPGA, GAG, generic address generator, genome analysis, hidden RAM, image compression, instruction stream, Kress-Kungmachine, MAC, microprocessor, Moore's law, MPI, multimedia, nano technology, oil and gas, paradox, parallelism, pipe networks, protein identification, radar, Reconfigurable Computing, reconfigurable main processor, reconfigurability overhead, remote reconfiguration, scalability, Smith-Waterman, software to configware migration, speedup, stereo vision, supercomputing, systolic array, terminology, transdisciplinary, wiring overhead, Xputer

Already decades ago the instruction-stream-based von Neumann (vN) paradigm [1, 2] lost its dominant role as the basic common model of computing systems. After many technology generations the microprocessor CPU, together with its software still the monopoly-like main focus point of CS education, often cannot meet even usual performance requirements. In most PCs it cannot

J. Henkel and S. Parameswaran (eds.), Designing Embedded Processors – A Low Power Perspective,
451–501.
© 2007 *Springer.*

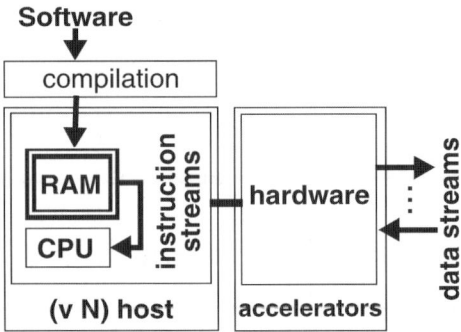

Fig. 20.1   Accelerators for embedded systems

even drive its own display: it needs a graphics accelerator. The vN paradigm as a common model has been replaced by a symbiosis of CPU and hardwired accelerators (Figure 20.1), not only for embedded systems. Most MIPS equivalents have been migrated from software to accelerators: software to ASIC migration [3], and more and more software to configware migration to be run on FPGAs (Field-Programmable Gate Arrays). The microprocessor has become the tail wagging the dog [4]. The basic model of most accelerators is data-stream-based (Figure 20.1, however, it is not a "dataflow machine"). It is not instruction-stream-based. For detailed explanations on this duality of paradigms see Section 2.

Accelerator design became difficult.   But soon also the design of accelerators has become increasingly difficult: the second VLSI design crisis (Figure 20.2).

With technology progress the mask cost and other NRE cost are rapidly growing (Figure 20.2a). Also the design cost is rapidly increasing, made worse by decreasing product life cycle length (Figure 20.2b). This explains the decreasing number of wafer starts (Figure 20.2c) as well as the rapid growth of the number of design starts based on FPGAs instead of ASICs (Figure 20.2d, [5]): the next paradigm shift, where the accelerator has become reconfigurable (Figure 20.3), so that the product life cycle can be extended by upgrading the product, even at the customers site by remote reconfiguration (Figure 20.2e). Now both paradigms are RAM-based: the instruction-stream-based CPU, and the data-stream-based FPGA or rDPA (reconfigurable Data-Path Array, see Section 4). The accelerator's RAM is no block RAM: it is hidden inside the reconfigurable interconnect fabrics: hRAM.

Most important silicon revolution after introduction of the microprocessor.
Such Reconfigurable Computing (RC) [4, 6–9] is the most important revolution to silicon application after the introduction of the microprocessor [10].

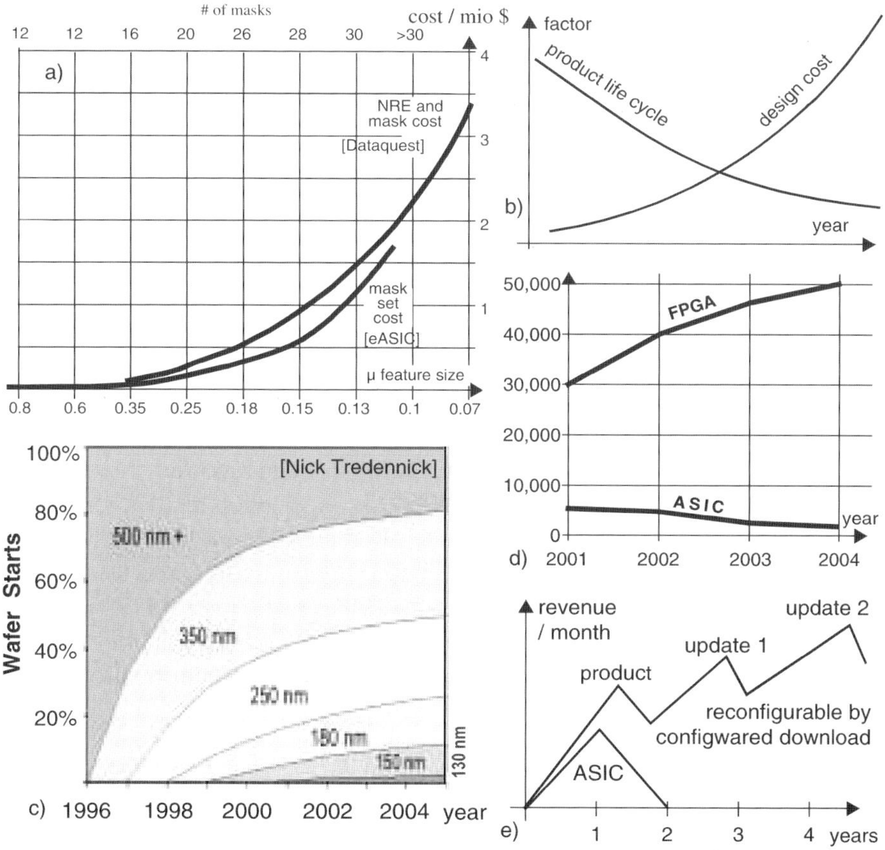

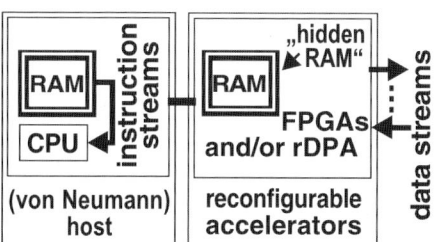

Fig. 20.2 The second design crisis (silicon crisis): a) increasing NRE and mask cost, b) growing design cost and decreasing product life cycle, c) hesitating new technology adoption, d) growing FPGA-based design starts vs. declining ASIC design starts [5] [90], e) reconfigurability extends product life cycles

Fig. 20.3 FPGAs replacing ASICs

It is the most important new machine paradigm with common model capability [11–16]. Emerging in the 1980s and now having moved from a niche market to mainstream we have with FPGAs a third class of platforms filling the gap between vN-type procedural compute engines and ASIC application-specific hardware [17]. FPGAs are the fastest growing segment of the semiconductor market.

Disruptive methodology.   When having been intel CEO Andy Grove claimed, that each technology providing a factor of 10 or more improvements over an established one, can be expected to become disruptive [18]. From CPU software to FPGA configware migrations speedup factors by up to more than three orders of magnitude have been published (Figure 20.4). This massively disruptive community is reporting a factor of 7.6 in accelerating radiosity calculations [19], a factor of 10 for FFT (Fast Fourier Transform), a speedup factor of 35 in traffic simulations [20]. For a commercially available Lanman/NTLM Key Recovery Server [21] a speedup of 50–70 has been reported. Another cryptology application reports a factor of 1305 [23]. A speedup by a factor of 304 is reported for a R/T spectrum analyzer [25]. In the DSP area [26] for MAC [26] operations a speedup factor of 100 has been reported compared to the fastest DSP on the market (2004) [27]. Already in 1997 vs. the fastest DSP a speedup between 7 and 46 has been obtained [28]. In biology and genetics (also see [29, 30]) a speedup of up to 30 has been shown in protein identification [31], by

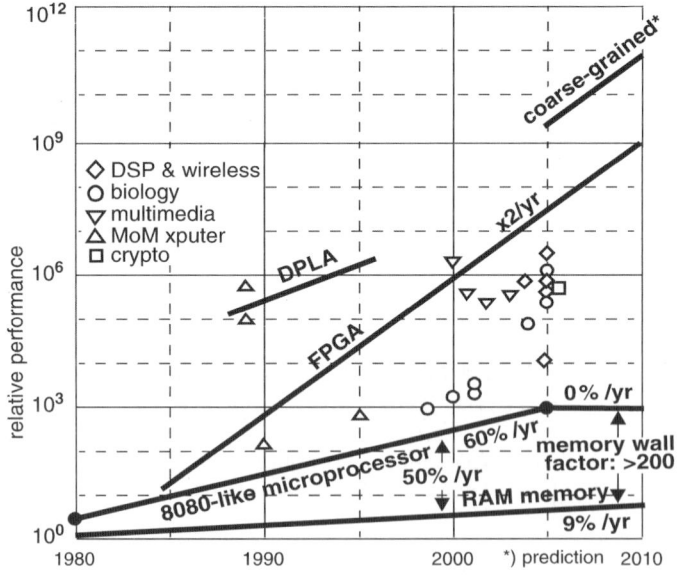

Fig. 20.4   Speedup by software to configware migration

133 [32] and up to 500 [33] in genome analysis, as well as 288 with the Smith–Waterman pattern matching algorithm at the National Cancer Institute [35]. In the multimedia area we find factors ranging from 60 to 90 in video rate stereo vision [36] and in real-time face detection [37], and of 457 for hyperspectral image compression [38]. In communication technology we find a speedup by 750 for UAV radar electronics [39]. These are just a few examples from a wide range of publications [41, 42, 44–46, 48, 50] reporting substantial speedups by FPGAs. Fortunately, in embedded systems, highest computational requirements are determined by a small number of algorithms, which can be rapidly migrated via design automation tools [51].

The first reconfigurable computing paradox. These disruptive performance results are a surprise. The first reconfigurable computing paradox [53–57]. The technology parameters are so bad that Andy Grove's rule is missed by several orders of magnitude. The effective integration density really being available to the FPGA application is much worse than this physical integration density. It is behind the Gordon Moore curve by about four orders of magnitude (Figure 20.5). Wiring overhead costs about two orders of magnitude and reconfigurability overhead costs about another two orders of magnitude (only about 1 out of about 100 transistors serves the application, whereas the other 99 transistors deserve reconfigurability). Another source of area inefficiency of FPGAs is routing congestion, which could cost a factor of 2 or more: not all desired connections can be routed because the supply of routing resources is exhausted. It is astonishing, how with such a bad technology such brilliant acceleration factors are obtained. For the explanation of

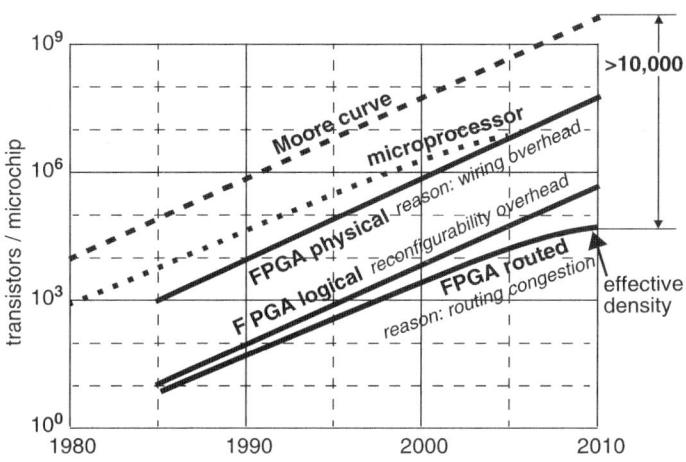

Fig. 20.5 The effective FPGA integration density

most speedup mechanisms involved see Section 4, since this is more easy at the higher abstraction level given by coarse-grained reconfigurable computing.

Earlier alternatives.   Let us have a brief look onto earlier alternatives to the FPGA. With the MoM-2, an early reconfigurable computer architecture, the following speedup factors have been obtained: 54 for computing a 128 lattice Ising model, >160 for Lee Routing, >300 for an electrical rule check, >300 for a 3by3 2D FIR filter [58], and between 2,300 and 15,000 for grid-based VLSI design rule check [4, 59–61]. Instead of FPGAs, which have been very small at that time, the MoM-2 used DPLA, a programmable PLA, which has been designed at Kaiserslautern and manufactured via the Mead-&-Conway-style multi-university VLSI project E.I.S [62]. The DPLA has been especially efficient for computing Boolean expressions. At the time it has been designed, a single DPLA replaced 256 state of the art FPGAs available commercially.

FPGAs became mainstream.   Already many years ago FPGAs have become mainstream throughout all embedded systems areas (Figure 20.6a). More recently, the FPGA-based pervasiveness of Reconfigurable Computing (RC) also spread over all application areas of scientific computing (a few example are listed in Figure 20.6b). As a side effect in addition to speedup factors of disruptive dimensions also a drastic reduction of the electric energy budget is provided down until about 10% [63] – along with a reduction of equipment cost by a similar dimension [63]. This has been reported from the supercomputing community. So far the immense electricity consumption has been considered one of the most severe obstacles on the classical way to the petaflop supercomputer. This side effect [64] extends the scope of the low power design community [65–68] beyond dealing with devices powered by a tiny battery.

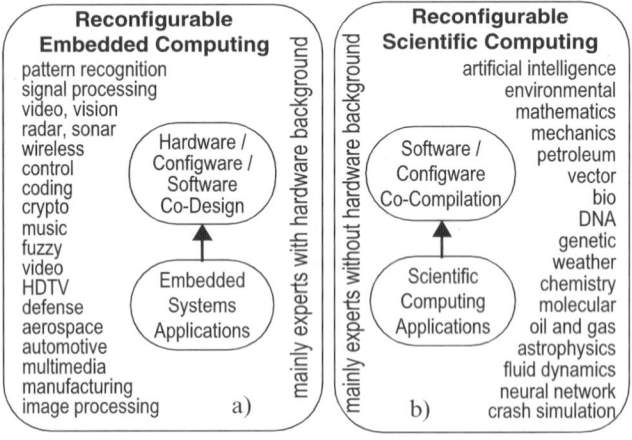

Fig. 20.6   Two different reconfigurable computing cultures

Low power design.   As leakage power and total power became a dramatic issue in very deep submicron technologies, the low power design community started in 1991 to take care of these power problems. As a spin-off of the PATMOS project funded by the Commission of the European Union [70], the annual PATMOS conference was founded as the first series on this topic area world-wide [71]. (PATMOS stands for Power and Timing Modelling, Optimization and Simulation.) Its annual sister conference series ISLPED (International Symposium on Low Power Electronics and Design) has been founded four years later in the USA [72]. The low power design community started exploring new design methodologies for designing leakage tolerant digital architectures, based on architectural parameters like activity, logical depth, number of transitions for achieving a given task and total number of gates. An early proposed design method selects the best architecture out of a set of architectures (baseline, sequential, parallel, pipelined, etc.) at optimal Vdd and threshold voltages. Another design method takes as constraints given Vdd and threshold voltages.

Hardware design on a strange platform.   Inside the embedded systems scene at first glance the use of reconfigurable devices like FPGAs has looked more like a variety of hardware design, but on a strange platform. But now we have two reconfigurable computing scenes (Figure 20.6). Meanwhile FPGAs are also used everywhere (even in oil and gas [73]) for high performance in scientific computing. This is really a new computing culture – not at all a variety of hardware design. Instead of HS co-design (Figure 20.1) we have here software/configware co-design (S/C co-design) (Figure 20.3), which is really a computing issue. This major new direction of developments in science will determine how academic computing will look in 2015 or even earlier. The instruction-stream-based mind set will loose its monopoly-like dominance and the CPU will quit its central role – to be more an auxiliary clerk, also for software compatibility issues, for running legacy code: a CPU co-processor serving to a reconfigurable main processor.

Educational deficits.   This new direction has not yet drawn the attention of the curriculum planners within the embedded systems scene. For computer science this is the opportunity of the century of decampment for heading toward new horizons [75–77]. This should be a wake-up call to CS curriculum development. Each of the many different application domains has only a limited view of computing and takes it more as a mere technique than as a science on its own. This fragmentation makes it very difficult to bridge the cultural and practical gaps, since there are so many different actors and departments involved. Only computer science can take the full responsibility to merge Reconfigurable Computing into CS curricula for providing Reconfigurable Computing Education from its roots. CS has the right perspective for a transdisciplinary unification in dealing with problems, which are shared across many different application domains. This new direction would also be helpful to reverse the current down trend of CS enrolment.

Unthinkable without FPGAs. The area of embedded systems is unthinkable without FPGAs [78]. This has been the driving force behind the commercial breakthrough of FPGAs. Almost 90% of all software is implemented for embedded systems [79, 81, 82] dominated by FPGAs usage, where frequently hardware/configware/software partitioning problems have to be solved. The quasi monopoly of the von Neumann mind set in most of our CS curricula prohibits such a dichotomic qualification of our graduates, urgently needed for the requirements of the contemporary and future job market. At a summit meeting of US state governors Bill Gates has drastically criticized this situation in CS education.

FPGAs and EDA. The pervasiveness of FPGAs also reaches the EDA (Electronic Design Automation) industry, where all major firms spend a substantial effort to offer a variety of application development tools and environments for FPGA-based product development [84]. Also FPGA vendors have EDA development efforts and cooperations with firms in the EDA industry and offer such tools and development environments. Since this is a highly complex market, this chapter does not go into detail because of a lack of space.

The Kress–Kung machine. After switch-on of the supply power the configuration code has to be downloaded to the FPGA's hRAM, which is a kind booting like known from the vN processor. But the source of this code for FPGAs is not software. It definitely does not program instruction streams. The advent of FPGAs provides a second RAM-based fundamental paradigm: the *Kress–Kung machine* [85], which, however, is data-stream-based, and not instruction-stream-based. Instead of organizing the schedule for instruction executions the compilation for FPGAs has to organize the resources by placement and routing, and, based on the result, to implement the data schedules for preparing the data streams moving through these resources (Figure 20.7d). FPGAs or the Kress–Kung machine, respectively, has *no "instruction fetch" at run time* (Figure 20.8). Not to confuse students and customers with the term *"software"* another term is used for these non-procedural programming sources of RC: the term *configware*. Not only FPGA vendors offer configware modules to their customers. Also other commercial sources are on the market: a growing configware industry – the little sister of the software industry.

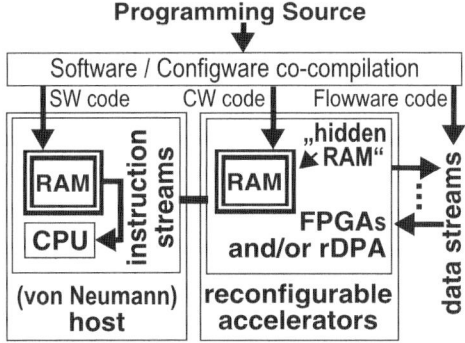

Fig. 20.7 The dual-paradigm model

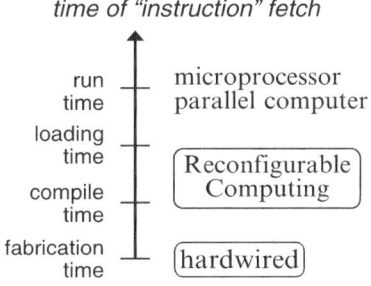

Fig. 20.8 "Instruction fetch"

1. Configurable Cells

Since their introduction in 1984, *Field-Programmable Gate Arrays* (FPGAs) have become the most popular implementation media for application-specific or domain-specific digital circuits. For a reading source on the role of FPGAs (providing 148 references) also see [78]. The technology-driven progress of FPGAs (for key issues see [86, 87]) is faster than that of microprocessors (Figure 20.5). FPGAs with 50 million system gates are coming, may be, soon [88]. The FPGA is an array of gate-level configurable logic blocks (CLB) embedded in a reconfigurable interconnect fabrics [87]. Its *configware* code (reconfiguration code [93]: Figure 20.10) is stored in a distributed RAM memory. We may also call it *hRAM* for *"hidden RAM"*, because it is hidden in the background of the FPGA circuitry.

Hidden RAM. The little boxes labelled "FF" in Figure 20.9 are bits of the hRAM. At power-on the FPGA is booted by loading the configware code down to the hRAM, mostly from an external flash memory. Usually the FPGA

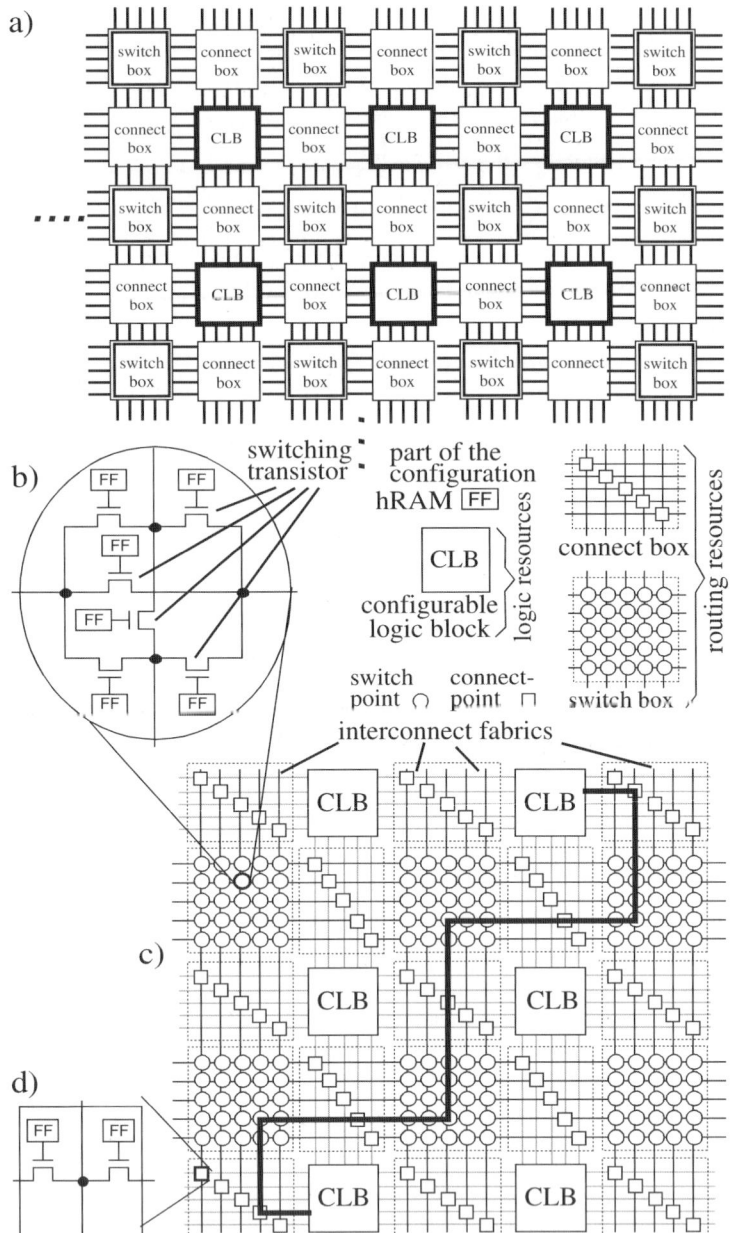

Fig. 20.9 Illustrating FPGA island architecture fine-grain reconfigurable resources: a) global view b) switch point circuit of a switch box, c) FPGA detailed view (only one configured "wire" shown), d) "connect point" circuit of a "connnect box"

platform		program source	machine paradigm
hardware		not programmable	(none)
recon-figurable[1]	fine-grained reconfigurable	*configware*	
	coarse-grain reconfigurable array (data-stream-based)	*configware* & *flowware*	antimachine
hardwired processor	data-stream-based computing	*flowware*	
	instruction-stream-based computing	software	von Neumann

[1] a good general term would be *morphware* [91], which, however, is a registered trade mark [92]

Fig. 20.10 Terminology (also see Figure 20.9 Illustrating FPGA island architecture)

includes the booting interface needed. There are efforts on the way to have the booting flash memory direct on the FPGA chip, or, to use flash technology directly for the hRAM [89], currently is less area-efficient than CMOS hRAM.

Fastest growing market segment. FPGAs are with 6 billion US-Dollars the fastest growing segment of the semiconductor market. Complex projects can be implemented on FPGAs, commodities off the shelf (COTS), without needing very expensive customer-specific silicon. The growth of the number of design starts is predicted to grow from 80,000 in 2006 to 115,000 in 2010 [90].

Two classes of reconfigurable systems. We may distinguish two classes of reconfigurable systems: *fine-grain* reconfigurable systems and *coarse-grain* reconfigurable systems (see Section 4). Reconfigurability of fine granularity means that the functional blocks have a datapath width of about only one bit. This means that programming at low abstraction level is logic design. Practically all products on the market are *FPGAs* (field-programmable gate arrays), although some vendors prefer different terms as a kind of brand names like, for instance, PLD (Programmable Logic Device, or rLD for reconfigurable logic device). Reconfigurable platforms and their applications have undergone a long sequence of transitions.

FPGAs replacing ASICs. First FPGAs appeared as cheap replacements of *MPGAs* (or *MCGAs*: Mask-Configurable Gate Arrays). Still today FPGAs are the reason for shrinking ASIC markets (Figure 20.2d), since for FPGAs no application-specific silicon is needed – a dominating cost factor especially in low production volume products. Although being cheaper than for fully hardwired solutions the ASIC fabrication cost (only a few specific masks needed) is still higher than for FPGA-based solutions. Later the area proceeded into a new model of computing possible with FPGAs. Next step was making use of the possibility for debugging or modifications the last day or week, which also lead to the adoption by the *rapid prototyping* community which also has lead to the introduction of *ASIC emulators* faster than simulators. Next step is direct *in-circuit execution* for debugging and patching the last minute.

Brand names.   Like for example, PLD (Programmable Logic Device) are often confusing. Reconfigurable platforms and their applications have undergone a long sequence of transitions. First FPGAs appeared as cheap replacements of *MPGAs* (or *MCGAs*: Mask-Configurable Gate Arrays). Still today FPGAs are the reason for shrinking ASIC markets (Figure 20.2c), since for FPGAs no application-specific silicon is needed – a dominating cost factor in low production volume products. Before FPGAs have been available, ASICs have been used to replace fully hardwired solutions (full mask set needed), because fabrication cost has been lower (only a few specific masks needed). Later the area proceeded into a new model of computing possible with FPGAs. Next step was making use of the possibility for debugging or modifications the last day or week, which also lead to the adoption by the *rapid prototyping* community which also has lead to the introduction of *ASIC emulator*, expensive (million-dollar range) and bulky machines filled with hundreds or thousands of FPGAs, used as accelerators for simulators used in traditional EDA environments. Because of high cost and poor performance the time for the ASIC emulator has been over around the end of last century [5]. Because of their enormous capacity and interesting built-in features meanwhile FPGAs directly offer a viable alternative for inexpensive FPGA-based prototypes.

Terminology problems.   A fine term would be "morphware" [91], which, however, is protected [92]. The historic acronyms *FPGA* and *FPL* are a bad choice, since "programming", i.e. scheduling, is a *procedural* issue in the *time domain*. For the same reason also the term *PLD* is a bad choice and should be replaced by *rLD (reconfigurable Logic Device)*. A program determines a *time sequence* of executions. In fact the FP in FPGA and in FPL, acronym for *field-programmable*, actually means field-reconfigurable, which is a structural issue in the *space domain: configuration in space*. For terminology also see Figure 20.24.

Island architectures.   Most important architectural classes of FPGAs are [94]: island architectures (Xilinx), hierarchical architectures (Altera), and row-based architectures (Actel). A more historic architecture is *mesh-connected*, sometimes also called *sea of gates* (introduced by Algotronix) [95]. For a survey on FPGA architectures see [96]. For illustration of FPGA fundamentals we use the historic simple island architecture as an example (Figure 20.9). Under the term "simple FPGA" we understand a pure FPGA, which unlike modern "platform FPGAs", do not include non-reconfigurable resources, like adders, memory blocks, etc.

An island style example.   A simple island style FPGA is a mesh of many CLBs (configurable logic blocks), embedded in a reconfigurable interconnect fabrics (Figure 20.9a). Most CLBs are LUT-based, where LUT stands for "look-up table". A simple example of a CLB block diagram is shown by Figure 20.12. Its functional principles by multiplexer implementation of the LUT are shown by Figure 20.13a and b, where in CMOS technology only 12 transistors are needed for the fully decoded multiplexer (Figure 20.13c). The island architecture is illustrated by Figure 20.9a. Figure 20.9b shows details of *switch boxes* and *connect boxes* being part of the reconfigurable interconnect fabrics. Figure 20.9d shows the circuit diagram of a *cross point* in a switch box, and Figure 20.9d from within a connect box. The thick wire example in Figure 20.9b illustrates how these interconnect resources are configured to

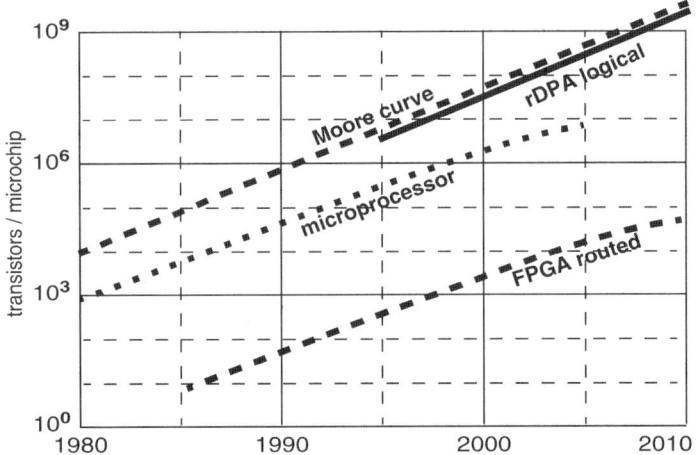

Fig. 20.11   Area efficiency of coarse-grained RC

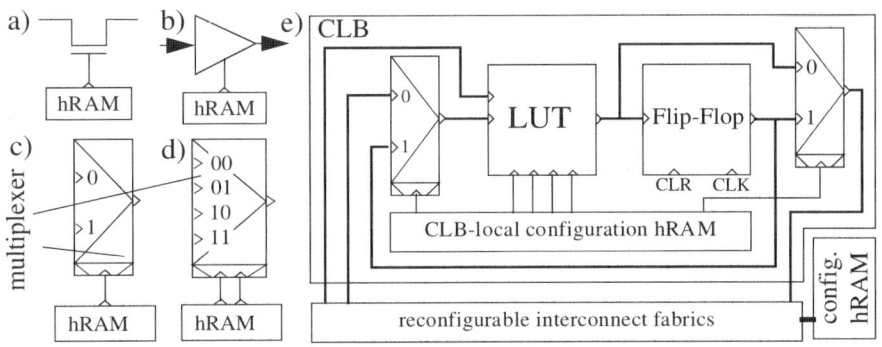

Fig. 20.12   Reconfiguration resources in FPGAs: a) pass transistor, b) tri-state buffer d) 2-way multiplexer, d) 4-way multiplexer, e) simplified CLB example

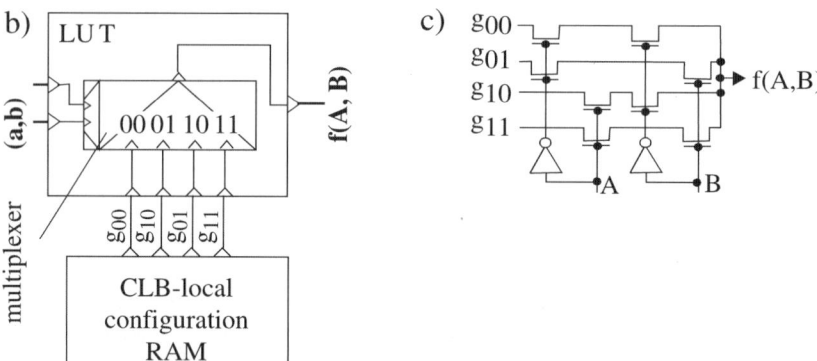

a)	configuration bits g_{ab}				f(A, B)	function
#	g_{00}	g_{01}	g_{10}	g_{11}		
0	0	0	0	0	0	constant 0
1	0	0	0	1	A and B	and
2	0	0	1	0	B disables A	if B then 0 else A
3	0	0	1	1	A	identity A
4	0	1	0	0	A disables B	if A then 0 else B
5	0	1	0	1	B	identity B
6	0	1	1	0	A exor B	antivalence
7	0	1	1	1	A or B	or
8	1	0	0	0	not(A or B)	nor
9	1	0	0	1	A coin B	equivalence
10	1	0	1	0	not(B)	negation of B
11	1	0	1	1	B implies A	if A then 1 else ¬ B
12	1	1	0	0	not(A)	negation of A
13	1	1	0	1	A implies B	if B then 1 else ¬ A
14	1	1	1	0	not(A and B)	nand
15	1	1	1	1	1	constant 1

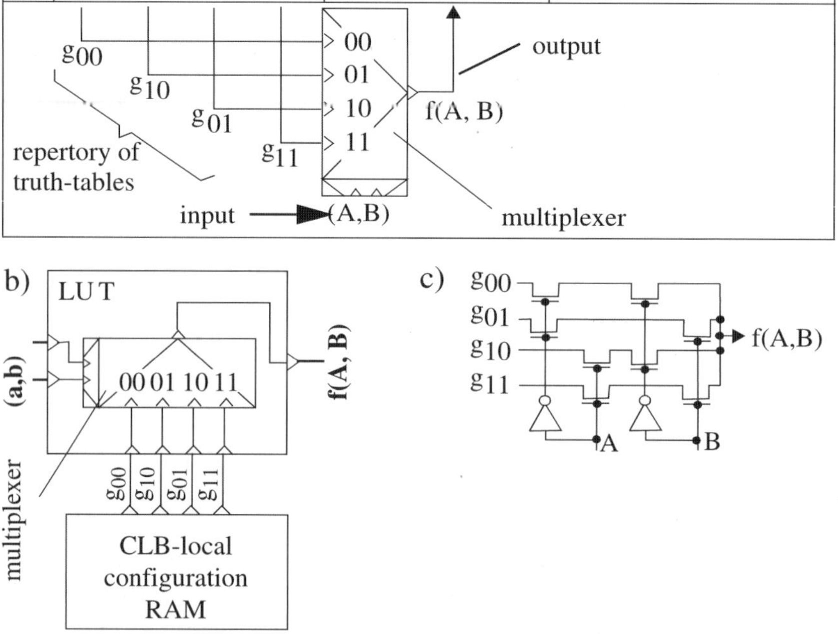

Fig. 20.13   Illustrating LUT implementation by multiplexer: example for functions of two variables: a) illustration of the function generator, b) illustration of LUT (look-up table) block use within CLB (compare Figure 20.12e), c) multiplexer circuit

connect a pin of one CLB with a pin of another CLB. The total configuration of all wires of an application is organized by a *placement and routing* software. Sometimes more interconnect resources are needed than are available, so that for some CLB not all pins can be reached. Due to such *routing congestion* it may happen that a percentage of CLBs cannot be used.

Fine-grain reconfigurable systems products. A wide variety of fine-grained reconfigurable systems products [94, 95] is available from a number of vendors, like the market leader Xilinx [97], the second largest vendor Altera [98], and many others. Also a variety of evaluation boards and proto-typing boards is offered. COTS (commodity off the shelf) boards for FPGA-based developments are available from Alpha Data, Anapolis, Celoxica, Hunt, Nallatech, and others, to support a broad range of in house developments. As process geometries have shrunk into the deep submicron region, the logic capacity of FPGAs has greatly increased, making FPGAs a viable implementation alternative for larger and larger designs. Deep submicron FPGAs are available in many different sizes and prices per piece ranging from 10 US-Dollars up to FPGAs with much more than a million usable gates for more than 1000 US-Dollars. For example, Xilinx offers the Virtex-4 multi-platform FPGA family on 90 nm process technology, the 550 MHz Virtex-5 family of FPGAs on 65 nm technology providing four platforms (high-performance logic, serial connectivity, signal processing, and embedded processing) and has pre-announced FPGAs with 50 million system gates for about 2005 [88]. Modern FPGAs support mapping entire systems onto the chip by offering on board all components needed, like several memory banks for user data, one or several microprocessors like ARM, PowerPC, MIPS, or others, a major number of communication interfaces (WAN, LAN, BoardAN, ChipAN, etc.) supporting contemporary standards, up to several GHz bandwidth, JTAG boundary scan circuitry to support testing, sometimes even multipliers. Low-cost FPGA development boards are available for universities [99]. A base version board is priced at 90 US-Dollars and contains a 400,000-gate Xilinx Spartan-3 FPGA. Other boards are available with up to 2 million gates. Each board comes bundled with free development software.

Low power dissipation and radiation tolerance. Also FPGAs featuring low power dissipation [66] or better radiation tolerance (for aerospace applications) are offered. Several major automotive corporations contracted FPGA vendors to develop reconfigurable devices optimized for this branch of industry. Some FPGAs commercially available also support partial column wise reconfiguration, so that different talks may reside in it and may be swapped individually. This may also support *dynamic reconfiguration* (*RTR*: run time reconfiguration), where some tasks may be in the execution state, whereas at the same time other tasks are being reloaded. Dynamic reconfiguration, however, tends to be tricky and difficult to understand and to debug. But static reconfiguration is straightforward and more easy to understand. Because reconfiguration is slow, also multi-context reconfigurable systems has been discussed, but is not yet available commercially. Multi-context reconfigurable systems features several alternative internal reconfiguration hRAM memory banks, for example 2 or 4 banks, so that reconfiguration can be replaced by an ultra fast context switch to another hRAM memory bank.

2.   von Neumann vs. Reconfigurable Computing Paradigm

It is well known that the growth rate of the integration density of microprocessors is slower than Moore's law. Because of the high degree of layout regularity the integration density of FPGAs, however, is going roughly by the same speed as with Moore's law [4] (Figure 20.5). But because of the high percentage of wiring area the transistor density of FPGAs are behind the Gordon Moore curve two orders of magnitude [4]. However, the number of transistors per chip on FPGAs has overhauled that of microprocessors already in the early 1990s and now is higher by two orders of magnitude [4].

The end of Moore's Law.   Increasing the architectural complexity and the clock frequency of single-core microprocessors has come to an end (e.g. see what has happened to the intel Pentium 4 successor project). Instead, multi-core microprocessor chips are emerging from the same vendors (e.g. AMD: 32 cores on a chip by 2010 [101]). But just more CPUs on the chip is not the way to go for very high performance. This lesson we have learnt from the supercomputing community paying an extremely high price for monstrous installations by having followed the wrong road map for decades. Such fundamental bottlenecks in computer science will necessitate new breakthroughs [103]. Instead of hitting physical limits we found that further progress is limited by a fundamental misconception in the theory of algorithmic complexity [104]. *Not processing* data is costly, *but moving* data. We have to rethink the basic assumptions behind computing. Such a change should take place, when the old paradigm is in crisis and cannot explain compelling new facts, like recently illustrated by the reconfigurable computing paradox. But this is not the first crisis of the old paradigm [105, 106]. However, the substantial investments of the past decades in the old paradigm causes resistance to a paradigm change, because knowledge accumulated up to that point may loose its significance ([107] as quoted by [108]). Those who have a vested interest in the old paradigm will always attempt to extend that paradigm to accommodate new facts. Thus the final victory of the new paradigm can only be guaranteed by a generation change [108]. But research on Innovation is less pessimistic [109].

Transdisciplinary approaches needed. Instead of the traditional reductionism we need transdisciplinary approaches, such as heralded by the current revival of Cybernetics, e.g. labelled as Integrated Design & Process Technology [110], or Organic Computing [111]. To reanimate the stalled progress in HPC for a breakthrough in very high performance computing we need a transdisciplinary approach for bridging the hardware/software chasm, which meanwhile has turned into a configware/software chasm. For much more successful efforts we need a transdisciplinary paradigm shift, over to a new fundamental model, such as available from the Reconfigurable Computing community dealing with configware engineering as a counterpart to software engineering.

Classical parallelism does not scale – [101, 104]. A very expensive lesson which we have already learnt from the supercomputing community with massively increasing the number of processors when going cheap COTS (commodity off the shelf). With the growing degree of parallelism, the programmer productivity goes down drastically ("The Law of More" [113]). It is an illusion to believe that scalability would get massively better, when all these processors will be resident on a single chip – as long as the reductionistic monopoly of the von Neumann mind set will not be relieved, where the classical fundamental paradigm is still based on concurrent sequential processes and message passing through shared memory, both being massively overhead-prone and extremely memory-cycle-hungry. Rescue should not be expected from threads, although intel pre-announced some tools intended to avoid that the programmers shy away. In his cover feature article [114] Edward A. Lee from UC Berkeley claims that for concurrent programming to become mainstream, we must discard threads as a programming model. Nondeterminism is the overhead-prone problem, not only hidden behind methodologies attributed "speculative". By the way, complications by threads perfectly illustrate the von Neumann-based software paradigm trap.

Escape the software development paradigm trap, said IRIS director Mark Bereit [116], who refutes the assumption that software development will always be difficult and bug-ridden, noting that this is due "solely to the software development paradigm that we have followed, unchallenged, for decades". One of the consequences of this paradigm trap crisis is demonstrated by (not only) the history of programming languages [117, 119], where 2500 different languages are seen as counted by Bill Kinnersley [120]. Diarmuid Piggott counts the total even higher with more than 8500 programming languages [121]. What an immense waste of manpower to make sure to stay caught inside the paradigm trap.

Bad scalability.   The software development paradigm is also the reason of bad scalability and bad programmer productivity in classical parallelism. Mark Bereit proposes reworking the model and studying other engineering disciplines for inspiration. He proposes to study mechanical engineering. But much better is studying Reconfigurable Computing. Tensilica senior vice president Beatrice Fu said that reconfigurable computing offers the option of direct processor-to-processor communications without going through memory nor through a bus. This paradigm shift is an old hat, but until recently mostly ignored, not only by the supercomputing community. Buses cause multiplexing overhead [123] and the dominating instruction-stream-based-only fundamental model is extremely memory-cycle-hungry [125]: the reason of the "memory wall". The alternative offered by reconfigurable computing is data stream parallelism by highly parallel distributed fast local memory.

More simple memory parallelism and more straightforward than, for example, interleaved memory access known from vector computers. Crooked Labelling. The difference between Parallel Computing and Reconfigurable Computing is often blurred by projects labelled "reconfigurable", which, in fact are based on classical concurrency on a single chip. To avoid confusion: switching the multiplexers or addressing the registers at run time is not "reconfiguration". At run time, real Reconfigurable Computing never has an instruction fetch: only data streams are moving around (This should not be confused with dynamically reconfigurable systems: a mixed mode approach switching back and force between reconfiguration mode and execution mode, which should be avoided for introductory courses.)

> The data stream machine paradigm has been around three decades as a niche. Software used it indirectly by inefficient instruction-stream-based implementations.

CPUs outperformed by FPGAs.   The world-wide total running compute power of FPGAs outperforms that of CPUs. Most total MIPS running worldwide have been migrated from CPUs to accelerators, often onto FPGAs. The FPGA market with almost 4 billion US-Dollars (2006) is the fastest growing segment of the integrated circuit market. Gartner Dataquest predicts almost 7 billion US-Dollars for the year 2010. Xilinx and Altera currently dominate this market with a share of 84%.

Bad Parameters.   However, FPGAs have even more comparably bad parameters: the clock frequency is substantially lower than that of microprocessors, FPGAs are power-hungry and expensive. We would like to call it the "Reconfigurable Computing Paradox" that despite of these dramatically worse technology parameters such enormous speedup factors can be obtained by software to configware migration (Figure 20.4). The explanation of this paradox is the machine paradigm shift coming along with such a migration.

The third reconfigurable computing paradox. It may be called the third paradox of Reconfigurable Computing that despite of its enormous pervasiveness, most professionals inside computer science and related areas do not really understand its paradigm shift issues. This massively affects implementer productivity. To support configware engineering projects often a hardware expert is hired who may be good implementer, but is not a good translator. From a traditional CS perspective most people do not understand the key issues of this paradigm shift, or, do not even recognize at all that RC is paradigm shift. A good approach of explanation is to compare the mind set of the software area vs. the one of the configware field. An dominant obstacle for understanding is also the lack of a common accepted terminology, which massively causes confusion.

Gradually wearing off. Currently the dominance of the basic computing paradigm is gradually wearing off with growing use of Reconfigurable Computing (RC) – bringing profound changes to the practice of both scientific computing and ubiquitous embedded systems, as well as new promise of disruptive new horizons for affordable very high performance computing: Due to RC the desk-top personal supercomputer is near [128]. To obtain the payoff from RC we need a new understanding of computing and supercomputing. To bridge the translational gap, the software/configware chasm, we need to think outside the box.

Going toward the dual-paradigm mind set is the current strong trend (Figure 20.7): the duality of the instruction-stream-based CPU paradigm, and its counterpart, the data-stream-based antimachine paradigm based on data counters instead of a program counter. The von Neumann paradigm using a program counter [1, 2] will not be explained by this chapter. Since about the early 1980s one of the two roots of the antimachine paradigm has been hidden inside the systolic array community [128, 129], mainly a group of mathematicians, who have nicely defined the concept of "data streams", coming along with a beautiful visualization by time/space diagrams (see Figure 20.14). Systolic arrays have been popularized in 1981 by H. T. Kung [129]. But for mathematicians a systolic array has only been a data path, a pipe network, but it has not been a computational machine, because the sequencer has been missing. But by the mathematicians, development of a methodology for generating these data streams at run time has been considered being somebody else's job. Refusing a transdisciplinary perspective this has been a typical result of the "formal" reductionism having been fashionable at that time.

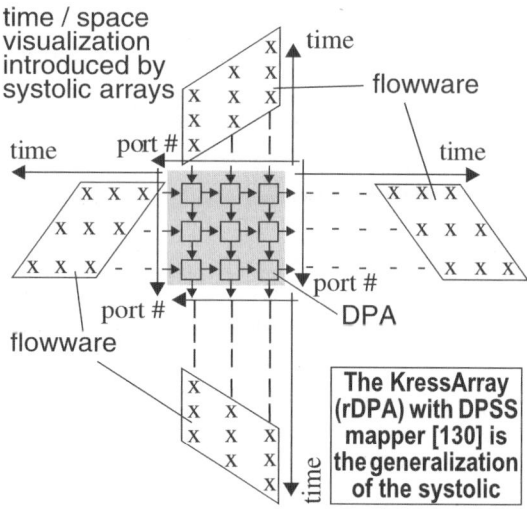

Fig. 20.14   Data stream definition (Flowware)

Mathematicians' paradigm trap.   The classical systolic array suffered from a severe restriction. It could be used only for applications with strictly regular data dependencies. For a hammer many things in the world look like a nail. About 25 years ago for the mathematicians working on systolic arrays all applications looked like algebraic problems. From this point of view their algebraic synthesis algorithms generating a systolic architecture from a mathematical formula have been based only on linear projections, which yield only uniform arrays with linear pipes: usable only for applications regular data dependencies.

Generalization of the systolic array.   Later Rainer Kress holding an EE degree discarded the mathematician's synthesis algorithms and used simulated annealing instead, for his DPSS (Data Path Synthesis System) [130]. This means the generalization of the systolic array: the Kress Array [131], or super-systolic array (we may also call it Kress–Kung Array), also supporting any wild forms of pipe networks, including any non-linear pipes like spiral, zigzag, branching and merging, and many other forms. Now reconfigurability makes sense. Figure 20.15 shows how the sequencing part is added to the systolic array data paths, so that we get a complete computational machine. Instead of a program counter we have multiple data counters supporting parallelism of data streams. The GAG (generic address generator [132, 134, 135]) is a data sequencer connected to a RAM block for data storage. GAG and RAM block are parts of the ASM (Auto-sequencing Memory), a generalization of the DMA (Direct Memory Access). The following paragraphs explain the machine

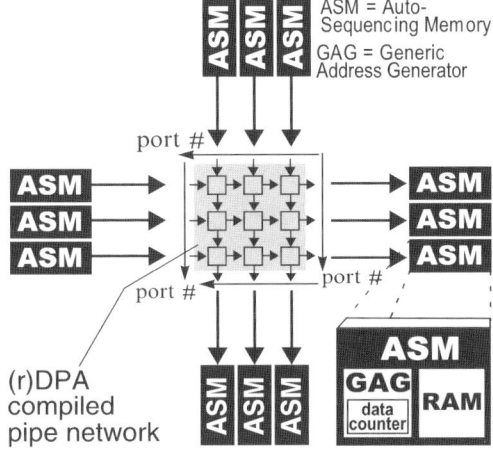

Fig. 20.15   Generating data streams by distributed memory

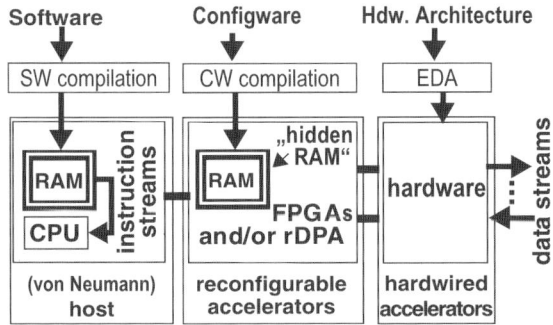

Fig. 20.16   De facto common model of contemporary computing systems: von Neumann-only is obsolete

principles and software/configware co-compilation techniques for the duality of machine paradigms.

Discussing software engineering vs. configware engineering.   In total we have three different kinds digital subsystems (Figure 20.16) and three different kinds of programming sources (Figure 20.16). The dual paradigm model can be illustrated by contrasting via Nick Tredennick's model of computer history (Figure 20.17). The algorithm of early machines like the Digital Differential Analyzer (DDA) could not be programmed: no source code was needed (Figure 20.17a). With von Neumann's classical software processor only the algorithm is variable, whereas the resources are fixed (hardwired), so that only one type of source code is needed: software (Figure 20.17b), from which the

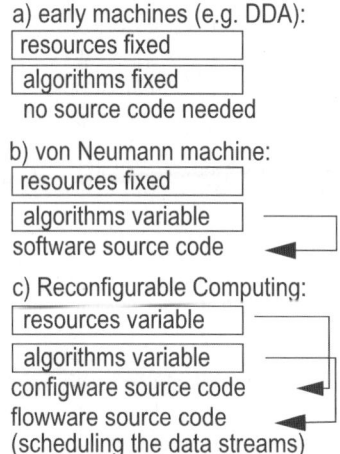

a) early machines (e.g. DDA):

| resources fixed |
| algorithms fixed |

no source code needed

b) von Neumann machine:

| resources fixed |
| algorithms variable |

software source code

c) Reconfigurable Computing:

| resources variable |
| algorithms variable |

configware source code
flowware source code
(scheduling the data streams)

Fig. 20.17   Nick Tredennick's machine classification scheme

#	Source	is compiled into:
1	Software	instruction schedule
2	Flowware	data schedule
3	Configware	a pipe network by placement and routing

Fig. 20.18   Sources by compilation targets

compiler generates software machine code to be downloaded into the *instruction RAM* – the instruction schedule for the *software processor* (Figure 20.17b). For the Kress–Kung machine paradigm, however, not only the algorithm, but also the resources are programmable, so that we need two different kinds of sources (Figure 20.18): *Configware* and *Flowware* (Figure 20.17d):

1. Configware [93] deserves structural programming of the resources by the "*mapper*" using *placement and routing* or similar mapping methods (for instance by simulated annealing [130, 131, 137–139] (Figure 20.19d).

2. Flowware [141] deserves programming of the data streams by the "*data scheduler*" (Figure 20.19b), which generates the flowware code needed for downloading into the generic address generators (GAG) within the ASM auto-sequencing memory blocks (Figure 20.15).

The dichotomy of language principles. These two different fundamental machine principles, von Neumann software machine vs. the Kress–Kung machine, the configware machine, are contrasted by the dichotomy of

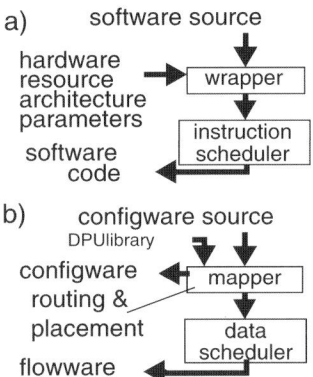

Fig. 20.19 Compilers: a) software compilation, b) configware/flowware co-compilation

languages (Figure 20.23) and by Figure 20.18 [56, 68]. Flowware Languages are easy to implement [141, 142]. A comparison with software programming languages is interesting [87]. Flowware language primitives for control instructions like jumps and loops can be simply adopted from classical software languages, however, for being used for manipulation of data addresses instead of instruction addresses. Flowware languages are more powerful than software languages and permitting parallel loops by using several data counters used simultaneously, such flowware language primitives are more powerful than these software primitives. Not handling instruction streams, flowware languages are much more simple (because at run time there is only "data fetch", however, no "instruction fetch".

Some remarks on Terminology:. Since the basic paradigm is *not* instruction-stream-based, necessarily the term "Configware" should be used for program sources, instead of the term "Software", which would be confusing (Figure 20.10). The term "software" must be unconditionally restricted to traditional sources of instruction-stream-based computing. In fact this paradigm relies on data streams, however, not on instruction streams.

Equivocalities of the term "data stream" are a problem, not only in education. In computing and related areas there is a babylonian confusion around the term "stream", "stream-based", or "data stream". There is an urgent need to establish a standards committee to work on terminology. For the area of reconfigurable computing the best suitable definition of "data stream" has been established around 1980 by the systolic array scene [128, 129], where data streams enter and leave a datapath array (a pipe network: illustrated by

Figure 20.14). In fact, there a set of data streams is a data schedule specifying, which data item has to enter or leave which array port at which point of time.

The tail is wagging the dog.   Because of their memory-cycle-hungry instruction-stream-driven sequential mode of operation microprocessors usually need much more powerful accelerators [4]: the tail is wagging the dog. The instruction-stream-based-only fundamental mind set (vN-only paradigm) as a common model often is still a kind of monopoly inside the qualification background of CS graduates. The real model practiced now is not the von Neumann paradigm (vN) handed down from the *mainframe age*. In fact, during the *PC age* it has been replaced by a symbiosis of the vN host and the non-vN (i.e. non-instruction-stream-based) accelerators. Meanwhile we have arrived at the (kind of post-PC) *reconfigurable system age* with a third basic model, where the accelerator has become programmable (reconfigurable).

What high level source language for HPC and supercomputing users?.
Useful for application development are *Co-compilers* (Figure 20.7), automatically partitioning from the programming source into software and configware and accepting, for instance a C-like language [143]. The methodology is known from academic co-compilers [143, 144], easy to implement since most of their fundamentals have been published decades ago [146]. Figure 20.18 contrasts the difference of intermediate sources to be generated by such a co-compiler. Although high level programming languages like C come closer to the such a user's mind set than classical FPGA synthesis environments, it is still confusing. Such imperative languages seem to have an instruction-stream-based semantics and do not exhibit any dual-paradigm features, because data-stream-based constructs are out of reach. It would be desirable to have a source language at one abstraction level higher than imperative languages, like the domain of mathematical formula. Such a system can be implemented by using a term rewriting system (TRS) [148] for dual-paradigm system synthesis [149].

FPGA main processor with auxiliary CPU.   There is a number of trend indications pointing toward an auxiliary clerk role of the CPU for running old software and taking care of compatibility issues. "FPGA main processor vs. FPGA co-processor" asks the CEO of Nallatech [150]: is it time for vN to retire? The RAMP project, for instance, proposes to run the operating system on FPGAs [151]. In fact, in some embedded systems, the CPU has this role already now. But often the awareness is missing.

The dichotomy of machine paradigms.   is rocking the foundation walls of computer science. Because of the lack of a common terminology this duality of paradigms is difficult to understand for people with a traditional

CS background. A taxonomy of platform categories and their programming sources, quasi of a terminology floor plan, should help to catch the key issues (Figure 20.10). The Kress–Kung machine is the data-stream-based counterpart of the instruction-stream-based von Neumann paradigm. The Kress–Kung machine does not have a program counter (Figure 20.24), and its processing unit *is not a CPU* (Figure 20.24). Instead, it is only a *DPU (Data Path Unit)*: without an instruction sequencer (Figure 20.24).

The enabling technology of the Kress–Kung machine. has one or mostly several *data counters* as part of the *Generic Address Generators (GAG)* [132, 134, 135] within data memory banks called *ASM (Auto-Sequencing Memory*, see Figure 20.15). ASMs send and/or receive data streams having been programmed from *Flowware* sources [133] (Figure 20.14). An ASM is the generalization of the DMA circuit (Direct Memory Access) [152, 153] for executing block transfers without needing to be controlled by instruction streams inside. ASMs based on the use of distributed memory architectures [154] are very powerful architectural resources, supporting the optimization of the data storage schemes for minimizing the number of memory cycles [135]. The MoM Kress–Kung machine based on such generic address generators has been published in 1990 [155, 156]. The use of data counters replacing the program counter has first been published in 1987 [157].

Hardwired Kress–Kung machines. There are also hardwired versions of the Kress–Kung machine. We may distinguish two classes of Kress–Kung machines (Figure 20.10): programmable ones (reconfigurable systems: reconfigurable Kress–Kung machine, needing two types of programming sources (see next paragraph and Figure 20.18): *Configware* for structural programming, and *Flowware*, for data scheduling. However, also hardwired Kress–Kung machines can be implemented for instance (the BEE project [158]), where the configuration is been frozen and cast into hardware before fabrication. The lack of reconfigurability after fabrication by not using FPGAs of such hardwired Kress–Kung machines substantially improves the computational density (Figure 20.5a) for much higher speedup factors and might make sense for special purpose or domain-specific applications. Since after fabrication a reconfiguration is impossible, only one programming source is needed: *Flowware*.

Dynamically reconfigurable architectures and their environment illustrate the specific flavor of *Configware Engineering* being able to rapidly shift back and force between run time mode of operation and configuration mode. Even several separate macros can be resident in the same FPGA. Even more complex is the situation when within partially reconfigurable FPGAs some modules are in run time mode, whereas at the same time other modules are in the configuration phase, so that an FPGA could reconfigure itself. Some macros can be removed at the same time, when other macros are active by being in the run time mode. *Configware operating systems* are managing such scenarios [159, 160]. On such a basis even *fault tolerance* by self-repair can be implemented [161, 162], as well as reconfigurable artificial neuronal networks [163, 164]. The electronics within the Cibola satellite [165] scheduled to be launched by October 2006 uses such fault tolerance mechanisms to cope with fault introduced by cosmic radiation [167]. Dynamic reconfigurability can be confusing for beginners and should be introduced not earlier than at graduate courses.

New educational approaches are needed.   Although configware engineering is a discipline of its own, fundamentally different from software engineering, and a configware industry is already existing and growing, it is too often ignored by our curricula. Modern FPGAs as COTS (commodities off the shelf) have all three paradigms on board of the same VLSI chip: hardwired accelerators, microprocessors (and memory banks), and FPGAs, and we need both, software and configware, to program the same chip. To cope with the clash of cultures we need interdisciplinary curricula merging all these different backgrounds in a systematic way. We need innovative lectures and lab courses supporting the integration of reconfigurable computing into progressive curricula.

3.   Future of FPGA (Technologies)

The scaling of CMOS technology will come to an end. It is unclear whether CMOS devices in the 10–20 nanometer range will find-a useful place in semiconductor products [168]. Timothy S. Fisher (Purdue) said, "But before we can even think about using nanotubes in electronics, we have to learn how to put them where we want them". [169]. New silicon-based technologies (e.g. silicon nanowires) and non-silicon based (e.g. carbon nanotubes) show the promise of replacing traditional transistors [170, 172, 174, 175]. However, there are multiple challenges to face, like the production of nanoscale CMOS with reasonable yield and reliability, the creation of newer circuit structures with novel materials as well as the mixing and matching of older and newer technologies in search of a good balance of costs and benefits [168]. Opportunities will be driven by the ability of designing complex circuits. Problems related to defect density, failure rates, temperature sensitivity can be expected. Our ability to define the right design technology and methodology will be key in the realization of products of these nanotechnology. It also will depend on the direction that the semiconductor road will take [168].

Lithographic patterning has long been the primary way of defining features in semiconductor processing [176]. Nanodesigns, however, will not simply be an extension of classical VLSI design. We may not be able to directly pattern complex features, but rather need self-assembly to create ordered devices, and post-fabrication reconfigurability. This creates new challenges for design and motivates different architectures than found on classical silicon. The crossbar may be to play a prominent role in these new developments. Its regular structure is well suited to nanotechnology-based assembly. A nanometer-scale crossbar is useful only if wires in each of its dimensions can be addressed individually. Research examines methods of stochastic addressing of nanowires [178] as well as the efficient data storage in crossbar-based memories [178].

Nanoscale features. For nanoscale features there seems to be a sufficient set of building blocks for fully nanoscale systems (Figure 20.20 shows a crossbar [170]). Scientists are developing a growing repertoire of techniques which allow us to define features (e.g. wire widths, spacing, active device areas) without lithography, requiring self-assembly for structures from individual atoms with tight dimensions, with wires just a few atoms wide [170]. Semiconducting nanowires with diameters down to 3 nm (about 6 atoms wide) have been grown [179, 180], composed of different materials or selectively doped [181] along their length using timed growth [182]. Diode junctions built by crossing P-doped and N-doped nanowires use field-effects to control conduction to implement switchable crosspoints or memory bits [183, 184]. Nanowires can be sandwiched into array blocks including programmable OR planes, memory planes, and signal restoration or inversion planes [178, 185, 186]. Flow techniques can be used to align a set of nanowires into a single orientation, close pack them, and then transfer them onto a surface [187, 188]. Switchable molecules can be assembled in a crossed array [183, 186], providing sublithographic scale, programmable junctions. It is very difficult to predict how many more years it will take to a large scale market introduction.

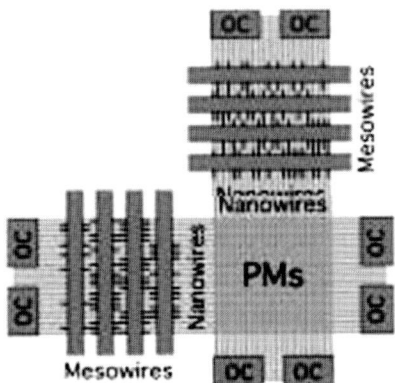

Fig. 20.20   Nanowire-based crossbar [John A. Savage] (courtesy IEEE)

4.   Coarse-Grained vs. Fine-Grained Reconfigurability

We may distinguish two classes of reconfigurable microchips [189]: *fine-grain* reconfigurable devices like FPGAs (see Section 1) and *coarse-grain* reconfigurable devices, rDPAs (reconfigurable DataPath Arrays) [190]. The table in Figure 20.27 gives a survey on differences between coarse-grained and fine-grained. Instead of the up to hundreds of thousands of about 1 bit wide CLBs (Configurable Logic Blocks) of modern FPGAs, an rDPA has a few (about up to 128 or 256 or much less) larger rDPU blocks (reconfigurable DataPath Units), for instance, 32 bits wide. For illustration a rDPU could be compared with the ALU of CPU (Figure 20.27).

Advantages of coarse-grained reconfigurability.   Coming along with functional level rDPUs as configurable blocks of rDPAs [190] – physically at a more convenient abstraction level much closer to a software user's mind set than that of FPGAs difficult to reach by design tools or compilation techniques, coarse-grained reconfigurability makes the educational gap smaller. To software people the configuration of FPGAs looked more like logic design on a strange platform, however, not like computation. Another advantage of coarse-grained reconfigurability is the much higher computational density than coming with FPGAs. In contrast to the very bad effective integration density of FPGAs (Figure 20.5), this density of coarse-grained reconfigurable arrays (rDPAs) almost reaches the Gordon Moore curve [4] (Figure 20.11, also see row 8 in Figure 20.27): it is better by four orders of magnitude. Because a coarse-grained array has only a few larger configurable blocks (about a few hundred rDPUs or much less) with very compact configuration codes the configuration time is reduced to microseconds – in contrast to milliseconds as known from FPGAs.

Free form pipe networks.   In contrast to a CPU, a DPU is not instruction-stream-driven and has no program counter and its operation is transport-triggered by the arrival of operand data. This new machine paradigm (the diametrical counterpart of von Neumann) is based on free form large pipe networks of rDPUs (without memory wall and compilation is easy), but not on concurrent sequential processes. There is no instruction fetch overhead at run time since these pipe networks, generalizations of the systolic array, are configured before run time. This new paradigm is based on data streams generated by highly parallel distributed on-chip local small but fast memory which consists of auto-sequencing memory (ASM) blocks (Figure 20.15) using reconfigurable generic address generators (GAG), providing even complex address computations not needing memory cycles [132]. This kind of memory parallelism is more simple and more straightforward than interleaved memory access known from vector computers. The ASM is reconfigurable. This means its address computation algorithm is not changed at run time. It does not need an instruction stream at run time. In embedded systems often a symbiosis of both models, instruction-stream-based and data-stream-based, practices a duality of machine paradigms (Figure 20.7). Figures 20.19 and 20.24 explain this by showing the differences.

Improved designer productivity. The even physically high abstraction level or coarse-grained arrays dramatically improves the designer productivity. This is utilized by the KressArray Xplorer [137, 138] design environment (Figure 20.29), including a design space explorer capable to compare and profile within a few hours a variety of different rDPU architectures and rDPA architectures, supporting the entire KressArray family concept [137, 138] featuring individual numbers of ports of individual path width at each side (example in Figure 20.29b) and many layers of background interconnect: mixture of buses and long distance point to pint. Why does this make sense? Since logic gates are general purpose elements, a basic FPGA is a universal device, whereas performance-optimized coarse-grained arrays tend to be more or less domain-specific, depending on the collection of functions available on a rDPU like, for instance, the physical availability of floating point operations. Figure 20.21 shows a rDPA example having been compiled by the KressArray Xplorer for the SSN filter of an image processing application, which computes the center pixel from the 3-by-3 pixel input as shown by Figure 20.22. This rDPA consists of 160 rDPUs, each 32 bits wide. The highly optimal solution is demonstrated by the following result properties: only two rDPUs are not used, and only one rDPU is used for rout-through-only. Almost all rDPUs are directly connected by nearest-neighbor interconnect, except six of them, which need a few additional backbus wires.

Data-stream-based. The Kress–Kung array non-von Neumann mode of operation, known from pipelines and from systolic arrays and Kress–Kung Arrays, is data-stream-based, instead of instruction-stream-based. The (r)DPU does not have an instruction sequencer, and its execution is kind of transport-triggered, i.e. it is triggered by the arrival of its operand data. The data streams are generated by local ASM distributed reconfigurable memory [132, 191] (Figure 20.15). The ASM (Auto-Sequencing Memory) is reconfigurable. This means that its address computation algorithm is not changed at run time. It does not need an instruction stream at run time. That is, the Kress–Kung machine paradigm is a non-von Neumann paradigm: it is the diametrical counterpart of the von Neumann principles.

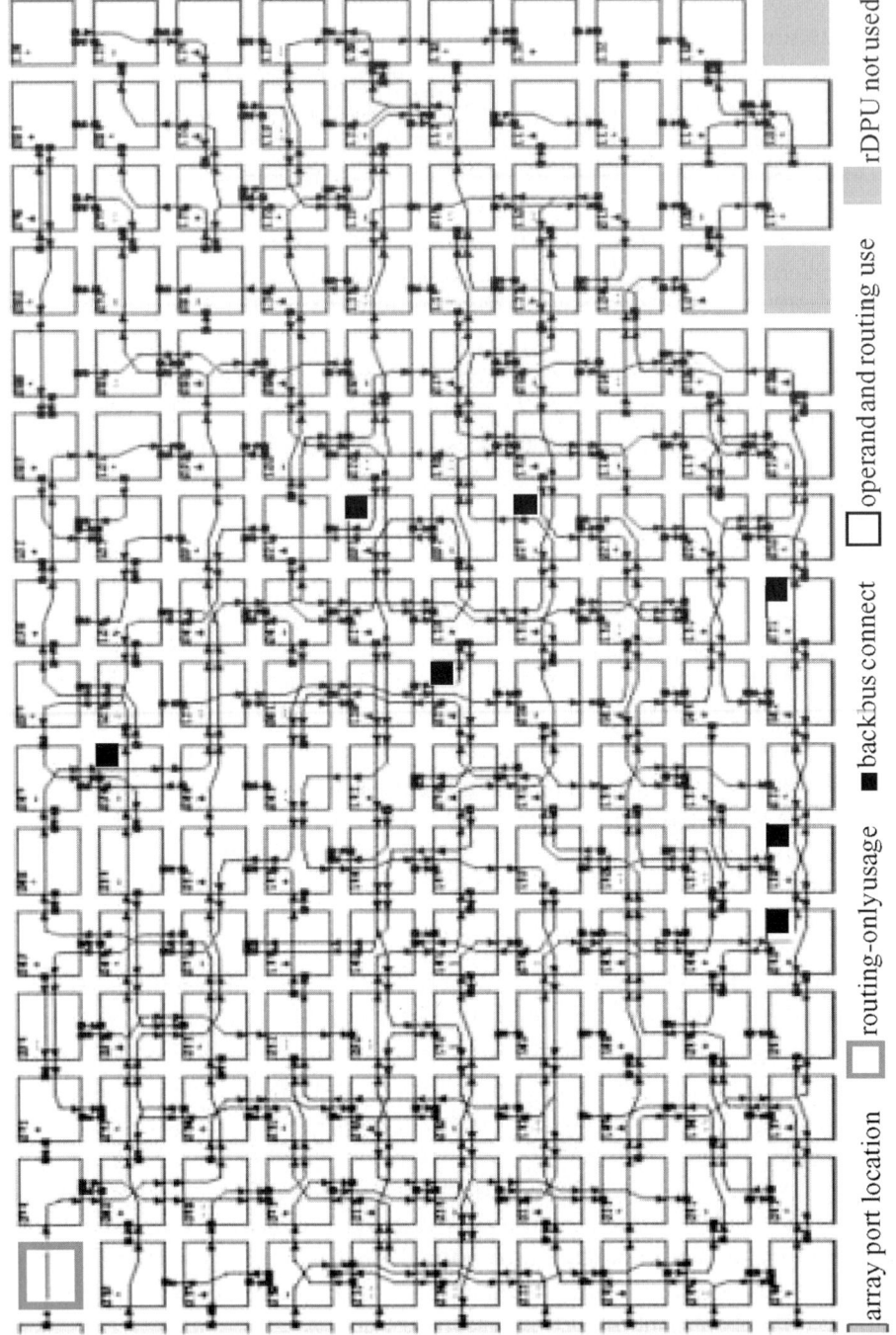

Fig. 20.21   Example of mapping an application (image processing: SNN filter) onto a (coarse grain) KressArray

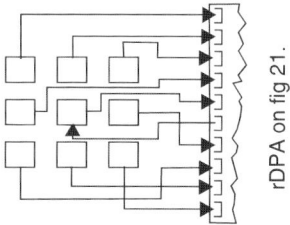

rDPA on fig 21.

Fig. 20.22   3-by-3 pixel window of SNN algorithm

Language category	Software Languages	Flowware Languages
sequencing managed by	read next instruction, goto (instruction address), jump (to instruction address), instruction loop, nesting, ***no parallel loops***, escapes, instruction stream branching	read next data item, goto (data address), jump (to data address), data loop, nesting, *parallel l∞ps*, escapes, data stream branching
data manipulation specification	yes	not needed (specified by configware only)
state register	program counter	multiple data counters
instruction fetch	memory cycle overhead	no overhead
address computation	massive memory cycle overhead	no overhead

Fig. 20.23   Software languages vs. flowware languages

type of functional unit	instruction sequencer included?	execution triggered by
CPU	yes	instruction fetch
DPU, or, rDPU	no	arrival of data[1]

1).transport-triggered

Fig. 20.24   Duality of paradigms: von Neumann vs. Kress–Kung machine

TTAs are von Neumann.   We should not be confused with "transport-triggered architectures" (TTA) [192–194] such as heralded by the TTA community: using buses driven by an (instruction-stream-based) move processor [195]. This is not really data-stream-based: it is not an antimachine architecture. It is instruction-stream-based instead of a pipe network driven by a locally distributed ASM memory architecture [132, 191]. TTAs are von Neumann architectures. In contrast to a pipe not changed at run time, the interconnect of a bus is usually frequently changed at run time – overhead-prone [123] and memory-cycle-hungry, since being under the control of instruction streams. For their data TTAs do not use (reconfigurable) ASM memory architectures. For this reason TTAs are von Neumann architectures.

FPGA-based speedup mechanisms better explained here.   The astonishingly high acceleration factors obtained by CPU to FPGA migration not only in embedded systems (Figure 20.4, also see Section 1), but also in supercomputing, are more easily explained within the higher abstraction level of coarse-grained reconfigurable systems here in this section. Figure 20.26a shows the execution (500 ns in total) of a conditional add operation on a simple hypothetical instruction set processor. Figure 20.26b illustrates its execution by a small coarse-grained array, where no memory cycles are needed and storing the final result to the rDPA's output buffer (slow technology) takes only

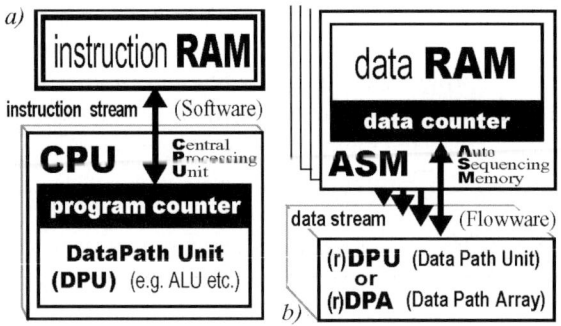

Fig. 20.25   Basic paradigms: a) von Neumann machine, b) antimachine (reconfigurable or hardwired)

S := R + (if C then A else B endif);

a)		memory cycles	nano seconds
	instruction fetch	1	100
if C then read A endif	decode instruction		
	operand fetch	1	100
	execute instruction		
if not C then read B endif	instruction fetch	1	100
	decode instruction		
	instruction fetch	1	100
addieren	decode instruction		
	execute instruction		
Speichern	store result	1	100
total		**5**	**500**

Fig. 20.26   Illustrating acceleration: a) instruction-stream-based execution on a simple hypothetical processor ($C = 1$), b) data-stream-based execution in a rDPA

5 ns. The speedup factor is 1000. We may summarize via this example that avoiding or minimizing memory access is the main secret of success of reconfigurable computing: with FPGAs or coarse-grained. Also the GAG reconfigurable generic address generator of the ASM helps a lot (Figure 20.15): at run time it does not need any memory cycle for address computation. To access bulk data storage the GAG methodology also helps to find storage schemes with minimal memory cycles [132, 135].

Communication effort at classical supercomputing. The communication complexity is growing exponentially because of typical bottle-necks, mainly determined by slow memory cycles [196] (Figure 20.28). Bus systems and other switching resources tend to require memory-cycle-hungry high administrative efforts: the von Neumann bottleneck [123]. Data transport at run time is a predominating problem, whereas we have an abundance of cheap CPU resources. Thinking in communicating concurrent processes like with the MPI (Message Passing Interface) standard is dominant. The communication effort tends to grow exponentially with the number of CPUs involved. The scalabil-

#		FPGA	rDPA
1	terminology	field-programmable gate array	reconfigurable datapath array
2	reconfiguration granularity	fine-grained	coarse-grained
3	data path width	~1 bit	e.g. ~32 bits
4	physical level of basic reconfigurable units (rU)	gate level	RT level
5	typical rU examples	LUT (look-up table): determines the logic function of the rU (and, or, not, etc. or flip-flop)	ALU-like, floating point, special functions, etc.
6	configuration time	milliseconds	microseconds
7	clock cycle time	about 0.5 GHz	about 1–3 GHz
8	typical effective integration density compared to the Gordon Moore curve	reduced by a factor of about 10,000 (Figure 18.5)	reduced only by a few percent (Figure 18.11)

Fig. 20.27   Fine-grained vs. coarse-grained reconfigurability

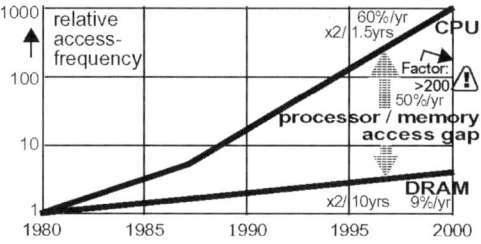

Fig. 20.28   The processor/memory access gap (Patterson's law [196])

ity of the performance of an application often drastically misses the theoretical
peak performance values, which the platform seems to offer [197]. Amdahl's
law explains only one of several reasons of this inefficiency [198]. However,
the Kress–Kung machine does not have a von Neumann bottleneck.

MPI (Message Passing Interface).   For reader outside the supercomputing
community this is explained as follows. MPI is a standard implementing the
distributed memory programming model by exchanging messages (also data)
between several processors of shared memory architectures. All functions are
available in a FORTRAN or a C version library with internal data structures,
which are mainly hidden from the user. MPI is based on the *Communicating
Sequential Processes* model (*CSP* [199, 200]) by Tony Hoare [201] for quasi-
parallel computing auf distributed, heterogeneous, loosely coupled computer
systems. The programming language Occam is an implementation of CSP.
JCSP combines CSP and Occam concepts in a Java-API. Herewith parallel
MPI programs are executable on PC clusters and dedicated parallel computers (e.g. communicating via common main memory). Most FPGA users are
not familiar with MPI and members of the supercomputing community do not
understand the languages used by FPGA-savvy hardware experts. But there are
efforts on the way to cope with the language problem [202]. For instance, also
co-compilation from a C-like language is an interesting approach [143, 144].

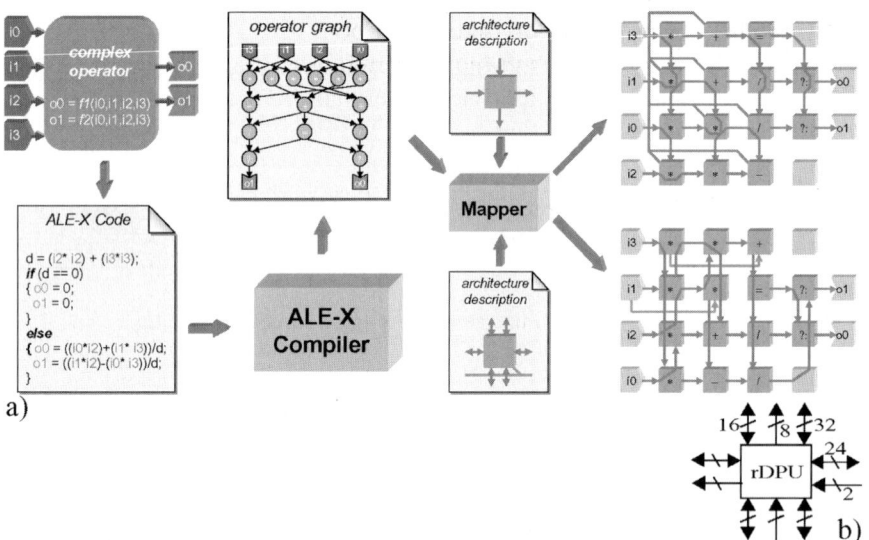

Fig. 20.29   KressArray Xplorer (design space explorer [137]): a) simplified example to illustrate platform space exploration by finding an optimized array depending on rDPU
architecture (1 or 2), b) KressArray family rDPU example architecture illustrating
flexibility

The stool is moved and not the grand piano.   With the instruction-stream-based mind set of the classical supercomputing, data transport often resembles moving the grand piano over to the stool of the pianist. However, the data-stream-based mind set of the reconfigurable computing community follows the inverse approach: the stool is moved and not the piano. Here MPI is unknown and concurrent processes are only a secondary aspect. Primarily the data are not moved by memory-cycle-hungry communication resources, but the locality of operations is assigned to the right places within the data streams, generated by ASM distributed memory (Figure 20.25b and 20.26). The communication paths between rDPUs as pipe networks – without any administrative overhead and usually without intermediate storage. Since the number of processors is usually much lower than the number of data items, there is not much to move – during the less precious compile time. At run time also no instructions are transported: another acceleration aspect.

Taxonomy of algorithms missing.   Now it is time to discuss time to space mapping methods for software to configware migration: with coarse-grained arrays and also in general. Some algorithms are easy to map, and some others are difficult to map. Let us look at examples from the embedded systems area. Classical DSP algorithms, for instance, are easy to map from time to space and the result does not need much interconnect. It is well known that some DSP algorithms even result just in simple a linear pipe. Other examples are extensively error-correcting (de)coding algorithms for a bad signal to noise ratio in wireless communication, like for turbo codes or the Viterbi algorithm and others [203], where the result requires an immense amount of interconnect resources: by far much more than available with only nearest neighbor connect within the (r)DPU array.

Missing transdisciplinary perspective.   On such samples of algorithmic cleverness we notice that each special application domain keeps its own trick box. But what is really missing in education and elsewhere is an all-embracing dual-paradigm taxonomy of algorithms. A reductionistic attitude forcing a fragmentation into many special areas keeps us from the benefit of transdisciplinary understanding the algorithms taxonomy problem. But an all-embracing taxonomy of algorithms is not even available from a single-paradigm perspective. Even in classic high performance computing and supercomputing there is a fragmentation into special application domains. For education also here a transdisciplinary perspective is missing.

The personal supercomputer is near.   The Munich-based PACT Corp. [204] with its coarse-grained reconfigurable XPA (Xtreme Processing Array) product (also see Figure 20.30) has demonstrated that a 56 core 16-bit rDPA

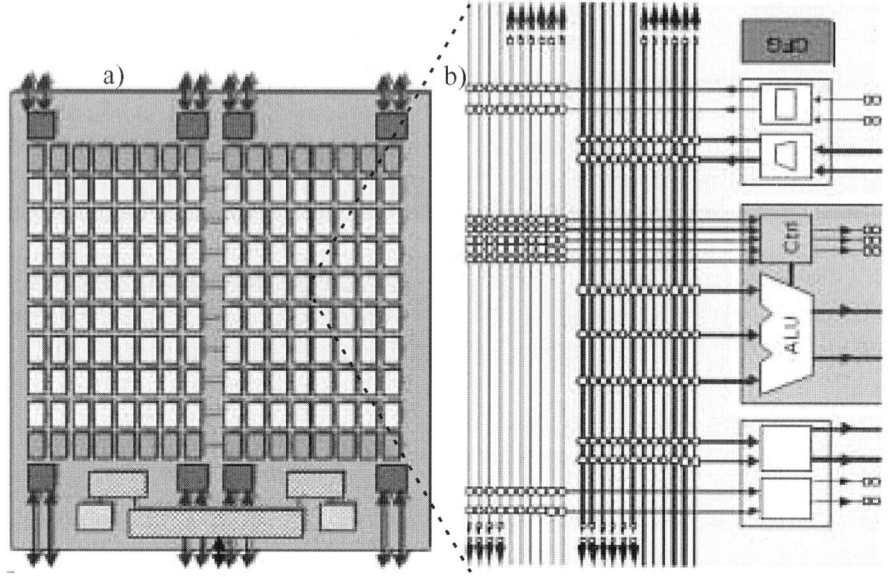

Fig. 20.30 PACT configurable XPU (xtreme processing unit): a) rDPA array, b) rDPU

running at less than 500 MHz can host simultaneously everything needed
for a world TV controller, like multiple standards, all types of conversions,
(de)compaction, image improvements and repair, all sizes and technologies of
screens, and all kinds of communication including wireless. More high performance by less CPUs, by reconfigurable units instead of CPUs. By this coarse-
grained methodology also a single-chip game console is feasible. A highly
promising vision would be a polymorphic multi-core super pentium with multiple dual-mode PUs, which under control of a mode bit could individually
run in CPU mode or in rDPU mode (not using the program counter). Choosing the right distributed on-chip memory architecture and the tight reconfigurable interconnect between these PUs is the key issue. But CPU vendors have
a lack of reconfigurable computing experience, whereas the rDPA vendors lack
concurrent computing experience. But because of educational deficits it is not
very likely that one of the major microprocessor vendors or one of the major
FPGA vendors will go toward multi-core microprocessors adopting the coarse-
grained reconfigurable computing array methodology. From a different point
of view this would also be a proposal for PACT Corp. to develop such a modified polymorphic version of the XPU array.

The second reconfigurable computing paradox.   Although for many
application areas again more orders of magnitude in speedup are expected
for coarse-grained arrays, compared to the orders of speedup having already

been obtained with FPGA-based solutions (see curve "coarse-grained" in Figure 20.4), there is no indication of efforts going into this direction, where future laptops reach a performance, which today requires a supercomputer. Another vision would be that, e.g., Xilinx would insert a rDPA onto a new platform FPGA targeting the scientific computing market. The RAMP project [151] having proposed to run the operating system on an FPGA would sound like: "Xilinx inside" replacing "intel inside". But there are also doubts whether Xilinx will understand such a vision of a future, where a dual-paradigm HPC and supercomputing will be found within embedded systems, achieved the performance, making unnecessary hangars full of monstrous expensive equipment unbelievably intensive guzzling electric power.

5.   History of FPGAs

About in the mid-1970s the field-programmable logic array (FPLA) was introduced by Signetics and Programmable array logic (PAL) was introduced by Monolithic Memories, Inc. (MMI), both for implementing combinational logic circuits – forerunners of the FPGA. A programmable logic array (PLA) is a programmable device used to implement combinational logic circuits. The PLA has a set of programmable AND planes, which link to a set of programmable OR planes, which can then be conditionally complemented to produce an output. This layout allows for a large number of logic functions to be synthesized in the sum of products (and sometimes product of sums) canonical forms. The PAL introduced 1978 and second sourced by National Semiconductor, Texas Instruments, and Advanced Micro Devices was a huge success. The programmable elements connect both the true and complemented inputs to AND gates, also known as product terms, which are ORed together to form a sum-of-products logic array. Before PALs were introduced digital designers used SSI (small-scale integration) components, such as 7400 series nand gates and D-flip-flops. But one PAL device replaced dozens of such 'discrete' logic packages in many products, such as minicomputers [210, 211], so that the SSI business went into decline.

First Xilinx FPGA.   Xilinx released the first FPGA in 1985, the XC2064 chip with a 1000 gate size. Two decades later, in the year 2004 the size of an FPGA was 10,000 times larger. In 1985 this was a new technology for implementing digital logic. These new FPGAs could be viewed either as small, slow gate arrays (MPGAs) or large, expensive PLDs [212]. The major deviation from PLDs was the capability to implement multi-level logic. Their main advantage was in low NRE cost for small volume designs and prototyping. In the late 1980s and early 1990s an explosion of different architectures was observed [96, 212]. A good reference on the introduction of FPGA architecture is [96].

Across the 100,000 gate threshold.   As long as design size remained within the range of several thousand gates, schematic design entry and a good gate-level simulator were enough to create and verify the entire design. Hardware descriptions languages started to sneak into schematic designs in the shape of HDL macros, and as designs migrated into tens of thousands of gates they gained importance. By the time FPGAs crossed the 100,000 gate threshold it became obvious that HDLs would have to eliminate schematic entry and gate-level simulators.

FPGAs in HPC.   mark the 'Beginning of a New Era' [127]. It is time to look for alternatives to the classical von Neumann computer architecture. The old recipe of increasing the clock speed, the number of CPUs, the cores per chip, and the threads per core just does not give us enough sustained computing performance. It seems that the von Neumann architecture has come to a dead end in HPC [127]. Things are now changing dramatically. FPGAs are being recognized as viable alternative to the massively power-consuming large scale computers [127].

Programming problems.   FPGAs have been mostly hidden in embedded systems [127]. Their programming was cumbersome and required specialists using some obscure programming languages like VHDL or Verilog. You might think of these languages as a kind of assembler language, but that is wrong. It is even worse: programming is done at the gate level, that is, at the very lowest level of information processing with NAND and NOR gates.

Educational deficits.   A less welcome side effect of the paradigm shift are educational deficits needing a training on the job, since typical CS or CE curricula ignore Reconfigurable Computing – still driving the dead road of the von Neumann-only mind set [126]. A new IEEE international workshop series on Reconfigurable Computing Education has been founded to attack this problem [41]. Within the several hundreds of pages of all volumes of the 2004 joint ACM / AIS / IEEE-CS curriculum recommendations [206] by the find and replace function found zero encounters of the term "FPGA" and all its synonyms or other terms pointing to reconfigurable computing. This is criminal. (For the 2005 version see [209].) These recommendations completely fail to accept the transdisciplinary responsibility of computer science to combat the fragmentation into many application-domain-specific tricky reconfigurable computing methodologies.

References

[1] A. Burks, H. Goldstein, J. von Neumann: Preliminary discussion of the logical design of an electronic computing instrument; US Army Ordnance Department Report, 1946

[2] H. Goldstein, J. von Neumann, A. Burks: Report on the mathematical and logical aspects of an electronic computing instrument; Princeton IAS, 1947

[3] M. J. Smith: Application Specific Integrated Circuits; Addison Wesley, 1997

[4] R. Hartenstein (invited): The Microprocessor Is No more General Purpose; Proceedings of IEEE International Symposium on Innovative Systems (ISIS), October 9–11, 1997, Austin, Texas

[5] L. Rizzatti: Is there any future for ASIC emulators? (or How FPGA-Based Prototypes Can Replace ASIC Emulators); EDA Alert e-Newsletter PlanetEE, February 18, 2003, www.planetee.com

[6] T. Makimoto (keynote): The Rising Wave of Field-Programmability; FPL 2000, Villach, Austria, August 27–30, 2000; Springer Verlag, Heidelberg/New York, 2000

[7] J. Becker (invited tutorial): Reconfigurable Computing Systems; Proceedings Escola de Microeletrônica da SBC – Sul (EMICRO 2003), Rio Grande, Brasil, September 2003

[8] J. Becker, S. Vernalde (editors): Advances in Reconfigurable Architectures – Part 1; special issue, International Journal on Embedded Systems, 2006

[9] J. Becker, S. Vernalde (editors): Advances in Reconfigurable Architectures – Part 2; special issue, International Journal on Embedded Systems, 2006

[10] F. Faggin, M. Hoff, S. Mazor, M. Shima: The history of 4004; IEEE Micro, December 1996

[11] R. Hartenstein (invited talk): Reconfigurable supercomputing: What are the Problems? What are the Solutions? Seminar on Dynamically Reconfigurable Architectures, Dagstuhl Castle, Wadern, Germany, April 2–7, 2006

[12] R. Hartenstein (opening keynote): From Organic Computing to Reconfigurable Computing; 8th Workshop on Parallel Systems and Algorithms (PASA 2006), March 16, 2006, Frankfurt/Main, Germany – in conjunction with the 19th International Conference on Architecture of Computing Systems (ARCS 2006), March 13–16, 2006, Frankfurt/Main, German

[13] R. Hartenstein (opening keynote): Supercomputing goes reconfigurable – About key issues and their impact on CS education; Winter International Symposium on Information and Communication Technologies (WISICT 2005), Cape Town, South Africa, January 3–6, 2005

[14] R. Hartenstein (plenum keynote address): Software or Configware? About the Digital Divide of Computing; 18th International Parallel and Distributed Processing Symposium (IPDPS), April 26–30, 2004, Santa Fe, New Mexico, USA

[15] R. Hartenstein (invited presentation): The Digital Divide of Computing; 2004 ACM International Conference on Computing Frontiers (CF04); April 14–18, 2004, Ischia, Italy

[16] R. Hartenstein (invited keynote address): The Impact of Morphware on Parallel Computing; 12th Euromicro Conference on Parallel, Distributed and Network based Processing (PDP04); February 11–13, 2004, A Coruña, Spain

[17] D. Chinnery, K. Keutzer: Closing the Gap between ASIC & Custom; Kluwer 2002

[18] A. Grove: Only the Paranoid Survive; Currence 1996

[19] A.A. Gaffar, W. Luk: Accelerating Radiosity Calculations; FCCM 2002

[20] M. Gokhale et al.: Acceleration of Traffic Simulation on Reconfigurable Hardware; 2004 MAPLD International Conference, September 8–10, 2004, Washington, D.C, USA

[21] F. Dittrich: World's Fastest Lanman/NTLM Key Recovery Server Shipped; Picocomputing, 2006 URL: [22]

[22] http://www.picocomputing.com/press/KeyRecoveryServer.pdf

[23] K. Gaj, T. El-Ghazawi; Cryptographic Applications; RSSI Reconfigurable Systems Summer Institute, July 11–13, 2005, Urbana-Champaign, IL, USA URL: [24]

[24] http://www.ncsa.uiuc.edu/Conferences/RSSI/presentations.html

[25] J. Hammes, D. Poznanovic: Application Development on the SRC Computers, Inc. Systems; RSSI Reconfigurable Systems Summer Institute, July 11–13, 2005, Urbana-Champaign, IL, USA

[26] *ASM* = Auto-Sequencing Memory; *DSP* = Digital Signal Processing; *EDA* = Electronics Design Automation; *ESL* = Electronic System-Level Design; *FIR* = Finite Impulse Response; *FPGA* = Field-Programmable Gate-Array; *MAC* = Multiply and Accumulate; *PU* = Processing Unit *rDPA* = reconfigurable Data Path Array; *rDPU* = reconfigurable Data Path Unit; *rE* = reconfigurable Element

[27] W. Roelandts (keynote): FPGAs and the Era of Field Programmability; International Conference on Field Programmable Logic and Applications (FPL), August 29–September 1, 2004, Antwerp, Belgium,

[28] J. Rabaey: Reconfigurable Processing: The Solution to Low-Power Programmable DSP; ICASSP 1997

[29] Y. Gu et al.: FPGA Acceleration of Molecular Dynamics Computations; FCCM 2004 URL: [30]

[30] http://www.bu.edu/caadlab/FCCM05.pdf

[31] A. Alex, J. Rose et al.: Hardware Accelerated Novel Protein Identification; FPL 2004

[32] N.N. Nallatech, press release, 2005

[33] H. Singpiel, C. Jacobi: Exploring the Benefits of FPGA-Processor Technology for Genome Analysis at Acconovis; ISC 2003, June 2003, Heidelberg, Germany URL. [34]

[34] http://www.hoise.com/vmw/03/articles/vmw/LV-PL-06-03-9.html

[35] N.N. (Starbridge): Smith-Waterman pattern matching; National Cancer Institute, 2004

[36] A. Darabiha: Video-Rate Stereo Vision on Reconfigurable Hardware; Master Thesis, University of Toronto, 2003

[37] R. McCready: Real-Time Face Detection on a Configurable Hardware Platform; Master thesis, University of Toronto, 2000

[38] T. Fry, S. Hauck: Hyperspectral Image Compression on Reconfigurable Platforms; Proceedings of FCCM 2002

[39] P. Buxa, D. Caliga: Reconfigurable Processing Design Suits UAV Radar Apps; COTS J., October 2005: [40]

[40] http://www.srccomp.com/ReconfigurableProcessing_UAVs_COTS-Journal_Oct05.pdf

[41] http://helios.informatik.uni-kl.de/RCeducation/

[42] R. Porter: Evolution on FPGAs for Feature Extraction; Ph.D.thesis; Queensland University, Brisbane, Australia, 2001: [43]

[43] http://www.pooka.lanl.gov/content/pooka/green/Publications_files/imge FPGA.pdf

[44] E. Chitalwala: Starbridge Solutions to Supercomputing Problems; RSSI Reconfigurable Systems Summer Institute, July 11–13, 2005, Urbana-Champaign, IL,USA

[45] S.D. Haynes, P.Y.K. Cheung, W. Luk, J. Stone: SONIC – A Plug-In Architecture for Video Processing; Proceedings of FPL 1999

[46] M. Kuulusa: DSP Processor Based Wireless System Design; Tampere University of Technology, Publ. #296: [47]

[47] http://edu.cs.tut.fi/kuulusa296.pdf

[48] B. Schäfer, S. Quigley, A. Chan: Implementation of The Discrete Element Method Using Reconfigurable Computing (FPGAs); 15th ASCE Engineering Mechanics Conference, June 2–5, 2002, New York, NY: [49]

[49] http://www.civil.columbia.edu/em2002/proceedings/papers/126.pdf

[50] G. Lienhart: Beschleunigung Hydrodynamischer N-Körper-Simulationen mit Rekonfigurierbaren Rechenystemen; 33rd Speedup/19th PARS Worksh.; Basel, Switzerland, March 19–21, 2003

[51] G. Steiner, K. Shenoy, D. Isaacs, D. Pellerin: How to accelerate algorithms by automatically generating FPGA coprocessors; Dr. Dobbs Portal, August 09, 2006 URL: [52]

[52] http://www.ddj.com/dept/embedded/191901647

[53] R. Hartenstein (keynote): The Transdisciplinary Responsibility of CS Curricula; 8th World Conference on Integrated Design & Process Technology (IDPT), June 25–29, 2006, San Diego, CA, USA

[54] R. Hartenstein (invited presentation): Reconfigurable Computing; Computing Meeting, Commission of the EU, Brussels, Belgium, May 18, 2006,

[55] R. Hartenstein (keynote): New horizons of very high performance computing (VHPC) – hurdles and chances; 13th Reconfigurable Architectures Workshop (RAW 2006), Rhodos Island, Greece, April 25–26, 2006

[56] R. Hartenstein: Configware für Supercomputing: Aufbruch zum Personal Supercomputer; PIK – Praxis der Informationsverarbeitung und Kommunikation, 2006; KG Saur Verlag GmbH, Munich, Germany – title in English: see [57]

[57] Configware for Supercomputing: Decampment for the Personal Supercomputer

[58] J. Becker et al.: An Embedded Accelerator for Real Time Image Processing; 8th EUROMI-CRO Workshop on Real Time Systems, L'Aquila, Italy, June 1996

[59] http://xputers.informatik.uni-kl.de/faq-pages/fqa.html

[60] W. Nebel et al.: PISA, a CAD Package and Special Hardware for Pixel-oriented Layout Analysis; ICCAD 1984

[61] J. Becker et al.: High-Performance Computing Using a Reconfigurable Accelerator; CPE Journal, Special Issue of Concurrency: Practice and Experience, Wiley, 1996

[62] http://xputers.informatik.uni-kl.de/staff/hartenstein/eishistory.html

[63] Herb Riley, R. Associates: http://www.supercomputingonline.com/article. php?sid=9095

[64] R. Hartenstein (opening keynote): The Re-definition of Low Power Design for HPC: A Paradigm Shift; 19th Symposium on Integrated Circuits and System Design (SBCCI 2006), August 28–September 1, 2006, Ouro Preto, Minas Gerais, Brazil

[65] Ch. Piguet (keynote): Static and dynamic power reduction by architecture selection; PATMOS 2006, September 13–15, Montpellier, France

[66] V. George, J. Rabaey: Low-Energy FPGAs: Architecture and Design; Kluwer, 2001

[67] R. Hartenstein (opening keynote address: The Re-definition of Low Power Digital System Design – for slashing the Electricity Bill by Millions of Dollars; The 19th Symposium on Integrated Circuits and System Design (SBCCI 2006), August 28–September 1, 2006, Ouro Preto, Minas Gerais, Brazil

[68] R. Hartenstein (invited contribution): Higher Performance by less CPUs; HPCwire, June 30, 2006, in conjunction with ISC 2006, Dresden, Germany – [69]

[69] http://xputers.informatik.uni-kl.de/staff/hartenstein/lot/HartensteinLess CPUs06.pdf

[70] R. Hartenstein, A. Nuñez et al.: Report on Interconnect Modelling: Second Periodic Progress Report; PATMOS Project, Universität Kaiserslautern, December 1991

[71] http://www.patmos-conf.org/

[72] http://www.islped.org/

[73] N.N.: R. Associates Joins Nallatech's Growing Channel Partner Program; Companies are Using FPGAs to Reduce Costs and Increase Performance in Seismic Processing for Oil and Gas Exploration; FPGA and Structured ASIC Journal BusinessWire, August 08, 2005 – URL: [74]

[74] http://www.fpgajournal.com/news_2005/08/20050808_03.htm

[75] R. Hartenstein (invited presentation): Reconfigurable Computing (RC) being Mainstream: Torpedoed by Education; 2005 International Conference on Microelectronic Systems Education, June 12–13, 2005, Anaheim, California, co-located with DAC (Design Automation Conference) 2005

[76] R. Hartenstein (invited opening keynote address): Reconfigurable HPC: Torpedoed by Deficits in Education? Workshop on Reconfigurable Systems for HPC (RHPC); July 21, 2004, to be held in conjunction with

HPC Asia 2004, 7th International Conference on High Performance Computing and Grid in Asia Pacific Region, July 20–22, 2004, Omiya (Tokyo), Japan

[77] R. Hartenstein (keynote): The Changing Role of Computer Architecture Education within CS Curricula; Workshop on Computer Architecture Education (WCAE 2004) June 19, 2004, in conjunction with the 31st International Symposiam on Computer Architecture (ISCA), Munich, Germany, June 19–23, 2004

[78] S. Hauck: The Role of FPGAs in Reprogrammable Systems; Proceedings of IEEE, 1998

[79] F. Rammig (interview): Visions from the IT Engine Room; IFIP TC 10 – Computer Systems Technology, URL: [80]

[80] http://www.ifip.or.at/secretariat/tc_visions/tc 10_visions.htm

[81] S. Hang: Embedded Systems – Der (verdeckte) Siegeszug einer Schlüsseltechnologie; Deutsche Bank Research, January 2001, [82], URL: [83]

[82] S. Hang ([81] title in English): Embedded Systems – The (subsurface) Triumph of a Key Technology;

[83] http://xputers.informatik.uni-kl.de/VerdeckterSiegeszug.pdf

[84] P. Gillick: State of the art FPGA development tools; Reconfigurable Computing Workshop, Orsay, France, September 2003

[85] http://antimachine.org

[86] V. Betz, J. Rose, A. Marquardt (editors): Architecture and CAD for Deep-Submicron FPGas; Kluwer, 1999

[87] R. Hartenstein: Morphware and Configware; (invited chapter) in: A. Zomaya (editor): Handbook of Innovative Computing Paradigms; Springer Verlag, New York, 2006

[88] S. Hoffmann: Modern FPGAs, Reconfigurable Platforms and Their Design Tools; Proceedings REASON Summer School; Ljubljana, Slovenia, August 11–13, 2003

[89] N.N.: Pair extend Flash FPGA codevelopment pact; http://www.electronicstalk.com/news/act/act120.html

[90] B. Lewis, Gartner Dataquest, October 28, 2002

[91] http://morphware.net

[92] http://www.morphware.org/

[93] http://configware.org

[94] D. Soudris et al.: Survey of existing fine grain reconfigurable hardware platforms; Deliverable D9, AMDREL consortium (Architectures and Methodologies for Dynamically Reconfigurable Logic); 2002

[95] J. Oldfield, R. Dorf: Field-Programmable Gate Arrays: Reconfigurable Logic for Rapid Prototyping and Implementation of Digital Systems; Wiley-Interscience, 1995

[96] J. Rose, A.E. Gamal, A. Sangiovanni-Vincentelli: Architecture of FP-GAs; Proceedings of IEEE, 81(7), July 1993

[97] http://www.xilinx.com

[98] http://www.altera.com

[99] N.N.: FPGA Development Boards and Software for Universities; Microelectronic System News, August 2, 2006, URL: [100]

[100] http://vlsil.engr.utk.edu/~bouldin/MUGSTUFF/NEWSLETTERS/DATA/1278.html

[101] S. Davidmann: Commentary: It's the software, stupid; El. Bus. online; August 15, 2006, URL: [102]

[102] http://www.reed-electronics.com/eb-mag/index.asp?layout=articlePrint&articleID=CA6360703

[103] Michio Kaku: Visions; Anchor Books, 1998

[104] U. Rüde: Technological trends and their impact on the future of supercomputers; in: H. Bungartz, F Durst, C. Zenger, C. (editors): High Performance Scientific and Engineering Computing: International FORTWIHR Conference on HPSEC, Munich, Germany, March 16–18, 1998 – URL: [112]

[105] F. Brooks: No Silver Bullet; COMPUTER, 20, 4 (April 1987)

[106] F. Brooks: The Mythical Man Month (Anniversary Edition with 4 new chapters) Addison Wesley, 1995

[107] T.S. Kuhn: The Structure of Scientific Revolution; University of Chicago Press, Chicgo, 1996

[108] V. Rajlich: Changing the Paradigm of Software Engineering; C. ACM 49, 8 (August 2006)

[109] E.M. Rogers: Diffusion of Innovations – 4th edition; The Free Press, New York, 1995

[110] http://www.sdpsnet.org/Other_PDFs/IDPT2006FinalProgram.pdf

[111] http://www.organic-computing.de/

[112] http://www10.informatik.uni-erlangen.de/Publications/Papers/1998/lncse_vol8_98.pdf

[113] V. Natoli: A Computational Physicist's View of Reconfigurable High Performance Computing; Reconfigurable Systems Summer Institute, Urbana, IL, July 2005, URL: [115]

[114] Edward A. Lee: The Problem with Threads; COMPUTER, May 2006

[115] http://www.ncsa.uiuc.edu/Conferences/RSSI/2005/docs/Natoli.pdf

[116] Mark Bereit: Escape the Software Development Paradigm Trap; Dr. Dobb's Journal (05/29/06)

[117] B. Hayes: The Semicolon Wars; American Scientist, July/August 2006 URL: [118]

[118] http://www.americanscientist.org/content/AMSCI/AMSCI/ArticleAlt Format/20066510252_307.pdf

[119] E. Levenez: Computer Languages History; http://www.levenez.com/lang/

[120] B. Kinnersley: The Language List – Collected Information on About 2500 Computer Languages, Past and Present; http://people.ku.edu/~nkinners/LangList/Extras/langlist.htm

[121] D. Piggott: HOPL: An interactive roster of programming languages; 1995–2006. URL: [122]

[122] http://hopl.murdoch.edu.au/

[123] G. Koch, R. Hartenstein: The Universal Bus considered harmful; in: [124]

[124] R. Hartenstein R. Zaks (editors): Microarchitecture of Computer Systems; American Elsevier, New York, 1975

[125] R. Hartenstein (opening keynote address): Why We Need Reconfigurable Computing Education: Introduction; The 1st International Workshop on Reconfigurable Computing Education (RCeducation 2006), March 1, 2006, Karlsruhe, Germany – in conjunction with the IEEE Computer Society Annual Symposiun on VLSI (ISVLSI 2006), March 2–3, 2006, Karlsruhe, Germany [126]

[126] http://hartenstein.de/RCedu.html

[127] A. Reinefeld (interview): FPGAs in HPC Mark "Beginning of a New Era"; HPCwire June 29, 2006 – http://www.hpcwire.com/hpc/709193.html

[128] N. Petkov: Systolic Parallel Processing; North-Holland, 1992

[129] H.T. Kung: Why Systolic Architectures? IEEE Computer, 15(1) 37–46 (1982)

[130] R. Kress et al.: A Datapath Synthesis System (DPSS) for the Reconfigurable Datapath Architecture; Proceedings of ASP-DAC 1995

[131] http://kressarray.de

[132] M. Herz et al. (invited p.): Memory Organization for Data-Stream-based Reconfigurable Computing; 9th IEEE International Conference on Electronics, Circuits and Systems (ICECS), September 15–18, 2002, Dubrovnik, Croatia

[133] http://flowware.net

[134] M. Herz et al.: A Novel Sequencer Hardware for Application Specific Computing; ASAP'97

[135] M. Herz: High Performance Memory Communication Architectures for Coarse-grained Reconfigurable Computing Systems; Dissertation, University of Kaiserslautern, 2001, URL: [136]

[136] http://xputers.informatik.uni-kl.de/papers/publications/HerzDiss.html

[137] U. Nageldinger et al.: Generation of Design Suggestions for Coarse-Grain Reconfigurable Architectures: International Conference on Field Programmable Logic and Applications (FPL) August 28–30 2000, Villach, Austria

[138] U. Nageldinger Coarse-grained Reconfigurable Architectures Design Space Exploration; Dissertation, University of Kaiserslautern, 2001, Kaiserslautern, Germany URL: [139]

[139] http://xputers.informatik.uni-kl.de/papers/publications/Nageldinger Diss.html

[140] R. Hartenstein (invited plenum presentation): Reconfigurable Super-computing: Hurdles and Chances; International Supercomputer Conference (ICS 2006), June 28–30, 2006, Dresden, Germany – http://www.supercomp.de/

[141] http://flowware.net

[142] A. Ast et al.: Data-procedural languages for FPL-based Machines; International Conference on Field Programmable Logic and Applications (FPL), September 7–9, 1994, Prag, Tschechia

[143] J. Becker et al.: Parallelization in Co-Compilation for Configurable Accelerators; Asian-Pacific Design Automation Conference (ASP-DAC), February 10–13, 1998, Yokohama, Japan

[144] J. Becker: A Partitioning Compiler for Computers with Xputer-based Accelerators, Dissertation, University of Kaiserslautern, 1997, URL: [145]

[145] http://xputers.informatik.uni-kl.de/papers/publications/BeckerDiss.pdf

[146] K. Schmidt: A Program Partitioning, Restructuring, and Mapping Method for Xputers, Dissertation, University of Kaiserslautern, 1994 URL: [147]

[147] http://www.shaker.de/Online-Gesamtkatalog/Details.asp?ID=930601 &CC=41546&ISBN=3-8265-0495-X&Reihe=15&IDSRC=4& LastAction=Search

[148] J.C. Hoe, Arvind: Hardware Synthesis from Term Rewriting Systems; VLSI'99, Lisbon, Portugal

[149] M. Ayala-Rincón, R. Hartenstein, C.H. Llanos, R.P. Jacobi: Prototyping Time and Space Efficient Computations of Algebraic Operations over Dynamically Reconfigurable Systems Modeled by Rewriting-Logic; accepted for ACM Transactions on Design Automation of Electronic Systems (TODAES), 2006

[150] A. Cantle: Is it time for von Neumann and Harvard to retire? RSSI Reconfigurable Systems Summer Institute, July 11–13, 2005, Urbana-Champaign, IL, USA

[151] http://bwrc.eecs.berkeley.edu/Research/RAMP/

[152] http://de.wikipedia.org/wiki/Direct_Memory_Access

[153] S. Heath: Embedded Systems Design, Elsevier, 2003

[154] F. Catthoor et al.: Data Access and Storage Management for Embedded Programmable Processors; Kluwer, 2002

[155] A.G. Hirschbiel et al.: A Novel Paradigm of Parallel Computation and Its Use to Implement Simple High Performance Hardware; Info-Japan'90 – International Conference memorating 30th Anniversary, Computer Society of Japan; Tokyo, Japan, 1990

[156] Invited reprint of [155] in Future Generation Computer Systems 7 91/92, pp. 181–198, North Holland

[157] W. Nebel et al.: Functional Design Verification by Register Transfer Net Extraction from Integrated Circuit Layout Data; IEEE COMPEURO, Hamburg, 1987

[158] C. Chang et al.: The Biggascale Emulation Engine (Bee); summer retreat 2001, UC Berkeley

[159] H. Simmler et al.: Multitasking on FPGA Coprocessors; International Conference on Field Programmable Logic and Applications (FPL), August 28–30, 2000, Villach, Austria

[160] H. Walder, M. Platzner: Reconfigurable Hardware Operating Systems: From Design Concepts to Realizations; Proceedings of ERSA 2003

[161] P. Zipf: A Fault Tolerance Technique for Field-Programmable Logic Arrays; Dissertation, University of Siegen, 2002

[162] M. Abramovici, C, Stroud: Improved BIST-Based Diagnosis of FPGA Logic Blocks; IEEE International Test Conference, Atlantic City, October 2000

[163] J.A. Starzyk, J. Guo, Z. Zhu: Dynamically Reconfigurable Neuron Architecture for the Implementation of Self-organizing Learning Array; International Journal on Embedded Systems, 2006

[164] M. Glesner, W. Pochmueller: An Overview of Neural Networks in VLSI; Chapman & Hall, London, 1994

[165] N. Ambrosiano: Los Alamos and Surrey Satellite contract for Cibola flight experiment platform; Los Alamos National Lab, Los Alamos, NM, March 10, 2004, URL: [166]

[166] http://www.lanl.gov/news/releases/archive/04-015.shtml

[167] M. Gokhale et al.: Dynamic Reconfiguration for Management of Radiation-Induced Faults in FPGAs; IPDPS 2004

[168] G. De Micheli (keynote): Nanoelectronics: challenges and opportunities; PATMOS 2006, September 13–15, Montpellier, France

[169] E. Venere: Nanoelectronics Goes Vertical; HPCwire, August 4, 2006, 15(3) – http://www.hpcwire.com/hpc/768349.html

[170] J.E. Savage: Research in Computational Nanotechnology; Link: [171]

[171] http://www.cs.brown.edu/people/jes/nano.html

[172] A. DeHon et al.: Sub-lithographic Semiconductor Computing Systems; HotChips-15, August 17–19, 2003, link: [173]

[173] http://www.cs.caltech.edu/research/ic/pdf/sublitho_hotchips2003.pdf

[174] L. Gottlieb, J.E. Savage, A. Yerukhimovich: Efficient Data Storage in Large Nanoarrays; Theory of Computing Systems, 38, pp. 503–536, 2005.

[175] J.E. Savage: Computing with Electronic Nanotechnologies; 5th Conference on Algorithms and Complexity, May 28–30, 2003, Rome, Italy

[176] International Technology Roadmap for Semiconductors. link: [177]

[177] http://public.itrs.net/Files/2001ITRS/, 2001.

[178] A. DeHon, P. Lincoln, J. Savage: Stochastic Assembly of Sublithographic Nanoscale Interfaces; IEEE Transactions in Nanotechnology, September 2003.

[179] Y. Cui et al.: Diameter-Controlled Synthesis of Single Crystal Silicon Nanowires; Applied Physics Letters, 78(15):2214–2216, 2001.

[180] A.M. Morales, C.M. Lieber: A Laser Ablation Method for Synthesis of Crystalline Semi-conductor Nanowires; Science, 279:208–211, 1998.

[181] M.S. Gudiksend et al.: Epitaxial Core-Shell and Core-Multi-Shell Nanowire Heterostructures; Nature, 420:57–61, 2002.

[182] M.S. Gudiksen et al.: Growth of Nanowire Superlattice Structures for Nanoscale Photonics and Electronics;. Nature, 415:617–620, February 7, 2002.

[183] C. Collier et al.: A Catenane-Based Solid State Reconfigurable Switch; Science, 289:1172–1175, 2000.

[184] C.P. Collier et al.: Electronically Configurable Molecular-Based Logic Gates. Science, 285:391–394, 1999.

[185] A. DeHon: Array-Based Architecture for FET-based Nanoscale Electronics; IEEE Transactions on Nanotechnology, 2(1):23–32, March 2003.

[186] C.L. Brown et al.: Introduction of [2] Catenanes into Langmuir Films and Langmuir-Blodgett Multilayers. A Possible Strategy for Molecular Information Storage Materials; Langmuir, 16(4):1924–1930, 2000.

[187] Y. Huang, X. Duan, Q. Wei, C.M. Lieber: Directed Assembley of One-Dimensional Nanostructures into Functional Networks. Science, 291:630–633, January 26, 2001.

[188] D. Whang, S. Jin, C.M. Lieber: Nanolithography Using Hierarchically Assembled Nanowire Masks; Nano Letters, 3, 2003. links: [205]

[189] R. Hartenstein (invited embedded tutorial): A Decade of Reconfigurable Computing: A Visionary Retrospective; Design, Automation and Test in Europe, Conference & Exhibition (DATE 2001), March 13–16, 2001, Munich, Germany

[190] http://en.wikipedia.org/wiki/RDPA

[191] R. Hartenstein et al.[1]: A High Performance Machine Paradigm Based on Auto-Sequencing Data Memory; HICSS-24 Hawaii International Conference on System Sciences, Koloa, Hawaii, January 1991

[192] J. Hoogerbrugge, H. Corporaal, H. Mulder: Software pipelining for transport-triggered architectures: Proceedings of MICRO-24, Albuquerque, 1991

[193] H. Corporaal: Microprocessor Architectures from VLIW to TTA; John Wiley, 1998

[194] http://en.wikipedia.org/wiki/Transport_Triggered_Architectures

[195] D. Tabak, G.J. Lipovski: MOVE architecture in digital controllers; IEEE Transactions on Computers C-29, pp. 180–190, 1980

[196] D. Patterson, T. Anderson, N. Cardwell, R. Fromm, K. Keeton, C. Kozyrakis, R. Thomas, K. Yelick: A Case for Intelligent RAM; IEEE Micro, March/April 1997.

[197] J. Dongarra et al. (editors): The Sourcebook of Parallel Computing; Morgan Kaufmann 2002

[198] G. Amdahl: Validity of the Single Processor Approach to Achieving Large-Scale Computing Capabilities; Proceedings of AFIPS 1967

[199] C.A.R. Hoare: Communicating Sequential Processes, Prentice-Hall, 1985 – URL: [200]

---

[1] Best paper award.

[200] http://www.usingcsp.com/cspbook.pdf

[201] http://de.wikipedia.org/wiki/Tony_Hoare

[202] Th. Steinke: Experiences with High-Level-Programming of FPGAs; International Supercomputer Conference (ICS 2006), June 28–30, 2006, Dresden, Germany

[203] N. Wehn (invited keynote): Advanced Channel Decoding Algorithms and Their Implementation for Future Communication Systems; IEEE Computer Society Annual Symposium on VLSI, March 2–3, 2006, Karlsruhe, Germany

[204] http://pactcorp.com

[205] http://www.cs.caltech.edu/research/ic/abstracts/sublitho_hotchips2003.html

[206] Joint Task Force for Computing Curricula 2004: Computing Curricula 2004. Overview Report; s. [207]

[207] http://portal.acm.org/citation.cfm?id=1041624.1041634

[208] http://hartenstein.de/ComputingCurriculaOverview_Draft_11-22-04.pdf

[209] http://www.acm.org/education/curric_vols/CC2005-March06Final.pdf

[210] Tracy Kidder: The Soul of A New Machine; Little Brown and Company, 1981 – reprint: [211]

[211] Tracy Kidder (reprint): The Soul of A New Machine; Modern Library, 1997

[212] S.-L. Lu: The FPGA paradigm; http://web.engr.oregonstate.edu/~sllu/fpga/lec1.html

Chapter 21

Dynamic Reconfiguration

Exploiting dynamic reconfiguration for power optimized architectures

Jürgen Becker and Michael Hübner
ITIV, Universität Karlsruhe (TH)

Abstract   The adaptivity of electronic systems to their environment and inner system status enables the processing of different applications in time slots on one reconfigurable hardware architecture. Adaptivity "On-Demand" related to non predictable requirements from the user or the environment means the optimization of power dissipation and performance at run-time by providing on-chip computation capacity. Traditional microprocessor based electronic systems are able to adapt the software to the required tasks of an application. The disadvantage in those system is the sequential processing of the software code and the fixed hardware architecture which doesn't allow to adapt internal structures in order to optimize the data throughput. Reconfigurable hardware, in this section Field Programmable Gate Arrays (FPGAs), allows adapting the hardware system architecture at design- and run-time as well as the integrated software within the included soft-core IP-Processing cores. This new system approach which enables the computing in time and space opens new degree of freedom in system design. Methods of the hardware- software co-design which were former developed for Application Specific Integrated System (ASIC) design and only can be exploited while design time now can be used while run-time in order to achieve an optimized system parameterization in terms of adapted IP and software.

Keywords: Field Programmable Gate Array (FPGA), Dynamic and partial reconfiguration, run-time adaptivity

J. Henkel and S. Parameswaran (eds.), Designing Embedded Processors – A Low Power Perspective,
503–512.

1.   Basis of Reconfiguration and Terminology

FPGA architectures, as described in Chapter 1, can be used to integrate digital systems as an architecture modeled in a HDL language. As mentioned, some of the presented FPGA architectures are one-time programmable or re-programmable. A special feature, the dynamic reconfigurability, can be exploited with a subset of actual available SRAM-based FPGAs. To understand the different ways to (re-)configure these devices, a terminology will be introduced in the following description.

In Figure 21.1, a configurable area of an FPGA is represented in the lower part of the schematic drawing. The area spans a 2-dimensional field with configurable blocks (see Chapter 1). On the time-axis, one configuration (application) can be load to the FPGA and runs as long the system is powered. In case of Fuse/Antifuse- and FLASH-based architecture, the application of this configuration starts immediately after power-up phase because of the non-volatile memory. In case of a SRAM-based architecture, the configuration must be reloaded from external memory to the FPGA. This traditional way of configuring an FPGA can be achieved with all types, Fuse/Antifuse-, FLASH- and SRAM-based FPGA.

In Figure 21.2 the previous described scenario is extended by different configurations which were loaded in different time frames. Certainly, this scenario is now valid only for FLASH- and SRAM-based FPGAs. To re-program the complete content of the FPGA, the device's memory must be erased. While SRAM-based FPGAs can be reseted and re-programmed, FLASH-based FPGA's memory must be erased before re-programming the device. Equally to the scenario in Figure 21.1, the reconfiguration method is possible in off-line mode. This means that the application has to be stopped in order to include the new functionality on-chip.

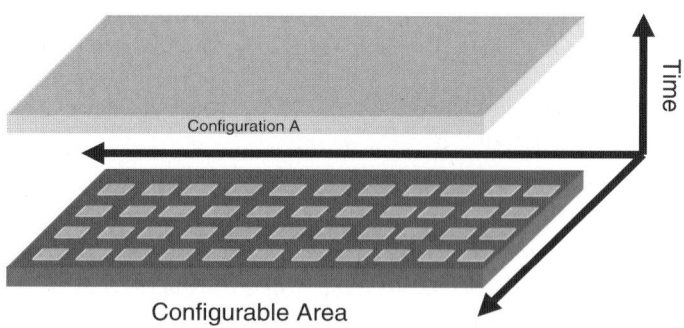

Fig. 21.1   Traditional configuration of FPGA

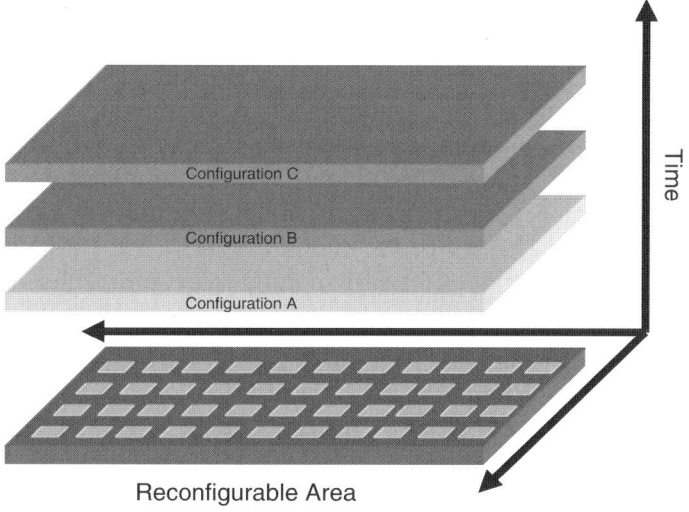

Fig. 21.2 Full reconfiguration of FPGA

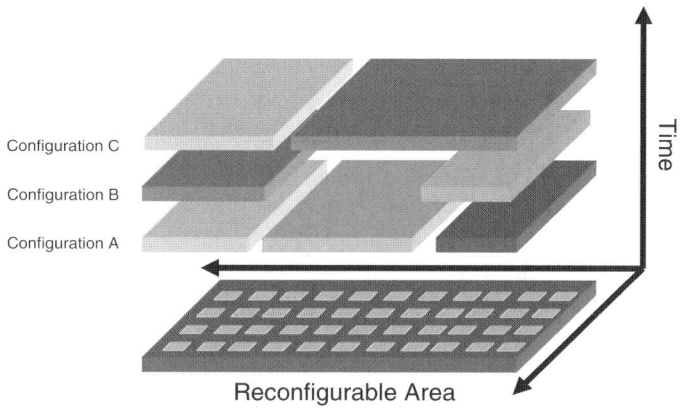

Fig. 21.3 Partial reconfiguration of FPGA

Figure 21.3 shows the scenario of partial reconfiguration of an FPGA. The difference to the process described with Figure 21.2 here is that not the complete configuration has to be written to FPGA's configuration memory. In the example of the schematic view in Figure 21.3, the left and right part of configuration A is substituted by configuration B. In the next step the configuration B is substituted by configuration C with two application parts. Partial reconfiguration is performed in off-line mode which means that active configurations were stopped for the time of reconfiguration. The described scenario only makes sense if the used FPGA architecture supports the re-programming of parts of the configuration memory. This is possible with both FLASH- and SRAM-based FPGA technology but in fact, no FLASH-based FPGA architecture is known until now which allows erasing only parts of configuration memory. Definitely the benefit here is that the complete configuration code does not have to be written to the device which leads to decreased configuration time.

The gap to the definition of the term "Dynamic and Partial Reconfiguration" can now be closed, if an FPGA architecture coupled with the scenario described with Figure 21.3 allows reconfiguring parts of the FPGA's configuration memory while run-time. Run-time in this context means that applications does not have to be stopped while reconfiguration phase. With the example of Figure 21.3, the medial part of configuration A stays active while the left and right parts were substituted. Consequently also here, the device must support the method of dynamic and partial reconfiguration. Until now only SRAM-based FPGAs enable this feature.

2.   Exploiting Dynamic and Partial Reconfiguration for Power-Reduction

In order to motivate the possibility of power-reduction by exploitation of dynamic and partial reconfiguration, a simple example starts with the data-flow graph in Figure 21.4.

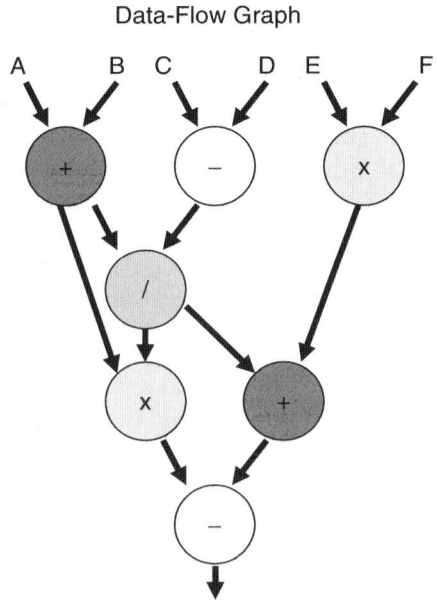

Fig. 21.4   Data-flow graph of a simple equation

$$Result = A + B/C - D + E * F - A + B/C - D + E * F \qquad (21.1)$$

The data-flow graph is the picture above, describes the equation in Equation 21.1. Lets assume that each operation has a latency of 1 which means that each time step an operand can calculate the related result. With this assumption it is obvious that the result can be calculated in four time steps.

Each operation utilizes logic-blocks by integration on a real FPGA. As described in Section 1, configurable logic-blocks with their programmable look-up tables (LUT) were used to implement the functions described in HDL language. Certainly also communication resources were utilized to enable the data-flow but neglected for the experiment described in the following text. In this example, the operators were integrated to a standard FPGA device to get an impression of the area amount. Therefore, Figure 21.5 displays real values for logic-block utilization on an FPGA. All operations were modeled with an bit-width of 32 bit.

By neglecting the data-dependencies of the several operators, Figure 21.6 pictures a schematic representation of the FPGA integrated equation mentioned above. The required value of logic-blocks for a full integration is the sum of all logic-blocks of each operator. In this example the amount of logic-blocks has the number of 3217.

From the data-flow graph pictured in Figure 21.4 can be recognized that some operations can be processed in parallel in each time step. In best case, three operations can be calculated in parallel within one-time frame (see Figure 21.7). Please notice that the addition operation in time frame two can be scheduled to time frame three.

In the left part of Figure 21.7, the sum of actual required logic-bocks in each time frame is listed. Significant here is that in the second time frame the highest amount of logic-blocks are required to support the necessary calculation. As mentioned before, the addition operation of time frame two can be sched-

Operation	Logic-Blocks
+ −	16
x	673
/	1807

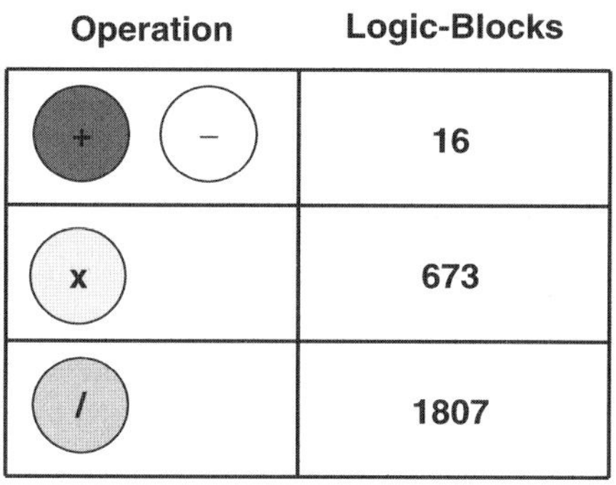

Fig. 21.5 Resource utilization

Configuration

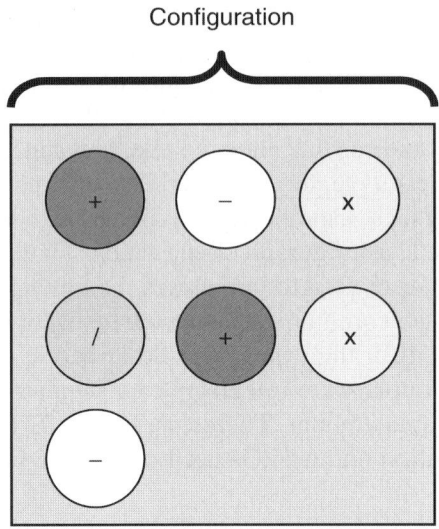

Fig. 21.6   Full integration of operators

Logic-Blocks   Configuration

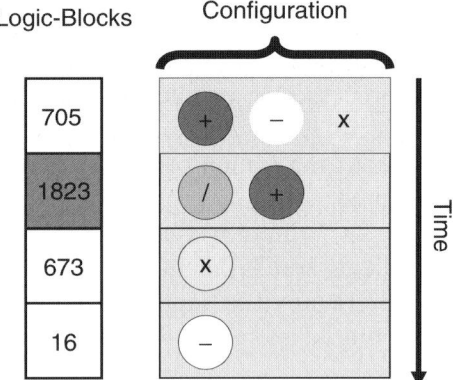

Fig. 21.7   Dynamic and partial reconfiguration of operators

uled to time frame three which has to be taken into account. By re-scheduling the addition operation to time frame three, the amount of logic-blocks in time frame two is reduced by 16 logic-blocks. Therefore the time frame with the highest utilization has the value of 1807 achieved by optimizing the scheduling of required operations. The example illustrates that dynamic reconfiguration can be exploited to reduce the amount of necessary logic-resources and therefore reduction of configuration area. In the example described above, a reduction of 56% can be achieved. Scheduling can affect the highest utilization in one time frame and has to be taken into account. In fact, this example integrated into a real system on FPGA would have a lower value for the relation between full configuration and the dynamic and partial method because of necessary control mechanisms for data-flow and reconfiguration. A real example integrated on FPGA will be discussed in [1].

With the example of Figure 21.7 was illustrated that scheduling, let it named here as "vertical scheduling", has an influence to the highest utilization of logic-blocks. By introduction of "horizontal scheduling" another important aspect of dynamic reconfiguration can be optimized. In reality, dynamic reconfiguration requires time because configuration data has to be transferred and stored to the configuration memory of the FPGA. To reduce time for this procedure it is profitable to minimize the amount of data is transferred to the configuration memory between the several time frames.

Figure 21.8 shows on the left part the example schedule as described before. On the right part of the figure, some operations were positioned on different physical positions on the reconfigurable area. The advantage here is that the number of substitutions allocated areas by operations is minimized as listed in the table on the left of both scheduling schemes. The result is a reduced data-transfer to the configuration memory.

The example in this subchapter should point out that dynamic and partial reconfiguration can be exploited to reduce power-consumption. A reduced utilization has a direct relation to the required chip-size and therefore to the power-consumption of the complete system. Ably scheduling methods allows to reduce utilization of logic-resource. A detailed introduction is described in [1]. Also the introduced "horizontal scheduling", merely a physical position of operations on the chip area, allows to reduce data-transfer to the configuration memory. The reduced data-transfer also reduces power-consumption of the complete system.

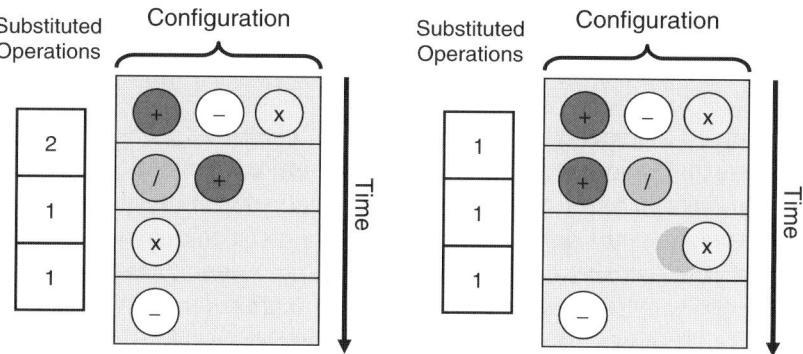

Fig. 21.8 Horizontal scheduling

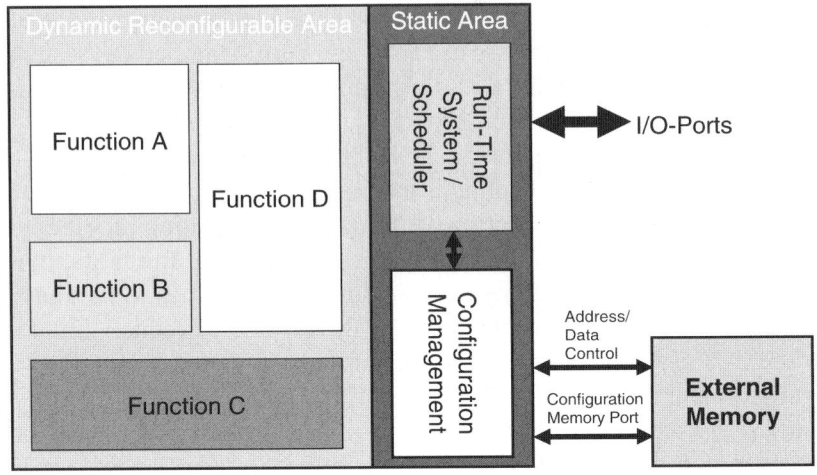

Fig. 21.9   Dynamic and partial reconfigurable system

Certainly additional external memory for loadable operations while run-time must be provided to build up a dynamic reconfigurable system.

Figure 21.9 shows a schematic view of a dynamic and partial reconfigurable system. As mentioned in the previous description, additional elements for scheduling mechanisms and the configuration management must be integration into the system. In the picture these two modules were fixed into a static area which is not affected by reconfiguration. In the left part the area for dynamic reconfiguration in established. While run-time, the scheduler enables the configuration management to load a new function from the external memory which is connected to the configuration memory of the FPGA. After finalizing this procedure, a function is loaded into the reconfigurable area and may substitute other functions.

In order to reduce the size and likewise power-consumption of external memory, the configuration code can be compressed before storing it for example to FLASH-memory. The usage of non-volatile FLASH memory has the advantage that the system is available after power-up phase however, it is also possible to connect other kind of memory, like SRAM-based memory for intermediate and more performant access to configuration data.

The introduction of run-time configuration code decompression on-chip has the most benefit, if the configuration memory port is available without external wiring. This means that a connection to the configuration management module is possible by establishing a physical connection within the FPGA.

Figure 21.10 shows the configuration management module extended by the decompressor unit. This decompressor unit is connected to an internal configuration memory port which is available in some actual FPGA architectures. It is

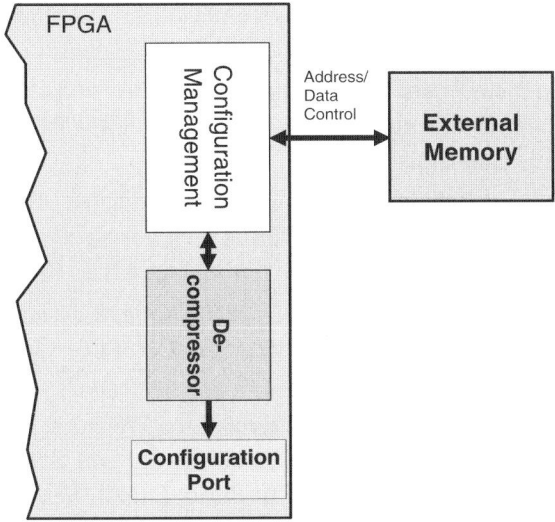

Fig. 21.10 Decompressor module of FPGA

necessary that the code decompression unit has a very low utilization of logic-blocks by simultaneously high performant data throughput to allow real-time execution of the system. Several approaches to compress and decompress data are available. Definitely the requirement of low utilization prefers to integrate an effective method for decompression. One approach with a high efficient method for FPGA configuration code compression/decompression is described in [2].

References

[1] Becker, Hübner, Ullmann: 'Real-Time Dynamically Run-Time Reconfiguration for Power-/Cost-optimized Virtex FPGA Realizations', VLSI-SoC 2003, Darmstadt, Germany

[2] Hübner, Ullmann, Weissel, Becker: 'Real-time Configuration Code Decompression for Dynamic FPGA Self-Reconfiguration', RAW Workshop at IPDPS 2004, Santa Fé, USA

[3] Thilo Pionteck, Carsten Albrecht, Roman Koch, Erik Maehle, Michael Hübner, Jürgen Becker: "Communication Architechtures for Dynamically Reconfigurable FPGA Designs", RAW 2007, Long Beach, California USA

[4] Alisson Brito, Matthias Kuehnle, Michael Huebner, Juergen Becker, Elmar Melcher: "Modelling and Simulation of Dynamic and Partially

Reconfigurable Systems using SystemC", ISVLSI 2007, Porto Alegre, Brazil

[5] Michael Huebner, Lars Braun, Juergen Becker: "On-Line Visualization of the Physical Configuration of a Xilinx Virtex-II FPGA", ISVLSI 2007, Porto Alegre, Brazil

[6] Jürgen Becker, Michael Hübner, Katarina Paulsson: "Physical 2D Morphware and Power Reduction Methods for Everyone", Dagstuhl Seminar, April 2006, Schloss Dagstuhl

[7] Jürgen Becker, Michael Hübner: "Run-time Reconfigurabilility and other Future Trends", Invited Talk, SBCCI 2006, Ouro Preto, Brazil, September

[8] Michael Hübner, Jürgen Becker: "Exploiting Dynamic and Partial Reconfiguration for FPGAs - Toolflow, Architecture and System Integration", Invited Tutorial, SBCCI 2006, Ouro Preto, Brazil, September

[9] J. Becker, M. Hübner, G. Hettich, R. Constapel, J. Eisenmann, J. Luka: "Dynamic and Partial FPGA Exploitation", to be published in the IEiEE Proceedings

[10] J. Becker, M. Hübner: "Dynamically Reconfigurable Hardware: "A Promising Way to On-Line Diagnosis, Fault-Tolerance and Reliability"", ISDS 2006, Romania

[11] J. Becker, M. Hübner, G. Hettich, R. Constapel, J. Eisenmann, J. Luka: "Dynamic and Partial FPGA Exploitation", Proceedings of the IEEE, Special Issue on Advance Automobile Technologies, February 2007, Volume 95, Number 2

Chapter 22

Applications, Design Tools and Low Power Issues in FPGA Reconfiguration

Adam Donlin
Xilinx Research

Abstract   Dynamic reconfiguration allows the circuit configured on an FPGA to be optimized to the required function and performance constraints. Traditionally, designers used dynamic reconfiguration to increase the speed or decrease the area of their system. This chapter considers a variety of use models for lowering the power requirements of dynamically reconfigurable processor systems. We begin with a review of FPGA reconfiguration and the applications that have been applied to it so far. We then describe the tool flow for reconfiguration and expand the discussion, first, to present experimental results from the literature that characterize the power-profile of reconfiguration itself and, second, to review system-level use models of dynamic reconfiguration that may improve the overall system power requirements.

Keywords: FPGA, dynamic reconfiguration, low power FPGAs, use models

1.   Introduction

Reconfiguration, the act of reprogramming the configuration memory of an FPGA, has been an active research topic since Xilinx first introduced FPGAs in 1985. The technique tantalizes design engineers with the prospect of trading off circuit area and execution time. Despite its intuitive appeal, reconfiguration has eluded widespread adoption. There are two major reasons for this. The first is that the design process for reconfigurable logic design is complex. Reconfiguration violates many of the basic assumptions of logic design, and, consequently, the majority of design tools have not supported it. The second reason is that it can be difficult to identify applications where the cost of reconfiguring

513

J. Henkel and S. Parameswaran (eds.), Designing Embedded Processors – A Low Power Perspective,
513–541.

the system is quickly and easily recovered. Estimating reconfiguration latencies requires the designer to have a detailed understanding of the FPGA's logic
and configuration architectures.

The aims of this chapter are threefold. First, it will present the reader with an
overview of FPGA reconfiguration and review some compelling examples of
dynamic reconfiguration from industry and academia. Second, it will review
the design flow required to actually create a reconfigurable application. This
will include a summary of the modern FPGA reconfiguration tool flow and the
major pieces of system IP that are required in such systems. Finally, the chapter
will present the opportunities and challenges of deploying FPGA reconfiguration to reduce system power.

2.   Applying Reconfiguration

In Chapter 21 definitions of FPGA reconfiguration were provided. This chapter
shows the configuration architecture of an FPGA in more detail. The configuration mechanism transfers configuration data to and from the FPGA's configuration memory. Configuration data defines the logical operation computed by
the FPGA's logic resources. Reconfiguration is the act of altering the functionality of the device during its duty cycle by re-writing the device's configuration
data. As discussed in Chapter 21 changing the configuration data can be done
while the device is in use or when it is offline. The amount of programming
data that is altered during each reconfiguration cycle can also vary. In some
cases, the entire configuration data is re-written (full-reconfiguration) whilst
in others, only a relevant subset of the configuration data need be changed
(partial-reconfiguration) (Figure 22.1).

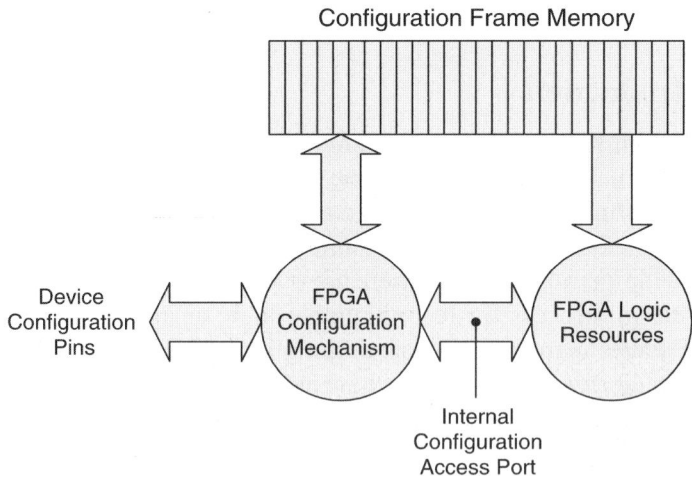

Fig. 22.1   FPGA configuration mechanism

Reconfiguration comes at a cost: it takes time to write the new data into the FPGA and some system resources must be dedicated to controlling the reconfiguration process. In early reconfigurable systems, reconfiguration controllers were external microcontrollers or secondary PLDs. Modern FPGAs provide an internal configuration access port (ICAP) allowing reconfiguration controllers to be implemented on the device that is being reconfigured. Such integration reduces the system's bill of materials and tightly integrated reconfiguration control typically has higher performance than an external controller.

To be used successfully, the cost of reconfiguration must be balanced against the projected benefits. The time required for each reconfiguration cycle to complete must be factored into the reconfiguration system design from the earliest stages of the design process. In Figure 22.2 we show how reconfiguration time is dependent on the proportion of the device being reconfigured. The organization of the configuration memory is influenced strongly by the FPGAs internal configuration control logic. In modern FPGAs, the configuration memory is segmented into configuration frames. A configuration frame will typically consist of some hundreds of bytes of configuration data. Each frame governs the implementation of a column of resources in the FPGA logic.

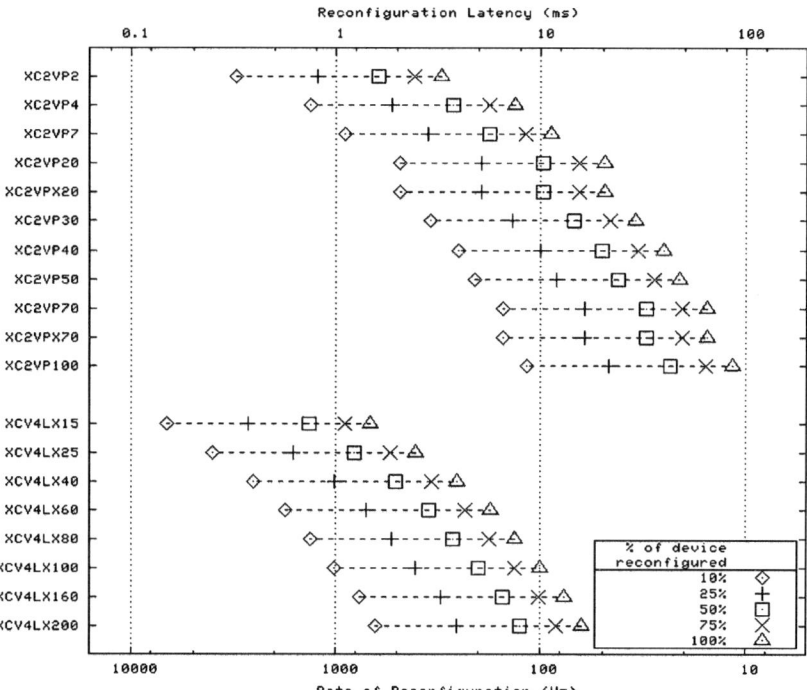

Fig. 22.2   Reconfiguration rates and latencies for Xilinx Virtex-II Pro and Virtex-4 devices

A configuration frame is the smallest unit of data that can be accessed in a single reconfiguration cycle. The exact amount of data in a configuration frame depends on the device being reconfigured. For example, in the Xilinx Virtex-II series of FPGAs, the column of logic resources governed by a configuration frame spans the full height of the device. Consequently, the size of a frame is proportional to the height of the device itself. Frames for "shorter" devices contain fewer data than frames for "taller" devices. In newer Xilinx Virtex-4 series FPGAs, configuration frames have a single fixed size, and, hence, no longer control logic resources that span the full height of the FPGA. This allows designers to have finer grain control over the resources in the FPGA that they are reconfiguring. Some early FPGA devices had configuration mechanisms that allowed very fine grain reconfiguration. For example, the Xilinx XC6200 series allowed reconfiguration of individual words, bytes or even single bits. Whilst this is clearly more flexible than a coarser, frame based configuration mechanism, it does come at a significant cost in silicon area.

Figure 22.2 indicates of the amount of configuration memory that can be reconfigured on a variety of modern FPGAs in a specific period of time. Each line on the graph represents a particular FPGA device. The points shown on each line indicate the percentage of the device configuration memory being reconfigured (10%, 25%, 50%, 75% and 100%, from left to right). The position of each marker point relative to the upper-x-axis indicates the reconfiguration latency (the time taken to complete the configuration for that proportion of the device's configuration memory). For example, reconfiguring 75% of a Xilinx XCV4LX160 incurs 10 ms of reconfiguration latency. Correspondingly, it is possible to reconfigure the device at a rate of 100 Hz. An application designer can use graphs like this early in the design process to correlate application events to the reconfiguration capabilities of a target FPGA.

2.1   Applications of FPGA Reconfiguration

In this section we review some interesting applications of FPGA reconfiguration. Our aim is to give the reader an overview of the diverse use cases of reconfiguration. Our discussion will not be exhaustive but it will highlight some of the most compelling reconfiguration applications that have been published in the literature.

Crossbar Switches.   A fully connected crossbar switch has M × N switches where any of M-inputs may be mapped exclusively to any one of the N-outputs. Over time, the mapping of inputs connected to outputs will change. Two of the most notable applications of reconfiguration are implementations of a crossbar switch on the FPGA programmable fabric. In 1996, Eggers et al. [1] reported the implementation of a 32 × 32 full crossbar for commercial deployment in an ultrasonic imaging application. The switch had a real-time switching requirement of 200 μs. Rather than using solely programmable multiplexor or logic structures in the FPGA architecture to implement the crossbar, a "skeleton" crossbar structure was created from the FPGA's routing resources. Dynamic reconfiguration was used to establish and break connections between orthogonal routing lines in the FPGA architecture. Indeed, the authors noted that direct use of logic structures actually would have lowered the performance of their design. Furthermore, the commercial viability of this design was not based solely on switching performance: alternative solutions had a significantly higher raw performance. The reconfigurable solution provided a better balance of price and performance with respect to the application's requirements.

In 2003, Young et al. [2] reported a crossbar switch design using a single Xilinx XC2V6000 device. Here, a major telecommunications company required a 928×928 port, fully connected crossbar. Each port supports 155 Mbps traffic with deterministic latency and a switching time on the order of a few hundred microseconds. In their design, the authors use dynamic reconfiguration to control fast, wide multiplexors within the routing structure of the FPGA. These multiplexors are only available to the designer through reconfiguration and were used to construct multiple 928-to-1 multiplexor structures. Compared to non-reconfigurable FPGA designs the reconfigurable crossbar achieved a 16 times gain in port density. Non-FPGA crossbar implementations, such as those available as discrete ASICs, target switching of far fewer, high-bandwidth data-streams. For this application, dynamic reconfiguration enabled the use of the FPGA fabric to provide a solution meeting the requirements for per-port bandwidth and overall port density.

In 2004, Lysaght and Levi [8] reported a reconfigurable multi-stage Clos-style switch. This design offered equivalent switching capacity to a fully connected switch but had better throughput and significantly reduced area. However, the reconfiguration control algorithm dynamically calculates configurations to be applied to the reconfigurable switch fabric. This is more complex than the earlier switch architecture which pre-computed and stored reconfiguration data in external memory. Table 22.1 summarizes the benefits of this reconfigurable switch design over its predecessor. Of particular note is the reduction in dynamic power consumed by the switch.

Software Defined Radio (SDR).   Software defined radios are complex and highly flexible digital communication systems. SDRs support multiple communication channels encoded with a variety of different communication protocols (also referred to as "waveforms"). The number of active channels and the protocols applied to those channels can change dynamically. For example, variations in atmospheric conditions or even the position of the remote

Table 22.1   Performance comparison of two reconfigurable crossbar designs

	1-stage Crossbar	3-stage Clos
Crosspoints	861,184	172,608
Size (number of configurable logic blocks required)	4,836	1,324
Max. Frequency	155 MHz	200 MHz
Max. Bandwidth	144 Gbps	186 Gbps
Path Latency	22 cycles	6 cycles
Path setup time (worst case)	0.66 ms	4 ms
Dynamic power (worst case, 155 Mhz)	11 Watts	6 Watts

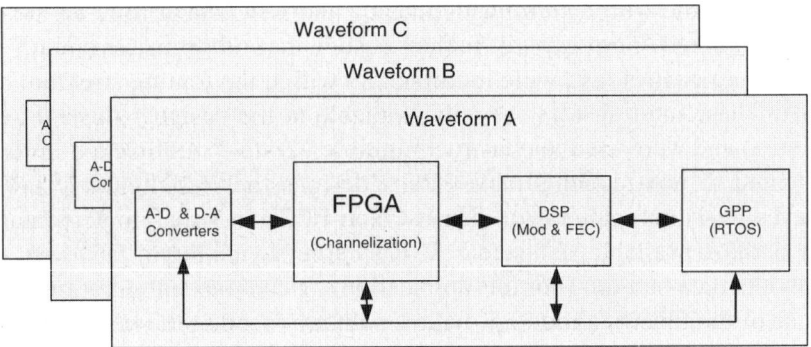

(i) SDR Dedicated Resource Model

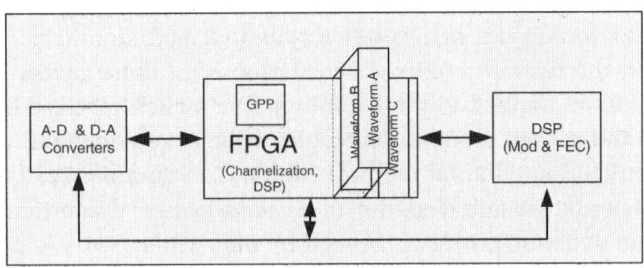

(ii) SDR Shared Resource Model

Fig. 22.3   Resource models for software defined ratio

communication terminals can provoke a change in the waveforms used by the SDR to maintain high-quality communication between the nodes. FPGAs are natural candidates for SDR systems because of their high-DSP performance (Figure 22.3).

The Joint Tactical Radio System is comissioned by the United States Military to serve as the standard, interoperable military software defined radio. Dynamic reconfiguration is specified as a key technology for the JTRS. Using reconfiguration, the JTRS can instantiate only the waveform circuits that are required by the currently active communication channels. This is commonly referred to as the JTRS "shared resource" model.

Without the ability to reconfigure, the JTRS designers would have to employ a "dedicated resource" model where the radio implements all possible combinations of protocols for every communication channel. If a radio supports M channels and N waveforms, the dedicated resource model requires $M \times N$ resources to implement all possible channel/waveform combinations. For anything other than trivial numbers of channels and waveforms, dedicated resources would be infeasable. The shared resource model allows the JTRS designers to use the minimum number of required logic resources. In 2004, ISR technologies and Xilinx demonstrated a working, reconfigurable software defined radio using the shared resources model [12]. Their system partially reconfigured a Xilinx Virtex-II Pro FPGA to implement only the waveform circuits that were required at any given point in time. *Dynamic* reconfiguration of the FPGA is a crucial requirement for SDR systems like the JTRS. Specifically, dynamic reconfiguration allows the JTRS to continue processing data for the other radio channels whilst reconfiguring a specific channel. Channels are handled independently, even during reconfiguration.

Automotive Reconfiguration.   The automotive sector has seen an astonishing increase in the use of integrated electronic control systems in the last two decades. These systems, referred to as Electronic Control Units (ECUs), typically manage discrete functions within the engine, chassis or passenger cabin. For example, fuel system and power train management in a modern vehicle will be controlled by two discrete ECUs. A complex wiring harness allows ECUs to exchange data and optimize the overall operation of the vehicle. For example, traction control systems in all-wheel drive vehicles exchange data over the wiring harness, allowing each wheel ECU to compensate a loss of traction. The exchange of data across a wiring harness is governed by a series of protocols. For example, the controller area network (CAN) and local interconnect network (LIN) protocols are commonly used in automotive electronics. The safety–critical design requirements are common in vehicle systems. Consequently, ECU communication protocols must offer guaranteed performance and the ability to detect and recover from failures safely.

Most vehicle electronic systems follow the same dedicated resource model outlined in the previous section. However, modern vehicles are quickly reaching the physical limits on the number of ECUs (and associated wiring harness) that can be included. A shared resource model using FPGA reconfiguration has been proposed to create multi-function ECUs [13, 14]. Here, the functions of each independent ECU are time multiplexed into the FPGA resources of a reconfigurable ECU. In [13], Becker et al. describe a multi-function door controller ECU. Since many door control functions are mutually exclusive, dynamic reconfiguration enables only the required subset functions to be implemented on the FPGA at any given point in time.

Genome Database Processing.   The three applications discussed in this section address the "supercomputer-class" problem of scanning genome data for matching (or near-matching) protein sequences. This problem maps well to the fine grain, data parallel compute model of FPGAs and, consequently, rapidly becomes I/O bound. The two main alternative compute architectures for implementing gene sequence matching are workstation clusters or arrays of ASICs. Whilst workstation clusters are dynamically reprogrammable, their compute and power density is less attractive than an FPGA solution. ASIC solutions have an attractive compute and power density, but cannot be dynamically reprogrammed.

The first paper to report a reconfigurable gene sequence matching design targeted the Xilinx XC3000 series FPGAs and was written by Lemoine and Merceron [3] in 1995. The style of reconfiguration described in this paper is unorthodox by modern definitions because a complete place and route cycle preceeds every reconfiguration. Still, this application can still be termed "run-time reconfigurable" given the relatively long run times involved in gene sequence matching. Careful logic circuit design also reduced the place and route tool run times to the order of 100 s. An optimized workstation solution took hours to a full day to complete a scan of the gene database. The reconfigurable FPGA solution, yielded a two-to-three order of magnitude gain in search speed.

In 2002, Guccione and Keller [5] and Yamaguchi et al. [4] both published gene sequence matching algorithms based on the Smith–Waterman dynamic programming algorithm. Both applications used modern FPGAs (a Xilinx XC2V6000 and XCV2000E respectively). In the system created by Yamaguchi et al. the search algorithm is segmented into two phases and the system periodically swaps between them. In the first phase, a circuit scans the database for near-matches of the search sequence. In the second phase, the circuit calculates and reports more detailed information about the optimal alignment of the search sequence within the database. The authors report that their system took 34 s to match a 2 Kb search sequence over a gene database with 64-million entries. In comparison to a high-end workstation system, this translated to a 330 times speed improvement. The authors of this system acknowledge an important assumption in their design was a sparse distribution of near-matching sequences within the gene database. This important simplification allows the application to amortize the reconfiguration overhead between the circuit phases. In the event that a search sequence occurs often within the database, the authors invoke sequential processing on a host computer.

The application developed by Guccione and Keller used the Xilinx JBits API to craft highly optimized implementations of the Smith–Waterman algorithm. The authors take advantage of transformations of the algorithm specification that allow a clean mapping to high-performance features of the FPGA architecture such as its carry chain. These enhancements are static in nature: dynamic reconfiguration is introduced by folding both algorithm constants and characters of the search string into lookup tables (LUTs) used as compute elements of the implementation. Changes in algorithm's parameters or string segments trigger a reconfiguration of the corresponding LUT values. The final design is reported to have performance in the range of 3.2 trillion comparisons per second with 11,000 processors on a single Xilinx XC2V6000. The authors report their single chip solution easily out-performed previous ASIC (0.2 trillion comparisons per second), and workstation cluster (0.2 trillion comparisons per second) solutions by more than an order of magnitude.

Other Applications.   Single event upsets (SEUs) are a phenomenon of the deep sub-micron semiconductor design. When transistor geometries are small enough, the energy deposited from neutron radiation passing through a device may cause latch-up in a memory cell. FPGAs are process technology drivers, and, as such, have become susceptible to SEUs before ASICs fabricated on older technologies. In [15], dynamic reconfiguration is used to overcome any corruption of FPGA configuration memory from an SEU. This process, known as bitstream "scrubbing", uses a small reconfiguration controller to read back the configuration data of the target FPGA. Errors are detected and corrected by writing the correct data back into the FPGA.

In 1993, Fawcett [6] reports early commercial uses of FPGA reconfiguration. The applications fall into two broad classes: reconfiguration in direct response to a physical event or reconfiguration for test and verification. The following three applications were designed to react to physical events. For example, an innovative computer monitor manufacturer designed its CRT to be rotated between portrait and landscape viewing modes. Each time a user physically changed the viewing orientation, Xilinx FPGAs within the monitor were reconfigured with different video addressing logic. Another manufacturer of 9-track $1/2$ inch tape drives produced a PC interface card with multiple Xilinx FPGAs. The devices on the interface card were reconfigured to accommodate processing of different data densities recorded onto the magnetic tapes. Finally, a printer company used reconfiguration in its printer controller cards. Each card had physical interfaces for two printers and used reconfiguration of a single Xilinx XC2000 to instantiate the control logic for the printer being used to print at the given moment in time. During idle periods, the same FPGA would be repeatedly reconfigured with control circuitry to query the printer status and execute test circuitry to execute diagnostics.

The second group of applications reported by Fawcett used reconfiguration for test and verification. One company developed voice compression modules and reconfigured two Xilinx XC2000 devices with built-in self-test (BIST) circuitry to apply diagnostics to other components in the system. A second company produced protocol analyzer hardware and used reconfiguration to switch between several pseudo-random bit error rate (BERT) pattern generators.

Conclusion.   Used carefully, reconfiguration can yield faster, smaller systems. However, these benefits do not come without careful consideration in the system's design and implementation phases. The costs of reconfiguration must be weighed carefully against the benefits. Looking to the future, one observation that we can make is that reconfiguration rates do not increase as quickly as system clock speeds. Consequently, larger numbers of compute cycles are sacrificed if the system's main application is suspended during reconfiguration. Increasingly applications will require dynamic reconfiguration to insulate the main application from the latencies of other styles of reconfiguration. Fortunately, most modern FPGAs support dynamic reconfiguration and many even allow self-controlled reconfiguration. However, because the FPGA is in operation during dynamic reconfiguration, a designer must take great care when designing, floorplanning and operating their system. In the next section we will review the design flow for reconfigurable systems and give particular focus to the design flow issues related to dynamic reconfiguration.

3.   Design Flow for Reconfigurable Systems

The standard FPGA design flow, as shown in Figure 22.4, transforms an FPGA design description into a configuration bitstream. Readers unfamiliar with the details of the standard FPGA design flow can find a comprehensive tutorial in [REF]. Typically, the FPGA design description is written in an HDL. Complex designs often use system integration tools like the Xilinx Embedded Development Kit or Xilinx System Generator to construct the initial design description from libraries of IP.

Each pass through the standard flow is independent of subsequent passes. The bitstream produced at the end of the flow is dependent only on the initial design specification. By contrast, the design flow for partial and dynamically reconfigurable designs require more than one pass through the standard design flow – each reconfigurable module passes through a "reconfiguration-aware" version of the standard design flow. Furthermore, the results of each pass are dependent on the output of at least one of the previous passes. This is a necessary precaution to avoid introducing resource conflicts between successive

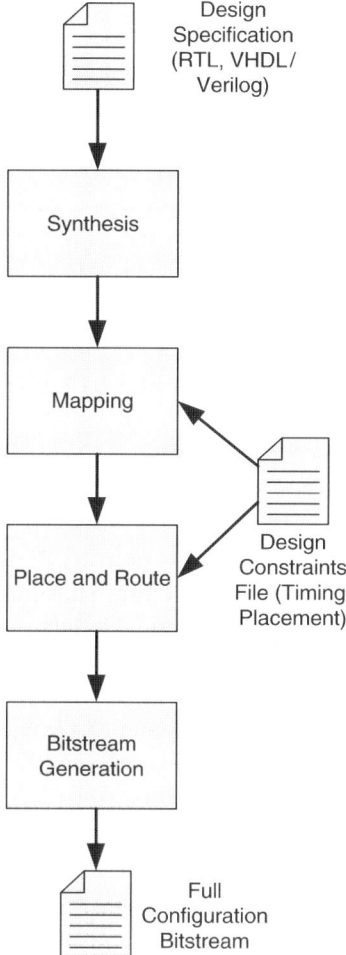

Fig. 22.4 Standard FPGA design flow (no reconfiguration)

configuration bitstreams when they are loaded onto the FPGA at runtime. A detailed description of the design flow for a dynamically reconfigurable system on Xilinx FPGAs was published by Lysaght et al. in [17]. Their discussion is summarized in the section below.

3.1 Dynamic Reconfiguration Design Flow

Figure 22.5 shows the logical structure of a simple reconfigurable system. An input circuit provides data to one of two dynamically reconfigured circuits. The configured circuit computes a result which is passed to the data sink module to be stored. Although simple, this example is representative of the

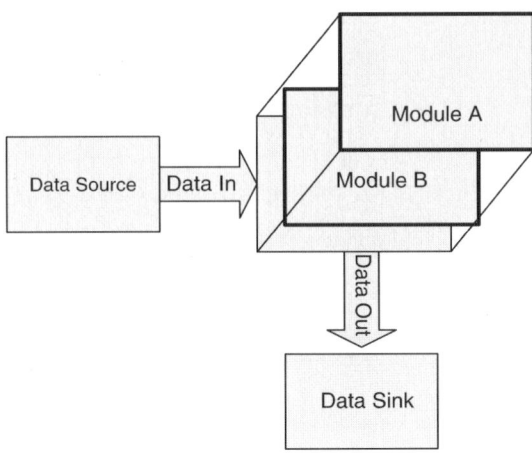

Fig. 22.5   Logical view of a simple reconfigurable design

shared resource model discussed in Section 2.1.2. The data source and data sink circuits will remain fixed for the entire runtime of the system. A single reconfigurable module region may contain either module A or module B as required by the system's reconfiguration controller. The reconfigurable flow produces three bitstream files for this example. The first is a full FPGA configuration containing just the fixed components of the design. This would include both of the fixed circuits and all of the routing paths leading up to and out of the reconfigurable region. The exact dimensions of the reconfigurable region must be specified in the design constraints file passed to the place and route tools. This allows the design tools to prevent logic or routes from fixed circuits entering the reconfigurable region.

Some specific terminology is used in [17] to refer to the different design components of our example.

- The fixed modules in the design hierarchy are referred to as the "static" design. The static design is further separated into a toplevel module (called "TOP") that contains all the global clock and IO pin resources.

- Inside the static design, the area reserved for a reconfigurable module is called a partially reconfigurable region (PRR).

- Each module that can be configured into a PRR is called a partially reconfigurable module (PRM).

Five main phases of the reconfigurable design flow are described in [17]. These are shown in Figure 22.6 and summarized as follows.

1. **Design partitioning**. A hierarchy of HDL design files is created. Logic for the static, top level design is gathered into one part of the design

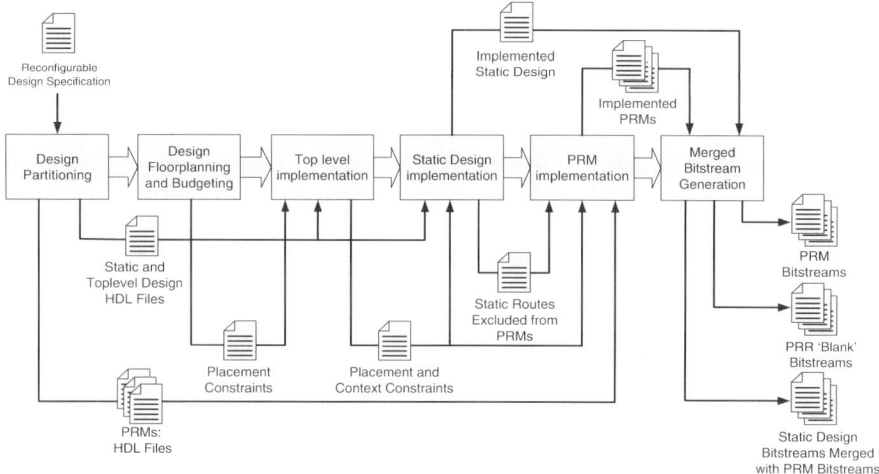

Fig. 22.6   Reconfigurable design flow

hierarchy. The source files for each PRM are also gathered into separate directories.

2. **Planning**. The appropriate area must be reserved for each PRR. The exact area reserved must accommodate the largest PRM to be configured into the PRR. Tools like Xilinx PlanAhead automate the detailed floor planning of the design and PRR. When this budgeting is complete, the area constraints are written into a file for use in later phases of the design flow.

3. **Top level context**. Information regarding clock, I/O and reconfiguration "bus macro" utilization must be available during implementation of the static and PRM design components. These resources are refered to as "top level context" and implemented separate from static design. The information gathered during this phase is passed to the later static design and PRM implementation phases to coordinate their use of the top level physical resources.

4. **Static design implementation**. Here, the static design is synthesized, reduced to a gate-level netlist and then implemented. The area constraints supplied from the planning phase prevent logic in the static design from being placed into a PRR. Static design routes may pass completely through a PRR. This helps reduce routing network congestion around each PRR. However, it is necessary for the design tools to now log exactly which routing resources have been used in the static design and prevent them from being used in the PRMs. This information

is logged into a file which is consulted during the implementation phase of the PRMs.

5. **PRM implementation.** During PRM implementation, the HDL descriptions of each PRM are placed in separate direcories. Copies of information extracted when building the top-level context and the routing resources used by the static design implementation are placed in each of the directories. The standard implementation tools are run on the HDL description of each PRM and use this information to ensure that no routes from the PRM traverse the boundary of the PRR. Routes from the static design passing through the PRR are similarly excluded from the place and route phase of each PRM impelemented in this stage of the design flow.

6. **Bitstream generation.** With the static design and each PRM implemented, this final stage of the design flow creates the bitstream files for use on the FPGA. Two scripts, described in more detail in [17], automate this process. The first script operates on the PRMs of designs with a single PRR. It generates a full bitstream containing the PRM already instantiated in the PRR and a partial bitstream containing only the PRM. The second script works on designs that contain more than one PRR. It creates a full bitstream where each PRR is populated with a prescribed PRM. Exactly which PRM is placed in which PRR is governed by a sequence of PRMs specified on the command line of the script. Alongside the full bitstreams, this script also generates a "blank" bitstream for each PRR. When written into the FPGA, a blank bitstream effectively removes any currently configured PRM from the given PRR.

HDL Design Structure.   As described above, the HDL specification of a dynamically reconfigurable design is different from a non-reconfigurable design. This is necessary because the static design and PRMs must be synthesized, placed and routed separately from each other. In this section we review the constraints on the structure of the reconfigurable design. Firstly, PRRs must be declared in the same top level of design hierarchy as the fixed modules. This limits the depth of design hierarchy where reconfiguration may be used to within the top level module only. Each of the PRMs and the fixed modules may have a complex hierarchy underneath them, but that hierarchy cannot contain any PRRs.

Figure 22.7 shows the design hierarchy for our simple design example. In it we can see the fixed modules of the top level design and the declaration of a PRR. The reader should note, however, that a series of "bus macros" have been declared at the interface points of the static design and the PRR. These macros play a crucial role during reconfiguration and are discussed in detail in the following section.

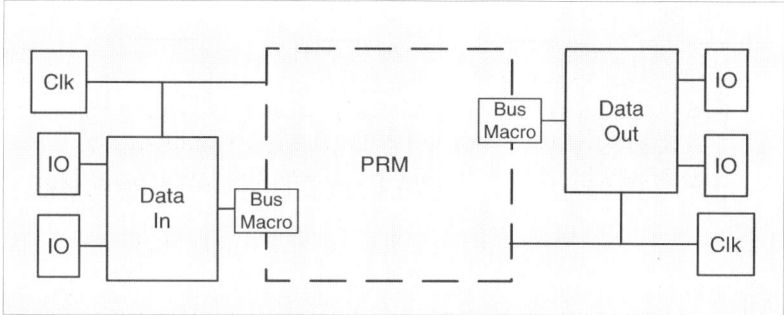

Fig. 22.7 Top level design hierarchy for simple reconfiguration example

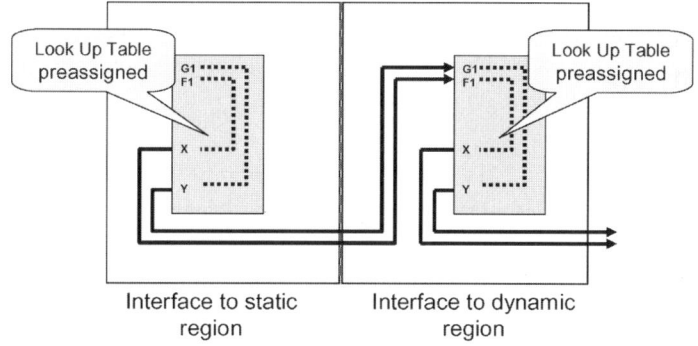

Fig. 22.8 Logical structure of a bus macro

Bus Macros.   In dynamic reconfiguration, we must be able to "zip" the routes that connect the static design with the interface of each PRM. To do this, the design tools constrain the exact, physical placement of the routing paths from the static design into the PRMs. Bus macros define that fixed interface between the static design and PRMs that are configured into the PRR. Every PRR must have a consistent and clear routing interface to the static design and every instance of a PRM that will be placed into a PRR must implement its interface. It is important to note that the interface is more than just the logical specification of the width and direction of signals. The *physical* routes used to implement those signals must be entirely consistent in every RPM and the static design. As we can see in Figure 22.8, one half of a bus macro resides in the static portion of the design. The other half resides in the dynamic portion. The design tools constrain these signals to be the same physical routing wires in the static design and all the PRMs when the modules are being implemented in the design flow. Consider, for example, a zipper: the alignment and size of the teeth on each side must be the same for the zipper to join the fabric on

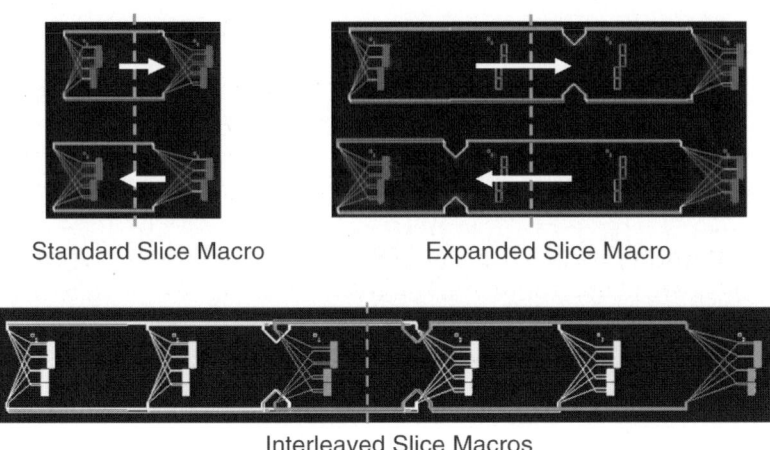

Standard Slice Macro   Expanded Slice Macro

Interleaved Slice Macros

Fig. 22.9   Physical view of LUT based slice macros

each side together. If there the two sides of the zipper do not match it will not operate correctly.

Early versions of the Xilinx reconfigurable design flow used the FPGA's tristate buffers and tristate routes to connect the static design to the configured RPM. However, tristate buffers became less common with successive FPGA device generations. In modern devices, there are no tristate logic resources whatsoever. A new form of bus macro using the lookup-table (LUT) resources of the FPGA was created [17, 18] to take its place. Figure 22.8 shows the logical structure of a LUT based bus macro and Figure 22.9 shows the physical layout of the macros on a FPGA. The most important characteristic of this new style of macro is that they completely constrain the routes between the static and dynamic parts of the design. The user's design entry tools must explicitly instantiate these LUT macros at the interface to each PRR, and on the interface of each PRM. This happens, typically, during the top level and planning phases of the dynamic reconfiguration flow.

LUT based bus macros are superior to the earlier tristate macros. For one, LUTs are relatively abundant in the FPGA architecture. This means the number of routes that can be exchanged between the static design and the PRMs is much larger when LUT based macros are used. LUT based bus macros also have fewer restrictions on where they can be placed in the FPGA. This allows macros to be packed more densely. They can also connect the static and dynamic regions horizontally or vertically. This turns out to be quite useful in FPGAs like the Xilinx Virtex-4 where there is a natural, vertical segmentation of the configuration memory into fixed height regions.

3.2   Deploying Dynamic Reconfiguration

Having successfully created the configuration bitstreams for the reconfigurable design, the loading and manipulation of the FPGAs configuration memory must be controlled. In this section we review two technologies for interfacing and controlling the configuration memory of an FPGA.

The Self Reconfiguring Platform (SRP).   The Xilinx Self Reconfiguring Platform (SRP) [11] is a piece of standard, low-level infrastructure that allows an FPGA to read and write its own configuration memory. The SRP comprises two parts: a hardware peripheral to drive the FPGA's internal configuration access port (ICAP) and a software API to fetch and manipulate configuration data through the ICAP.

Figure 22.10 gives an overview of the SRP hardware architecture. The SRP hardware interfaces directly to a Xilinx embedded processor using the IBM CoreConnect bus protocol. The dedicated dual-port RAM included in the SRP hardware allows the control circuit to buffer configuration frames as they are copied to and from the FPGA configuration memory. The authors in [11] use this to implement a "Read-Modify-Write" protocol when modifying the FPGA's configuration memory. Rather than sending an entire frame to the embedded processor one byte at a time, the SRP hardware manipulates the configuration frame data in its buffer directly. This approach is very efficient when only small bit manipulations to the configuration data are required.

The software architecture of the SRP is presented in Figure 22.11. The two most significant layers are its ICAP interface and the Xilinx Partial Reconfiguration Toolkit (XPART) layer. The ICAP interface implements the basic configuration frame read and write operations. By supplying an appropriate frame

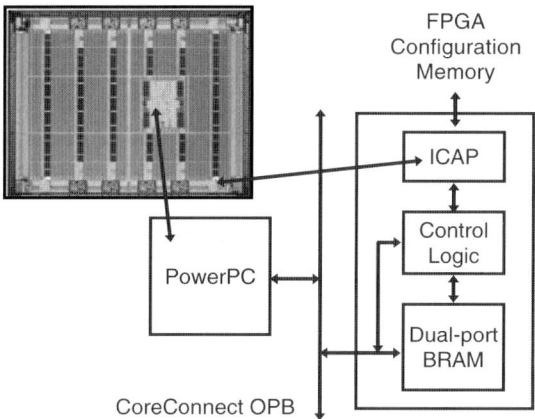

Fig. 22.10   SRP configuration control hardware

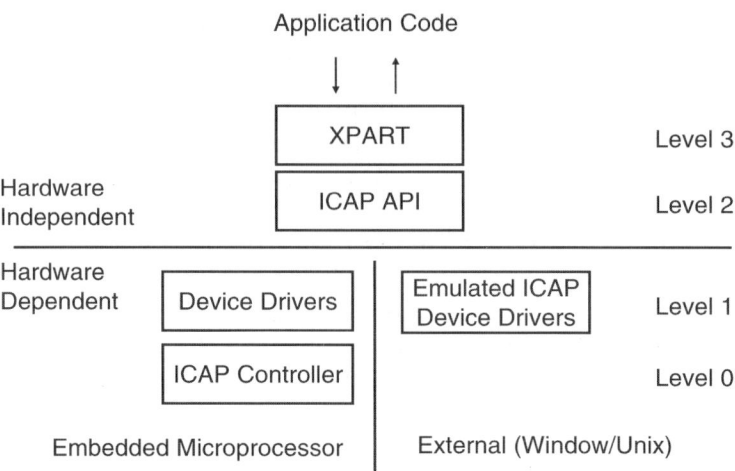

Fig. 22.11 The Xilinx SRP software APIs

address, the ICAP interface's read method reads the corresponding frame configuration data into the SRP hardware's buffer memory. Similarly, the ICAP interface's write method writes data directly from the SRP hardware buffer into the supplied configuration frame address.

XPART raises the programmer's level of abstraction above that of configuration frames. Instead, XPART methods automate the manipulation of the configuration bits that control distinct logic resources. For example, the XPART API provides a method to alter the configuration of a particular lookup table. Now, the user simply supplies the co-ordinates of the required resource in the FPGA logic and the data they wish to write into it. The XPART API translates the resource's logical co-ordinates into the device specific frame addresses. It then fetches, modifies and writes back the appropriate sequence of configuration frames.

The Xilinx Virtual File System (XVFS). The SRP software is an effective, low level programming interface to the FPGA configuration. However, requiring every interaction with the FPGA's configuration memory to be written with a low level programming API can become cumbersome. The Xilinx Virtual File System (XVFS) [19] is a standard, higher-level interface to FPGA configuration memory (Figure 22.12).

Instead of a programming API, the XVFS represents FPGA configuration data with the common file system interface metaphor of hierarchies of directories and files. The XVFS directories in a "physical" view contain files that represent the subset of the FPGA configuration data that controls each logic resource in the FPGA. For example, the LUTs of the FPGA architecture are

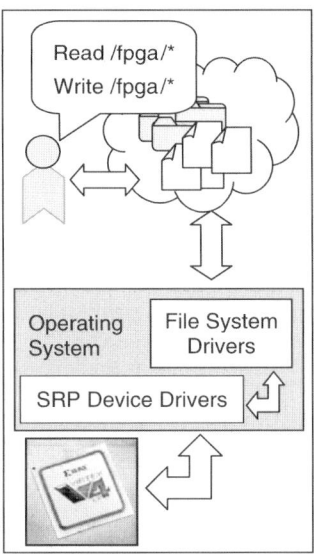

Fig. 22.12   The Xilinx Virtual File System

presented to the user of the XVFS as files. These files are organized under two levels of directory hierarchy that represent, respectively, the X and Y co-ordinates of each LUT in the device. Similar organizations exist for other device resources such as Flip-Flops and BRAMs. The benefits of the XVFS as a configuration interface come from from its intuitiveness. Files and directories in the XVFS share most of the semantics of files and directories on a hard drive. Representing the configuration of FPGA resources as files also frees the user to process those files with their choice of programming or scripting language.

Other representations of the configuration data, beyond the basic "physical" view, are also possible. For example, an "IP core" view of the FPGA is also defined. This view contains directories that represent selected regions of configuration data for each of the IP cores used in the system architecture. This XVFS view is particularly useful for diagnosing faults as many of its files are dedicated to presenting the state of the IP core.

A "floorplan" view takes the XVFS design abstraction one step further by allowing directories to represent large geometric regions of the FPGA configuration memory. Each directory maps to a specific hierarchical region in the system's logical floor plan. The files found in each floor plan directory allow the configuration memory of the region to be manipulated. For example, writing new configuration data into a directory of the floor plan view allows new functions to be configured into the system. Furthermore, the existing regions of the floor plan may be archived by simply copying the requisite files and directories to a hard-drive. When the archive is to be restored, all the user

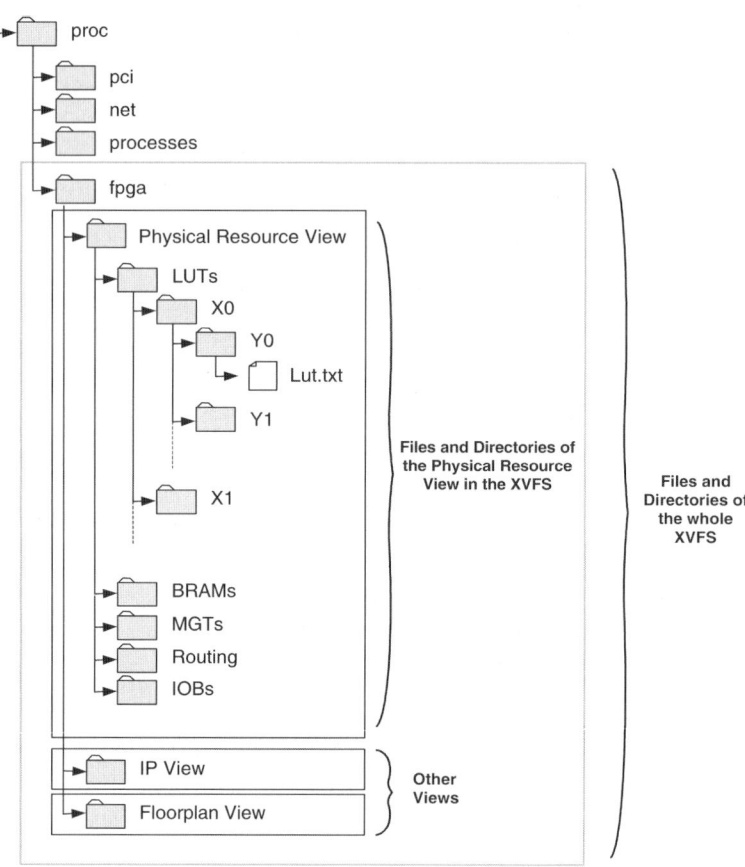

Fig. 22.13 : Files and directories in the Xilinx Virtual File System

need to is copy the same files back into the correct floor plan view directory (Figure 22.13).

4.   Low Power and FPGA Reconfiguration

4.1   The FPGA Low Power Landscape

Functions implemented in FPGA gates consume more power than the same functions implemented in comparable technology ASIC gates. However, FPGAs are commonly used in applications where power sources are in relative abundance. And, when compared to programmable processing architectures like microprocessors or DSPs, FPGAs *do* have much lower power requirements.

Table 22.2   Power requirements of portable consumer devices versus mainstream FPGAs

Power characteristic	Market requirement	Consumer FPGA (XC3S1000) [20]	Mainstream FPGA (XCV4LX60) [21]
Operating power (dynamic + leakage)	~100 mW	~250 mW @ 150 MHz	~4 W @ 150 Mhz[1]
Standby power	10–100 uW	~100 mW @ 25 C	250 mW @ 25 C

[1]The reader should note that the Virtex-4 design in [21] has a higher switching rate than the Spartan design in [20].

Recently, interest in low-power FPGAs has been inspired by the rapid increase in portable consumer electronics. These systems demand high-compute performance within stringent power budgets. This poses some interesting challenges to designers. ASICs clearly have the power and volume pricing characteristics to address the market but their design cycles are hostile. Also, consumer designs do not necessarily mean high-volume designs: rapid product cycles are common in consumer, portable products. Microprocessors and DSPs have very low unit prices and high programmability but their performance and power characteristics are far from optimal for portable systems. Consequently, there has been a surge of interest in researching low-power FPGA architectures.

In [20] it is stated that portable electronic products have an operating power budget in the range of 100 mW. To preserve their energy source, these products also require a standby mode with a power budget in the range of 10 s or 100 s of μW. Table X uses data presented in [20] to compare the market requirements for low power with mainstream FPGA power characteristics. As we can see, a standard FPGA does not meet the proposed power requirements (see Table 22.2).

It is interesting to consider the breakdown of the power consumption of different resources in the FPGA. Table 22.3 summarizes the percentage of static and dynamic power consumed by a Xilinx Spartan device.

The aim of this section is to explore the role of reconfiguration in reducing FPGA power consumption. We can see from the data in Table 22.3 that there are actually two ways to do this. Since the FPGA configuration memory accounts for nearly half of the static power consumed by the FPGA, the underlying device architecture can be modified to lower its static power. Alternatively, the user may employ reconfiguration as design technology to optimize their running application for low power operation.

Table 22.3   Power requirments of FPGA logic resources

Resource	Static power	Dynamic power
Clock	–	19%
Logic	20%	19%
Routing	36%	62%
Configuration Memory	44%	–

Modifying the FPGA architecture has the advantage of being largely transparent to the user. In [20], a low power FPGA named *"Pika"* was reported. The authors used three techniques to reduce the power requirements of their architecture: voltage scaling, power gating and low leakage configuration memory cells. Overall, the authors reduced the configuration memory leakage by two orders of magnitude. However, this comes at a cost of higher fabrication complexity to use the lower leakage mid-oxide transistors. The technique also increased the area of the configuration memory by 8% compared to the non-optimized memory. The performance of configuration memory was also likely to be lower with this technique although exact reduction was not reported.

Reconfiguration Power Requirements.   Optimizing the FPGA device architecture is not mutually exclusive with reconfiguration in the users design flow. However, reconfiguration for low power introduces the question of how much additional power is required to modify the configuration memory? As noted in Table 22.3, the static power consumption of the memory is high. Intuitively, a design concern is that the configuration memory cells would consume equally large amounts of power during a reconfiguration cycle.

In reality the power required during a reconfiguration cycle is not significant. Firstly, during a reconfiguration cycle, only the configuration controller and a small portion of the configuration memory is active. Studies in single event upset mitigation [22] show that a relatively small proportion of the FPGA configuration bitstream is used in any given user design. Writes to configuration memory are even further reduced when using partial reconfiguration to load new circuits into the device. These issues, combined with the fact that reconfiguration usually happens infrequently, mean that the power required during reconfiguration is small relative to the operating power of the user design. Lab measurements of the power consumed during reconfiguration were reported by Becker et al. in [10]. They show that a Xilinx Virtex-II device draws 60 mW additional power during a reconfiguration cycle of 350 ms for a full configuration and 100 ms for a partial configuration. The design's baseline operating power was more than 900 mW.

4.2   Low Power Use Cases for Dynamic Reconfiguration

In this section we review some interesting use cases for lowering system power with FPGA reconfiguration. Whilst these techniques are still under research, they indicate the possiblities of reconfiguring for low power.

Use Case 1: Dynamic Power Optimized Configurations.   Dynamic power consumption is directly related to a circuit's switching activity. Reconfiguration can be used to select a system implementation that has lower switching activity for a given set of input stimulus. There are two variants of this approach: the first selects between multiple, power optimized hierarchical blocks of the system and the second dynamically *removes* inactive blocks from the system.

Configuring optimized functions.   Consider the idealized example in Figure 22.14. Here, a design block computes a function, F(x), of which there are two alternative implementations (F′(x) and F″(x)). Each implementation is profiled to determine its power efficiency over a range input stimulus. The resulting graph, shown on right side of figure X, demonstrates that neither block is power efficient for the entire input range. The most power efficient implementation of the system is one where F′(x) processes data in the half of the input range and F″(x) is used for the second half.

Reconfiguration is used to ensure that the most power efficient circuit is loaded into the FPGA at any given point in time. Clearly this requires some degree of data coherence in the input data to the circuit. In situations where the input data varies widely and indeterminately, this technique will not reduce system power. Fortunately, modern algorithms already exploit data coherence. An MPEG encoder, for example, is typically benchmarked against

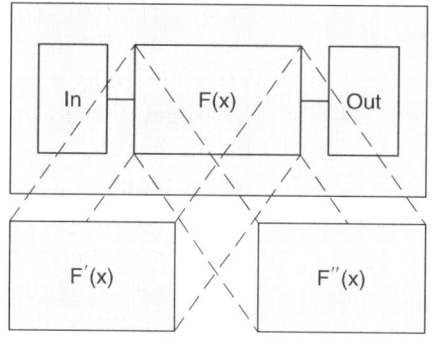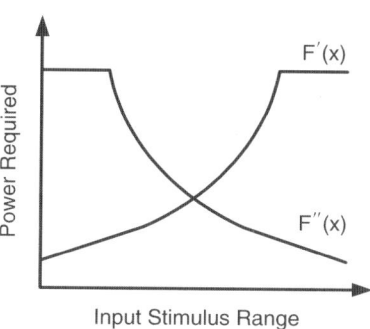

Fig. 22.14   Power optimization of design blocks

multiple video sequences with different amounts and rates of motion in the image sequence.

Profiling the various implementations of F(x) requires detailed RTL timing simulations. This yields trace files that power analysis tools such as Xilinx XPower use to calculate the power requirements of circuit for the given input stimulus. The designer then balances the overheads of reconfiguring the system to have the lowest power implementation of F(x) according to the degree of data coherence in the input stream.

De-configuring inactive functions. Section 2.1.0 introduced multi-modal systems as excellent candidates for reconfiguration. These systems comprise multiple, mutually exclusive subsystems where each subsystem implements a distinct operating mode. Reconfiguration reduces the operating power of these systems by intentionally *removing* the subsystems that are not required at a given point in time. The power reduction comes from the ability to map such designs to smaller FPGAs. Also, because the "inactive" circuits are not necessarily "quiescent", removing the inactive circuits also eliminates any dynamic power that is drawn by them.

Consider the simple example system in Figure 22.15. Here, three modes are implemented by distinct functions, F(x), G(x) and H(x). In the leftmost system, static power is higher because a large FPGA is required to implement all three independent modes simultaneously. The system's dynamic power is also higher because the synchronous "inactive" modes still draw some power on each clock cycle. The system shown on the right uses reconfiguration to implement only the correct circuit for the current operating mode. Here, static power is lower because a smaller FPGA is used to implement the system. Dynamic power is also lower because only the operating mode circuit is implemented at any given time. The power required to implement the reconfiguration is not shown in this figure. To recall Section 4.1.0, the reconfiguration power is small with respect to the operating power of a circuit. Because the operating modes of a circuit are assumed to change infrequently, the actual power consumed for each configuration cycle is quickly amortized.

Clock Gating with Reconfiguration. Clock gating is a standard low power design technique. It reduces system power by enabling and disabling the clock for regions of the design that are not required at that time. This forces the region into a quiescent state, eliminating its dynamic power consumption. In constrast to de-configuring the region, the region logic is not actually removed from the system. Instead, its ability to react to input stimulus has been removed. As a result, only the modules with active clocks contribute to dynamic power. Clock gating is useful when the reconfiguration time for an entire module is prohibitively long.

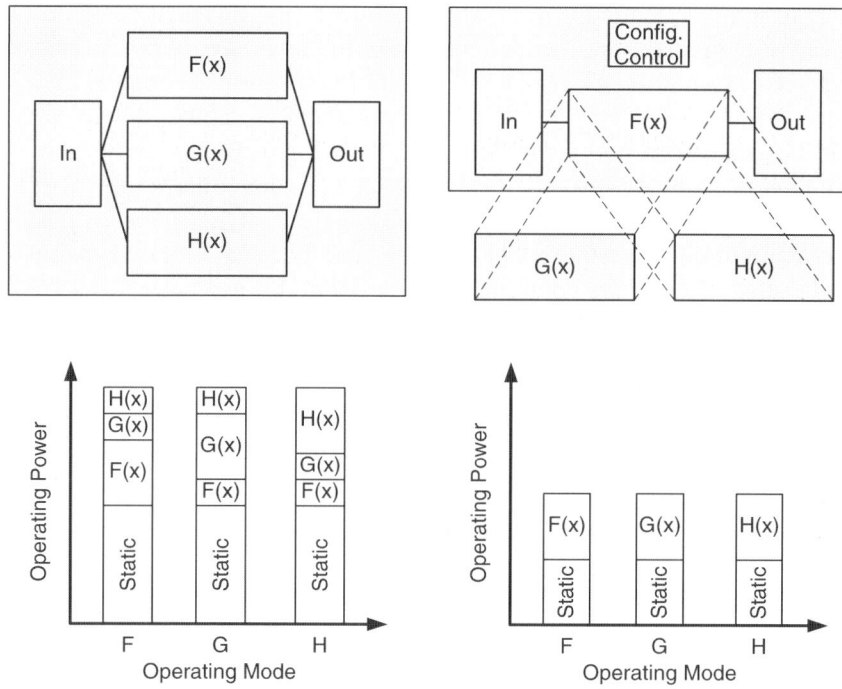

Fig. 22.15 Deconfiguring inactive circuits

Clock gating in ASIC designs requires programmable pass gates be inserted in the clock tree. The number of pass gates required depends on the level of granularity that is required. Fewer pass gates means coarser regions of the design are affected by each gate. Since each pass transistor controlling the clock influences the performance of the system, the technique cannot be used indiscriminately. Determining the size of the regions that can be clock gated is challenging because it depends on the user's final design. The system architect may decide this in advance or tools may attempt to automatically profile and extract the set of clock gated modules during the physical design steps.

Modern FPGAs provide multiple independent clock trees. Which clocks are routed to which regions of the device is decided during physical implementation. The configuration bitstream for a module contains configuration frames that program the FPGA's clock resources to route the correct clock signals to user logic. Unlike ASIC clock gating, FPGA clock controls are not optimized to a specific user design. The clock tree of an FPGA has a fixed, regular pattern. Consequently, it can be enabled and disabled only at fixed, regular intervals. The area of programmable logic that is driven by a segment in the clock tree is similarly constrained.

The design challenge for the system architects and physical design tools is to insure that modules are mapped into the optimal number of clock controlled regions. Gayasen et al [23] report the use of FPGA region constraints to reduce the area of a user design. This allowed the authors to "power down" areas of the FPGA that were not occupied by any user logic. By clustering the logic from each module around a dedicated clock tree, the authors may also have gated the clock to that module.

Reconfiguration is required in this use case because clock segments can only be controlled from configuration memory. If we want to dynamically alter which segments of the FPGA clock trees are enabled, the corresponding configuration memory cells must be altered. Careful floor planning of the user design around the clock tree can minimize the number of clock gates that are modified. This reduces the amount of configuration data required for each clock activation or deactivation. Configuration controlled clock gating can, therefore, be applied more rapidly and at a finer granularity than the other reconfiguration based power management techniques.

Low Leakage Configurations.   The techniques presented so far influence dynamic power more than static power. In this final use case, a technique to lower static power is presented. In the *"Pika"* architecture, static power is reduced by adding pass gates to the power supply of different regions of the device. This allows static power consumption to be addressed in much the same way that clock gating reduces dynamic power. The regions of logic that are not required at a particular point in time are simply powered down. Clearly, care must be taken to allow state and configuration data to be restored correctly when the region is reactivated.

Modern FPGAs (*"Pica"* excepted) do not implement power gating. To reduce static power in a standard FPGA, a different technique is required. In [24], Tuan and Lai state the static power consumed by a programmable structure depends on the values of its configuration memory cells and inputs. For example, they state the static power consumption of a simple, programmable 2-input multiplexor can vary up to $4\times$ with respect to its configuration and input values. This property can be exploited to construct "low-leakage" configuration bitstreams where the data values written to memory cells and driven to circuit inputs minimize the overall static power consumption of the device or region.

Figure 22.16 shows a slight variation to the simple multi-modal system used as an example in Section 4.2.0.0. Clock gating can reduce the dynamic power required by this system, but this time its static power cannot be lowered by moving it to a smaller FPGA. The I/O requirements of the F(x) and H(x) modules mandate the use of a large FPGA with the appropriate number of I/O resources. Reconfiguration allows the system to overwrite the functions not

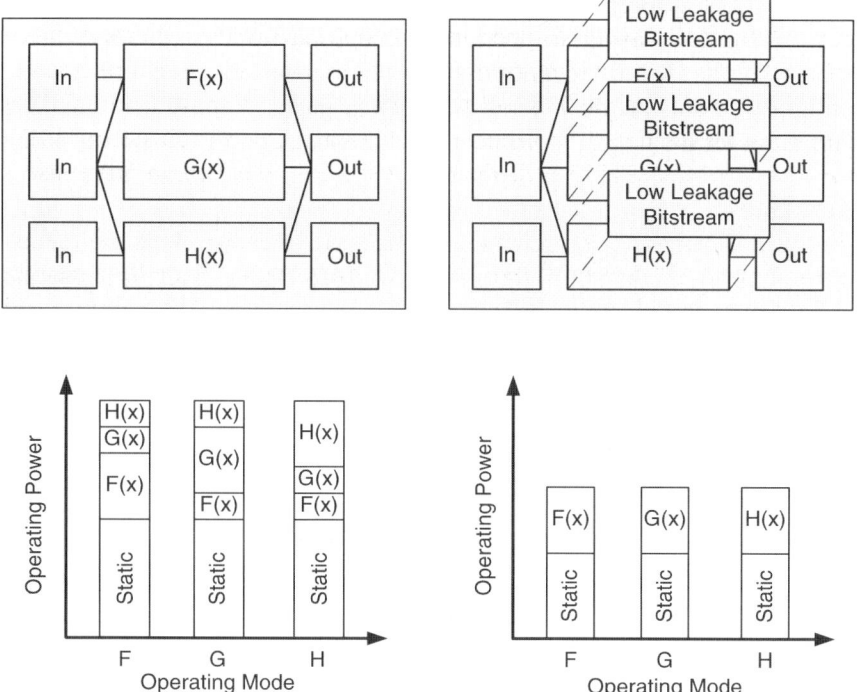

Fig. 22.16 Lower static power with low leakage bitstreams.

used by the system's current mode with low-leakage bitstreams. This is indicated in the two graphs at the bottom half of Figure 22.16.

5. Conclusion

FPGA reconfiguration has been a compelling and somewhat elusive design technique for almost 20 years. In this chapter we reviewed some of the main applications of reconfiguration and explored the typical design flow required to create such systems. Certainly there is much work to be done, including a greater exploration of the use of reconfiguration to lower other system parameters like dynamic and static power consumption. Indeed, many of the design considerations for low power systems are equally applicable to dynamically reconfigurable systems. Today, the path of a reconfigurable systems designer may be a complex one but we can see that, with insight and careful design, FPGA reconfiguration yields great rewards.

Acknowledgments

I would like to express my great thanks to Jürgen Becker and Michael Hübner for their invitation to participate in this book. Also, Brandon Blodget, Tobias

Becker (now at Imperial College, London), Jeff Mason, Jay Young, James Anderson, Brendan Bridgeford, Manuel Uhm, Patrick Lysaght, and the other members of the Xilinx reconfiguration team invested great efforts on the reconfiguration design flow discussed in this paper. Finally, my thanks also to Tim Tuan for his helpful comments and feedback on FPGA power issues.

References

[1] H. Eggers, P. Lysaght, H. Dick, G. McGregor, Fast Reconfigurable Crossbar Switching in FPGAs, in 6th International Workshop on Field Programmable Logic and Applications (FPL), 1996, Springer LNCS 1142.

[2] S. Young, P. Alfke, C. Fewer, S. McMillan, B. Blodget, D. Levi, A High I/O Reconfigurable Crossbar Switch, IEEE Symposium on FPGAs for Custom Computing Machines, 2003. To appear.

[3] E. Lemoine, D. Merceron, Runtime Reconfiguration of FPGA for Scanning Genomic Databases, in IEEE Symposium on FPGAs for Custom Computing Machines (FCCM), 1995.

[4] Y. Yamaguchi, Y. Miyajima, T. Marugama, A. Konagaya, High Speed Homology Search Using Runtime Reconfiguration. In 12th International Workshop on Field-Programmable Logic and Applications, 2002, Springer, LNCS 2438.

[5] S. Guccione, E. Keller, Gene Matching Using JBits. In 12th International Workshop on Field-Programmable Logic and Applications, 2002, Springer, LNCS 2438.

[6] B. Fawcett, Applications of Reconfigurable Logic in More FPGAs, Editors Will Moore and Wayne Luk, Abingdon EE & CS Books, 1993.

[7] S. Guccione, D. Levi, P. Sundararajan, JBits: A Java-Based Interface for Reconfigurable Computing, 2nd Annual Military and Aerospace Applications of Programmable Devices and Technologies Conference (MAPLD).

[8] P. Lysaght, D. Levi, Of Gates and Wires, p. 132a, 18th International Parallel and Distributed Processing Symposium (IPDPS'04) – Workshop 3, 2004.

[9] J. Becker, M. Hübner, K.D. Müller-Glaser, R. Constapel, J. Luka, J. Eisenmann, Automotive Control Unit Optimisation Perspectives: Body Functions on-Demand by Dynamic Reconfiguration, Date 2005, Munich, Germany.

[10] Becker, Hübner, Ullmann: Power Estimation and Power Mesurement of Xilinx Virtex FPGAs: Trade-offs and Limitations, SBCCI 03.

[11] B. Blodget, S. McMillan P. Lysaght, A lightweight approach for embedded reconfiguration of FPGAs, Design, Automation and Test in Europe Conference and Exhibition, 2003, pp. 399–400

[12] M. Uhm, Making the adaptivity of SDR and Cognative Radio Affordable: Going beyond flexibility and adaptivity in FPGAs, Xilinx DSP Magazine, May 2006.

[13] M. Hübner, J. Becker, Seamless Design Flow for Runt-Time Reconfigurable Automotive Systems, DATE Workshop on Future Trends in Automotive Electronics and Tool Integration, March 2006, pp. 47–51.

[14] A. Donlin, FPGA Reconfiguration in Automotive Systems, DATE Workshop on Future Trends in Automotive Electronics and Tool Integration, March 2006, pp. 52–58.

[15] C. Carmichael, Correcting single-event upsets through Virtex partial reconfiguration, Xilinx Application Note, XAPP216, June 1, 2000.

[16] P. Kane, Digital Electronics with PLD Integration, Prentice-Hall, 2001.

[17] P. Lysaght, et al, Enhanced Architectures, Design Methodologies and CAD Tools for Dynamic Reconfiguration on Xilinx FPGAs. In Proceedings of International conference on Field Programmable Logic and Applications, 2006.

[18] J. Becker, M. Huebner, M. Ullmann, Real-Time Dynamically Run-Time Reconfiguration for Power-/Cost optimized Virtex FPGA Realizations, Proceedings of the IFIP International Conference on Very Large Scale Integration (VLSI-SoC), Darmstadt, Germany, December 1–3 2003, pp. 129–134.

[19] A. Donlin, et al, A virtual file system for dynamically reconfigurable FPGAs, Field Programmable Logic and Applications (FPL), 2004.

[20] T. Tuan, et al, A 90nm Low-Power FPGA for Battery-Powered Applications, In proceedings of International Symposium on FPGAs, February 2006.

[21] A. Telikepalli, Power vs. Performance: The 90nm Inflection Point, Xilinx White Paper (WP223), April 2005.

[22] P. Sundararajan, et al, Estimation of Single Event Upset Probability Impact of FPGA Designs. In Proceedings of Military Applications of Programmable Logic Devices, 2003.

[23] A. Gayasen, et al, Reducing Leakage Energy in FPGAs using Region Constrained Placement. In Proceedings of International Symposium on FPGAs, 2004.

[24] T. Tuan B. Lai, Leakage Power Analysis of a 90nm FPGA. In Proceedings of Custom Integrated Circuits Conference (CICC), 2003.

Index